ENCYCLOPEDIA
of FOODS
A GUIDE TO HEALTHY NUTRITION

A Guide To
HEALTHY
NUTRITION

PART I

The latest research shows that the foods we choose to eat—or not to eat—may increase our life span or the quality of our lives. Not a day goes by, it seems, without feature news stories about food and its impact on health. The message that we can reduce our chances of developing cancer, high blood pressure, diabetes, and other diseases by maintaining a healthy weight, decreasing the fat and calories in our diets, eating more vitamin- and mineral-rich fruits and vegetables, and getting fit is becoming a familiar one. As more research is done, the link between diet and the risk of developing common diseases such as heart disease or cancer is becoming clearer and clearer. Thanks to this research, we are beginning to understand the dietary and lifestyle factors that are most likely to ensure a long, healthy life. And the good news is that we can incorporate these factors into our own lives without sacrificing taste or giving up the foods we enjoy, by discovering and eating tastier, nutritious fruits, vegetables, and whole grains.

Despite the well-publicized connections among diet, weight, and health, statistics show that the prevalence of overweight and obesity is increasing at an alarming rate, particularly in children and teens. So why do we cling to our unhealthful habits? For many, the nutrition and fitness guidelines published by the government and by health organizations may seem over-whelming. We may have questions about why we should eat what the experts recommend, what foods are or are not nutritious, and whether it is better to obtain some nutrients from food or from a multivitamin or nutritional supplement. Finally, for most of us, eating is pleasurable, and familiar foods are comforting. The idea of making a major change in the kinds and amounts of food that we eat is daunting.

Part I of this book provides you with an overview of the principles of good nutrition, provides you with insight as to why the experts recommend what they recommend, and then gives you some practical tips on how to change the way you eat while still enjoying good food. Chapter 1 begins by reviewing the current guidelines for nutrition and fitness. Chapter 2 provides an overview of the basics of nutrition, including a description of the known nutrients and other food components. Chapter 3 discusses the role of good nutrition in the prevention and treatment of common diseases. Chapters 4 and 5 provide suggestions for planning and preparing meals and selecting healthful foods. They give tips on eating out, shopping, reading food labels, and modifying family favorites with healthful recipe makeovers. Sample recipes are provided to show that following a well-balanced, nutritious diet can be an enjoyable undertaking, and 2 weeks of menus are given to help you get started.

In this chapter, you will be introduced to the basic principles of nutrition. You will learn about the value of a diet rich in fruits, vegetables, and grains as the foundation for good health and how to select nutritious foods that contribute to a healthful diet. You will then be equipped to discern the best approach for your nutritional well-being.

You will also learn about:

- The new Dietary Reference Intakes and how they are used
- Health Goals—The importance of nutrition
- Dietary Guidelines—Your gateway to nutrition knowledge
- The Food Guide Pyramid: A guide to eating well

OPTIMIZING HEALTH

We are surrounded by a vast array of foods to eat and activities to pursue. Every day we make choices among those foods and activities based on our cultural background, knowledge, experiences, and goals. Each choice may have an impact on our overall health and quality of life. Our ancestors' food choices were limited by what they could gather, catch, cultivate, and harvest. Physical pursuits were determined by the work that needed to be done. Today, advances in agriculture, transportation, food preservation, and storage bring nearly every type of food from every country of the world to our local supermarkets throughout the year. With such a limitless array of foods, choosing the ones that promote health is easier than ever, but making these choices requires knowledge and motivation. This chapter explains how the guidelines established by nationally recognized health and nutrition authorities can be used to help you understand the food choices that promote health, choose the foods that contain needed nutrients, and select appropriate serving sizes.

THE DIETARY REFERENCE INTAKES (DRIs)

We all need the same nutrients, but the amounts we need depend on our age, sex, and a few other factors. For example, women who are pregnant or breastfeeding need more of most nutrients. The Food and Nutrition Board of the Institute of Medicine, National Academy of Sciences, a group of nutritional scientists from the United States and Canada, has established the Dietary Reference Intakes (DRIs), a set of recommendations for nutrient intake. The DRIs are age- and sex-specific. With the exception of fats and carbohydrates (whose requirements depend only on our calorie needs), a separate DRI is set for each of the known nutrients for each of 10 different age groups. From the age of 9 years, males and females have separate DRIs, and additional DRIs are set for women who are pregnant or breastfeeding.

How did the nutrient recommendations originate? Concerned with the need to provide proper nutrition for newly drafted World War II soldiers, many of whom were undernourished, the Department of Defense commissioned the first set of nutrient recommendations (called the Recommended Dietary Allowances) in 1941. Since then, nutrient recommendations

have undergone periodic revision based on advances in our understanding of nutrition.

Today, nutrition research addresses not only the prevention of nutritional-deficiency diseases but also the role of nutrients in reducing the long-term risk for diseases such as heart disease and cancer. Taking into consideration the resulting expansion of scientific knowledge about the roles of nutrients in health since the first recommendations were established, the latest revision was begun in 1997. The new Dietary Reference Intakes (DRIs) include the Recommended Dietary Allowance, the Estimated Average Requirement, the Adequate Intake, and the Upper Limit.

The Recommended Dietary Allowance (RDA) is the amount of each nutrient that is sufficient to prevent nutritional deficiencies in practically all healthy people. The Estimated Average Requirement (EAR) is the amount of a nutrient that is estimated to meet the requirement of half the population of an age- and sex-specific group. For some nutrients, too little is known about them to establish an RDA. For these, an Adequate Intake (AI) is determined. This is the intake that should be adequate to meet the needs of most people. A safe Upper Limit (UL) has been established for some nutrients. Establishment of this value reflects our growing recognition that some nutrients may help promote health and prevent disease in amounts that exceed the RDA. The UL is the maximal daily intake of a nutrient that is likely to be free of the risk of adverse health effects in almost all individuals in the designated group.

How are the DRIs used? They are the basis for all nutritional plans used by health care facilities and providers, food services, food manufacturers, and others who plan diets. As you will learn below, the Food Guide Pyramid, the research-based food guide developed by the government, is based on the DRIs. In addition, the Daily Values, the information on food labels that helps you determine how a food contributes to your total nutrient intake, are based on the DRIs (see the Appendix: Dietary Reference Intakes, page 421).

AMERICA'S HEALTH GOALS

According to recent statistics, our eating habits—the foods we eat and drink and those we avoid—play a major role in preventing 4 of the 10 leading causes of death in the United States. These include heart disease, cancer, stroke, and diabetes (see sidebar: Top 10 Causes of Death, this page). In addition, one in four adults has high blood pressure, a leading contributor to stroke, heart attack, kidney failure, and premature death. (See Chapter 3, page 47, for the important role of diet.) We didn't always have this knowledge. But now that we do, experts in nutrition working with the federal government have provided us with nutrition and physical activity guidelines for staying healthy and preventing disease.

Many government and health care associations focus their efforts on helping Americans eat well. Chief among them is the U.S. Department of Health and Human Services, which created a set of national health goals entitled Healthy People 2010.

Healthy People 2010

The ultimate goals of Healthy People 2010 are to improve the nation's health status and to eliminate health disparity among segments of the U.S. population. One of the priorities of this initiative is to foster a change in America's eating

---

TOP 10 CAUSES OF DEATH* (U.S. POPULATION)

Many of the leading causes of death in the United States are directly related to diet and excessive alcohol consumption.

RANK	CAUSE OF DEATH
1†	Heart disease
2†	Cancer
3†	Stroke
4	Chronic obstructive pulmonary disease (emphysema)
5‡	Accidents/injuries
6	Pneumonia and influenza
7†	Diabetes mellitus
8‡	Suicide
9	Kidney disease
10‡	Chronic liver disease, cirrhosis

*Top 10 causes of death according to the National Center for Health Statistics, 1997.
†Causes of death in which diet plays a part.
‡Causes of death in which excessive alcohol consumption plays a part.

HEALTHY PEOPLE 2010

Overall Health Goals: Increase quality and years of healthy life and eliminate health disparities among different segments of the population

FOCUS AREA: NUTRITION AND OVERWEIGHT*

Objective:

Promote health and reduce chronic diseases associated with diet and weight

Weight status and growth:

Increase proportion of people who are of healthy weight

Decrease obesity in adults

Decrease overweight or obesity in children and adolescents

Reduce growth retardation in children

Food and nutrient consumption:

Increase fruit intake (2+ servings daily)

Increase vegetable intake (3+ servings daily)

Increase grain product intake (6+ servings daily)

Decrease saturated fat intake (less than 10% of calories)

Decrease total fat intake (no more than 30% of calories)

Food and nutrient consumption (continued):

Decrease sodium intake (2,400 milligrams or less daily)

Meet dietary needs for calcium

Reduce iron deficiency and anemia

Schools, worksites, and nutrition counseling

Meals and snacks at school should contribute to overall dietary quality

Employers promote nutrition education and weight management at the worksite or through health plans

Nutrition counseling for medical conditions

Include nutrition counseling in physician office visits

Food security

Increase access to nutritionally adequate and safe foods for an active, healthy life

*Nutrition and Overweight is one focus area (of 28) that targets interventions designed to increase quality and years of healthy life and to eliminate health disparities among different segments of the population.

habits. The primary means for achieving these goals is through various nutrition initiatives (see sidebar: Healthy People 2010, above).

Mounting scientific evidence supports a link among diet, health promotion, and disease prevention. Improved nutrition has the potential to prevent or delay many diseases often associated with advancing age. With prevention of illness comes the possibility of reducing health care costs. Therefore, one of the main nutrition objectives is to promote health and reduce chronic diseases associated with diet and obesity. This includes reducing the number of people who die of heart disease, reducing the number of cancer deaths, reducing the prevalence of overweight and diabetes, and reducing growth retardation in children.

To help achieve these health goals, specific nutrition targets were set. These include:

- increasing the proportion of the population who are at a healthy weight
- optimizing food and nutrient consumption, emphasizing fruits, vegetables, and whole grains
- improving nutrition and nutrition education at schools and at worksites
- including nutrition counseling as a regular part of health care
- increasing access to a healthful and safe food supply

Coalitions of government health agencies and the food industry are working collaboratively to provide consistent messages that emphasize the importance of eating a diet rich in plant foods—fruits, vegetables, and grains—and containing less fat (see Other Voices, page 14).

For example, the government has required that labels on foods provide clear and concise information on nutrient

content and truthful health claims based on scientific fact. The food industry is developing healthier lower-fat and lower-calorie products, restaurants are identifying healthful choices on menus, and educational efforts on the importance of good nutrition have been stepped up.

5 a Day for Better Health

The National Cancer Institute of the United States and the Produce for Better Health Foundation (a nonprofit consumer education foundation representing the fruit and vegetable industry) collaborated in a unique partnership in 1991 to develop the 5 a Day for Better Health program. This is a nationwide educational effort to encourage Americans to eat 5 or more servings of fruits and vegetables every day for better health. A minimum of 5 servings of fruits and vegetables a day provides the RDA for many of the vitamins and minerals (see Chapter 2, page 29). The recommendation that we eat 5 to 9 servings of fruits and vegetables each day also is based on the results of numerous studies showing the positive effects of fruits and vegetables on health as a result of their ability to reduce the risk of cancer and other diseases. Ample consumption of fruits and vegetables forms the basis of some of the Dietary Guidelines for Americans and the guidelines of the American Cancer Society and others, outlined below. The 5 a Day program works through state public health departments, retail food stores, school classrooms and cafeterias, the military, and various media. The goal of the program is to educate the

public about the benefits of fruits and vegetables and to demonstrate easy and delicious ways to fit more of them into your diet (see Chapters 4 and 5, pages 79 through 149).

Have We Made Progress?

The explosion of health information and nutrition education programs has led to good progress on several fronts. Deaths from heart disease have declined and, to a slight degree, so have deaths from some cancers. On average, the intake of total fat and saturated fat has decreased. Food labeling provides much more useful information now. Restaurants offer more low-fat and low-calorie options on their menus.

Although consumption of grain products is on the rise, many grains are in the form of snacks such as corn chips and popcorn. Fewer than one-third of American children and less than one-half of adults eat the recommended 5 servings of fruits and vegetables. Overall, fat intake is decreasing (from 40 percent of calories in the late 1970s to 33 percent in the mid-1990s). However, only about a third of adults meet the "30 percent or fewer calories from fat" recommendation of nutrition experts.

Nutritionists are now assessing our progress in meeting the goals of Healthy People 2010. These efforts will include evaluating healthful behaviors in the areas of fitness and nutrition, ensuring a safe food supply, and reducing and preventing diseases such as osteoporosis, cancer, diabetes, heart disease, and stroke.

Of course, national goals are met one person at a time. Fortunately, there is a road map for achieving fitness and health. Scientists and nutrition experts have mapped out a sound plan for healthful eating and exercise based on the most current findings about nutrition.

THE DIETARY GUIDELINES FOR AMERICANS

Aim for Fitness

Aim for a healthful weight

Research clearly shows that being overweight greatly increases your risk for many diseases, including heart disease, cancer, and diabetes. If you are overweight, combining a healthful eating plan with regular physical activity is the most effective way to lose weight and to sustain

the loss (see sidebar: Diet and Exercise—The Perfect Pair, page 10). If you are at a healthy weight, your goal is to maintain that weight. Chapter 3 (page 47) provides further information on weight control.

Be physically active each day

Everyone—young and old—can improve their health by being more active. Choose activities that you enjoy and can do regularly. Although you will gain more health benefits with high-intensity exercise that lasts 30 minutes or more, low-to-moderate activities can be part of your routine. For some people, this means fitting more activity of daily living into your usual routine. This could include using the elevator less and using the stairs more, parking farther from rather than closer to your destination, gardening, or golfing without a cart. For others, a more structured program might be preferred, such as at a worksite or health club. Whichever you choose, the goal is to include at least 30 minutes of activity every day.

The need for regular physical activity is so important that the Surgeon General of the United States has issued a report entitled Physical Activity and Health, which has its own guidelines for achieving activity. They are the following:

- Physical activity should be performed regularly. Include a minimum of 30 minutes of moderate physical activity (such as brisk walking) on most, if not all, days of the week. For most people, greater health benefits can be obtained by engaging in activity that is more vigorous or of longer duration.
- Previously sedentary people should start with short durations of moderate activities and gradually increase duration or intensity.
- Physical activity should be supplemented with strength-enhancing exercises at least twice a week to improve musculoskeletal health, maintain independence in performing the activities of daily life, and reduce the risk of falling.
- Consult with a physician before beginning a new physical activity program if you have—or are at risk for—a medical condition (such as heart disease, high blood pressure, or diabetes), or if you are a man older than 40 years or a woman older than 50.

Build a Healthy Base

Let the Pyramid guide your food choices

Your body needs more than 40 nutrients and other substances for good health. No one food can give you all the nutrients your

---

DIETARY GUIDELINES FOR AMERICANS

The ABCs of good health

Aim for Fitness
- Aim for a healthful weight
- Be physically active each day

Build a Healthy Base
- Let the Pyramid guide your food choices
- Choose a variety of grains daily, especially whole grains
- Choose a variety of fruits and vegetables daily
- Keep foods safe to eat

Choose Sensibly
- Choose a diet that is low in saturated fat and cholesterol and moderate in total fat
- Choose beverages and foods to moderate your intake of sugars
- Choose and prepare foods with less salt
- If you drink alcoholic beverages, do so in moderation

---

body needs, no matter how much you enjoy it or how nutritious the food is. By eating a wide variety of foods each day, you will keep your meals exciting and you will achieve the balance of nutrients that best ensures good health. (See page 11 for more information on the Food Guide Pyramid.)

Choose a variety of grains daily, especially whole grains

Choosing a diet rich in grains, especially whole grains, reduces your risk of many diseases. These foods provide different types of vitamins, minerals, and fiber, as well as phytochemicals—important plant substances that may be beneficial to health. Rely on a wide variety of these foods rather than supplements as your source of nutrients, fiber, and phytochemicals. Aim for 6 servings each day—more if you are very active—and include several servings of whole-grain foods. (See Chapter 2, page 33.)

Choose a variety of fruits and vegetables daily

Fruits and vegetables are essential in your diet. They provide many vitamins, minerals, phytochemicals, and fiber, and they are low in calories and provide no fat. The goal is to have at least 2 servings of fruit and 3 servings of vegetables every day. Variety is important. Choose different colors and

DIET AND EXERCISE—THE PERFECT PAIR

Diet along with exercise is the most effective way to lose weight.
A deficit of 500 calories daily can add up to a loss of 1 pound a week.
Here are some suggestions for skimming approximately 250 to 300 calories off your daily diet
and for burning an additional 250 calories through increased physical activity.

DIET	EXERCISE
Decrease usual meat intake by 3 to 4 ounces daily	Walk at a moderate pace for 60 minutes daily
Cut butter, margarine, or oil by 2 tablespoons daily	Garden for 50 minutes daily
Eliminate two 12-ounce cans of sweetened carbonated beverages daily	Swim laps for 30 minutes daily
Decrease beer intake by two 12-ounce cans daily	Jog 25 minutes daily
Do not eat a candy bar	Bike briskly for 25 minutes daily

kinds of fruits and vegetables. (See Chapters 4 and 5, pages 79 through 149, for ideas on how to include these important foods as regular features in your meals. Part II, page 150, also describes the bounty from which to choose.)

Keep foods safe to eat
Food safety is vital. It starts well before you purchase food. However, the steps you control also make a difference. They include making sure you have clean hands and work surfaces—before and during the handling of food. Take care to separate raw, cooked, and ready-to-eat foods at all times. Keep hot foods hot and cold foods cold. Make sure to cook food to the proper temperature. Refrigerate perishable foods and leftovers promptly. Follow the dates on containers. And finally, when in doubt, throw it out. (For further discussion on food safety issues, see Chapter 5, page 148.)

Choose Sensibly

Choose a diet that is low in saturated fat and cholesterol and moderate in total fat
Fat is a nutrient that is essential for health, but too much fat in your diet, especially saturated fat, increases your risk of several diseases, including heart disease. Most important, learn to identify the sources of fats, saturated fats, and cholesterol, and make healthful food choices. (See Chapter 2, Fats, page 26, and Chapter 3, Coronary Artery Disease, page 61.)

Choose beverages and foods to moderate your intake of sugars
Some foods that contain natural sugar (such as fruits, vegetables, and milk products) also contain essential nutrients. Others, such as table sugar, sugar-sweetened carbonated beverages, candy, and some baked goods, supply calories but few other nutrients. When consuming sugar, moderation is key. (See Chapter 2, Carbohydrates, page 18.)

Choose and prepare foods with less salt
Sodium, a nutrient, is a major part of table salt (sodium chloride). It is found naturally in many foods in small amounts. Salt and sodium compounds are also added to processed foods, and salt may be used in cooking or added at the table. Reducing sodium intake lowers high blood pressure in some individuals. Moderation in sodium intake is recommended. (See Chapter 3, High Blood Pressure, page 53.)

If you drink alcoholic beverages, do so in moderation
Alcoholic beverages (beer, wine, and hard liquor) are a source of extra calories. When consumed in excess, alcohol can impair judgment, result in dependency, and lead to several serious health problems. However, evidence suggests that a moderate intake of alcohol is associated with a lower risk of disease of the heart and blood vessels (cardiovascular disease) in some individuals. Discuss the consumption of alcohol with your health care provider. (See sidebar: Alcohol and Health, page 387.)

THE POWER OF THE FOOD GUIDE PYRAMID

The Food Guide Pyramid, the triangular symbol you see on many food packages, was developed by nutrition experts at the U.S. Department of Agriculture (USDA) (see below). The Pyramid is an educational tool that translates nutrient requirements into the foods you need to eat and helps you put into action the advice offered by the Dietary Guidelines. In graphic form, the Pyramid displays the variety of food choices and the correct proportions needed to attain the recommended amounts of all the nutrients you need without consuming an excess of calories. The Pyramid divides all foods into six categories, based on the nutrients they contain.

THE FOOD GUIDE PYRAMID

Fats, oils, and sweets (eat sparingly)

Milk products: Skim or low-fat milk, yogurt, low-fat or nonfat cheeses or cottage cheese

Meats and other high-protein foods: Lean meats, poultry, fish, eggs (3 to 4 yolks per week), cooked dry beans, peas, lentils, peanut butter, nuts, seeds, tofu

Vegetables: Fresh or cooked vegetables, vegetable sauces, or juices

Fruits: Fresh fruit (apple, apricots, banana, berries, dates, figs, grapefruit, grapes, guava, kiwi, mango, melon, nectarine, orange, pineapple), canned fruit, juices

Grains: Whole-grain breads, bagels, English muffins, breakfast cereals (whole-grain, cooked, or ready-to-eat), crackers, tortillas, pancakes, pasta, rice

The Food Guide Pyramid was developed by the U.S. Department of Agriculture. The pyramid incorporates many principles that emphasize a plant-based diet that is low in fat, high in fiber, and rich in important vitamins, minerals, and other nutrients. All of these factors contribute to optimal health and help you to control your weight and to reduce the risk of heart disease and some types of cancer. The arrangement of the food groups in a pyramid shape calls attention to the kinds of foods to eat more of and those to eat in moderation.

SIZING UP YOUR SERVINGS

Even if you eat a variety of foods, serving sizes are an important part of maintaining a healthful weight.
Knowing them can help you gauge if you are eating enough food — or too much.

ONE SERVING EQUALS

GRAINS
1 slice bread
1 ounce ready-to-eat cereal (large handful or check the package label)
1/2 cup cooked cereal, rice, or pasta (similar to the size of an ice cream scoop)

FRUITS
1 medium apple or orange (size of tennis ball)
1 medium banana
1/2 cup cut-up, canned, or cooked fruit
3/4 cup 100% fruit juice

VEGETABLES
1 cup raw leafy vegetables (the size of your fist)
1/2 cup other vegetables, chopped (raw or cooked)
3/4 cup vegetable juice

MILK PRODUCTS
(**choose low-fat varieties**)
1 cup milk or yogurt
1 1/2 ounces natural cheese (the size of a pair of dice
  or pair of dominoes)
2 ounces low-fat processed cheese

MEATS
2 to 3 ounces of cooked lean meat, poultry, or fish
  (about the size of a deck of cards or the palm of your hand)
THE FOLLOWING ALSO EQUAL 1 OUNCE OF MEAT:
1/2 cup cooked dry beans or legumes (ice cream scoop)
1 egg (3 to 4 yolks per week)
2 tablespoons peanut butter
1/3 cup nuts
1/2 cup tofu

FATS, OILS, AND SWEETS
(These foods add calories and are usually low in nutrients. Eat them sparingly.)

How Many Servings Do You Need Each Day?

This table tells you how many servings to aim for from each food group. The number of servings you need depends on your age, sex, and how active you are. The table also indicates how much fat (in grams) should be your limit. This includes the fat you find in foods and the fat that you add to foods.

	MOST WOMEN, OLDER ADULTS	CHILDREN, TEEN GIRLS, ACTIVE WOMEN, MOST MEN	TEEN BOYS, ACTIVE MEN, VERY ACTIVE WOMEN
Calorie level	About 1,600	About 2,200	About 2,800
		SUGGESTED NUMBER OF SERVINGS	
Grain group	6	9	11
Fruit group	2	3	4
Vegetable group	3	4	5
Milk group	2–3*	2–3*	2–3*
Meat group	2 (for a total of 5 ounces)	2 (for a total of 6 ounces)	3 (for a total of 7 ounces)
Total fat (less than 30% of calories)	53 grams or less†	73 grams or less†	93 grams or less†

*3 servings are recommended for women who are pregnant or breastfeeding, teenagers, and young adults up to age 24.

†Values are rounded off.

The six categories of the Pyramid are:
- Grain products (bread, cereal, rice, and pasta)
- Fruits
- Vegetables
- Milk products (milk, yogurt, cheese)
- Meats and other high-protein foods (lean meats, poultry, fish, dry beans, eggs, and nuts)
- Fats, oils, and sweets

The shape of the Pyramid, widest at the base and narrowest at the tip, makes it easy to visualize the contribution that each group of foods should make to your overall eating plan when you follow the Dietary Guidelines. The emphasis of the Pyramid is on increasing the proportion of fruits, vegetables, and grains—those foods that form the base of the Pyramid—and decreasing the proportion of higher-fat foods—the ones at the very top—in our diets.

The grain group, which includes bread, cereal, rice, and pasta, forms the broad foundation of the Pyramid to emphasize that grains should be a major contributor to our overall diet. As often as possible, our choices of grain foods should be those made from whole grains, for the most nutritional value.

As illustrated by the Pyramid, in addition to grains, our diet should include ample servings of fruits and vegetables. If our daily need is to be met for vitamins, minerals, fiber, and other important phytochemicals (plant chemicals that are believed to play a role in preventing disease), the bulk of our diets must come from plant foods.

Because of the saturated fat they contain, meats, poultry, and seafood (the high-protein foods) and dairy products (high in protein, calcium, and other minerals) should make a smaller contribution to our daily fare. Foods that occupy the tip of the Pyramid, pure fats (cooking oil, butter, and margarine) and high-fat, high-sugar sweets, are the ones to include only sparingly, like the proverbial icing on the cake. The Pyramid is designed to promote and encourage

a plant-based diet, one that is based primarily on grains, fruits, and vegetables. Yet, by including all types of foods, the Pyramid emphasizes the need for us to choose a variety of foods and the fact that there are no "bad" foods.

The Pyramid is designed to address the needs of all persons older than 2 years by providing a range of recommended servings for each food group (see sidebars: Sizing Up Your Servings, page 12, and How Many Servings Do You Need Each Day? page 13). The number of servings that you should choose from each food group depends on your calorie needs, which in turn depend on your age, size, sex, and activity level. The lower number of servings provides a total daily energy intake of about 1,600 calories. This calorie level meets the needs of most sedentary women and some older adults. The higher number of servings, which provides approximately 2,800 calories, is recommended for physically active men, teen boys, and some very active women. The middle range of servings is designed to provide about 2,200 calories, sufficient for children, teen girls, active women, and most men. These calorie estimates assume that you choose lean meats, lower-fat dairy foods, and vegetables and grains prepared and eaten with minimal added fat and sugar.

In 1999, the USDA released a Children's Pyramid. The Children's Pyramid was designed to address the needs of 2- to 6-year-olds. It has proportionally smaller portions and numbers of recommended servings from each group of foods except fruits and vegetables. This emphasizes that children, too, need at least 5 servings of fruits and vegetables a day.

Similar to the Dietary Guidelines for Americans, the Food Guide Pyramid undergoes periodic updates to reflect what we have learned about the role of nutrition in disease prevention. To get an idea of the changes you might see in the next Pyramid, let's take a look at some other Pyramids that have been constructed.

Other Pyramids

The risk for heart disease and some types of cancer among people who live in the Mediterranean region—southern Italy, France, Spain, and Greece—is significantly lower than the risk in Americans. Nutritional scientists have uncovered strong evidence that the eating, drinking, and exercise habits of the Mediterranean people play a major role in their low risk for disease. The Mediterranean diet has been illustrated as a "Mediterranean Pyramid," based on our own Food Guide Pyramid. It is built on a foundation of pasta, bread, rice, and other grains, with large contribu-

tions of vegetables and legumes (beans and peas) and small portions of meat, poultry, seafood, and dairy products. The fat used in cooking and for dressings is olive oil, rather than butter. Desserts consist of fresh fruits, and meals are accompanied by wine. This plant-based diet is naturally low in saturated fat, higher in monounsaturated fats (from olive oil), and rich in fiber, vitamins, minerals, and phytochemicals (see Chapter 2, page 33). In addition to adhering to a plant-based diet (which includes generous servings of legumes such as kidney beans, peas, and lentils), Mediterranean people have a more physically active lifestyle than most Americans. This factor also may contribute to their lower risk of heart disease. The major difference between the Mediterranean diet and the USDA Food Guide Pyramid lies in the distinction between the recommendation to lower total fat, which places all high-fat foods at the tip of the Pyramid, and the Mediterranean practice of including monounsaturated fats but limiting saturated fats (see Chapter 2, page 26).

You also may have heard of other Pyramids, such as a Vegetarian Pyramid and an Asian Pyramid. Like the Mediterranean Pyramid, these pyramids were constructed to illustrate dietary practices of groups whose risk of heart disease and some types of cancer is lower than that of people who consume a typical Western diet. Not surprisingly, these pyramids also illustrate diets that are plant-based and low in saturated fat. Still other pyramids are designed merely to showcase foods that are native to particular regions or produced by particular companies. The Mayo Clinic Healthy Weight Pyramid and the California Pyramid, however, have one unique feature that we may see incorporated into a future Food Guide Pyramid, that is, the inclusion of fruits and vegetables, rather than grains, as the foundation of the Pyramid. By replacing grains with fruits and vegetables at the base, the critical need to increase our intake of these foods is emphasized.

OTHER VOICES: GUIDELINES OF HEALTH ORGANIZATIONS

In addition to the Food Guide Pyramid and the Dietary Guidelines for Americans, several private health organizations (the American Heart Association, the American Cancer Society, and the American Institute for Cancer Research) have issued their own nutritional guidelines.

When the guidelines of all the major health organizations are compared, they are similar. Recognizing this, experts from the American Heart Association, the American Cancer

Society, the American Dietetic Association, the American Academy of Pediatrics, the American Society for Clinical Nutrition, and the National Institutes of Health met in 1999 to review scientific evidence and to identify practices that are effective against major diseases. The dietary recommendations that these groups have in common include:

- **Total fat**: no more than 30% of calories
- **Saturated fat**: less than 10% of calories
- **Monounsaturated fat**: no more than 15% of calories
- **Polyunsaturated fat**: no more than 15% of calories
- **Cholesterol**: no more than 300 milligrams daily
- **Carbohydrates**: 55% or more of calories
- **Salt (sodium chloride)**: less than 6 grams daily (4 grams of sodium)

These recommendations can be achieved by following the Dietary Guidelines for Americans (page 9) and by using the Food Guide Pyramid (page 11).

THE BOTTOM LINE: OPTIMIZING HEALTH

The message is clear. Nutrition experts agree that when you lower the total fat, saturated fat, and added sugar in your diet and increase the vitamins, minerals, and fiber by eating more fruits, vegetables, and grains, you can improve your quality of life and help prevent many of the diseases that are the leading causes of death. Now that you know the goals and guidelines for healthful eating, we will provide you with the nutrition and food selection knowledge you need to put those guidelines into practice.

Chapter 2 explains the nutrients we all need, the roles they play in promoting health, the best food sources for these nutrients, and how your nutritional needs change throughout your life. Chapter 3 describes how your risk for serious illnesses is influenced by your diet. Chapters 4 and 5 help you learn to use the Food Guide Pyramid to choose the most nutritious foods, and they give you guidance about planning and preparing healthful, appetizing meals.

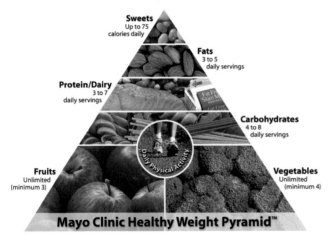

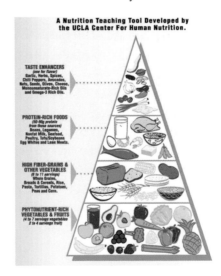

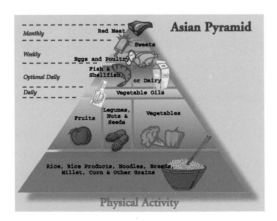

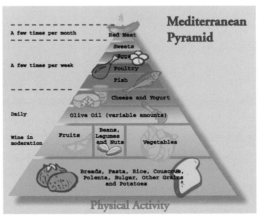

In this chapter, you will be introduced to the nutrients your body needs. You will learn the role each nutrient plays in your body, how much you need and how often, ideal food sources of each nutrient, and how all the nutrients work to optimize health.

You will also learn about some recently identified substances in plant foods, the phytochemicals, that may promote health and help prevent disease.

THE NUTRIENTS AND OTHER FOOD SUBSTANCES

There is no one perfect food. We need an assortment of nutrients that can be obtained only by eating a wide variety of foods. What is it that our bodies need? Scientists have identified more than 40 different nutrients in food. These substances are essential for growth and for the chemical reactions and processes that keep us alive and functioning (metabolism).

Except for an extremely small number of foods that consist almost entirely of one nutrient, the vast majority of the foods we eat are mixtures of many nutrients. Nevertheless, each group of foods included in the Food Guide Pyramid (grains, fruits and vegetables, milk products, and meats) (see Chapter 1, page 11) is unique in the types of nutrients it contributes to our diets. For example, fruits and vegetables are the main source of many vitamins, minerals, and complex carbohydrates in our diets, and the meat group (including dry beans and legumes, eggs, poultry, and fish) is the main source of protein for most people.

It can be difficult to understand the difference between the nutrients themselves and the foods that contain them. For example, when you hear nutrition experts talk about the need to get more complex carbohydrates, what do they mean and what foods contain those nutrients? In this chapter, we focus on the nutrients themselves—how they are digested, what happens to them in the body, and what they do for you. We also say a little about the best food sources of each nutrient, because, after all, when you go to the supermarket, you don't look for protein, starch, fiber, and antioxidants, you look for chicken, rice, raisin bran, and orange juice.

Nutrients are sorted into categories on the basis of their chemical structures and functions. Carbohydrates, proteins, and fats contained in foods are known as the macronutrients, because they are required in the largest quantities. In addition to their other functions, macronutrients provide energy in the form of calories. Vitamins and minerals are known as the micronutrients. They are required by your body in much smaller quantities. Although the micronutrients help your body use the energy in macronutrients, they provide no energy (calories) themselves. Water is also an essential, calorie-free nutrient. The work our bodies do each day causes us to deplete some of our stores of these essential nutrients. Only by maintaining a diet that is rich in various nutrient-containing foods can we replace those lost nutrients.

In addition to the known nutrients, substances in foods of plant origin, called phytochemicals or phytonutrients (*phyto* is the Greek word for plant), have been identified in recent studies. These phytochemicals may promote health and help prevent certain diseases. Hundreds of such compounds are being identified in the fruits, vegetables, nuts, beans, and grains we eat, although only a few have been thoroughly studied. How these various phytochemicals influence our health is a promising new area of research for nutrition experts.

THE MACRONUTRIENTS: CARBOHYDRATES, PROTEINS, AND FATS

Each of the macronutrients—carbohydrates, proteins, and fats—plays various roles in the function of our bodies. In addition to their unique functions, all of the macronutrients supply calories. When we eat more protein, carbohydrate, or fat than we need to replenish what we have used, the excess is converted to and stored as fat. Calories are used to support all muscular activity, to carry out the metabolic reactions that sustain the body, to maintain body temperature, and to support growth. But when we consistently take in more calories than we use, we gain weight. Weight is maintained when energy (calorie) intake balances energy output (see Chapter 3, page 48).

CARBOHYDRATES

The carbohydrates are a vast and diverse group of nutrients found in most foods. This group includes simple sugars (like the sugar you add to your morning coffee) and complex forms such as starches (contained in pasta, bread, cereal, and in some fruits and vegetables), which are broken down during digestion to produce simple sugars. The main function of the simple sugars and starches in the foods we eat is to deliver calories for energy. The simple sugar glucose is required to satisfy the energy needs of the brain, whereas our muscles use glucose for short-term bouts of activity. The liver and muscles also convert small amounts of the sugar and starch that we eat into a storage form called glycogen. After a long workout, muscle glycogen stores must be replenished. Both simple sugars and starches provide about 4 calories per gram (a gram is about the weight of a paper clip). Because carbohydrates serve primarily as sources of calories (and we can get calories from other macronutrients), no specific requirement has been set for them (see Chapter 1, The Dietary Reference Intakes [DRIs], page 5). But health experts agree that we should obtain most of our calories (about 60 percent) from carbohydrates. Our individual requirements depend on age, sex, size, and activity level.

In contrast to the other carbohydrates, fiber (a substance contained in bran, fruits, vegetables, and legumes) is a type of complex carbohydrate that cannot be readily digested by our bodies. Even though it isn't digested, fiber is essential to our health. Nutrition professionals recommend 25 to 30 grams of fiber daily.

Simple Sugars

Simple sugars make foods sweet. They are small molecules found in many foods and in many forms. Some simple sugars occur naturally in foods. For example, fructose is the sugar that naturally gives some fruits their sweet flavor.

Table sugar, the sugar that we spoon onto our cereal and add to the cookies we bake, also called sucrose, is the most familiar simple sugar. A ring-shaped molecule of sucrose actually consists of a molecule of fructose chemically linked to a molecule of another simple sugar called glucose. Sugars such as fructose and glucose are known as monosaccharides, because of their single (mono) ring structure, whereas two-ringed sugars such as sucrose are known as disaccharides. Another disaccharide, lactose, the sugar that gives milk its slightly sweet taste, consists of glucose linked to yet another simple sugar called galactose. The inability to digest lactose to its constituent sugars is the cause of lactose intolerance, a condition common to adults of Asian, Mediterranean, and African ancestry.

The table sugar that we purchase is processed from sugar cane or sugar beets. As an additive to many different types of prepared or processed foods, sucrose adds nutritive value (in the form of calories only), flavor, texture, and structure, while helping to retain moisture. Today, sucrose is most often used to sweeten (nondietetic) carbonated beverages and fruit drinks (other than juice), candy, pastries, cakes, cookies, and frozen desserts. One of the most commonly consumed forms of sugar is called high-fructose corn syrup. High-fructose corn syrup is also commonly used to sweeten sodas, fruit drinks (not juices), some ice creams, and some manufactured pastries and cookies. Other forms of sucrose include brown sugar, maple syrup, molasses, and turbinado (raw) sugar.

Foods that are high in added sugar are often low in essential nutrients such as vitamins and minerals. Unfortunately, these foods are often eaten in place of more nutrient-rich foods such as fruits, vegetables, and low-fat whole-grain products, and they may prevent us from obtaining essential nutrients and lead to weight gain.

Nutritionists are concerned by the enormous increase in sugar consumption by Americans during the past 30 years, particularly because much of this sugar is in the form of soft drinks. On average, teens today drink twice as much soda as milk, and young adults drink three times as much soda as milk. As a result, their intake of calcium-rich foods is low, a factor that is thought to contribute to lower bone

mass. This can lead to an increased risk of bone problems as we grow older (see Chapter 3, Osteoporosis, page 67).

The increase in sugar consumption also has been attributed to the increasing availability of low-fat versions of such dessert and snack foods as cookies, cakes, and frozen desserts. Often, the sugar content of these foods is high because sugar is used to replace the flavor lost when the fat is decreased. Sugar promotes tooth decay, when consumed in forms that allow it to remain in contact with the teeth for extended periods (see sidebar: "Hidden" Sugar in Common Foods, this page).

Thus, foods that are high in sugar, or sugar and fat, and have few other nutrients to offer appear at the top of the Food Guide Pyramid because they should be eaten sparingly. In contrast, choosing fresh fruits, which are naturally sweetened with their own fructose, or low-fat yogurt, which contains lactose (natural milk sugar), allows us to get the vitamins and minerals contained in those foods as well as other food components that contribute to health but may not have yet been identified.

On the positive side, there is no credible evidence to demonstrate that sugar causes diabetes, attention deficit-hyperactivity disorder, depression, or hypoglycemia. No evidence has been found that sugar-containing foods are "addictive" in the true sense of the word, although many people report craving sweet foods, particularly those that are also high in fat.

"Hidden" Sugar in Common Foods

Some foods contain sugar that has been added during processing. The following foods contain a large amount of sugar. The sugar content is shown in grams, and its equivalents in teaspoons are also given.

Try to eat high-sugar foods less frequently or in smaller amounts. Check labels and compare similar foods—choose those that are lower in sugar content. Go easy on adding sugar to food.

Food	Amount	Grams of Sugar	Teaspoons of Sugar
Fruit punch	12 ounces	40	10
Carbonated soft drink, sweetened	12 ounces	40	10
Ice cream	1 cup	40	10
Yogurt with fruit	1 cup	35	9
Candy bar	1 average	30	8
Apple pie	1 slice	15	4
Sweetened cereal	1 cup	15	4
Jam, jelly	1 tablespoon	10	2 1/2
Donut	1	10	2 1/2
Honey	1 teaspoon	5	1
Brown sugar	1 teaspoon	5	1
Table sugar	1 teaspoon	5	1

Complex Carbohydrates

Found almost exclusively in foods of plant origin, complex carbohydrates are long chains of molecules of the simple sugar glucose. The complex carbohydrates in plant foods can be divided into two groups: starch and fiber.

Starch is the form of carbohydrate that is found in grains, some fruits and vegetables, legumes, nuts, and seeds. It provides energy for newly sprouting plants. Fiber is the tougher material that forms the coat of a seed and other structural components of the plant (see illustration on page 20). Starches are digested by our bodies into their constituent glucose molecules and used for energy, whereas fiber is not. Starch, like simple sugars, provides 4 calories per gram, whereas fiber (sometimes called nonnutritive fiber) provides no calories. Like simple sugars, the role of starches in our diets is mainly to provide energy.

Fiber is actually a family of substances found in fruits, vegetables, legumes, and the outer layers of grains. Scientists divide fiber into two categories: those that do not dissolve

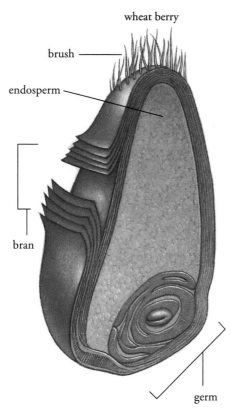

wheat berry

brush

endosperm

bran

germ

The wheat kernel (or seed) consists of the fiber-rich outer bran layer; the inner endosperm, which is composed of starch, proteins, and B vitamins and is made into flour; and the germ, which is ground and sold as wheat germ, a rich source of vitamin E.

in water (insoluble fiber) and those that do (soluble fiber). Insoluble fiber, also called roughage, includes cellulose, hemicellulose, and lignin, found in vegetables, nuts, and some cereal grains. Soluble fibers include pectin, found in fruits, and gums, found in some grains and legumes (see sidebar: Plant Fibers: Insoluble and Soluble, below).

Fiber-rich diets, which include ample amounts of whole-grain foods, legumes, and fresh vegetables and fruits, have been linked with a lower risk of several diseases. Nutrition scientists are just beginning to understand the role of dietary fiber in maintaining health. Fiber appears to sweep the digestive system free of unwanted substances that could promote cancer and to maintain regularity and prevent disorders of the digestive tract. Fiber also provides a sense of fullness that may help reduce overeating and unwanted weight gain. Diets that are rich in fiber and complex carbohydrates have been associated with lower serum cholesterol and a lower risk for high blood pressure, coronary artery disease, and some types of cancer. But does this mean that it's okay just to take a fiber pill? No! Rather, the studies that have shown the beneficial effects of a high-fiber diet (containing 25 to 30 grams of fiber per day) have been those in which the dietary fiber is in the form of fruits, vegetables, whole grains, and cereals. These and other studies suggest that not only the fiber in these foods but also

PLANT FIBERS: INSOLUBLE AND SOLUBLE

Insoluble Fiber—also known as roughage
These fibers hold onto water, add bulk, and promote movement through the intestine.

CELLULOSE	HEMICELLULOSE	LIGNIN
Whole-wheat flour	Bran	Woody portions of plants:
Bran	Whole grains	such as stem of broccoli, carrots
Nuts	Vegetables	Wheat
Vegetables	Fruits	Fruits with edible seeds, such as raspberries and strawberries

Soluble Fiber—dissolves in water and becomes gummy
These types of fibers can help lower blood cholesterol and blood glucose.

GUMS	PECTIN
Oats	Apples
Legumes and dried beans	Citrus fruits: oranges, grapefruits
Barley	Carrots

WHERE ARE THE WHOLE GRAINS?

Whole-grain bread products are labeled as whole grain, whole wheat, or rye.

In contrast, bread products labeled as *made with* wheat, cracked wheat, seven-grain, multi-grain, stone-ground wheat, or any of several other names contain *mostly* refined flour and lack the health-promoting effects of a whole-grain product.

the vitamins, minerals, and other compounds they contain contribute to their health-promoting effects.

The Dietary Guidelines for Americans (see Chapter 1, page 8) recommend that we obtain most (about 60%) of our calories from carbohydrates, preferably complex carbohydrates, in the form of foods such as whole grains, fruits, vegetables, and legumes. These foods are good sources of fiber, essential vitamins, minerals, and other phytochemicals and are also more likely to be low in fat.

The average American today consumes only about a third of the recommended amount of fiber. To obtain as many of the potential benefits as possible, you need to obtain complex carbohydrates and fiber from various food sources. Although studies indicate that our intake of carbohydrates is increasing, the contribution of whole-grain foods remains small, partly because identifying whole-grain foods can be confusing. For ideas on what whole-grain foods to look for in your supermarket, see the sidebars Where Are the Whole Grains? and Finding Fiber, this page.

Foods that are naturally good sources of fiber or have fiber added are allowed to make claims on their labels regarding their fiber content. What do the terms used to describe fiber content mean? When you see the phrase "high fiber" on a food label, it means that 1 serving (defined on the Nutrition Facts panel) of the food contains 5 grams of fiber or more per serving. A food that contains 2.5 to 4.9 grams of fiber in a serving is allowed to call itself a "good source" of fiber, and a food label that says "more fiber" or "added fiber" has at least 2.5 grams more fiber per serving.

Sugar Substitutes

For the same reason that people have recently sought substitutes for fat, noncaloric sugar substitutes became popular in the 1960s as people began to try to control their weight. Sugar substitutes are of two basic types: intense sweeteners and sugar alcohols.

FINDING FIBER

How much fiber will you find in the foods you eat?

FOOD	SERVING	FIBER (GRAMS)
Apple with skin	1 medium	4
Banana	1 medium	3
Orange juice	3/4 cup	Less than 1
Orange	1 medium	3
Strawberries	1 cup halves	4
Broccoli, raw	1/2 cup	1
Broccoli, cooked	1/2 cup	2
Carrot, raw	1 medium	2
Potato, baked with skin	1 medium	4
Spinach, cooked	1/2 cup	2
Spinach, raw	1 cup	1
Tomato	1 medium	1
Lentils, cooked	1/2 cup	8
Beans, baked	1/2 cup	7
White bread	1 slice	Less than 1
Whole-wheat bread	1 slice	2
Pumpernickel bread	1 slice	2
Rice, white	1/2 cup	Less than 1
Rice, brown	1/2 cup	2
Bran cereal, 100%	1/3 cup	8
Corn flakes	1 cup	1
Oatmeal, cooked	from 1/3 cup dry	3
Popcorn, air popped	1 cup	1
Spaghetti, cooked, regular	1/2 cup	1
Spaghetti, cooked, whole wheat	1/2 cup	3
Hummus	2 tablespoons (chickpea dip)	2

Intense sweeteners are also called non-nutritive sweeteners, because they are so much sweeter than sugar that the small amounts needed to sweeten foods contribute virtually no calories to the foods. These sweeteners also do not promote tooth decay. Currently, four such intense sweeteners are available, both for use in processed foods and for home consumption. The U.S. Food and Drug Administration (FDA) has set "acceptable daily intakes" (ADI) for these sweeteners. The ADI is the amount that can be consumed daily over a lifetime without risk.

One of these sweeteners is aspartame (NutraSweet brand). It is manufactured by chemically modifying the naturally occurring amino acid phenylalanine. This sweetener can't be used by people with phenylketonuria (a rare congenital disorder that disrupts the body's ability to metabolize phenylalanine and can result in severe nerve damage). Despite extensive safety testing showing aspartame to be safe, its use has been implicated by the popular press in everything from headaches to loss of attentiveness. At this time, there is no scientific validity to these claims. Aspartame is not heat stable, so it can't be added to foods that will be cooked or baked, although it can be added to some foods (such as coffee) after heating.

Saccharin, a second non-nutritive sweetener, was associated with cancer in mice when it was fed in very large amounts. However, further studies have found no links between saccharin and human cancer. This recently led the U.S. government to remove it from its list of potential cancer-causing chemicals. Although saccharin is heat stable,

in some cases it cannot satisfactorily be used in baking because it lacks the bulk of sugar.

Acesulfame K (Sunnett), a third intense sweetener, was approved by the FDA in 1998 for use in soft drinks, although it was used in various food products before that. About 200 times sweeter than sugar, this noncaloric product has been extensively tested for safety. After reviewing more than 90 studies, the FDA deemed the sweetener safe in amounts up to the equivalent of a 132-pound person consuming 143 pounds of sugar annually (an Acceptable Daily Intake of 15 milligrams per kilogram of body weight; 1 kilogram is about 2.2 pounds). Because it is not metabolized, acesulfame K can be used safely by people with diabetes. The sweetener is more heat stable than aspartame, maintaining its structure and flavor at oven temperatures more than 390° Fahrenheit and under a wide range of storage conditions. Like saccharin and aspartame, acesulfame K lacks bulk, so its use in home baking requires recipe modification. The flavor of acesulfame K has been described as

Foods contain a variety of carbohydrates, from simple to complex. Fruits, vegetables, grains, and dairy products all contain carbohydrate.

clean and quickly perceptible, although disagreement exists about whether it leaves an aftertaste.

A fourth intense sweetener, sucralose (Splenda), was approved by the FDA in 1998 for sale and use in commercial food products. Sucralose is made by chemically modifying sucrose (table sugar) to a non-nutritive, non-caloric powder that is about 600 times sweeter than sugar. Before approving sucralose, the FDA reviewed more than 110 research studies conducted in both human and animal subjects. It concluded that the sweetener is safe for consumption by adults, children, and pregnant and breast-feeding women in amounts equivalent to the consumption of about 48 pounds of sugar annually (an Acceptable Daily Intake of 5 milligrams per kilogram of body weight). People with diabetes may also safely consume the sweetener, because it is not metabolized like sugar. In addition, sucralose is highly stable to heat and so will not lose its sweetness when used in recipes that require prolonged exposure to high temperatures (such as baking) or when stored for long periods. The product is currently available in the form of a powdered sugar substitute and in some commercial baked goods, jams and jellies, sweet sauces and syrups, pastry fillings, condiments, processed fruits, fruit juice drinks, and beverages, and its use is approved for various additional products. However, use of sucralose in home baking is expected to be limited by its low bulk in comparison with table sugar.

Foods containing intense sweeteners should not be given to infants or children, who need energy to grow and to sustain their high activity levels. Foods that contain intense sweeteners and lack any nutritive value also should not replace nutrient-dense foods in your diet.

The sugar alcohols xylitol, mannitol, and sorbitol contain less than 4 calories per gram. These sugar alcohols are digested so slowly that most are simply eliminated. Unfortunately, excessive consumption can cause diarrhea or bloating in some people.

So-called "natural" sweeteners provide the same number of calories as sugar and have acquired the reputation, albeit incorrectly, of being healthier than sugar, because they seem more natural than processed table sugar. These include honey, maple syrup and sugar, date sugar, molasses, and grape juice concentrate. In reality, these sweeteners contain no more vitamins or minerals than table sugar. Honey may harbor small amounts of the spores of the bacteria that produce botulism toxin and should never be given to babies younger than 1 year.

The Bottom Line on Carbohydrates

Carbohydrates—sugars and starches—are the main source of fuel for our bodies. When we choose carbohydrate-rich foods, our best bets are fruits, vegetables, whole grains, and legumes, because these foods are also rich sources of health-promoting vitamins, minerals, phytochemicals, and fiber. But like all calories, extra calories from carbohydrates beyond those we need to replenish the energy we burn are converted to fat and stored in our fat cells. Non-caloric sweeteners seem to be a safe alternative to sugar for most people, but the foods that contain them are often nutritionally empty and their use in home cooking is limited. The so-called natural sweeteners are no better for you than sugar.

PROTEIN

Protein is an essential part of our diets. Proteins are large, complex molecules resembling tangled strings of beads. Each of the "beads" on the string is one of a group of smaller molecules called amino acids. Amino acids are composed of carbon, oxygen, hydrogen, and nitrogen, and some contain sulfur.

Using the amino acids from the protein you eat, the body makes more than 50,000 different proteins. These proteins are the main structural elements of our skin, hair, nails, cell membranes, muscles, and connective tissue. Collagen, the main protein in our skin, provides a barrier to the invasion of foreign substances. Proteins in cell membranes determine what substances can enter and exit cells. Our muscles, which contain some 65 percent of the body's total protein, give our bodies their shape and strength. Proteins in connective tissues such as tendons, ligaments, and cartilage enable our skeletons to function, form internal organs, and hold the organs in place. Proteins in the blood carry oxygen to all cells and remove carbon dioxide and other waste products. The proteins in muscle, connective tissue, and blood make up most of the protein in the body. Other proteins called enzymes accelerate metabolic processes, and still other proteins and amino acids are hormones and neurochemicals, the substances that deliver signals throughout the body and regulate all metabolic processes.

During periods of growth, our bodies must manufacture and store large amounts of protein. Therefore, the requirement for protein in our diets is higher during growth. But even when we are not growing, each of the unique

proteins in the body has a finite lifespan and must be replaced continuously. So the need for protein never ends.

Dietary Protein and Body Protein

The thousands of proteins that make up our bodies are assembled on demand from some 20 different amino acids. What are these amino acids, and where do they come from? The protein from the meat we ate last night is not directly incorporated into our muscles. The proteins in the foods we eat are digested first into small "peptides." Some of these peptides are further digested into their constituent amino acids. Only amino acids and small peptides are actually absorbed by the small intestine into the bloodstream. They are then delivered to the liver, muscles, brain, and other organs, where they are used to make new proteins or converted to other amino acids needed by those organs.

Of the 20 amino acids that make up all proteins, 9 are considered "essential" because they cannot be made in our bodies and must be obtained from the foods we eat. Of the remaining 11, some are essential for infants and persons with certain diseases (see sidebar: Amino Acid Classification, page 25). The rest of the amino acids are considered "nonessential," because our bodies can make them in adequate amounts, if necessary. Nevertheless, they are easily supplied by eating a well-balanced diet that includes a variety of foods.

Most foods contain protein. Some foods are better sources of protein than others. "Complete" proteins are those that contain all the essential amino acids in amounts needed to synthesize our body's proteins. The best sources of complete protein are lean meats and poultry, fish, low-fat dairy products, and eggs (see Part II, High-Protein Foods, page 291, and Dairy Foods, page 345).

Lean meats (including poultry and fish) and dairy products aren't the only foods that contain protein. By eating a variety of foods, including grains and legumes, you are likely to meet your needs.

The grains and cereals group of foods, which form the base of the Food Guide Pyramid, are excellent sources of protein, but because these proteins often lack one or more essential amino acids, they are called "incomplete" proteins. For example, the proteins in corn are low in the essential amino acids lysine and tryptophan, and wheat is low in lysine. In contrast, legumes tend to be rich in lysine but a bit low in methionine. Among the legumes, soybeans contain the most complete protein.

Does this mean you must eat meat, eggs, and dairy products (foods of animal origin) to get all the amino acids you need? Not at all. By eating a variety of different foods, including grains and legumes, you are likely to get all the amino acids you need and in the correct amounts. People of many cultures and vegans (vegetarians who eat no foods of animal origin) get adequate amounts and types of protein by eating various combinations of plant proteins including beans, corn, rice, and other cereal grains. Although it was once thought necessary to combine these foods at the same meal, nutrition experts now agree that they can be eaten at various times throughout the day.

When we eat grains and legumes, rather than foods of animal origin (a more frequent source of protein in our diets), we gain additional health benefits. Whole-grain foods and legumes are rich in vitamins, minerals, fiber, and other substances that optimize health. If that does not seem like reason enough to make the trade, grains and legumes lack the high levels of saturated fat present in foods of animal origin, which, as you will learn below, are linked to many diseases.

Contrary to popular belief, simply eating more dietary protein, in excess of recommended amounts, will not result in bigger muscles. Our bodies do not store excess protein. If we eat more protein than our bodies need to replenish the amino acids we have used during the day, the excess amino acids are converted to, and stored as, fat. Dietary protein, like carbohydrates, supplies about 4 calories of energy per gram. Because our requirements for protein mainly depend on our body's size, our need for protein increases during times of rapid growth. Therefore, the recommendations for protein are age-dependent and are slightly higher for pregnant and breastfeeding women than for other adults (see the Appendix: Dietary Reference Intakes, page 421). The recommended allowances ensure an adequate protein intake by nearly all healthy people. Nevertheless, many Americans typically consume twice this amount, often in the form of meat and dairy products that are high in

AMINO ACID CLASSIFICATION

Of the 20 amino acids that make up all proteins, only 9 are considered "essential" in our diets because they cannot be made by our bodies and must be obtained from the foods we eat.

ESSENTIAL IN OUR DIETS	ESSENTIAL IN OUR DIETS UNDER SOME CIRCUMSTANCES	NONESSENTIAL IN OUR DIETS
Histadine	Arginine	Alanine
Isoleucine	Cysteine*	Asparagine
Leucine	Tyrosine†	Aspartic acid
Lysine		Glutamic acid
Methionine		Glutamine
Phenylalanine		Glycine
Threonine		Proline
Tryptophan		Serine
Valine		

Cysteine can be synthesized from the essential amino acid methionine.
†*Tyrosine can be synthesized from phenylalanine when supplied in adequate amounts.*

saturated fat, which increases the risk for coronary artery disease and some forms of cancer.

What if we eat too little protein? Few Americans are at risk of eating too little protein. However, individuals on severely restricted diets, those who are unable to eat, and those whose needs are increased because of illness or trauma may experience protein deficiency. To replenish the pools of essential amino acids that have been depleted, in order to make critical proteins such as enzymes and hormones, the body of a protein-deficient person begins to rob protein from muscle by digesting that protein to its constituent amino acids. Because muscle is needed for various vital functions (for example, diaphragm muscles for breathing and heart muscles for pumping our blood), the loss of large amounts of muscle protein can be fatal. Fortunately, the vast majority of people, even those who engage in regular, rigorous endurance exercise, can easily meet their need for protein by eating a balanced diet based on the Food Guide Pyramid.

The Bottom Line on Protein

Adequate protein is critical for growth, metabolism, and health, but eating more protein than we need will not build bigger muscles. Conversely, excess protein is converted to fat. Foods of animal origin are high in protein but may also be high in total and saturated fat. Lean meats and dairy products, fish, legumes, and grains are the best sources of protein.

FATS

It's difficult to read a newspaper or listen to the evening news without hearing something new about fat and its connection with disease. Diets that are high in fat are strongly associated with an increased prevalence of obesity and an increased risk of developing coronary artery disease, high blood pressure, diabetes mellitus, and certain types of cancer. Health authorities recommend that we reduce our total fat intake to about 30 percent of total calories. They also recommend that we limit our intake of saturated fat (the type of fat most often found in meat and dairy products) to less than 10 percent of our fat calories and try to be sure that the fat we do eat is mostly the monounsaturated or polyunsaturated type. These changes have been shown to decrease our risk for several diseases.

Fat as a Nutrient

Fat is an essential nutrient, because our bodies require small amounts of several fatty acids from foods (the so-called essential fatty acids) to build cell membranes and to make several indispensable hormones, namely, the steroid hormones testosterone, progesterone, and estrogen, and the hormone-like prostaglandins. Dietary fats also permit one group of vitamins, the fat-soluble vitamins (A, D, E, and K), to be absorbed from foods during the process of digestion. Fats help these vitamins to be transported through the blood to their destinations. The fat in our bodies also provides protective insulation and shock absorption for vital organs.

As a macronutrient, fat is a source of energy (calories). The fat in food supplies about 9 calories per gram, more than twice the number of calories as the same amount of protein or carbohydrate. As a result, high-fat foods are considered "calorie-dense" energy sources. Any dietary fat that is not used by the body for energy is stored in fat cells (adipocytes), the constituents of fat (adipose) tissue (see

Chapter 3, Obesity, page 47). The Dietary Guidelines for Americans recommend that no more than 30 percent of our calories should come from fat, and only a third of that should be saturated fat.

Sorting Out the Fats

Our health is influenced by both the amount and the type of fat that we eat. Fats are molecules; they are classified according to the chemical structures of their component parts. But you don't need to be a chemist to understand the connection between the various fats in foods and the effect these fats have on the risk for disease. Some definitions will help.

Dietary fats, or triglycerides, are the fats in foods. They are molecules made of fatty acids (chain-like molecules of carbon, hydrogen, and oxygen) linked in groups of three to a backbone called glycerol. When we eat foods that contain fat, the fatty acids are separated from their glycerol backbone during the process of digestion.

Fatty acids are either saturated or unsaturated, terms that refer to the relative number of hydrogen atoms attached to a carbon chain. Fat in the foods that we eat is made up of mixtures of fatty acids—some fats may be mostly unsaturated, whereas others are mostly saturated (see sidebar: A Comparison of Fats, page 27).

Monounsaturated fatty acids are fatty acids that lack one pair of hydrogen atoms on their carbon chain. Foods rich in monounsaturated fatty acids include canola, nut, and olive oils; they are liquid at room temperature. A diet that provides the primary source of fat as monounsaturated fat (frequently in the form of olive oil) and includes only small amounts of animal products has been linked to a lower risk of coronary artery disease. This type of diet is commonly eaten by people who live in the region surrounding the Mediterranean Sea (see Chapter 1, page 14).

Polyunsaturated fatty acids lack two or more pairs of hydrogen atoms on their carbon chain. Safflower, sunflower, sesame, corn, and soybean oil are among the sources of polyunsaturated fats (which are also liquid at room temperature). The essential fatty acids, linoleic and linolenic acid, are polyunsaturated fats. Like monounsaturated fats, polyunsaturated fats lower blood cholesterol levels and are an acceptable substitute for saturated fats in the diet.

Saturated fatty acids, or saturated fats, consist of fatty acids that are "saturated" with hydrogen. These fats are found primarily in foods of animal origin—meat, poultry, dairy products, and eggs—and in coconut, palm, and

A COMPARISON OF FATS

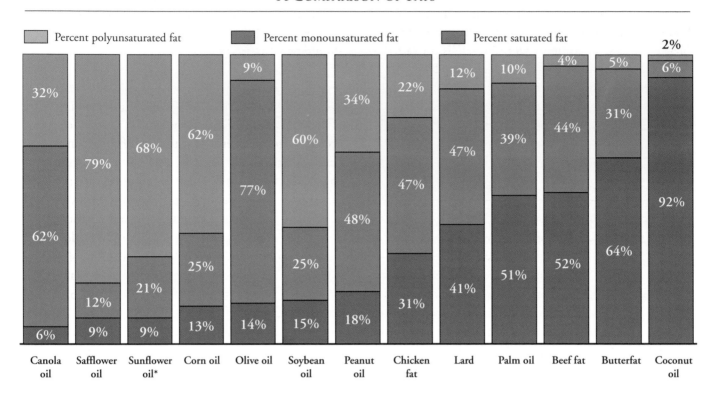

| | Percent polyunsaturated fat | | Percent monounsaturated fat | | Percent saturated fat |

Canola oil: 32%, 62%, 6%
Safflower oil: 79%, 12%, 9%
Sunflower oil*: 68%, 21%, 9%
Corn oil: 62%, 25%, 13%
Olive oil: 9%, 77%, 14%
Soybean oil: 60%, 25%, 15%
Peanut oil: 34%, 48%, 18%
Chicken fat: 22%, 47%, 31%
Lard: 12%, 47%, 41%
Palm oil: 10%, 39%, 51%
Beef fat: 4%, 44%, 52%
Butterfat: 5%, 31%, 64%
Coconut oil: 2%, 6%, 92%

Total is not 100% because of the presence of other, minor fat compounds.

palm kernel oil (often called "tropical oils"). Foods that are high in saturated fats are firm at room temperature.

Because a high intake of saturated fats increases your risk of coronary artery disease, nutrition experts recommend that less than 10 percent of your calories should come from saturated fats. To find out how to calculate your fat allowance, see the sidebar Recommended Fat Intake, page 28, and to determine the total and saturated fat contents of some foods, see the sidebar Where's the Fat? page 28.

Omega-3 fatty acids are a class of polyunsaturated fatty acids found in fish (tuna, mackerel, and salmon, in particular) and some plant oils such as canola (rapeseed) oil. These fatty acids have made the news because of the observation that people who frequently eat fish appear to be at lower risk for coronary artery disease. Omega-3 fatty acids also seem to play a role in your ability to fight infection.

Hydrogenated fats are the result of a process in which unsaturated fats are treated to make them solid and more stable at room temperature. The hydrogenation process, which involves the addition of hydrogen atoms, actually results in a saturated fat. Trans-fatty acids are created by

hydrogenation. An increase in consumption of these fats is a concern because they have been associated with an increased risk of coronary artery disease. Hydrogenated fat is a common ingredient in stick and tub margarine, commercial baked goods, snack foods, and other processed foods.

Cholesterol is a waxy, fat-like substance that is a necessary constituent of cell membranes and serves as a precursor for bile acids (essential for digestion), vitamin D, and an important group of hormones (the steroid hormones). Our livers can make virtually all of the cholesterol needed for these essential functions. Dietary cholesterol is found only in foods of animal origin, that is, meat, poultry, milk, butter, cheese, and eggs. Foods of plant origin, that is, fruits, vegetables, nuts, seeds, legumes, grains, and the oils derived from them, do not contain cholesterol. Eggs are the food most often associated with cholesterol, because the average large egg contains about 210 milligrams of cholesterol (only in the yolk), and the recommended daily cholesterol intake is 300 mg or less. However, for most people, meat contributes a higher proportion of cholesterol to the diet than do eggs, because cholesterol is found in both the lean and fat portions

RECOMMENDED FAT INTAKE

*Health experts recommend consuming no more than 30 percent of our calories from fat—with no more than
10 percent from saturated fat. Here is a chart to help you determine how much fat you should eat.*

	CALORIES	TOTAL FAT (GRAMS)	SATURATED FAT (GRAMS)
Most women & older adults	About 1,600	53	Less than 17
Most men, active women, teen girls & children	About 2,200	73	Less than 24
Active men & teen boys	About 2,800	93	Less than 31

Note: These are guidelines for healthy individuals. If you have heart disease or high blood cholesterol or triglyceride levels, or if you are overweight, ask your health care professional to recommend the types of fat and amounts most appropriate for you.

of meat. Shellfish have acquired an undeserved reputation for being high in cholesterol. Their cholesterol and total fat contents are actually comparatively low (see sidebar: Cholesterol Content of Foods, page 29).

Fat Substitutes

To appeal to our desire for lower-fat substitutes for our favorite high-fat foods, the commercial food industry has developed low- or lower-fat versions of many foods using various fat replacers. Until recently, fat replacers always consisted of proteins or carbohydrates, such as starches or gels, but the kinds of foods that could be prepared with these fat replacers were limited by their inability to withstand the high temperatures of frying. In 1996, after a long period of development, safety testing, and governmental review, the first non-caloric fat, olestra, was approved by the FDA for use in the manufacture of savory (non-sweet) snacks (such as crackers and chips). Because olestra is a modified fat, it is the first heat-resistant fat substitute, which allows it to be used to make fried foods. In addition, olestra gives foods the flavor and creamy "mouth feel" of high-fat foods.

FDA approval of olestra was controversial for two reasons. First, this artificial ingredient, if approved and accepted, would be the first in history to be consumed in quantities comparable to the quantities of fat, carbohydrates, and proteins we currently consume from food sources. In other words, these novel, previously unknown substances could become major parts of the diets of some people, and there

WHERE'S THE FAT?

Foods That May Pack More Fat Than You Think

FOOD	TOTAL FAT (GRAMS)	SATURATED FAT (GRAMS)	CALORIES
Butter (2 Tblsp)	23	14	200
Porterhouse steak (3 oz)	21	8	270
Peanuts (1/4 cup)	18	3	210
Cake donut (1 large)	16	3	300
Blue cheese dressing (2 Tblsp)	16	3	150
Peanut butter (2 Tblsp)	16	3	190
Chocolate/peanut/ nougat candy bar (2 oz)	14	5	270
Batter-fried chicken thigh (3 oz meat)	14	4	240
Croissant, cheese (2 oz)	12	6	235
Croissant, plain (2 oz)	12	7	230
Ice cream (1/2 cup rich vanilla)	12	7	180
Cream cheese (2 Tblsp)	10	6	100
Milk, whole (1 cup)	8	5	150
Milk, 2% (1 cup)	5	3	120
Sour cream (2 Tblsp)	5	3	50
Half-and-half (2 Tblsp)	3	2	40

would be no historical experience to tell us what the substances might do in our bodies. Some scientists predicted that the substance would cause serious gastrointestinal complaints despite controlled studies demonstrating its safety. However, in the first year of availability of olestra-containing foods, the predicted intestinal problems were not significant. Tests in which volunteers ate large quantities of olestra-containing potato chips or regular potato chips without knowing which type they were eating showed no differences in gastrointestinal complaints between the two groups. Second, tests of olestra showed that it inhibits the absorption of fat-soluble compounds (vitamins A, D, E, and K and some carotenoids) from foods eaten at the same time as the olestra-containing foods, whereas it has no effect on the absorption of other nutrients or on the body's stores of fat-soluble vitamins. To compensate for this effect of olestra on fat-soluble vitamin absorption, foods prepared with olestra have small amounts of these vitamins added to them. At this writing, the range of foods that can include olestra as a fat substitute is quite narrow. Some questions do remain about the long-term safety of the product, although long-term studies in young, growing animals and several studies in humans have shown no negative effects.

How should you decide whether to include foods with fat replacers in your eating plan, and how much of these foods do you include? From a health standpoint, small amounts of olestra-containing foods appear to be harmless. But from a purely nutritional standpoint, most foods that contain fat replacers are snack foods essentially devoid of nutritional benefit. In addition, these foods are not calorie-free. Many remain high in calories, and some foods that contain carbohydrate fat replacers are even higher in calories than their higher-fat counterparts, so they are still calorie-dense, nutritionally poor foods. It's fine to choose small amounts of these foods occasionally, but better low-fat snack choices include fruits, vegetables, nonfat yogurt, and whole-grain pretzels and breads.

The Bottom Line on Fats

Dietary fat is a source of energy, but high-fat diets, especially diets high in saturated fat, increase the risk of gaining excessive amounts of weight and of developing diabetes, coronary artery disease, high blood pressure, and several types of cancer. This increased risk is the reason that health experts encourage us to reduce our intake of total and saturated fats by:

CHOLESTEROL CONTENT OF FOODS

Cholesterol is found only in animal products. Below is a list of common foods and their cholesterol content.

FOOD	SERVING SIZE	CHOLESTEROL (IN MILLIGRAMS)
Chicken without the skin, lean beef or pork	3 oz	70-80
Egg	1	About 210
King crab	3 oz	45
Lobster	3 oz	60
Organ meat (liver)	3 oz	330
Shrimp	3 oz	165

- increasing our intake of fruits, vegetables, and whole-grain foods, which are naturally low in fat, and preparing them with a minimum of added fats
- consuming low-fat dairy products such as nonfat milk and yogurt and reduced-fat cheeses
- limiting our intake of red meat, poultry, and fish to 5 to 7 ounces daily
- choosing lean cuts of red meat and poultry, removing the skin before eating poultry, and preparing the meat with a method that uses little or no additional fats
- choosing some fish that is high in omega-3 fatty acids and preparing it with little or no added fat

THE MICRONUTRIENTS: VITAMINS AND MINERALS

Vitamins and minerals are required by every process in your body. Unlike the macronutrients, vitamins and minerals by themselves do not contain energy. Instead, they work with the energy-rich macronutrients—carbohydrates, protein, and fats—and with each other to help your body to release, use, and store the energy from those macronutrients.

Vitamins

Vitamins are small but complex molecules. In addition to helping us to use and store energy from macronutrients, they assist the molecules responsible for vision to perform their function, they serve as regulatory hormones for bone formation, and they act as antioxidants to preserve cellular

function (see Vitamins and Minerals as Antioxidants, page 31).

Each of the vitamins was discovered and its requirement determined by its ability to cure and prevent a particular disease or group of symptoms. For example, the discovery that a substance in limes could cure and prevent the disease called scurvy led to the discovery that our bodies require vitamin C and that scurvy is the result of vitamin C deficiency. Dietary Reference Intakes (DRIs) are being established for each vitamin. They describe the amount of the vitamin that should prevent symptoms of deficiency in most people, with a little extra added.

Today, the diseases that result from vitamin deficiencies rarely occur except in severely malnourished individuals or in those with certain medical conditions. Rather, in a virtual nutrition revolution, research has progressed beyond the identification and treatment of simple nutrient deficiencies to the realization that some of the vitamins (or foods that contain them) may help maintain health by preventing the development of chronic diseases such as cancer and heart disease in otherwise well-nourished people.

The Food Guide Pyramid is based on the DRIs. It tells us the number of servings, in each group of foods, that will supply us with the recommended allowance of most of the vitamins. Nutrition research also has begun to support the idea that a few of the vitamins and minerals, notably those referred to as antioxidants (see Vitamins and Minerals as Antioxidants, page 31), may provide even more benefit if taken in quantities somewhat greater than the recommended amounts. This idea raises some questions. Is there such a thing as too much of a vitamin? Should these extra vitamins come from food, or is it okay to take a supplement if you just can't eat that much? And, should the recommended amounts for these vitamins be increased? Although there really is no answer to the last question yet, the answers to the first two questions depend on the type of vitamin. As we begin our discussion of vitamins, we want to emphasize that although it is virtually impossible to overdose on vitamins from food alone, some vitamin supplements definitely offer too much of a good thing (see sidebar: What Are Daily Values? this page, and Supplement Sense, page 38).

The 14 essential vitamins can be classified into two groups: water-soluble and fat-soluble. They are classified on the basis of their molecular structure, which determines the way the vitamins are carried in food and in the bloodstream and the manner in which they are stored in your body. The text that follows describes which vitamins fall into which category and what that means for your health.

Water-Soluble Vitamins

There are 10 water-soluble vitamins. The B complex vitamins have various roles, some of which involve their action, in concert, to regulate the body's use of energy from food. Folic acid is an important factor in the regulation of growth. During the early stages of pregnancy, folic acid is important for preventing a type of birth defect known as a neural tube defect. Vitamin C, also known as ascorbic acid, functions in various ways, many of which seem to be related to its antioxidant properties (see the Appendix: A Quick Look—Vitamins, Their Functions and Food Sources, page 430).

As their name implies, water-soluble vitamins dissolve in water. The body strives to maintain the optimal level of each of the water-soluble vitamins for its immediate needs. Surplus water-soluble vitamins are excreted in the urine and through perspiration, because they are not stored in the body to any appreciable extent. Water-soluble vitamins must be replenished almost daily, preferably by eating foods that are rich in these vitamins. Fruits, vegetables, grains, and beans are excellent sources of the water-soluble vitamins (with the exception of vitamin B_{12}, which is found only in foods of animal origin). However, if you choose to use

WHAT ARE DAILY VALUES?

Have you ever wondered what "% Daily Value" means on the Nutrition Facts label? How does it relate to the Dietary Reference Intakes (DRIs)? As you may recall from Chapter 1, the DRIs, set by the Food and Nutrition Board of the National Academy of Sciences, are the amounts of each nutrient recommended for most healthy people. Because DRIs are both sex- and age-specific, each nutrient has a range of DRIs. To make it easier to show how a food meets your recommended allowance for some of the more critical nutrients, the Food and Drug Administration has established a Daily Value for each nutrient, which is approximately the highest recommended amount for that nutrient. The "% Daily Value" reported on the Nutrition Facts label is based on a maintenance calorie level of 2,000 calories daily. If your maintenance calorie level is 1,500 calories, your daily values may be a bit lower, so the nutrient contents of the food satisfy a higher percentage of your daily value.

multivitamins or single-nutrient supplements to get your water-soluble vitamins, moderation is advised (see Supplement Sense, page 38), because high doses of several of the B vitamins can have harmful effects, and high doses of vitamin C may contribute to the formation of kidney stones.

Fat-Soluble Vitamins

The fat-soluble vitamins, A, D, E, and K, are found in the food you eat, absorbed into your bloodstream, and carried throughout your body attached to fat molecules. Because fat-soluble vitamins can be stored in the body, they do not need to be replenished on a daily basis. Vitamins A and D are stored in the liver, and reserve supplies may be sufficient for as long as 6 months. Reserves of vitamin K, however, may be sufficient for only a few weeks, and the supply of vitamin E can last somewhere between several days and several months.

If taken in excess, usually in the form of a supplement, fat-soluble vitamins can accumulate in the body. Large stores of vitamins A and D can actually become harmful. Fortunately, it is difficult to get an excess of fat-soluble vitamins from food. For example, beta-carotene, the molecule found in some foods of plant origin that gives carrots and squash their yellow-orange color, is converted to vitamin A in the body. But because the chemical reaction that converts beta-carotene to vitamin A is carefully regulated, it is nearly impossible to get vitamin A toxicity from eating fruits and vegetables. Foods that provide vitamins D, E, and K would need to be consumed in such quantity that achieving toxic levels of these nutrients is highly unlikely. However, a high intake of vitamin K-containing foods may contribute to abnormal bleeding in people who are receiving blood-thinning (anticoagulant) medication. Therefore, it is important to keep foods that are high in vitamin K relatively constant. Vitamin E supplements also are not recommended for these individuals.

The recent development of diet drugs that work by inhibiting your body's absorption of fat has raised concerns about the possibility that the use of such drugs could lead to deficiencies in the fat-soluble vitamins. Preliminary results suggest that when taken as directed, these drugs may interfere with the absorption of beta-carotene and vitamin D from foods. If these findings are confirmed, persons who take these prescribed drugs to manage their weight also will be required to take a daily multiple vitamin that contains the fat-soluble vitamin.

As mentioned in the discussion of fat substitutes, early research on the fat substitute olestra showed that eating olestra-containing snack foods interferes with the body's ability to absorb some of the fat-soluble vitamins from the other foods we eat. To compensate for this effect, the manufacturers of snack foods that contain olestra were required to fortify these products with fat-soluble vitamins. Thus, eating olestra-containing snack foods does not seem to create a significant risk for deficiency of fat-soluble vitamins (unless you consistently choose to eat them in place of more nutritious foods).

To learn about the many roles of vitamins, how much of each of them you need, and the best food sources for each one, see the Appendix: A Quick Look—Vitamins, Their Functions and Food Sources, page 430.

Minerals

Minerals are just what the term indicates—elements found in the earth. Like the vitamins, minerals play a multitude of roles in our bodies. Unlike the vitamins, some minerals—calcium and phosphorus—have a structural function. These minerals are the main components of our bones and teeth. Calcium has an additional critical role. Along with several other major minerals—sodium, chlorine, potassium, and magnesium—calcium is a regulator of cell function. The minerals sodium, chloride, and potassium (also referred to as electrolytes) are responsible for maintaining the balance of fluids inside and outside of cells and, along with calcium, controlling the movement of nerve impulses.

Trace minerals are those that your body needs in smaller amounts, usually less than 20 milligrams daily. These include iron, chromium, cobalt, copper, fluoride, iodine, manganese, molybdenum, selenium, and zinc. The mineral iron forms the active part of hemoglobin, the protein in your blood that delivers oxygen to different sites in your body and picks up carbon dioxide. Although DRIs have been established for some of the trace minerals, those for which too little is known to establish precise DRIs have a recommended Adequate Intake (AI) (see the Appendix: A Quick Look—Minerals, Their Functions and Food Sources, page 432).

Vitamins and Minerals as Antioxidants

Several vitamins and minerals are considered antioxidants. These include vitamins E and C, beta-carotene (which can be converted to vitamin A), other carotenoids (some may be converted to vitamin A and also play a role in cell devel-

opment), and the minerals selenium, copper, zinc, and manganese. What are antioxidants? What do they do?

Every cell in our body needs oxygen to use the nutrients that food provides. However, when oxygen is used by cells, by-products called free radicals are formed. If allowed to accumulate, these free radicals can damage tissues, cells, and deoxyribonucleic acid (DNA, the genetic material of cells). The process of oxidative damage can be observed as the browning that occurs when sliced apples or potatoes are exposed to the air or the rancid flavor that butter and cooking oils develop when stored for long periods. Environmental pollutants such as cigarette smoke and ultraviolet light from the sun also contribute to the formation of free radicals in our bodies. Although not proved, studies suggest that excess free-radical production can increase the risk of cancer, heart disease, cataracts, and the other types of cell deterioration that are associated with aging.

Just as the vitamin C in lemon juice can prevent sliced apples from browning, antioxidants scavenge and neutralize the effects of free radicals in our bodies. Each antioxidant has its own unique effect. Vitamin C, which is water-

FOOD SOURCES OF ANTIOXIDANTS

CAROTENOIDS
Beta-carotene—carrots, broccoli, sweet potatoes, greens (dandelion, turnip, beet, spinach), squash (butternut, Hubbard), red bell peppers, apricots, cantaloupe, mango

Alpha-carotene—greens (see above), carrots, squash (see above), corn, green peppers, potatoes, apples, plums, tomatoes

Lycopene—tomatoes, watermelon, pink grapefruit

VITAMIN C
Bell peppers (red and green), guavas, greens (see above), broccoli, Brussels sprouts, cauliflower, strawberries, papayas, oranges and grapefruits and their juices

VITAMIN E
Polyunsaturated vegetable oils, seeds, nuts, fortified cereals, greens (see above), tomato products

SELENIUM
Wheat germ, Brazil nuts, whole-wheat bread, bran, oats, turnips, brown rice, orange juice

Some of the many food sources of antioxidants.

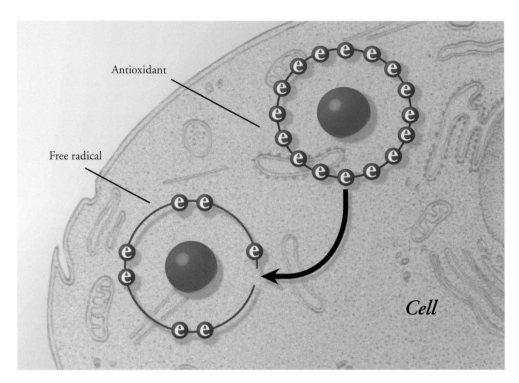

Antioxidant

Free radical

Cell

When oxygen is used by cells, by-products called free radicals are naturally formed. Free radicals are molecules with a missing electron. Simply put, free radicals "want" their full share of electrons. They will take electrons from vital cell structures, causing damage and leading to disease. Antioxidants are able to donate electrons. Nutrients such as vitamin C, vitamin E, or beta-carotene are antioxidants that block some of this damage by donating electrons to stabilize and neutralize the harmful effects of free radicals.

soluble, removes free radicals from body fluids and cell structures composed mainly of water. Beta-carotene and vitamin E are fat-soluble. They seem to be active primarily in fat tissues and cell membranes throughout the body. The mineral selenium is an antioxidant that assists vitamin E.

What is the best source of antioxidants? With the possible exception of vitamin E, the best source of antioxidants is food (see Supplements: Foods, or Functional Foods? page 34). Fruits, vegetables, and grains provide a wide variety of both known and yet to be discovered antioxidants that appear to protect your body's vital functions.

WATER

As an essential nutrient, water is the most often overlooked and taken for granted. Yet 75 percent of our body weight is water. Water contributes to nearly every major process in our bodies. It keeps our body temperature stable, maintains body chemicals at their proper concentrations, carries nutrients and oxygen to cells, and removes waste products. Water also cushions joints and protects organs and tissues.

An insufficient intake of water or excessive loss of water can result in dehydration and heat exhaustion, a condition characterized by dizziness, vomiting, muscle cramps,

fatigue, and confusion. Fortunately, under most circumstances our bodies are good at telling us when we are nearing dehydration. When we feel thirsty, our bodies need water. However, as we get older, our bodies' ability to sense dehydration decreases, and older adults often lose the sense of thirst. Our need for water increases with exercise; exposure to hot or even warm, dry, or extremely cold conditions; pregnancy and breastfeeding; and the use of some medications. Nutritionists recommend that we drink 8 or more 8-ounce glasses of water daily. Many fruits and vegetables are 80 to 90 percent water. Therefore, in addition to the vitamins and minerals they supply, fruits and vegetables also contribute to our total water intake. For more information on water, see Part II, Encyclopedia of Foods, page 150.

ON THE NUTRIENT HORIZON: PHYTOCHEMICALS

Carotenoids. Isoflavones. Capsaicins. You may have heard these words in radio advertisements for the latest supplement or seen them in last week's newspaper. But what are they? All of them fall into the category of substances called phytochemicals or phytonutrients. A phytochemical is, literally, any chemical found in a plant (*phyto* is the Greek word for

plant). Armed with the knowledge that people whose diets are rich in foods of plant origin are at lower risk for many serious diseases, nutritional scientists have begun to try to isolate the actual chemicals in foods that may be responsible for promoting health and preventing disease. Nutritionists have adopted the term "phytochemical" or "phytonutrient" to refer to any one of a growing list of substances they have isolated that appear to prevent disease in laboratory animals.

The phytochemicals identified so far are known to have various roles in the plants from which they originate, including capturing the energy from sunlight and conferring resistance to infection by fungi, bacteria, and viruses. How they function in our bodies, and how they may be responsible (along with the known vitamins, minerals, and fiber found in plant foods) for the health-promoting effects, is just beginning to be understood.

The antioxidant beta-carotene is one of a group of phytochemicals known as carotenoids. Beta-carotene, the substance that gives carrots their orange color and their name, is converted to vitamin A (retinol) in our body. Other carotenoids include lutein and zeaxanthin (from green vegetables) and lycopene (from tomatoes). Diets rich in foods containing carotenoids have been associated with a reduced risk of cancer and heart disease. Researchers are also investigating whether lutein-rich diets may be linked to a lower risk for macular degeneration, a disease of the retina that may lead to blindness.

The isoflavones found in soybeans are associated with lower blood cholesterol and a decreased risk for coronary artery disease. In addition, isoflavones, which are also referred to as phytoestrogens (estrogen-like molecules isolated from plants), appear to reduce some of the symptoms of menopause and may confer a lower risk for breast and other cancers (see sidebar: Soy What? this page). For health benefits, 20 to 25 grams of soy protein per day is recommended. For some of the other phytochemicals that have been identified, see the sidebar Phytochemicals: A Food Pharmacy That May Fight Disease, page 35.

Although some phytochemicals are now available in pill form, we do not yet know enough about how they function to assume that any one, on its own, will promote health and prevent disease without the presence of the vitamins, minerals, fiber, and other, as yet unidentified, substances in plant foods. Scientists are making progress in determining how phytochemicals work, but the best way to ensure an adequate intake of all potentially health-promoting substances in foods of plant origin is to eat the foods themselves—fruits, vegetables, beans, nuts, seeds, and whole grains. See the Appendix: Phytochemical Contents of Selected Foods, page 484.

SOY WHAT?

There is increasing evidence that isoflavones, plant estrogens found in soy foods, may have some of the same effects as estrogen. Beneficial effects may include:

- Lower blood lipid levels—Soy may decrease total cholesterol, low-density lipoprotein cholesterol, and triglycerides.

- Decreased cancer risk—Cancer population studies show a decreased risk for hormone-related cancers of the breast, ovary, endometrium, and prostate in countries where plant-based diets with a high content of phytoestrogens are consumed.

- Diminished menopausal symptoms—Population studies show that women who eat soy as their main protein source may have far fewer distressing menopausal symptoms.

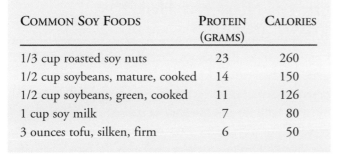

COMMON SOY FOODS	PROTEIN (GRAMS)	CALORIES
1/3 cup roasted soy nuts	23	260
1/2 cup soybeans, mature, cooked	14	150
1/2 cup soybeans, green, cooked	11	126
1 cup soy milk	7	80
3 ounces tofu, silken, firm	6	50

SUPPLEMENTS: FOODS, OR FUNCTIONAL FOODS?

Supplements are concentrated forms of vitamins, minerals, fiber, amino acids, fatty acids, herbal products, enzymes, plant or animal tissue extracts, or hormones. Some supplements contain one or two known nutrients or a small group of nutrients such as B vitamins or antioxidants.

PHYTOCHEMICALS

A Food Pharmacy That May Fight Disease

PHYTOCHEMICAL	FOOD	POTENTIAL BENEFIT
Allyl sulfides	Garlic, onions, leeks	May protect against coronary artery disease, abnormal blood clotting, cancer
Alpha-linoleic acid	Flaxseed, soybeans, walnuts	Decreases inflammation
Anthocyanosides	Eggplant, blood oranges, blueberries	May protect against cancer
Capsaicins	Chili peppers	Topical analgesic
Carotenoids, including lycopene, lutein	Orange, red, yellow fruits; many vegetables, including tomatoes	May protect against coronary artery disease, macular degeneration, and cancer
Catechins	Tea (especially green tea)	May protect against cancer
Cellulose (fiber)	Whole-wheat flour, bran, cruciferous and root vegetables, legumes, apples	May protect against colon cancer, coronary artery disease
Coumarins	Carrots, citrus fruits, parsley	May protect against blood clots
Ellagic acid	Strawberries, raspberries, blueberries	May protect against cancer
Flavonoids (including resveratrol)	Citrus fruits, onions, apples, grapes, wine, tea	May protect against cancer
Gums (fiber)	Oats, barley, legumes	May prevent colon cancer
Hemicellulose (fiber)	Bran, whole grains	May prevent colon cancer
Indoles	Cruciferous vegetables	May protect against cancer
Isoflavones	Soybeans and soy products	May diminish menopausal symptoms; may protect against cancer, may lower blood lipid levels, may improve bone health
Isothiocyanates	Cruciferous vegetables	May prevent lung, esophageal cancer
Lignins (fiber)	Whole grains, oranges, pears, broccoli, flaxseed	May protect against cancer
Monoterpenes	Citrus fruits	May protect against cancer (pancreatic, breast, prostate tumors)
Pectin (fiber)	Apples, citrus fruits, strawberries	May protect against cancer, coronary artery disease, diabetes
Phenolic acids	Brown rice, green tea	May protect against cancer
Protease inhibitors	Soybeans, all plants	May protect against cancer
Phytosterols	Legumes, cucumbers	May help prevent coronary artery disease and breast cancer
Saponins	Garlic, onions, licorice, legumes	May protect against cancer

Others, like the well-known multivitamins, may contain an entire panel of vitamins and minerals in amounts close to the Dietary Reference Intakes (DRIs). Still other supplements may not contain any substances yet identified as nutrients or even demonstrated as beneficial to health.

With few exceptions, foods are better sources than supplements for the nutrients we need. A diet based on the Food Guide Pyramid, especially one that is rich in fruits, vegetables, whole-grain foods, and legumes, will provide most of the nutrients we know we need and the ones we have not yet identified. When we build our diet on a foundation of whole foods, we reap the added benefit of phytochemicals (known and unknown) and all the types of fiber we have begun to realize are important for health.

A Closer Look: Supplements and the Law

When a dietary supplement is available in the store, it is naturally assumed that it is safe. Some government agency has checked to make certain it is not harmful, right? Not anymore.

In 1994, the Dietary Supplement Health and Education Act removed dietary supplements from pre-market safety evaluations required of food ingredients and drugs. Drugs and food ingredients still undergo a lengthy Food and Drug Administration (FDA) safety review before they can be marketed, but the 1994 legislation eliminated the FDA's authority to regulate the safety of nutritional supplements before they go on the market. Now, the FDA can intervene only after an illness or injury occurs.

Claims may not be made about the use of dietary supplements to diagnose, prevent, treat, or cure a specific disease. For example, a product may not carry the claim "cures cancer" or "treats arthritis."

The FDA can still restrict the sale of an unsafe dietary supplement when there is evidence that the product presents a significant or reasonable safety concern. But the agency must wait for complaints about a product before acting.

The legislation also changed guidelines for marketing supplements. Because the nutritional supplement industry is now largely unregulated, there is no guarantee of product purity or of the amount of active ingredient in a supplement—even from one package to the next of the same product.

Dietary supplements can be enticing. However, because of a law passed in 1994, the Dietary Supplement Health and Education Act, the Food and Drug Administration (FDA) is more limited in what it can do to regulate the safety, purity, and labeling of supplements than what it does for drugs or even foods. Supplement manufacturers are required to list the ingredients of their products but are not accountable for the validity of those lists. Supplements may contain more or less of the active ingredient than they claim or may contain various impurities. Moreover, manufacturers are not required to list possible side effects of supplements on labels or in promotional materials (see sidebar: A Closer Look: Supplements and the Law, this page). Finally, we don't yet know the active ingredient or ingredients in many herbal supplements and plant foods, so we have no way of knowing whether the commercially available extracts of those herbs or foods will have the same benefits as the foods themselves.

Herbal Supplements

The popularity of herbal products, those made from extracts of plants and believed to have medicinal properties, continues to increase. Americans spend $700 million a year on herbal remedies. The use of some plant remedies dates back thousands of years, and plant materials are the basis for many of our most helpful medications, including aspirin and morphine. Scientists continue to investigate and discover new medicinal uses for substances in plants (see sidebar: Herbal Products, page 37).

Even though some herbal remedies may show beneficial effects, most show little evidence of providing any health benefits. In fact, some may have serious health risks and may interfere with the action of some medicines. Because herbal products are considered dietary supplements rather than drugs, the FDA is limited in its ability to regulate these substances. Ongoing studies continue to investigate selected herbs for their safety and effectiveness, so that more information will be available to consumers in the future. There is no guarantee of quality control. In the meantime, follow these precautions when considering a supplement:

- Do not use herbal remedies for treatment of serious illnesses.
- Do not give herbal (or other) supplements to infants and children.
- Avoid all herbal supplements if you are pregnant or trying to become pregnant.

HERBAL PRODUCTS

HERB	POSSIBLE USE	PRECAUTIONS
Chamomile *Matricaria chamomilla*	Internally for indigestion and as an anti-inflammatory. Externally for skin inflammations	Weak potential for allergic sensitization
Comfrey *Symphytum officinale*	Externally for bruises and sprains (where skin is intact)	Taken internally, it contains traces of alkaloids, which can cause liver damage and cancer risk
Purple coneflower *Echinacea*	Protects against colds	Possibility of an excessive immune response. Should not be used by persons with multiple sclerosis, AIDS, or tuberculosis
Ephedra-containing compounds Ma-huang	Stimulant	Headache, irritability, sleeplessness. High doses can cause dangerous increase in blood pressure and heart rate. Addictive over long periods
Garlic *Allium sativum*	May improve blood lipid levels (equal to 2 to 4 cloves a day)	Large quantities can lead to stomach complaints, rare skin allergic reaction
Ginger *Zingiber officinale*	For loss of appetite, travel sickness	Should not be taken with gallstone conditions
Ginkgo *Ginkgo biloba*	Improved blood flow to the brain and peripheral circulation; improvement in concentration and memory deficits	Mild gastrointestinal complaints, allergic skin reactions. Can change blood clotting; therefore, consult with physician if taking anticoagulant therapy
Ginseng *Panax ginseng*	Tonic for fatigue and declining work capacity	High doses can cause sleeplessness, high blood pressure, and edema
Guarana *Paullinia cupana*	Stimulation due to caffeine-like content. Diuretic	Caution advised for persons with sensitivity to caffeine, renal diseases, hyperthyroidism, panic anxiety
Indian tobacco *Lobelia inflata*	Asthma treatment, stimulates respiratory center to open airways and ease breathing	High doses can cause convulsions, respiratory problems, and even death
Kava *Piper methysticum*	Nervousness and insomnia	Gastrointestinal complaints, eye pupil dilation, and disorders of visual equilibrium. May interact with (enhance) other central nervous system substances, such as alcohol and barbiturates
St. John's wort *Hypericum perforatum*	For mild depression and anxiety	Digestive complaints of fullness or constipation. Sensitivity to sunlight in large doses. Consult with physician if taking antidepressant medications
Saw palmetto *Serenoa repens*	Urinary problems in men with non-cancerous (benign) prostate enlargement. Does not reduce the enlargement	Use under a physician's supervision, not as a substitute for medical treatment
Yohimbe bark *Pausinystalia yohimba*	Used for sexual disorders, as an aphrodisiac, although it appears to be ineffective	Do not use if liver or kidney disease is present. Side effects include anxiety, increase in blood pressure, rapid heartbeat, tremor, and vomiting. High doses can cause heart failure

- If you have a medical condition, check with your health care provider before taking herbal supplements.
- In addition, if you are taking medications, do not take herbal supplements before discussing them with your health care provider.

Supplement Sense

In general, high-dose vitamin or mineral supplements add little to our health and may in themselves cause illness. Those that contain more than 100 percent of your estimated daily needs may result in serious nutrient imbalances or even toxicity. Such imbalances do not occur when your source of

USING SUPPLEMENTS

Supplements are not substitutes. Supplements do not replace the hundreds of nutrients in whole foods needed for a balanced diet, and they will not fix poor eating habits. If you are considering taking a supplement, heed the following:

- Do not self-prescribe.

- See your doctor or health care provider if you have a health problem.

- Discuss any supplement that you are taking. Supplements may interfere with medications.

- Read the label. Supplements can lose their potency over time, so check the expiration date on the label. Also, look for the initials "USP." They stand for the testing organization, US Pharmacopeia, which establishes testing standards for compounds. The Food and Drug Administration (FDA) does not analyze supplements before they enter the marketplace.

- Stick to the Daily Value (DV). Choose supplements that are limited to 100 percent or less. Take no more than what is recommended by your doctor or health care provider. The toxic levels are not known for some nutrients.

- Do not waste dollars. Synthetic supplements are the same as so-called natural types. Generic and synthetic brands are less expensive and equally effective.

- Store in a safe place. Iron supplements are the most common cause of poisoning deaths among children.

vitamins comes from foods rather than supplements, because foods contain safe amounts of multiple nutrients, and if you follow the Food Guide Pyramid's recommended number of servings you will likely meet the recommended amounts for most nutrients.

As discussed earlier, most Americans, including athletes, consume considerably more protein than recommended, and more than their bodies can use. Protein or amino acid powders provide no benefit and are a poor substitute for protein-rich foods that contain necessary vitamins and minerals. Similarly, pills that promise to deliver all the fiber we need daily are a bad risk, because these pills invariably provide only one type of fiber, whereas each type of fiber found in foods of plant origin appears to confer unique health-promoting benefits. The fatty acids we need also are available in more than adequate amounts in various foods.

Who Needs a Vitamin or Mineral Supplement?

Eating a variety of foods, especially those of plant origin, allows most of us to acquire all the known nutrients, food substances, and as yet unidentified nutrients that our bodies need. Nevertheless, supplements may be appropriate for some individuals. Who are these people?

- Pregnant or breastfeeding women have an increased need for most vitamins and minerals. Folic acid is especially important early in pregnancy. Women who are capable of becoming pregnant should ensure that their daily intake of folic acid from supplements and fortified foods is 400 micrograms. These vitamins and minerals are contained in the prenatal supplements that are prescribed by your health care provider.
- Older adults may absorb some nutrients poorly, particularly folate, vitamin B_{12}, and vitamin D. They therefore may require supplements.
- People on restricted diets may require supplements of some vitamins and minerals.
- People with diseases of the digestive tract or other serious illnesses that limit their absorption of some vitamins and minerals may require supplements.
- People taking prescription medications may have altered needs for a variety of nutrients.
- Smokers appear to have an increased need for antioxidants, especially vitamin C. (However, even this increased requirement for vitamin C is easily satisfied by eating nutrient-rich foods.)
- People who drink alcohol to excess may require supplements.

- Vegetarians who eat no animal products may not get adequate vitamin B_{12}, iron, and zinc. Vegetarians who avoid dairy products are at a greater risk for calcium deficiency than are those who do eat dairy products.
- Some evidence suggests that a daily supplement of vitamin E may reduce the risk of heart disease.

Before taking any supplement, discuss it with your health care provider, and be sure to mention any medications you are taking (see sidebar: Using Supplements, page 38).

The Future Is Here: Functional Foods

If your breakfast this morning included calcium-fortified orange juice and toast made with folate-enriched flour, you are a consumer of functional foods. Just what is a functional food? As the fastest growing category of new food products, these are foods or food components to which manufacturers have added ingredients that are known or believed to promote health and prevent disease (see Chapter 4, What Do the Terms "Fortified" and "Enriched" Mean? page 92). Although the name "functional foods" is new, the concept is not: when it was discovered in the early part of the 20th century that some thyroid disease was caused by a deficiency of the mineral iodine, manufacturers began enriching table salt with iodine. Since that time, we have also seen vitamin D-fortified milk, breakfast cereals fortified with a variety of vitamins and minerals, and the addition of preservatives that are themselves antioxidants to almost all processed foods (see Chapter 3, Cholesterol-Lowering Margarine, page 66). The past few years, though, have seen a virtual explosion of functional foods, some based on careful research and supported by nutrition experts and some with questionable, if any, potential benefits.

The Bottom Line on Supplements

With a few exceptions, most of us should get all the vitamins, minerals, fiber, essential fatty acids, protein, and phytochemicals we need from the food we eat, rather than from supplements. Always consult your health care provider before trying a supplement. A few of the functional foods that have recently appeared on the market have proven benefits, although most have yet to demonstrate their value.

If your goal is to eat foods that deliver on their promise of providing all the necessary vitamins, minerals, fiber, and other yet to be identified health-enhancing substances, you need to:

- increase your intake of whole grains, fruits, and vegetables as sources of vitamins, minerals, fiber, and phytochemicals
- decrease your intake of foods of animal origin (meats, dairy products, eggs), particularly those that are high in saturated fat, substituting lean alternatives and plant sources of protein
- limit your use of fats and cooking oils when preparing, serving, and eating food

NUTRITION AND YOUR STAGE OF LIFE

Choosing to eat wisely throughout life is one of the most important components of living a healthful lifestyle. Using the Dietary Guidelines for Americans and Food Guide Pyramid as standards for your food choices and amounts, you will be well on your way to establishing healthful eating habits for your entire life. Unfortunately, many people become discouraged by nutrition advice because they mistakenly think they cannot eat their favorite foods.

A more positive and encouraging approach is to consider that no food is forbidden. Good health comes from eating a variety of foods—meats, dairy products, and especially

vegetables, fruits, and grains—in sensible amounts. Keep in mind that your sex, age, weight, and health status are also important considerations when it comes to determining your nutritional needs and maintaining good health throughout life. In other words, what is nutritionally right for one person may not be the same for you.

In the following pages, you will explore several natural transition points during a lifetime which prompt variances in nutritional needs. These specific times of life, from infancy to preschooler, school-age to adolescence and teen years, to young and then older adulthood, are important to understanding the changes you can expect as you age and the strategies needed to meet your need for specific nutrients and optimal health.

Infancy

No human being grows more rapidly than an infant. On average, a baby triples his or her weight during the first year and grows taller by 50 percent. A newborn may grow from 7 pounds to 14 pounds in just 6 months, then to 20 or 21 pounds by the end of the first full year. It is easy to

Breastfeeding provides the right balance of nutrients—and more.

understand why proper nourishment for infants can provide a healthy head start on life.

Breast milk contains just the right balance of nutrients such as protein, carbohydrates, and fat. It also provides the infant with antibodies to fight some common childhood illnesses, and it may be unique in that it decreases the risk of food allergies. Breast milk is also easy for the baby to digest.

Mothers who cannot or choose not to breastfeed can still provide good nourishment to their infants with bottle feeding of commercial formulas. Careful preparation is required for each feeding, and formulas must be stored safely.

As a baby becomes more hungry, it is best to increase the frequency of the breast- or bottle-feedings. Most nutrition experts recommend that solid food should not be started until after the fourth month. Many suggest waiting until your baby is at least 6 months old.   Although babies may be ready for solid foods in a few weeks or months, the decision to start giving solid foods should be based on a baby's daytime behavior and eating habits and coordinated with a baby's increasing nutritional needs. Even if solid foods are started, breast milk or formula should continue to be included in a baby's diet for at least 1 year.

A baby's first solid food will probably be cereal. Cereal is a versatile food because it can be mixed very thin for babies just starting on solid foods and can be thickened as babies work on chewing and swallowing. After cereals, fruits and vegetables are easiest for babies to digest.

From 9 months on, babies make a gradual transition into toddlerhood and develop a feeding schedule that mimics the family's mealtimes. Sometime after 12 months, babies may be on a schedule of three meals a day with the family and breast milk, formula, or snacks between meals.

By 1 year of age, babies may have as many as four to six teeth and be developing a more defined and stronger chewing motion. With these developments, babies can handle foods of thicker consistency, such as lumpy or chopped foods. Continue offering new solid foods at the rate of one new food a week. One of the most important considerations at this age is offering baby food in a form that is appropriate to the baby's development. Any solid food should be tender and soft enough to be easily squashed with your fingers.

The typical meal for a 1-year-old includes 1 tablespoon from each of the major food groups: milk, meat, vegetables, fruits, breads, and cereal grains. That menu may translate into a tablespoon of cooked carrots, two bites of rice, a taste

INFANT NUTRITION AT A GLANCE

Breastfeeding is preferred

- Meets baby's special nutrient needs
- Protects against infection
- May protect against food allergies
- To be major part of baby's diet for at least 1 year

What about milk?

- Whole milk may replace breast milk or formula at 1 year
- Continue whole milk until age 2, then may switch to lower-fat milk

When is baby ready for solid foods?

- Baby can sit with support
- Shows interest in foods others are eating
- Can move food from front to back of mouth
- Weight has doubled since birth
- Nurses eight or more times or drinks more than 32 ounces of formula in 24 hours
- No earlier than 4 to 6 months of age

Which foods, what order?

- Iron-fortified, single-grain cereals
- Vegetables
- Fruits
- Strained meats
- Soft, mashed table foods
- Finger foods

Other tips about solid foods and eating in general

- Offer one new food about every week
- Offer small amounts frequently throughout the day
- Avoid encouraging baby to eat too much or too fast
- Infants generally have an inborn sense to consume an adequate amount of calories
- Egg whites are highly allergenic. Wait until baby is 1 year or older to introduce eggs
- Babies younger than 1 year old should not be given honey. It may contain botulism spores, which can produce a highly toxic poison. After age 1, the digestive tract is able to render the botulism spores harmless

---

of meat, and a couple bites of pear. Try to feed your child at regular intervals while paying attention to cues that may suggest that your child is hungry.

Good eating habits begin early (see sidebar: Infant Nutrition at a Glance, above).

Preschool Years: Ages 1 to 5

As babies become toddlers, they make the transition into eating food the rest of the family eats. The rapid weight gain characterized during the first year levels off during the second, with an average gain of 5 to 6 pounds. Because a child is not growing at the same rate as during infancy, he or she may not want the same quantities of food that were once enjoyed. Some preschoolers may be uninterested in eating, whereas others seem finicky. Sometimes preschoolers are reluctant to try new foods or expand their food repertoire beyond three or four favorites. There are several things you can do to help overcome these challenges without forcing a child to eat (see sidebar: Tips for Feeding Toddlers, page 42).

By this time, a child should be eating foods from each of the food groups represented in the Food Guide Pyramid. However, do not expect a preschooler to eat a completely balanced diet every day. When allowed to choose from a selection of nutritionally sound foods, most children tend to select diets that, over several days, offer the necessary balance.

Until age 2, fat should not be limited in a child's diet. Dietary fat and cholesterol are important for an infant's growth. After age 2, children can begin to consume fats in moderation just as the rest of the family does. This type of diet includes grain foods, vegetables, fruits, low-fat dairy products, lean meats, and their substitutes.

Keep in mind that every child's energy needs are different. Thus, snacks are often appropriate for children, especially for smaller preschoolers who cannot eat enough to satisfy their energy needs all at once. Small amounts of various foods eaten frequently over the course of the day as a snack are healthful and normal. However, completely uncontrolled snacking can diminish a child's appetite for meals.

School-Age Children: Ages 6 to 12

The increasing independence of a school-age child may be a welcome contrast to the constant demands of a preschooler. By early school age, a child should be well on the way to establishing healthful eating habits and regular physical activity to maintain a healthful weight.

A normal school-age child will gain about 7 pounds a year, and his or her height will increase by approximately 2.5 inches a year. As children approach their teen years, boys and girls differ distinctly in growth patterns. Puberty generally begins about age 10 in girls and age 12 in boys and normally lasts for 2 to 3 years. During these years, growth spurts occur.

It is during the school-age years that the guidance of parents is especially important to formulating good nutrition habits. Modeling healthful eating practices such as

TIPS FOR FEEDING TODDLERS

- Make it comfortable for a child to eat—comfortably seated, foot support, finger foods.
- Develop a routine—schedule regular meals and snacks.
- Plan one familiar food at each meal.
- Offer various textures and colors—separate them on the plate.
- Serve small portions and let the child ask for more.
- Allow a child to eat at his or her own pace.
- Teach and reinforce good table manners.
- Remain calm if a meal is left untouched.
- Set a good example.

Foods with color and crunch are interesting—and fun—for youngsters.

eating foods that are low in fat and high in complex carbohydrates and fiber is important. Emphasize breakfast as an important meal. Pack healthful lunches that include fruit, vegetables, bread or some other form of starch, a meat or other protein, and low-fat milk. If a child participates in the school lunch program, talk about how to make nutritious food choices. Provide fruits, vegetables, whole-grain breads, or low-fat yogurts as after-school snacks. If a child participates in vigorous physical activity, more calories may be needed.

One of the major challenges for some school-age children is controlling weight. If a child eats more calories than are used, the pounds will add up, particularly if the child is inactive. Besides the social and emotional stresses that may result from peers who make fun of a child's excess weight, a higher than desirable weight at this age can increase the risk for later health problems, such as diabetes, high blood pressure, and increased blood cholesterol or triglyceride values.

Still, overweight children have the same nutrient needs as other children. The goal should be to stop or slow the rate of weight gain and allow height (growth) to catch up. Do not allow a child to restrict certain foods or to try fad diets. Instead, provide healthful foods in lesser amounts.

The best way to teach a child about good nutrition is to set a good example in your own eating habits.

Teenage Years: Ages 13 to 19

The teen years and the arrival of puberty are the second period of remarkable growth for youngsters. It is a period of profound development that has important nutritional implications. As a result, requirements increase for energy and all nutrients.

The growth and energy requirements of the teen years nudge daily calorie needs upward. On average, boys 11 to 14 years old need to have approximately 2,500 calories per day. From age 15 to 18, daily calorie requirements increase to 2,800 calories. Teenage girls also require more calories, but in the neighborhood of 2,200 calories a day.

Most of the calories a teenager consumes should take the form of the complex carbohydrates found at the bottom of the Food Guide Pyramid. It is also a good idea for teens to have 3 servings of calcium-rich foods a day (milk, yogurt, cheese, certain vegetables) to make certain that needs are met for growing bones. Iron is also important to the expanding volume of blood in the body and for increasing muscle mass. Teenage girls can be at risk for a shortage of iron as a result of iron loss through menstruation. To ensure ample dietary iron, encourage teens to eat fish, poultry (especially dark meat), red meat, eggs, legumes, potatoes, broccoli, rice, and iron-enriched grain products.

Growing, active teenagers have a real need to snack between meals. Encourage healthful snacks such as fresh fruits and raw vegetables, low-fat yogurt, low-fat milk, whole-grain bread, popcorn, pretzels, and cereals.

Several factors challenge the ability of teenagers to eat well. The typical teen has a busy school schedule, extracurricular activities, and often part-time employment. These may lead to skipping breakfast and other meals in favor of more meals from vending machines and fast-food restaurants, and more snacking on convenience items. For example, adolescence is an especially important time to get adequate calcium. Inadequate calcium may make teens more prone to the development of osteoporosis and other diseases in the future.

When weight gain accompanies these habits, many teenagers turn to fad diets for quick weight loss. All of these pressures may lead to nutritional excesses, deficiencies, and, at the extreme, eating disorders.

Excessive weight concerns can have more serious implications. Extremes in eating patterns, either severe undereating or excessive overeating, may result in serious—even life-threatening—health risks. These extremes may impair some bodily functions, including decreased hormone production, and thereby slow sexual maturation in both girls and boys. Consult a health care professional if an adolescent has a problem with weight.

Adulthood

The adult body is dynamic, changing subtly as the years march by. Therefore, what was a good diet for you in your second or third decade of life may no longer be a good fit at age 50 or 60.

For example, your metabolism—the way in which your body converts the food you eat into energy—slows. This means that you gradually need less food for a similar activity level. Fewer and fewer calories are needed as you grow older—about 10 percent less per decade from age 50 onward. This slowing of the metabolism is perfectly natural and occurs because you lose muscle mass (which utilizes most of the energy you produce) as you age. Exercise helps to maintain muscle mass and helps you burn calories. However, most Americans still need to reduce calories.

Variety and moderation remain the keys to a healthful diet. A balanced diet ensures proper intake of vitamins, minerals, proteins, carbohydrates, and other nutrients. Moderation controls calories and is especially important with regard to consumption of alcohol. Drinking plenty of water, eating fiber-rich foods, and staying as active as possible can help to stave off constipation.

Pregnancy and Breastfeeding

The best time to start thinking about good nutrition is before a woman becomes pregnant. Then she can be certain that her baby will have all the essential nutrients from the moment of conception.

Babies born at low weights (less than 5.5 pounds) have a greater likelihood for development of health problems. Mothers-to-be can help prevent this from occurring by eating the well-rounded diet that women their age would ideally eat, as well as achieving an appropriate weight for their height. Women who are 15 percent or more underweight present a special risk for a difficult pregnancy and childbirth.

Before becoming pregnant, talk to your health care provider about your need for a folic acid (a B vitamin) supplement. Doctors and scientists agree that use of folic acid supplements can reduce the occurrence of a birth defect called a neural tube defect. One form of this defect is spina bifida, an incomplete closure of the spine. Folate levels also may be affected by consuming alcohol and

TIPS FOR PREGNANT AND BREASTFEEDING WOMEN

"Morning" sickness

"Morning" sickness is not necessarily unique to mornings. About half of all pregnant women experience this during the first 12 weeks of pregnancy.

- Eat dry toast or crackers.
- Eat smaller amounts—more frequently—so that your stomach isn't too empty or too full.
- Limit or avoid spicy and fried foods.
- Sip liquids. Try crushed ice or frozen ice pops to avoid dehydration if you are vomiting.
- Cook in a well-ventilated kitchen.

Heartburn

During the later part of pregnancy, hormonal changes and the pressure due to an expanding uterus pushing on your stomach slow the rate at which food leaves the stomach. To diminish heartburn:

- Eat smaller meals more frequently.
- Limit or avoid chocolate and spicy and fried foods.
- Limit caffeine.

Constipation

Increased pressure from the growing baby can slow down the movement of the contents in your bowels.

- Drink plenty of liquids—2 to 3 quarts daily.
- Eat whole-grain breads and cereals, fresh fruits, and raw vegetables.
- Exercise moderately and regularly.

Anemia

Pregnancy increases the risk for anemia because of an inadequate intake of iron and folate, greater needs for these nutrients because of the baby, or both.

- Eat iron-rich foods (red meat, eggs, liver, dried fruit, iron-fortified cereals).
- Eat foods with folate (leafy green vegetables, oranges and grapefruit, dry beans, and cereals fortified with folic acid).
- Follow your health care provider's recommendations about supplements.

Breastfeeding

- Drink plenty of liquids—2 to 3 quarts daily.
- Choose calcium-rich foods (milk, yogurt, cheese, pudding, tofu).
- If you have a limited diet, check with your health care provider about continuing your prenatal vitamin-mineral supplement.
- Exercise regularly and moderately.

Toxins

- If you drink alcohol, stop. If you don't, don't start.
- If you smoke, stop. If you don't, don't start.
- Avoid excessive salt, which may cause fluid retention and increase blood pressure.

---

smoking cigarettes, both of which are toxic to the developing fetus and should not be used when you are pregnant. Taking oral contraceptives is also associated with low levels of folate.

It is essential to talk with your health care provider about folic acid because too much of any supplement can harm your health. The best dietary sources of folic acid include fortified breakfast cereals and enriched grain products. Folate (the natural form of this vitamin found in foods) is found in leafy green vegetables, oranges and grapefruit, black-eyed peas, kidney beans, and other cooked dried beans. Even if you eat a well-balanced diet, prenatal vitamins are recommended.

Research has proved that women who have a normal weight at the time of conception have the healthiest pregnancies and babies if they gain 25 to 35 pounds. Women who are underweight may need to gain additional weight. Even women who are overweight should plan on gaining about 15 to 20 pounds. During the first trimester, additional calories may not be needed. It is important, however, to make sure that your diet provides the best nutrition possible for you and your unborn baby. In the second and third trimesters, you will need about 300 extra calories per day beyond your normal diet. Concentrate on foods such as lean meat, low- or no-fat dairy products, and dark green vegetables, all of which provide generous amounts of

During pregnancy, choose snacks wisely—with health in mind. This snack includes 1 cup frozen yogurt topped with granola, raisins, apples, and cinnamon. (See Chapter 5, page 132, for recipe.)

vitamins, minerals, and protein. If you enjoy milk and dairy products, it will be easy to meet your need for calcium, a crucial mineral during pregnancy. If you do not, ask your doctor or registered dietitian about calcium-fortified foods. In some cases, a supplement may be recommended.

Even with the increased need for calories, it is nearly impossible to get sufficient amounts of iron, which is needed in double the amounts recommended for non-pregnant women. The added iron is needed for the expanded blood volume that accommodates the changes in your body during pregnancy. Iron is also needed for the

formation of tissue for both your baby and the placenta. At birth, a newborn needs enough stored iron to last for the first 6 months of life. The best advice is to eat a healthful diet and take only the supplements recommended by your caregiver.

Once a baby is born, the mother who chooses to breast-feed still needs extra calories—typically about 500 calories per day. Continue to concentrate on eating nutrient-rich foods (see sidebar: Tips for Pregnant and Breastfeeding Women, page 44).

Senior Years

Maintaining a healthful diet into older adulthood can be a challenge, particularly if part of what makes eating a pleasurable experience—the senses of taste and smell—decline. Tooth loss or mouth pain can further complicate the act of eating. The medicines that must be taken to treat chronic diseases may affect appetite, when meals are eaten, and, in some cases, how food tastes.

Other life changes, including the loss of a spouse, fewer social contacts, economic hardships, and increasing dependence on others for day-to-day care, combine to make getting a healthful diet more difficult, but more important than ever.

As people become older, they can expect to need fewer calories than they did when they were younger—about 10 percent less per decade from age 50 onward. Persons age 70 should be eating about 80 percent of the calories they were eating 20 years earlier. Nonetheless, older people continue to need other nutrients in the same quantities as ever.

This decline in calorie intake translates into a need for food that is rich in nutrients, to provide what you used to get with more calories when you were younger. Keeping a lower-calorie intake in mind, you can simply follow the basics of the Food Guide Pyramid as you get older: heavy on the grain foods, fruits, and vegetables; adequate meats and dairy products; light on the fats, oils, and sweets.

Watch for signs of poor nutrition in older adults and consult a physician, registered dietitian, or other health care professional to address their nutritional needs and to determine whether supplementation with vitamins is needed.

In this chapter, you will read about the importance of nutrition in maintaining or improving your health. Whether you are healthy, have a medical problem, have a family history of one, or know a family member or friend with an illness, nutrition plays a major role in prevention and treatment of common diseases.

THE FOOD-HEALTH CONNECTION

Knowing what nutrients comprise a well-balanced diet, in what foods to find them, and in what quantities to eat them are some of the first steps to good health. Applying this knowledge by eating nutrient-rich foods and incorporating physical activity into your schedule at any stage of life are the greatest investments you can make in sustaining good health.

Although healthful eating may lower your risk for certain diseases, there are no guarantees that adhering to the tenets of good nutrition will prevent an illness from developing. Science has shown that not all diseases or disorders are associated with what you eat. However, statistics do show that lifelong food selections may influence the risk for some diseases.

Research continues to evaluate and clarify the role that diet and nutrition play in the promotion of health and in the development of obesity, high blood pressure, diabetes mellitus, coronary artery disease, osteoporosis, cancer, and other illnesses. The nutritional recommendations for prevention of many diseases are similar (see Chapter 1).

In this chapter you will learn about selected conditions in which nutrition plays an important role

and how proper nutrition may affect and even alter the course of these conditions.

OBESITY

If obesity were merely a matter of aesthetics, it would be of less concern. But obesity is a health issue. It is associated with an increased risk of diabetes, lipid abnormalities, coronary artery disease, high blood pressure, certain cancers (such as breast, colon, and gallbladder in women and colon and prostate in men), stroke, degenerative arthritis, respiratory problems, sleep disturbances, and gallbladder disease.

Obesity places a huge burden on society in terms of lost lives, ongoing illnesses, emotional pain, discrimination, and economic cost (nearly $100 billion annually). The most ominous burdens posed by being overweight are reduction of the quality of life and shortening of life span. The likelihood of dying early (compared with the average age at death of all people in the population) progressively increases the more overweight you are. Diseases caused by obesity are the second leading cause of preventable deaths in the United States.

The Battle of the Bulge

With countless diet programs and products promising to help you shed pounds, losing weight should be easy. Simply eating too much and not being active enough are the causes of most overweight problems. But you also know it is hard to lose weight and even harder to keep it off. The cause of overweight and obesity is a chronic imbalance of calories ingested and calories burned. Genetic and environmental factors also contribute to obesity.

Americans spend more than $33 billion a year on weight-loss products and services, but they are losing the "battle of the bulge." Despite the great desire of Americans to be thinner, they have become more obese. Some have even declared that the United States has an "obesity epidemic." It is estimated that more than 50 percent of adult Americans are overweight. The prevalence of obesity also is increasing in several other countries.

Get the Terms Straight

"Overweight" and "obesity" are terms that often are used interchangeably, but they have different meanings. "Overweight" refers to having excess body weight compared with the norm for a person's height, but the term does not account for what tissue is making up the weight. For example, athletes are often overweight according to weight-for-height tables because they have increased muscle mass. However, for most people, overweight means having too much fat.

"Obesity" refers to body fat in excess of what is healthful for an individual. In healthy women, an acceptable level of body fat ranges from 25 to 35 percent. In contrast, an acceptable range of body fat in men is from 10 to 23 percent.

How Your Body Uses Food

The number of calories used by an individual is determined by three factors: basal metabolic rate, the thermic effect of the food eaten, and the calories used during physical activity. The basal metabolic rate is the amount of energy needed to maintain bodily functions when an individual is at rest. This component accounts for 60 to 75 percent of the daily calorie requirement in sedentary adults. The major determinant of the basal metabolic rate is the amount of fat-free mass in the body. Muscle is one example of fat-free mass. Resistance (strength) training can increase the amount of muscle and therefore increase the basal metabolic rate. Resistance training also can help prevent the loss of lean mass that normally occurs with aging. Men tend to have more muscle than women and therefore burn more calories.

The thermic effect of food is the energy required to digest, metabolize, and store nutrients. The thermic effect of food accounts for about 10 percent of the total daily calorie use. The number of calories burned during exercise can vary tremendously depending on the amount of exercise performed. For most so-called sedentary persons, the activities of daily living (such as walking, talking, and sitting) account for 15 to 20 percent of the daily calorie use.

Should You Lose Weight?

How do you determine whether you are overweight or obese? Scientists can use sophisticated tests to measure body composition. However, these are not necessary for most individuals. You can measure your change in weight over time. Alternatively, you can calculate your body mass index (BMI) and determine its relationship to health risks.

Pinpointing Your Risk

There are risk factors that indicate a predisposition for obesity. Among these risk factors are the following:

Body mass index (BMI)—BMI is defined as your weight (in kilograms) divided by the square of your height (in meters). Simpler still, look it up in the Body Mass Index Table on page 49. The advantage of BMI over bathroom scales and weight-for-height tables is that it normalizes weight for height and helps determine whether you have a healthful or unhealthful percentage of total body fat.

People who should not use the BMI for determining health risks include competitive athletes and body builders. Their BMI will be high because they have a larger amount of muscle. BMI is also not predictive of health risks for growing children, women who are pregnant or lactating, and frail, sedentary older adults.

A BMI from 19 to 24.9 is associated with a minimal to low health risk. A BMI from 25 to 29.9 is considered overweight and is associated with moderate health risks. A BMI of 30 or higher is considered obese and is associated with a substantially greater risk for development of various diseases. Extreme obesity is a BMI of more than 40.

Body shape—Increasing attention has been focused on the distribution of body fat as a potential indicator of health risk. Specifically, excess fat in the abdomen is associated with an increased risk for development of various metabolic illnesses, including diabetes mellitus, increased blood lipid levels, and high blood pressure. In contrast,

BODY MASS INDEX TABLE*

To use the table, find your height in the left-hand column. Move across to find your weight.
The number at the top of the column is your BMI.

	HEALTHY		OVERWEIGHT					OBESE				
BMI	**19**	**24**	25	26	27	28	29	30	35	40	45	50
HEIGHT						BODY WEIGHT (POUNDS)						
4′10″	91	115	119	124	129	134	138	143	167	191	215	239
4′11″	94	119	124	128	133	138	143	148	173	198	222	247
5′0″	97	123	128	133	138	143	148	153	179	204	230	255
5′1″	100	127	132	137	143	148	153	158	185	211	238	264
5′2″	104	131	136	142	147	153	158	164	191	218	246	273
5′3″	107	135	141	146	152	158	163	169	197	225	254	282
5′4″	110	140	145	151	157	163	169	174	204	232	262	291
5′5″	114	144	150	156	162	168	174	180	210	240	270	300
5′6″	118	148	155	161	167	173	179	186	216	247	278	309
5′7″	121	153	159	166	172	178	185	191	223	255	287	319
5′8″	125	158	164	171	177	184	190	197	230	262	295	328
5′9″	128	162	169	176	182	189	196	203	236	270	304	338
5′10″	132	167	174	181	188	195	202	209	243	278	313	348
5′11″	136	172	179	186	193	200	208	215	250	286	322	358
6′0″	140	177	184	191	199	206	213	221	258	294	331	368
6′1″	144	182	189	197	204	212	219	227	265	302	340	378
6′2″	148	186	194	202	210	218	225	233	272	311	350	389
6′3″	152	192	200	208	216	224	232	240	279	319	359	399
6′4″	156	197	205	213	221	230	238	246	287	328	369	410

* This table has been converted to use weight in pounds and height in feet and inches.

The body has a nearly unlimited capacity to store fat. Excess fat in the abdomen can lead to illnesses, including diabetes, high blood lipid levels, and high blood pressure. It is also associated with an increased risk for coronary artery disease, stroke, and certain cancers. Losing weight will reduce these risks and lessen the strain on the lower back, hips, and knees.

Measuring your waist circumference can be helpful to determine how your body distributes fat. Fat in your abdomen increases your risk for high blood pressure, coronary artery disease, diabetes, stroke, and certain cancers.

people whose excess fat is located in their lower body (hips, buttocks, and thighs) seem to have minimal or no increased risk of these diseases. Upper-body obesity also is associated with an increased risk for coronary artery disease, stroke, and certain cancers.

Therefore, it can be helpful to assess your health risk by measuring your waist circumference. A measurement of more than 35 inches in women and 40 inches in men is associated with increased health risks, especially if you have a BMI of 25 or more.

What Is There to Lose? To Gain?

Although no one is without health risk—even the fittest person can have a heart attack, diabetes, or cancer—health and well-being are apt to be in less jeopardy if BMI, body shape, and family health history do not indicate problems. However, if your BMI is 25 or more, if your fat is primarily located in your upper body, and if you have a personal or family history of diabetes, heart disease, high blood pressure, or sleep apnea, losing weight can greatly improve your health.

Keep in mind that BMI and waist circumference are just starting points. Other factors also are important. When in doubt, seek a medical evaluation by your physician. A thorough history, examination, and blood studies can clarify whether your weight is having adverse effects on your health. The appropriate plan of action then can be tailored to meet your individual needs.

Getting Started

Losing body fat and keeping it off are not easy. Losing weight and then maintaining a healthful weight require collaboration with knowledgeable health care professionals. Obesity is not only a medical issue but also is a lifestyle issue. Your habits can help you maintain a desirable body weight or they can hamper your efforts to lose weight or even cause you to gain further weight. The types and amounts of food you eat and the exercise you perform will determine whether you gain, lose, or maintain your weight. Therefore, experts recommend that any weight loss program should consist of three main components: nutrition, exercise (or activity), and behavior modification. (See sidebars: Weight Management Programs: Look for These Criteria, and Special Situations, this page.)

Nutrition

Liquid meals, over-the-counter diet pills, and special combinations of foods promising to "burn" fat are not the answers to long-term weight control and better health. Learning to eat differently—to enjoy a well-balanced diet of fewer calories—is the best strategy to achieve health and weight goals. You should begin by substituting the words "healthful nutrition program" for "diet."

Most people try to lose weight by eating 1,000 to 1,500 calories a day. In many instances, eating fewer than 1,400 calories makes it difficult to eat a balanced diet containing the recommended levels of nutrients. Therefore, nutrition programs that are too low in calories may be hazardous to your health.

You can lose weight by eating fewer calories or by increasing exercise. A caloric deficit of 3,500 calories is required to lose 1 pound of fat. Over 7 days, this can be

WEIGHT MANAGEMENT PROGRAMS: LOOK FOR THESE CRITERIA

Nearly 8 million Americans enroll in some kind of structured weight-loss program each year. Choose a safe and helpful program that meets these criteria.

- **Physician participation**—Consult a doctor if you have a health problem or it is necessary to lose more than 15 to 20 pounds.

- **Qualified staff**—Registered dietitians can design a nutrition plan. Specialists in behavior modification and exercise are also recommended.

- **Nutritionally balanced diet**—Any weight management plan should ensure adequate nutrition.

- **Exercise**—At a minimum, there should be instructions for starting a safe exercise program.

- **Reasonable weight loss goals**—Weight loss should be slow and steady. A 1- to 2-pound weight loss per week is a good start. A loss of 10 percent of starting weight is also a reasonable initial goal.

- **Help in changing lifestyle**—Programs should help improve life-long eating and exercise habits. There should be opportunity for follow-up support—even after goals are reached.

achieved by cutting 500 calories each day from your usual food intake or by cutting 250 calories each day (such as one or two fewer cookies) and burning an additional 250 calories with exercise (such as by walking briskly for 30

SPECIAL SITUATIONS

- The only way to lose weight and keep weight off is to combine a healthful nutrition program ("diet") with regular physical activity. Prescription medications for weight loss may be considered in moderately overweight or obese people with weight-related health problems. Use of these medications does not eliminate the need for healthful nutrition and activity or behavior modification strategies. Medical supervision is essential because there are potential risks, in addition to benefits.

- Weight-loss surgery may be considered only in clinically severe obesity (BMI of more than 35) with weight-related medical problems or in extreme obesity (BMI more than 40).

- Over-the-counter diet products are a huge financial success. Many of them promise to help you lose weight by increasing your metabolism or decreasing your appetite. However, these effects can be dangerous, and any weight loss isalmost always temporary.

minutes). The good news is that a relatively small loss of weight can make a big difference in reducing the risk of health complications from obesity. Even a 10 percent weight loss can lead to improvement in your blood sugar level, lipid values, and blood pressure. Once this degree of weight loss has been achieved, further weight loss goals can then be set, if needed. Rather than aiming for an "ideal" weight, which may not be achievable or desirable, focus on achieving and maintaining a healthful weight.

A healthful diet for controlling weight includes foods from all food groups in the Food Guide Pyramid—ensuring balance and variety—in sensible amounts (see Chapter 1, Food Guide Pyramid, page 11. Also see Chapters 4 and 5 for planning and preparing healthful meals). It is helpful to review the energy density of the food consumed. Fat contains 9 calories/gram, protein 4 calories/gram, and carbohydrates 4 calories/gram. Alcohol contributes 7 calories/gram.

For most people, the volume of food consumed determines how full you feel. Therefore, eating a small amount of an energy-dense food (such as fat) is usually not filling, whereas eating a large enough amount to fill you up results in a very large calorie intake. To lose weight, decrease your total calories by cutting back on the fat while filling up on low-calorie high-nutrient foods such as vegetables, fruits, and grains. You also can eat lower-fat versions of foods. However, be careful, because low-fat is not always low-calorie (see sidebar: Low-Fat Is Not Necessarily Low-Calorie, this page).

Healthful eating habits also avoid the feast or famine phenomenon. Distributing food selections throughout the day provides nourishment to support daily activities and can help to eliminate energy highs and lows. Three meals and occasional snacks also keep one's appetite in check.

It is also important to avoid the hazards of repeatedly losing and gaining weight. Although repeated dieting is still a matter of debate, some studies suggest that it may lower the rate at which calories are burned. When a person is off the diet and more food is eaten, the body stores fat faster and more efficiently. This effect causes regain of the lost weight. In this circumstance, the amount of weight regained often is more than that lost in the first place.

LOW-FAT IS NOT NECESSARILY LOW-CALORIE

FOOD (AMOUNT)	CALORIES	FAT (GRAMS)
Cookie (1 cookie)		
Chocolate chip		
Regular	80	4.5
Reduced-fat	70	3
Sandwich-type		
Regular	50	2
Reduced-fat	55	1.5
Crackers		
Soda (5 crackers)		
Regular	60	2
Fat-free	50	0
Wheat (15 crackers)		
Regular	130	6
Fat-free	180	0
Ice cream (1/2 cup)		
Regular	135	7
Low-fat	120	2
Potato chips		
(1 ounce or 1 individual bag)		
Regular	150	10
Reduced-fat	135	6
Pretzels		
(1 ounce or 1 individual bag)		
Regular	110	1
Fat-free	110	0

Physical Activity

Improved eating habits in combination with decreased calorie intake and calorie-burning exercise is the best way to lose weight and maintain the results. In addition to enhancing weight loss efforts, physical activity promotes loss of body fat, increases muscle mass, and increases cardiovascular fitness. Regular exercise not only helps you lose weight by increasing the number of calories you burn but also makes it easier to keep off the weight that you have already lost.

Walking is a good choice for getting started on an exercise plan. A daily 2-mile walk burns approximately 1,000 to 1,200 calories per week. In addition, do not discount the physical activity that is part of ordinary activities of daily living—housework, climbing stairs, gardening—all important forms of exercise that can contribute to weight loss. It may be easier to maintain a schedule if you exercise with a friend.

Many experts recommend that you also participate in resistance (or strength) training. Resistance training can increase your muscle mass (which, as discussed previously, increases your basal metabolic rate), your bone density (which can help protect against osteoporosis), and your balance and coordination (which can lower the risk of injury), and it can help improve your posture.

People who are overweight, have been inactive, or have medical problems should check with their physician before starting an exercise program. An exercise stress test may be helpful because it can measure the response of the heart to exercise and help establish a safe starting level. (See Chapter 1, pages 8 to 10, for further discussion about the benefits of exercise.)

Behavior Modification

Behavior modification programs help identify triggers and treatments for unhealthful eating habits—such as uncontrolled snacking or late night eating. Such programs often explore the reasons behind an inactive lifestyle. Food diaries and activity logs frequently are used to heighten your awareness of what and how much is eaten and the type and amount of exercise done daily. These journals often provide insight into what triggers undesirable eating and what can be done to avoid or change this behavior.

Other key behavioral elements needed for success include:

Commitment—There must be personal motivation to change—to eat healthfully and to get regular physical activity.

Priorities—It takes significant mental and physical energy to change deeply ingrained habits. If other major issues or life stresses are present, it is important to seek help for dealing with them. When balance in life is regained, you can focus on healthful nutrition more effectively.

Realistic goals—The best goals are those that can be achieved. Set daily or weekly goals that allow progress to be measured and rewarded. Aim to lose about 1 to 2 pounds weekly. It is important to know that small losses of weight— or improvements in physical activity—can improve health. Measurements of blood pressure, blood sugar level, and blood cholesterol and triglyceride values are more important in terms of health than the number on the scale. Accept the fact that there will be setbacks. Instead of becoming frustrated or angry, resume your health program once again.

Group support—Joining with people facing the same challenges can promote sharing of ideas and facilitate commitment.

Physical activity enhances weight loss efforts by promoting loss of body fat, increasing muscle mass, and increasing cardiovascular fitness.

HIGH BLOOD PRESSURE

High blood pressure, or hypertension, is the most common major health condition in the United States. It is one of the leading causes of heart attack, heart failure, stroke, kidney failure, and premature death. It also can damage parts of the circulatory system—including blood vessels in the heart, brain, eyes, and kidneys.

Sometimes called the "silent killer," high blood pressure is a disease that can go undetected for years. The higher the blood pressure or the longer it goes undiagnosed, the worse the outlook. More than 50 million Americans have high blood pressure. Of that number, an estimated one-third do not know they have the disease.

There has been remarkable progress in detecting, treating, and controlling high blood pressure. Recently, there has been a substantial increase in the number of individuals who are aware of their high blood pressure and who are being treated for it. At the same time, the incidences of coronary artery disease and stroke have significantly decreased, partially as a result of progress in the detection, treatment, and control of high blood pressure. Despite these efforts, most people with high blood pressure do not have adequately controlled blood pressure. Unfortunately, high blood pressure remains a serious medical problem.

Approximately 5 percent of cases of high blood pressure can be traced to underlying diseases, such as kidney disorders or conditions that cause narrowing of the arteries. This is called "secondary hypertension." But most cases have no known cause. This type is called "primary" or "essential" hypertension.

Risk Factors for High Blood Pressure

There are risk factors that indicate a predisposition for the development of high blood pressure. Among these risk factors are the following:

Family history—A family history of high blood pressure is a risk for hypertension.

Age—High blood pressure becomes more common among all people as they grow older. After all, nearly two-thirds of Americans aged 70 or older have the disease.

Sex—Although men and women are typically equally affected, women who use birth control pills and smoke cigarettes have a greater chance of having high blood pressure.

Race—As an example, high blood pressure is more common among blacks than whites.

You either have or don't have the above risk factors. There are also other factors that increase your risk for high blood pressure which you can do something about:

- obesity—control your weight
- lack of exercise—get moving
- alcohol consumption—take control or abstain
- excessive salt (sodium) intake—cut down

CLASSIFICATION OF HIGH BLOOD PRESSURE

	SYSTOLIC (MM HG)		DIASTOLIC (MM HG)
Optimal*	120 or less	and	80 or less
Normal	129 or less	and	84 or less
High-normal	130-139	or	85-89
Hypertension			
Stage 1†	140-159	or	90-99
Stage 2†	160-179	or	100-109
Stage 3†	180 or higher	or	110 or higher

*Optimal pressure with respect to cardiovascular risk.
†Based on the average of two or more readings taken at each of two or more visits after an initial screening.

Measuring Blood Pressure

Blood pressure is the force that the blood exerts on the artery walls as the heart pumps it through the body. The amount of force depends on various factors, including how hard the heart pumps and the volume of blood it pumps. Another factor is the amount of resistance blood encounters in the arteries—the thicker or more clogged the blood vessels, the greater the resistance. The elasticity of arteries also plays a role in blood pressure. The less elastic arteries are, the less they expand under the force of a heartbeat and the higher the resistance.

The standard way to measure blood pressure is in millimeters of mercury (mm Hg). This unit of measurement refers to how high the pressure inside the arteries is able to raise a column of mercury. Each blood pressure measurement has two numbers. The top number is the systolic blood pressure, or the highest pressure within the arteries that occurs during "systole," when the heart is contracting. The bottom number is the diastolic blood pressure, or the lowest pressure within the arteries that occurs during "diastole," when the heart is relaxed and filling with blood.

For most people, a blood pressure of 120/80 mm Hg or lower is considered healthy.

Effects of High Blood Pressure

Repeated blood pressure measurements that are more than 130/85 mm Hg indicate a potential for problems, a greater risk of progressing to definite high blood pressure, and the need to have regular blood pressure checks. A measurement of 140 systolic or more, 90 diastolic or more, or both indicates high blood pressure, which requires treatment. Recent guidelines stress the importance of treating to achieve a pressure less than 130/85. Indeed, treating to these lower levels of blood pressure further reduces the incidence of stroke, heart and kidney damage, and other vascular problems.

When high blood pressure occurs, it means that the force of the blood in the arteries is excessive and many health complications can occur (see illustration on page 55).

Heart

High blood pressure forces the heart to work harder than it should to pump blood to distant tissues and organs. Another way to understand blood pressure is to think of it as a weight or a load that the heart muscle must "push" against. Like any muscle, the heart gets larger with pushing heavy

weights. Eventually, the heart's pumping efficiency decreases when the muscle can no longer adapt to the excessive workload that the high blood pressure demands. When this occurs, the heart muscle may weaken and the heart can fail.

Arteries

High blood pressure can also accelerate the development of plaque within the arteries, a condition known as atherosclerosis. With the narrowing of the artery walls, the risk of heart attack is increased. High blood pressure also can lead to bulges (aneurysms) in the arteries. If an aneurysm in a major artery ruptures, the results can be catastrophic and possibly fatal.

Brain

High blood pressure increases the risk of having a stroke, which occurs when a blood vessel within the brain either ruptures or is blocked.

Kidneys

The kidneys filter waste products from the blood and maintain proper blood minerals and blood volume. When these functions are impaired or compromised, so is their role in helping to maintain blood pressure. These effects can produce a destructive cycle that results in increasing blood pressure and a gradual failure of the kidneys to remove impurities from the blood.

Eyes

High blood pressure often causes problems with the eyes. Examination of the retina can reveal narrowing of the arteries, small hemorrhages, and accumulations of protein that have leaked from affected blood vessels (exudates). Although it is unusual for high blood pressure to impair vision, it can occur as a result of severe constriction of the retinal arteries and swelling (edema) of the retina during episodes of increased blood pressure. Treating the high blood pressure is the only way to reverse this vision loss.

Treating High Blood Pressure

Fortunately for many people, high blood pressure is preventable. Even those who already have high blood pressure or are at increased risk may be able to reduce the number and doses of medications needed to control it and minimize other health complications by the following lifestyle modifications.

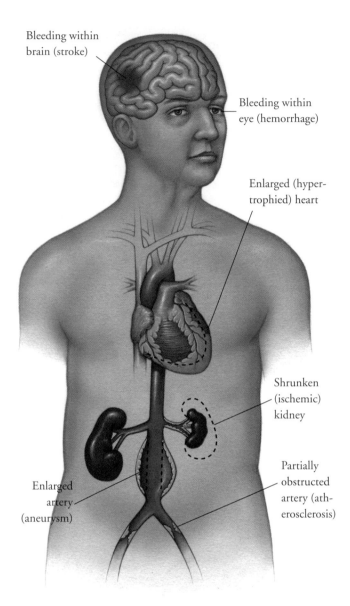

Bleeding within brain (stroke)

Bleeding within eye (hemorrhage)

Enlarged (hypertrophied) heart

Shrunken (ischemic) kidney

Partially obstructed artery (atherosclerosis)

Enlarged artery (aneurysm)

High blood pressure can damage vital organs and result in many life-threatening medical problems.

Lose weight—If you are overweight, losing weight is the most effective non-drug method for lowering blood pressure. A weight loss of as little as 10 pounds can significantly reduce blood pressure in many overweight people with high blood pressure. In some people, weight loss alone is sufficient to avoid the need for blood pressure medication.

Exercise—When compared with more active and fit peers, sedentary individuals with normal blood pressure have a 20 to 50 percent increased risk for development of high blood pressure. Regular aerobic exercise such as walking

or biking for 30 to 45 minutes most days of the week is a very effective means of lowering blood pressure.

Limit alcohol—Excessive alcohol intake is a risk factor for high blood pressure and stroke. It also can interfere with the effects of blood pressure medications. Men who drink should limit their intake to no more than 2 drinks a day; women should have no more than 1 drink daily (see Chapter 4, A Drink Defined, page 87).

Do not smoke—Smoking a cigarette temporarily increases blood pressure for up to 30 minutes. Smoking is also a major risk factor for cardiovascular disease. Everyone, especially people with high blood pressure, needs to quit smoking or never start.

Limit or avoid high-sodium foods—A high intake of sodium in the diet increases blood pressure in some people. The average American consumes about 4,000 milligrams or more of sodium a day. People with high blood pressure should limit their sodium intake to less than 2,400 milligrams a day, and many experts recommend the same limit for everyone.

THE COMBINATION DIET FROM THE DASH STUDY

This eating plan is from the "Dietary Approaches to Stop Hypertension" (DASH) study from the National Institutes of Health. This diet is rich in vegetables, fruits, and low-fat dairy foods and low in saturated fat, total fat, and cholesterol. It is also high in potassium, calcium, and magnesium.

FOOD GROUP	DAILY SERVINGS	SERVING SIZES	SIGNIFICANT NUTRIENTS IN EACH FOOD GROUP
Grains and grain products (emphasis on whole grains)	7-8	1 slice bread 1/2 cup dry cereal 1/2 cup cooked rice, pasta, or cereal	Energy and fiber
Vegetables	4-5	1 cup leafy vegetable 1/2 cup cooked vegetable 6 ounces vegetable juice	Potassium, magnesium, and fiber
Fruits	4-5	1 medium fruit 1/2 cup fresh, frozen, or canned fruit 1/4 cup dried fruit 6 ounces 100% juice	Potassium, magnesium, and fiber
Low-fat or nonfat dairy products	2-3	8 ounces milk 1 cup yogurt 1.5 ounces cheese	Calcium and protein
Lean meats, poultry, fish	2 or less	3 ounces cooked meats, poultry, or fish	Protein and magnesium
Nuts, seeds, and legumes	4-5 per week	1.5 ounces (1/3 cup) nuts 2 tablespoons (1/2 ounce) seeds 1/2 cup cooked legumes	Magnesium, potassium, protein, and fiber; high in calories

Note: This plan provides approximately 2,000 calories. The number of daily servings from a food group should vary with individual needs.

Follow a balanced nutrition program—A low-fat, high-fruit and vegetable diet can lower blood pressure impressively, all by itself. A large percentage of people with high blood pressure may be able to decrease their need for blood pressure medication if they follow the recommendations of the Food Guide Pyramid. A diet following this plan promotes weight loss and is high in minerals such as calcium, potassium, and magnesium, which have been associated with lower blood pressure. Chapters 4 and 5 (pages 79 through 149) provide ideas for planning and preparing healthful meals.

The National Heart, Lung and Blood Institute recently sponsored a study testing the effects of different diets on blood pressure, called the "DASH" (Dietary Approaches to Stop Hypertension) study. Participants in the study ate one of three diets: an average American diet, a diet rich in fruits and vegetables, or a "combination" diet that emphasized fruits, vegetables, and low-fat dairy products and was low in fat and saturated fat. Sodium consumption was the same in all three diets. Participants were asked to limit obviously salty food, to rinse canned vegetables, and to not add extra salt to food.

Both the fruit-and-vegetable diet and the "combination" diet lowered blood pressure. However, the combination diet was the most effective. Within that group, people with above-normal blood pressure (more than 129/80 mm Hg) and those with high blood pressure (more than 140/90 mm Hg) experienced reductions of their blood pressure similar to those achieved with some blood pressure medications.

Researchers believe that people following the combination diet fared better because of the low saturated fat and high fruit and vegetable mixture that provided adequate potassium, magnesium, and calcium. For people with normal blood pressure, the combination diet may help to avoid blood pressure problems. If blood pressure is only slightly increased, following this diet may actually eliminate the need for medication. For people with severe high blood pressure, the diet may allow reduction in blood pressure medication (see sidebar: The Combination Diet From the DASH Study, page 56).

When Medications Are Needed

When lifestyle changes alone are not effective for lowering high blood pressure, medications may be required. Medications vary in the way they control blood pressure. Some types help the kidneys to eliminate sodium and water, some make the heart beat more slowly and less forcefully, and others enable the blood vessels to relax and decrease the resistance to blood flow. Your physician will determine which drug or combination of drugs is best suited for you.

DIABETES MELLITUS

Each year, the words "you have diabetes" are spoken with greater frequency, often to unsuspecting individuals. Among Americans, the prevalence of diabetes has grown dramatically during the 20th century. Today, more than 16 million Americans have diabetes—90 percent of them are older than 40 years. Interestingly, half of these adults do not even know they have the disease because symptoms develop gradually and, at first, are hard to identify. Early diagnosis is important, though, because the longer diabetes goes untreated, the greater your risk for serious complications.

What Is Diabetes?

Diabetes mellitus is a disorder of metabolism—the way the body uses digested food for energy and growth. The origin of the name "diabetes mellitus" is Greek, referring to sweetness or honey (mellitus) that passes through (diabetes). After a meal, food is broken down into simpler forms and absorbed by the body. Simple sugars, amino acids, and fatty acids are used by the body or converted by the liver into sugar (glucose), the preferred fuel the body burns for energy. For cells to use this form of sugar, insulin—a hormone that is produced by the pancreas—must "unlock" the cells to allow glucose to enter.

The pancreas, a long, thin organ that is about the size of a hand, is located behind the stomach. Normally the pancreas produces the right amount of insulin to accommodate the amount of sugar that is in the blood. Diabetes is actually not a single condition but a group of diseases with one thing in common—a problem with insulin. In a person with diabetes, the pancreas does not produce sufficient insulin to meet the body's needs. This insufficiency may develop if the pancreas stops producing the right amount and quality of insulin, if the rest of the body's cells do not respond properly to insulin, or a combination of both. Insulin is required for glucose to be metabolized properly. If there is not enough insulin, excess glucose builds up in the blood, and the resulting condition is called hyperglycemia.

Persistent hyperglycemia causes almost all the symptoms that may alert an individual to the development of diabetes.

These symptoms can include frequent urination, extreme thirst, blurred vision, fatigue, unexplained weight loss, recurrent infection, tingling or loss of feeling in the hands or feet, and hunger.

With diabetes, the excess glucose spills into the urine, and the urine output increases. Consequently, the person with diabetes becomes dehydrated and thirsty. Fatigue results when the supply of glucose to cells is not available, causing energy levels to decline. To compensate for the lost fuel, the body burns stored fat, and weight loss and hunger may occur. Left untreated, persistent hyperglycemia is also responsible for most of the long-term complications of diabetes (see Long-Term Complications of Diabetes, this page).

There are several types of diabetes—different disorders with different causes. The two most common types are described here.

Type 1 Diabetes

Type 1 diabetes occurs in 1 in 10 people with diabetes. In these individuals, the pancreas produces little or no insulin. To control blood sugar, insulin must be taken. Most people whose diabetes is diagnosed before age 30 have the insulin-dependent type. It used to be called insulin-dependent diabetes mellitus or juvenile-onset diabetes. Type 1 diabetes affects both sexes equally. In most cases it is due to an autoimmune disease in which the body's immune system attacks and destroys the insulin-producing beta cells in the pancreas. This type of diabetes commonly develops in childhood, but it can occur at any age. It can develop unnoticed for several years and then suddenly become apparent, often after an illness. Some people—particularly children and teenagers—may first become aware of the disease when they develop ketoacidosis. This is a serious complication in which the blood becomes more acidic because of severe insulin deficiency.

Type 2 Diabetes

Type 2 diabetes accounts for about 85 to 90 percent of diabetes in people older than 30. It was previously called non-insulin-dependent or adult-onset diabetes. Most persons with type 2 diabetes are overweight or obese.

For people with this type of diabetes, total absence of insulin is not the problem. The problem is that the body does not make enough insulin to meet its needs and the insulin does not work normally to control glucose levels. This is termed "insulin resistance" and leads to hyperglycemia.

Excess weight is by far the greatest risk factor for development of type 2 diabetes. Most people who develop type 2 diabetes are overweight, a condition that appears to impair insulin action. Someone who is overweight and has diabetes may be able to achieve a normal blood sugar without medication simply by losing weight. Losing as little as 10 percent of body weight has been shown to lower blood glucose. Surprisingly, persons who are not overweight by traditional criteria also may be at risk for diabetes. Excess body fat distributed mostly in the abdomen increases the chance for development of type 2 diabetes. Other risk factors include age, race, heredity, and lack of physical activity.

Type 2 diabetes may go undetected for many years, because hyperglycemia develops slowly and the disease may not immediately produce the classic symptoms of diabetes. Unfortunately, even without symptoms, there are hidden dangers, including damage to major organs such as the heart and kidneys. Because the disease usually develops after age 40, and the incidence increases more steeply after age 55, it is important that people in middle age be screened for the disease and see their physician if symptoms develop.

Type 2 diabetes is more common among Native Americans, Hispanics, African-Americans, and westernized Asians than among people of European ancestry.

Long-Term Complications of Diabetes

Numerous studies have shown that keeping the blood glucose level close to normal delays the onset and prevents the progression of eye, kidney, and nerve diseases caused by diabetes. Even if blood glucose has not been controlled in the past, any improvement in diabetes control may help to avoid or delay complications of diabetes and their progression. Treatment of high blood lipid values and high blood pressure, which are commonly associated with diabetes, is also important.

Eye disease—In the absence of good glucose control, eye disease develops in nearly everyone with diabetes. Diabetic retinopathy occurs because high blood sugar (especially coupled with high blood pressure) can damage the small blood vessels in the retina (the light-sensitive area within the eye). Diabetes also can lead to cataracts, damage the macula (the area in the eye where the optic nerve is located), and increase the risk of glaucoma.

In addition to keeping your glucose levels under control, it is important to have regular eye examinations with an ophthalmologist, who dilates the eyes and does a thorough

Blood sugar can be kept within normal levels by balancing the main treatment tools for diabetes—nutrition, weight control, exercise, and medication. In some people, blood sugar may be controlled by a combination of loss of excess weight, good nutrition, and regular exercise. Others may need medication.

examination of them. A non-dilated eye test is not adequate for screening.

Kidney failure—In the absence of good glucose control, a person with diabetes is 20 times more likely to develop kidney failure than someone who does not have the disease. Kidney disease results when chronic high blood sugar damages the small vessels in the kidneys which are responsible for filtering waste from the blood. Ultimately, kidney failure may occur, requiring dialysis or a kidney transplant. People with diabetes should have the function of their kidneys evaluated routinely.

Nerve damage—Also called neuropathy, nerve damage occurs in 30 to 40 percent of people with diabetes. Nerve damage can cause numbness and tingling, pain, insensitivity to pain and temperature, and extreme sensitivity to touch. Experts think the damage results from the effect of chronic high blood sugar on blood vessels that supply nerve cells. The feet are especially vulnerable to neuropathy.

Cardiovascular disease—Chronic high blood sugar is associated with narrowing of the arteries (atherosclerosis), high blood pressure, heart attack, and stroke. It is also associated with increased blood levels of triglycerides (a type of blood fat) and decreased levels of HDL ("good") cholesterol. Unless appropriately treated, an individual with diabetes is 5 times more likely to have a stroke and 2 to 4 times more likely to have coronary artery disease. Also,

smoking dramatically accelerates the development of these cardiovascular complications. Anyone with diabetes should stop smoking.

Infections—High blood sugar impairs the function of immune cells and increases the risk of infections. The mouth, gums, lungs, skin, feet, bladder, and genital area are common sites of infection. Nerve damage in the legs and feet can make someone with diabetes less aware of injuries or infection, increasing the risk of amputation. With proper care of the feet, foot complications can be minimized or avoided.

Can Diabetes Be Prevented?

Although an area of active research, there is currently no proven means of preventing type 1 diabetes. However, these lifestyle changes minimize the risk of, and may actually prevent, type 2 diabetes:

Maintain a healthful weight—Most people who develop type 2 diabetes are overweight. Aggressive efforts aimed at achieving and maintaining a healthful weight may be beneficial, especially in combination with exercise.

Eat a balanced diet—A diet low in saturated fat and sugar and high in complex carbohydrates and dietary fiber also has been linked to a reduced risk of diabetes.

Exercise—People who exercise regularly have a significantly lower incidence of type 2 diabetes.

How Is Diabetes Diagnosed?

It is important that diabetes be diagnosed early, before too much damage is done. As a first step, the American Diabetes Association recommends that everyone should visit a physician and get a fasting blood glucose test at age 45. A sample of blood is drawn after fasting overnight, and its glucose level is measured. A normal fasting result is between 70 and 110 milligrams of glucose per deciliter of blood (mg/dL).

A retest every 3 years is advised. If an individual has any risk factors for diabetes or any symptoms, testing at more frequent intervals or at an earlier age is recommended. Diabetes is diagnosed when a fasting blood glucose value is more than 125 mg/dL or a glucose value is more than 200 mg/dL accompanied by symptoms of diabetes. The diagnosis should be confirmed by repeat testing on a different day.

Tools for Controlling Diabetes

Persons with diabetes can live a full life by following a few basic principles to control their disease. Diabetes can be managed with at-home blood glucose tests, healthy nutrition habits, weight control, routine exercise, and medications (if needed).

The most important step is to learn to control the blood sugar value, which means maintaining it as near to normal as possible or in the goal range determined by your physician. Vigilant control of blood sugar levels may dramatically reduce the risk of eye, kidney, and nerve damage. This also lowers the risk of heart attack, stroke, and limb amputation, in addition to pro moting a more desirable level of blood lipids.

Blood sugar can be kept within normal levels by balancing the main treat ment tools for diabetes—nutrition, weight control, exercise, and medication. In some people, blood sugar may be controlled by a combination of weight loss, good nutrition, and regular exercise. Others may need medication.

Blood Sugar Levels

The level of glucose in the blood depends on several factors: when meals are eaten, how many calories are consumed, activity level, and the dose of medication prescribed. The stress of an illness may also alter the level of blood glucose. Successful daily management of diabetes may prevent or minimize emergencies that may result when blood sugar levels are too high or too low. Because these emergencies can cause mental confusion or loss of consciousness, people with diabetes should wear a medical alert identification and acquaint family members, friends, neighbors, and coworkers with the signs and symptoms of an emergency and steps to a proper response.

Blood sugar levels can be monitored with home testing meters. Testing not only measures the amount of sugar in the blood but also enables the person to identify the reasons behind high and low values so that adjustments can be made in the dose of diabetes medications. Frequently the person can learn how to adjust the dose of oral medication or insulin to achieve the desired glucose value. The health care team caring for the person with diabetes can determine reasonable blood sugar goals.

Nutrition

The old restrictive diets for diabetes no longer apply. It is now known that the best diet for diabetes control is consistent with what everyone should eat for good health. How much you eat is just as important as what you eat in controlling blood sugar. To keep blood sugar on an even keel, people with diabetes should not eat large meals or skip meals; they should eat smaller servings at regular intervals instead. As for the menu, choices should follow Food Guide Pyramid recommendations—emphasize whole grains, legumes, and vegetables. These foods are higher in complex carbohydrate and fiber and can help control blood sugar. Even though fruit contains sugar, it should not be avoided.

Because people with diabetes are at higher risk for cardiovascular disease, it's important to keep the fat intake to about 30 percent of total daily calories and to limit cholesterol-containing foods. Items at the top of the Food Guide Pyramid—such as fats and sweets—should be eaten sparingly. Research has shown that the total amount of carbohydrate consumed at a given time, rather than the type of carbohydrate, is the most important factor in control of blood sugar. Therefore, if not present in excess, sugar can be included as a part of a well-balanced meal.

Many tools are available to help with meal planning. For most people, eating three meals at regular times and avoiding excessive sweets are enough to control blood sugar. Dietitians may provide a simple method, such as the Food Guide Pyramid, to encourage variety, proportion, and

moderation in food selections and to also ensure healthful food choices. Chapters 4 and 5 provide ideas for planning and preparing healthful meals. Other techniques may be needed and recommended by your physician, registered dietitian, or certified diabetes educator.

Weight Control

Because many people who have diabetes are overweight, maintaining a healthful weight and level of activity is the key to keeping the disease under control and minimizing the risk for serious complications. The more overweight people are, the more resistant their cells become to their own insulin; losing weight decreases that resistance. Often, weight loss of just 10 percent can improve blood sugar and have lasting beneficial effects.

Exercise

With exercise, some people with type 2 diabetes may even reduce or eliminate their need for insulin or oral diabetes medication. Anyone who has diabetes should check with a physician before starting an exercise program. Studies have shown that those at high risk for diabetes who exercise have a 50 percent lower incidence of type 2 diabetes. Exercise helps control weight, makes cells more sensitive to insulin, increases blood flow, and improves circulation in even the smallest blood vessels. A leaner body also helps burn calories more efficiently. Moreover, exercise lowers your risk of cardiovascular disease.

Exercise can affect blood sugar levels up to 24 hours. So if insulin is a part of the treatment plan, check with a physician to receive guidelines for frequency of blood sugar testing and insulin adjustment.

Medications

Along with nutrition, weight control, and exercise, medications may be necessary to achieve a desired glucose level. Persons with type 1 diabetes must take insulin by injection. Insulin cannot be taken by mouth because it breaks down in the digestive tract. The type of insulin and number of daily injections depend on individual needs. Insulin also can be administered by pump.

In some persons, type 2 diabetes can be managed by healthful nutrition and exercise alone. If the desired glucose level is not achieved, oral medications may be prescribed. There are several classes of oral medications. Some stimulate the pancreas to produce more insulin, and some help insulin to work more effectively in the body by decreasing the sugar made by the liver and by increasing the sugar removed from the blood to the cells. If oral medications do not achieve the glucose goal, insulin injections may be required.

CORONARY ARTERY DISEASE

Knowing your blood lipid (fat) levels—the various forms of cholesterol and triglycerides—is important. But what these numbers mean can be confusing, and how the different types of cholesterol—not to mention other blood fats—relate to coronary artery (heart and blood vessel) disease also can be a puzzle.

Coronary artery disease is the number one killer of Americans. Studies point to certain abnormalities in cholesterol and triglyceride levels as a major contributor to this problem. There is plenty of good news to suggest that public awareness of heart disease and its risk factors is making a difference. Deaths from cardiovascular disease continue to decrease. Much credit for this encouraging trend goes to improved treatments and modification of the risk factors for heart disease, including lowering cholesterol levels. Despite these substantial improvements, the American Heart Association reports that cardiovascular disease still kills almost 1 million Americans each year. This is more than all cancer deaths combined. More than 6 million Americans experience symptoms due to coronary artery disease. As many as 1.5 million Americans will have a heart attack every year, and about half a million of them will die.

Blockage of the coronary arteries supplying the heart muscle (which can lead to heart attack) causes more deaths, disability, and economic loss than any other type of heart disease.

Atherosclerosis

The coronary arteries are the heart's own circulatory system. They supply the heart with blood, oxygen, and nutrients. The heart uses this blood supply for energy to perform its continuous task of pumping. Coronary artery disease can take many different forms, but each has essentially the same effect: the heart muscle does not get enough blood and oxygen through the coronary arteries. Consequently, its own demands for oxygen and nutrients are not met. This condition can be either temporary or permanent.

Most coronary artery disease is caused by atherosclerosis (also known as "hardening of the arteries"). The term

"atherosclerosis" comes from the Greek *ather* (meaning "porridge") and *sklerosis* (meaning "hardening"). Healthy arteries are flexible, strong, and elastic. The inner layer of arteries is smooth, enabling blood to flow freely.

Atherosclerosis can be a silent, painless process in which cholesterol-containing fatty deposits accumulate in the walls of the arteries. These accumulations occur as lumps called plaques. As plaque deposits enlarge, the interior of the artery narrows, and the flow of blood is then reduced (see the illustration below). If reduced flow occurs in the coronary (heart) arteries, it can lead to a type of chest pain called angina pectoris.

As a plaque enlarges, the inner lining of the artery becomes rough. A tear or rupture in the plaque may cause a blood clot to form. Such a clot can block the flow of blood or break free and plug another artery. If the flow of blood to a part of the heart is stopped, a heart attack results. If the blood flow to a part of the brain stops, a stroke occurs.

A Closer Look at Blood Lipids

Many factors influence the clogging of arteries, but cholesterol is a primary one. Cholesterol is a waxy, fat-like substance (a lipid). Although it is often discussed in negative terms, it is an essential component of the body's cell membranes. It also serves to insulate nerves and is a building block in the formation of certain hormones. The liver uses it to make bile acids, which help digest food.

Confusion about cholesterol is due in great part to the all-purpose use of the term. Cholesterol has two sources: the foods we eat (about 20%) and the cholesterol that is made by the body (about 80%). Dietary cholesterol is found only in animal products, such as meat and dairy products, or foods made with animal products. Examples include all meats, fish, and poultry, eggs, and milk products. In addition, both the amount and the type of fat eaten influence the blood cholesterol level. Both saturated (primarily from animals) and trans-saturated (oils that have been processed to make them more solid) fats increase the amount of cholesterol made by the liver. (See Chapter 2, page 26, for further description of dietary fats.)

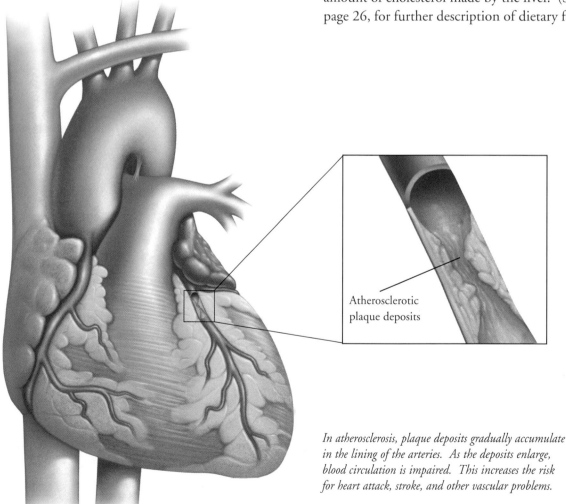

Atherosclerotic plaque deposits

In atherosclerosis, plaque deposits gradually accumulate in the lining of the arteries. As the deposits enlarge, blood circulation is impaired. This increases the risk for heart attack, stroke, and other vascular problems.

MINI-GLOSSARY OF LIPID-RELATED TERMS

- **Apolipoprotein**—Proteins that combine with lipids to make them dissolve in the blood.

- **Cholesterol**—A soft, waxy substance in the blood and in all your body's cells. It is used to form cell membranes, some hormones, and other needed tissues. Dietary cholesterol is found only in food derived from animal sources.

- **HDL cholesterol**—About 20 to 30 percent of blood cholesterol is carried by high-density lipoproteins (HDL). HDL carries cholesterol away from the arteries and back to the liver, where it is removed from the body. HDL seems to protect against heart attack. This is why HDL cholesterol is referred to as the "good" cholesterol.

- **LDL cholesterol**—Low-density lipoprotein (LDL) cholesterol is the main cholesterol carrier in the blood.

When there is too much LDL cholesterol circulating in the blood, it can slowly build up in the walls of the arteries that feed the heart and brain. This is why LDL cholesterol is often called the "bad" cholesterol.

- **Lipoproteins**—Lipids combined with apoproteins.

- **Triglycerides**—Triglycerides are derived from fats eaten in foods or made in the body from other sources such as carbohydrates. Calories ingested in a meal and not needed immediately by tissues are converted to triglycerides and transported to fat cells to be stored. Hormones regulate the release of triglycerides from fat tissue to meet the daily need for energy between meals.

- **VLDL cholesterol**—In the fasting state, very low-density lipoproteins (VLDL) contain 15 to 20 percent of the total blood cholesterol, along with most of the triglycerides.

Unique Roles: LDL and HDL Cholesterol and Triglycerides

Cholesterol and triglycerides are fats and are insoluble in the blood. However, when they combine with protein they become lipoproteins and are able to dissolve in and be carried by blood throughout the body. (See sidebar: Mini-Glossary of Lipid-Related Terms, above.)

Low-density lipoprotein (LDL) cholesterol is the main cholesterol carrier in the blood. There is a direct relationship between the level of LDL cholesterol (or total cholesterol) and the rate of coronary artery disease. When there is too much LDL cholesterol circulating in the blood, it can slowly build up in the walls of the arteries that feed the heart and brain. For this reason, LDL is often referred to as the "bad" cholesterol.

If there are too many LDL particles in the blood, or the liver (the normal site of metabolism) does not remove LDL quickly enough from the blood, it builds, particularly in blood vessels. It is the role of high-density lipoprotein (HDL) to counteract this effect.

About a third to a fourth of blood cholesterol is carried by HDL. HDL carries cholesterol away from the arteries and back to the liver, where it is removed from the blood. It is therefore often referred to as the "good" cholesterol. A high level of HDL seems to protect against atherosclerosis and heart attack. The opposite is also true: a low HDL level

indicates an increased risk of atherosclerosis. Thus, the goal is to have a high HDL cholesterol level and a low LDL cholesterol level.

Triglycerides in the blood are derived from fats eaten in foods or produced when the body converts excess calories, alcohol, or sugar into fat. Most triglycerides are transported through the bloodstream as very low-density lipoprotein (VLDL). Some cholesterol is also present in VLDL.

A certain amount of triglycerides in the blood is normal. Hormones regulate the release of triglycerides from fat tissue to meet the body's needs for energy between meals. However, at high levels, triglycerides may contribute to the development of atherosclerosis. Increased triglyceride levels also may be a consequence of other diseases, such as untreated diabetes mellitus.

Calories ingested at a meal and not used immediately by tissues are converted to triglycerides and transported to fat cells to be stored.

Blood Testing

The only way to determine whether cholesterol and other blood lipids are in a desirable range is to have them measured by a blood test. The National Cholesterol Education Program guidelines recommend that total cholesterol, HDL cholesterol, and triglycerides be measured at least once every 5 years in all adults age 20 or older. However, your

physician also may recommend that the screening include LDL cholesterol and triglycerides.

Triglycerides must be measured after an overnight fast because eating can have a marked effect on blood triglyceride levels. Therefore, fast for at least 12 hours before blood is drawn. Do not drink alcohol for 24 hours before the test. If you have a risk factor(s) for heart disease, consult your physician regarding the optimal frequency of testing. (See sidebar: Risk Factors for Coronary Artery Disease Other Than LDL Cholesterol, below.)

In a sense, it is incorrect to think of a cholesterol (or triglyceride) level as being strictly abnormal or normal. Although ranges of cholesterol levels have been identified which are considered "too high," there is no "magic number" that separates risky levels from safe levels. Actually, the ranges for adults are based on a consensus of experts. They have identified lipid levels in the blood above which the risk for development of coronary complications is high enough to warrant medications or lifestyle changes.

People with cholesterol or triglyceride levels in the higher-risk zones are said to be hypercholesterolemic or hypertriglyceridemic (*hyper* means "high," and *emic* means "in the blood"). But, as with all risk factors, being in the "high" range does not guarantee that coronary artery disease will develop, nor does being in the "low" range guarantee avoiding it.

Blood test numbers are only guidelines. If the numbers stray from the desirable range, a physician can provide advice on what to do. Remember that each number takes on greater meaning in light of the other lipid results and in the presence of other cardiovascular disease risk factors (see sidebar: Your Blood Lipid Test Results—What Do Those Numbers Mean? page 65).

RISK FACTORS FOR CORONARY ARTERY DISEASE OTHER THAN LDL CHOLESTEROL

- Age: males 45 years or older and females age 55 years or older, or females who have had premature menopause without estrogen replacement
- Family history of early coronary artery disease
- Current cigarette smoking
- High blood pressure
- Low HDL cholesterol (less than 40 mg/dL)
- Diabetes mellitus

The Reasons for High Blood Lipid Levels

Why do some people have high cholesterol and triglycerides? High levels may result from genetic makeup or lifestyle choices or both. Heredity may endow people with cells that do not remove LDL or VLDL cholesterol from the blood efficiently, or with a liver that produces too much cholesterol as VLDL particles or too few HDL particles. Lifestyle factors such as a high-fat diet, obesity, smoking, and physical inactivity also can cause or contribute to high cholesterol levels, increasing an individual's risk for atherosclerosis.

For a more complete picture of cardiovascular health, other risk factors—beyond cholesterol and triglycerides—must be considered. The more risk factors an individual has in combination with undesirable lipid levels, the greater the chances for development of cardiovascular disease.

The risk factors for cardiovascular disease are divided into those that can be changed and those that cannot.

Risk Factors That Can Be Changed or Treated

Smoking cigarettes damages the walls of the blood vessels, making them more receptive to the accumulation of fatty deposits. Smoking also may lower the HDL by as much as 15 percent. Quitting smoking may return the HDL to a higher level.

High blood pressure damages the walls of the arteries, thus accelerating the development of atherosclerosis. Some medications for high blood pressure increase LDL and triglyceride levels and decrease HDL levels. Others do not. Blood pressure that is properly managed decreases the progression and risk for cardiovascular disease.

Sedentary lifestyle is associated with a decrease in HDL. Aerobic exercise is one way to increase HDL. Aerobic activity is any exercise that requires continuous movement of the arms and legs and increases the rate of breathing. Even 30 to 45 minutes of brisk walking every other day helps protect the cardiovascular system.

Obesity is a risk to cardiovascular health. Excess body fat increases total cholesterol, LDL cholesterol, and triglyceride levels. It also lowers the HDL cholesterol level. Obesity increases blood pressure and the risk for diabetes, which can increase the chances of heart disease developing. Losing just 10 percent of excess body weight can improve triglyceride and cholesterol levels.

Diabetes can increase the triglyceride level and decrease the HDL cholesterol level. Good control of blood sugar helps reduce increased triglyceride levels.

YOUR BLOOD LIPID TEST RESULTS—WHAT DO THOSE NUMBERS MEAN?

Use this table as a general guide. The importance of each number varies according to your sex, health status, and family history. For example, if you already have heart disease, you will want to lower your LDL cholesterol level to less than 100 mg/dL. Your health care provider can help clarify your specific risk.

TEST	LEVEL (IN MG/DL)		
	OPTIMAL	BORDERLINE	UNDESIRABLE
Total cholesterol	Less than 200	200-240	More than 240
HDL cholesterol Considered "good"—the higher, the better	60 or more	–	Less than 40
LDL cholesterol[*] Considered "bad"—the lower, the better	Less than 100	130-160	More than 160
Cholesterol/HDL ratio	Less than 4.5	4.5-5.5	More than 5.5
LDL/HDL ratio	Less than 3	3-5	More than 5
Triglycerides	Less than 150	150-200	More than 200

[*]*LDL may be measured directly or may be estimated from the other numbers if your triglyceride level is lower than 400 mg/dL. You can estimate LDL yourself by using this equation: LDL = Total Cholesterol - HDL + triglyceride level divided by 5.*

Estrogen deficiency and menopause increase the risk of heart disease. Conversely, estrogen replacement lowers the risk in certain groups of estrogen-deficient women.

Risk Factors That Cannot Be Changed

Aging—Aging usually increases the level of LDL cholesterol, although the reasons are not understood. It could be the aging process itself that causes this, or an increase in body fat with advancing age.

Sex—Cardiovascular disease is not only a man's disease, as once thought. It is the number one killer of women, claiming the lives of 500,000 women every year. Cardiovascular disease occurs in women almost as often as it does in men. It just happens later in life. Before menopause, a woman's risk of coronary artery disease is lower than that of a man. Menopause results in an increase in LDL cholesterol and a decrease in the protective HDL cholesterol. After menopause, a woman's risk of heart disease is the same as that of a man. Treatment with estrogen helps to return the risk to premenopausal levels.

Family history—A family history of abnormal lipid levels or early heart disease increases the risk of heart attack and stroke.

Lifestyle Changes to Reduce Risk

There is significant opportunity to reduce the risk of getting cardiovascular disease. Changes in nutrition along with increased physical activity and learning to decrease stress can improve blood cholesterol and triglyceride levels. Making dietary changes to improve blood cholesterol and triglyceride levels involves these steps:

Maintain a desirable weight—A diet that is high in fat also can be unnecessarily high in calories and contribute to an unhealthy weight. Decrease the total amount of fat eaten. Limit fat—saturated, polyunsaturated, and monounsaturated—to less than 30 percent of your total daily calories (see Chapter 2, pages 26 to 29, for a description of these types of fat). Some individuals may need to restrict fats even more. Because all foods with fats contain a combination of these fats, it is important to reduce total fat.

Reduce saturated fat—The major dietary culprit in an increased blood cholesterol level and increased risk for coronary artery disease, saturated fat is typically solid or waxy at room temperature. Minimize your intake of saturated fat. Foods high in saturated fat include red meats and dairy products as well as coconut, palm, and other tropical oils (check the ingredient portion of the food label).

Replace saturated fat with unsaturated fat—Polyunsaturated and monounsaturated fats should make up the remaining fat allowance. In the recommended amounts, polyunsaturated fats reduce LDL cholesterol, but at the expense of the protective HDL cholesterol, whose levels also may decrease. Polyunsaturated fats are usually liquid at room temperature and in the refrigerator. Vegetable oils such as safflower, corn, sunflower, soy, and cottonseed oil are high in polyunsaturated fat. Monounsaturated fats tend to have the same effects on LDL cholesterol without lowering HDL cholesterol. Monounsaturated fats are liquid at room temperature but may start to solidify in the refrigerator. Olive, canola, and nut oils are sources of monounsaturated fats. (See Chapter 2, pages 26 to 29, for further discussion of unsaturated fats, and see sidebar: Cholesterol-Lowering Margarine? below.)

Limit trans fat—This fat is also called partially hydrogenated vegetable oil. This type of fat may be as harmful to your health as saturated fat because it increases blood cholesterol levels, among other effects. Major sources are hardened vegetable fat, such as margarine or shortening, and products made from these fats, such as cereals, cookies, and crackers.

CHOLESTEROL-LOWERING MARGARINE?

Who would have thought a person could consume margarine and possibly lower cholesterol? This functional food was approved for use in the United States and was introduced to the grocery shelves in 1999. Two types are available: Benecol and Take Control. Benecol is made with a refined form of plant sterol called stanol ester, which is derived from wood pulp. Take Control contains sterol esters, which are made from vegetable oils, soybean, and corn. These new margarines may help lower LDL cholesterol when used as directed by a physician. The margarines may lower LDL cholesterol 7 to 10 percent. Therefore, it is important that they be used in conjunction with a healthful diet full of whole grains, fruits, and vegetables and one that is low in total fat, saturated fat, and cholesterol.

Reduce dietary cholesterol—The daily limit for dietary cholesterol is 300 milligrams. Dietary cholesterol is found only in foods made from or containing animal products. A good way to lower dietary cholesterol is to limit the amount of meat and dairy products. Organ meats and egg yolks are also high in cholesterol. (See Chapter 2, page 27.)

Eat a plant-based diet—A diet that has generous amounts of grains, vegetables, and fruits is naturally lower in fat and has good sources of soluble fiber and antioxidants, which may protect blood vessels from damage and plaque buildup. Chapters 4 and 5 present ideas for planning and preparing healthful meals.

Fruits and vegetables and whole-grain products are also natural sources for folate—a B vitamin that controls the amount of homocysteine in the blood. Homocysteine is an amino acid (a building block of protein) normally found in your body. Your body needs homocysteine to manufacture protein to build and maintain tissue.

Problems arise when there is too much homocysteine, which can cause the tissues lining the arteries to thicken and scar. Cholesterol builds up in the scarred arteries, leading to clogged vessels and blood clots. Adequate intake of this vitamin can help normalize homocysteine levels and may reduce the risk for cardiovascular disease (see sidebar: Folate and Heart Disease, page 67). There is also accumulating evidence that vitamin E may reduce the risk of heart attack.

Cardiovascular Disease and Physical Activity

Unfortunately, most of the population of the United States is sedentary. Sedentary people have nearly twice the risk of having a fatal heart attack as active people of the same age when other factors—such as smoking and high cholesterol—are equal. Consult a physician before embarking on an exercise program. Then, follow these tips for maximal results:

- Choose an aerobic activity. It can be something like walking, jogging, bicycling, or swimming.
- Gradually increase the time and frequency of the exercise. Work up to exercising for 30 minutes daily.

When Are Medications Necessary?

If changes in lifestyle have not brought lipid values into the goal range, medication may be necessary. Before recommending a medication, your physician will use careful judgment and weigh many variables—sex, age, current health, family history of early heart disease or abnormal lipids, and the side effects of medication.

FOLATE AND HEART DISEASE

Homocysteine is an amino acid that is made from dietary protein. Too much homocysteine can damage arterial walls, allowing fatty plaque deposits to clog arteries and promote blood clotting.

FOOD	SERVING SIZE (CUP)	FOLATE AMOUNT (MICROGRAMS)	% DAILY VALUE (BASED ON 400 MICROGRAMS)	
Breakfast cereals	1/2 to 1	100-400* (check label)	25-100	Folate can decrease homocysteine levels. About 400 micrograms a
Lentils (cooked)	1/2	180	45	day is enough to lower your blood
Chickpeas	1/2	140	35	concentration of homocysteine. To
Asparagus	1/2	130	33	get more folate, eat plenty of beans,
Spinach	1/2	130	33	fruits, and vegetables—preferably
Black beans	1/2	130	33	raw or lightly cooked. Half the folate
Kidney beans	1/2	115	29	in foods can be lost in cooking.

*Manufacturers of grain-based foods now fortify their products with folic acid—a synthetic form of the vitamin.

OSTEOPOROSIS

The word "osteoporosis" literally means "porous bones." With osteoporosis, bones become weak and brittle—so brittle that even mild stresses, such as bending over to pick up a book, pushing a vacuum, or coughing, can cause a fracture.

The strength of your bones relates to their mass or density and is in part due to the calcium, phosphorus, and other mineral levels. In osteoporosis, the strength is decreased as calcium and other minerals are slowly depleted and bone density is undermined.

Bone is living tissue that is continually changing—new bone is made and old bone is broken down, a process called "remodeling" or "bone turnover." The cells called osteoclasts dissolve or "resorb" old bone cells, leaving tiny cavities. Another type of bone cells, called osteoblasts, line or fill these cavities with a soft honeycomb of protein fibers that become hardened by mineral deposits.

A full cycle of bone remodeling takes 2 to 3 months. When you are young, your body makes new bone faster than it breaks down old bone, and bone mass increases. Peak bone mass is reached in your mid-30s.

The mineral-hardened honeycomb, which accounts for bone strength, depends on an adequate supply of calcium. Estrogen also plays a key role in bone health by slowing the resorption of old bone and promoting new growth.

With aging, bone remodeling continues, but people lose slightly more than they gain. At menopause, when estrogen levels decrease, bone loss accelerates to 1 to 3 percent per year. Around age 60, bone loss slows again but it does not stop. Men can also have osteoporosis. By an advanced age, women have lost between 35 and 50 percent of their bone mass, and men have lost 20 to 35 percent (see illustration on page 70).

Risk Factors for Osteoporosis

Despite the gloomy statistics, osteoporosis is not an inevitable part of aging. With identification of the major causes of the disease and their risk factors, osteoporosis can be detected early and treated. Moreover, a greater understanding of the role of nutrients and hormones and new and continually emerging medications are raising hopes for prevention of the disease.

How do you assess your personal chances for getting osteoporosis? Listed below are several risk factors that should be considered and evaluated:

Sex—One's sex is the most significant indicator of risk. Fractures from osteoporosis are about twice as common in women as in men. Women build less bone than men by early adulthood. Women also generally consume less calcium than men. Prolonged calcium deficiency is a risk. Moreover, studies have documented a tendency for low

calcium intake among adolescent girls—a time at which calcium is especially needed for bone development.

Family history—Having a mother or sister with the disease may increase your risk.

Race—Whites are at greatest risk, followed by Hispanics and Asians. African-Americans have the lowest risk. Whites have a higher risk because they generally attain a lower peak bone mass than the others.

Age—The older an individual, the higher the risk for osteoporosis.

Small body frame—In general, the smaller the body frame, the thinner the bone.

Lifestyle choices—Smoking increases bone loss, perhaps by decreasing the amount of estrogen the body makes and reducing the absorption of calcium in the intestine. In addition, women smokers tend to enter menopause earlier than nonsmokers—a significant risk factor in itself. Consumption of too much caffeine or alcohol can lead to bone loss. A sedentary lifestyle is a risk factor. Weight-bearing physical activity strengthens bones.

Prolonged calcium deficiency does not merely mean that newly consumed calcium is not going into the bones. Because the body also needs calcium circulating in the blood, it will "rob" calcium from the bones to provide adequate calcium in the blood.

Estrogen deficiency—The less a woman's lifetime exposure to estrogen, the higher her risk for osteoporosis. For example, a woman will have a higher risk if she has an early menopause or began menstruating at a later age. Early menopause due to surgical removal of the ovaries also increases the risk for osteoporosis.

Women generally experience a sudden drop in estrogen at menopause, which accelerates bone loss. Men experience a much more gradual decline in the production of testosterone, and therefore they do not experience as rapid a loss of bone mass. Recent evidence suggests that estrogen also may play an important role in bone metabolism in men.

There also can be a deficiency of estrogen as a result of very low weight caused by eating disorders such as anorexia nervosa or excessive physical activity.

HOW YOU GROW SHORTER

In osteoporosis, bones become porous and weak.
Bones in the spine can compress, causing loss in height.

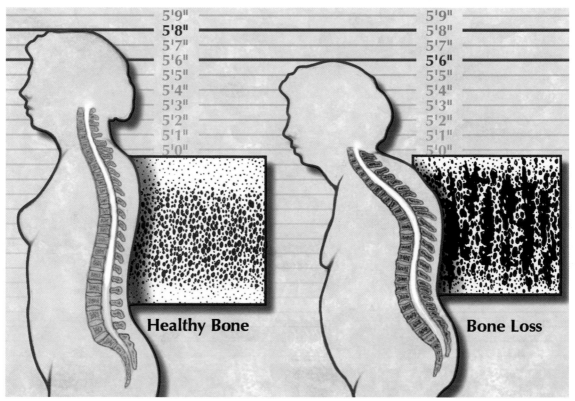

Healthy Bone

Bone Loss

Immobilization—This robs bones of the weight-bearing exercise that can help to build bone mass. Someone who is bedridden or otherwise off their feet for any extended period could have such a problem.

Medications—Some medications can contribute to osteoporosis. Long-term use of corticosteroids, such as prednisone, cortisone, prednisolone, and dexamethasone, is very damaging to bone. People who need to take such medications for other conditions (for example, asthma, rheumatoid arthritis, or psoriasis) should have their bone density monitored because they may require treatment to slow the rate of bone loss. Too much thyroid hormone caused by an overactive thyroid gland or excess thyroid replacement also can cause bone loss. In addition, other medications also can adversely affect bone health. If you are at risk for osteoporosis, your physician will take this into account when prescribing medications.

Bone Health

An individual's risk for the development of osteoporosis depends on how much bone mass has been attained by age 25 to 35 (peak bone mass) and how rapidly it is lost afterward. The higher the peak bone mass, the more bone a person has "in the bank" and the less likely that person is to develop osteoporosis as bone is lost during normal aging or during menopause.

The decrease in bone mass and the microscopic deterioration of the skeleton can lead to an increased risk of bone fragility and fractures, back pain, and a loss of height (see illustration on page 70). The most common fractures resulting from osteoporosis include vertebra, hip, forearm, and wrist. Fractures can also easily occur in other bones.

Fractures can severely compromise lifestyle and limit mobility. If an older woman breaks a hip, for instance, she has only a 25 percent chance of ever resuming her former level of physical activity. A broken hip can easily be the end of an independent way of living or even contribute to an early death from complications of operation or immobilization. Once you have had a fracture, there is an increased risk of having another.

Although osteoporosis generally is considered a disease that plagues elderly women, it can begin in others in their 40s and 50s. This potential crippler affects 8 to 10 million American women and about 2 million older American men. The cost of osteoporosis to the American economy is about $13 billion per year. Given current aging trends, these costs will increase dramatically in the coming decades.

Testing for Osteoporosis

Physicians can detect early signs of osteoporosis with a simple, painless bone density test. This test uses x-ray or ultrasound technology to measure bone density at likely sites for fractures. The test also can predict the risk of future fractures. Measuring bone density helps identify individuals who have osteoporosis or the threat of osteoporosis. (See sidebar: Recommendations for Measuring Bone Mineral Density in Women, below.) In addition, other tests for osteoporosis are being developed and tested.

Reducing the Risk for Osteoporosis

Although osteoporosis may not be fully preventable, researchers are agreeing that the following steps can delay its onset or improve the treatment outlook and thereby the quality of life.

Maximal Bone Mass

Building maximal bone mass will make bone fractures later in life less likely. Maximal bone density depends partly on the inherited ability to make bone, the amount of calcium consumed, and the level of exercise.

Estrogen Replacement Therapy

Estrogen replacement therapy is the single most important way for women to reduce their risk for osteoporosis during and after menopause, provided there is understanding of the

RECOMMENDATIONS FOR MEASURING BONE MINERAL DENSITY IN WOMEN

The American Association of Clinical Endocrinologists recommends a bone density test to assess the risk of osteoporosis in women—around the time of menopause or after menopause—who are concerned about osteoporosis and willing to accept available treatment. This includes women who:

- have x-ray findings that suggest the presence of osteoporosis
- are beginning or receiving long-term glucocorticoid therapy (such as prednisone)
- are perimenopausal or postmenopausal and have parathyroid disease
- are undergoing treatment for osteoporosis, as a way of monitoring the effects of therapy

OSTEOPOROSIS IN MEN

Although osteoporosis mainly affects women, 2 million men have the disease, including one-third of men older than 75 years. Warning signs in men include a change in posture or sudden back pain. But the most common way osteoporosis is diagnosed in men is because of a loss of height or a fracture.

Smoking and drinking excessively are significant risk factors for osteoporosis in men. In addition, men have most of the same risk factors as women—use of medications that accelerate bone loss, lack of exercise, smoking, excessive alcohol use, and inadequate calcium intake or absorption. Low testosterone also may increase the risk. Your physician can determine whether a bone mineral density test is needed.

Prevention strategies include getting adequate calcium, vitamin D, and weight-bearing exercise and consuming no more than 2 alcoholic drinks a day.

risks and benefits of this type of therapy and it is closely monitored by a physician. Estrogen receptors are present in bone, and estrogen inhibits bone breakdown. Estrogen replacement therapy can decrease or prevent bone loss and reduce the risk of spine and hip fractures by as much as 50 percent. For women who already have osteoporosis (and who have no conditions prohibiting the use of estrogen), starting estrogen replacement therapy can increase bone density by as much as 10 percent in the spine and 5 percent in the hip.

However, there are some risks associated with estrogen replacement therapy. Taking estrogen alone, without its natural balancing hormone progesterone, increases the risk of cancer of the uterus. And there may be a small increase in the risk of breast cancer from long-term use of estrogen. However, estrogen treatment may help decrease the risk of heart disease in certain groups and may also decrease the risk of other diseases, such as dementia.

Women who have breast or uterine cancer, uncontrolled high blood pressure, or a tendency to form blood clots should avoid estrogen. If blood triglyceride levels are high, women should consult their physicians to see if an alternative form of estrogen is an option. There are new types of estrogen (estrogen analogs) that may prevent or slow the rate of bone loss without the increased risk of breast cancer. However, these drugs may not be as effective as estrogen and

may not have all of the beneficial effects. In addition to estrogen and estrogen analogs, other types of medications are available and new ones are under development to prevent and treat osteoporosis.

Calcium and Vitamin D

Adequate amounts of calcium and vitamin D are critical for building peak bone mass in younger years and for slowing bone loss in later years (see sidebar: How Much Vitamin D Is Enough? page 71).

Calcium is a vital mineral in the body. In addition to being one of the essential building blocks of bone, calcium is essential to the function of the muscles (including the heart) and the function of the nerves, and it helps the blood to clot in case of injury. If people fail to get enough calcium in their diets, their bodies will take calcium from the bones to keep the blood calcium level constant. Recommendations for calcium vary a bit according to age and medical status. For ages 9 to 18, 1,300 milligrams of calcium daily is recommended. For ages 19 to 50, the recommendation is 1,000 milligrams. Finally, for ages 51 or older, 1,200 milligrams is recommended. During pregnancy and lactation, the recommendation is 1,300 milligrams.

BONE DENSITY OVER TIME

Bone density, which varies by sex and race, peaks in your mid-30s and then slowly declines with age. In general, the higher your peak bone mass, the lower your risk of having fractures caused by osteoporosis.

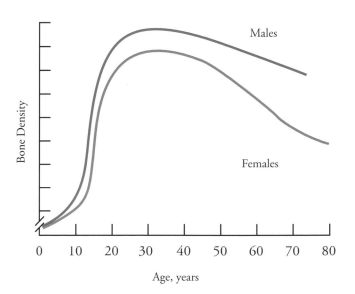

The richest dietary sources of calcium are milk, cheese, and yogurt. Other sources of calcium are broccoli, turnip greens, canned fish with bones, and calcium-fortified orange juice and tofu. Of course, if you are trying to get extra calcium through your diet, you should monitor your weight; foods such as whole milk, certain cheeses, and ice cream are rich in calories and fat (see sidebar: Food Sources of Calcium, below). Chapters 4 and 5 also present ideas for planning and preparing healthful meals.

Depending on dietary choices and habits, food alone can provide the recommended amounts of calcium. If you are unsure of the calcium content of foods, follow this simple formula for estimating intake of dietary calcium: assign

How Much Vitamin D Is Enough?

Recommended Intake for Vitamin D

Age (years)	Daily Amount (micrograms)	Daily Amount (International Units)
9-50	5	200
51-70	10	400
71+	15	600
Pregnancy and lactation	5	200

Note: 1 microgram = 40 international units (IU) of vitamin D.

Food Sources of Calcium

Food Item	Amount	Calcium (milligrams)	Calories
Milk			
Skim	1 cup	300	85
2 percent	1 cup	300	120
Whole	1 cup	300	150
Nonfat, dry	1/3 cup	285	90
Yogurt			
Plain, low-fat	1 cup	400	145
Fruited, low-fat	1 cup	300	225
Frozen, low-fat	1 cup	200	200
Pudding			
Skim milk	1/2 cup	150	105
Ice cream	1/2 cup	90	130
Ice milk	1/2 cup	90	90
Cheese			
Swiss	1 ounce	270	110
Cheddar	1 ounce	205	115
Mozzarella (part skim)	1 ounce	185	70
Cottage cheese			
Whole milk	1 cup	125	220
2 percent milk (low-fat)	1 cup	155	205
American cheese (processed)	1 ounce	105	175
Salmon (canned with bones)	3 ounces	205	130
Sardines (canned with bones)	4 sardines	185	100
Papaya	1 medium	75	120
Orange	1 medium	50	60
Macaroni and cheese	1 cup	360	430
Pizza with cheese	1 slice	230	280
Tofu (calcium-fortified)	1/2 cup	130	90
Almonds	1 ounce	75	175

Weight-bearing exercise increases bone density, muscle strength, and coordination, which can lower your risk for osteoporosis, falls, and broken bones.

approximately 300 milligrams for the baseline diet, and add 300 milligrams for each serving of a dairy product (cup or slice) and 160 milligrams for each serving of a calcium-fortified food that you eat. If your diet is not adequate in calcium, a supplement may be indicated (see sidebar: Tips for Selecting and Taking a Calcium Supplement, below).

Dietary absorption of calcium can be assessed by your physician. A 24-hour urine collection can measure the calcium content to determine how well or how poorly calcium is being absorbed from your diet.

Vitamin D helps the body absorb and metabolize calcium and deposit it in the bones. People can get vitamin D from vitamin D-fortified milk, liver, fish, egg yolks, and exposure to sunshine. Getting 10 to 15 minutes of

TIPS FOR SELECTING AND TAKING A CALCIUM SUPPLEMENT

If a calcium supplement is indicated, here are tips to help decide which type of calcium to take and how to take it.

Read labels—Various types of calcium supplements contain different amounts of elemental calcium, which is the actual amount of calcium available for absorption.

Recommendations are for elemental calcium, so look for how much elemental calcium is in each tablet. If elemental calcium is not listed on the label, the amount can be calculated as follows:
- Calcium carbonate (40 percent calcium):
    multiply total amount of calcium carbonate by 0.4
- Calcium citrate (21 percent calcium):
    multiply total amount of calcium citrate by 0.21
- Calcium lactate (13 percent calcium):
    multiply total amount of calcium lactate by 0.13
- Calcium gluconate (9 percent calcium):
    multiply total amount of calcium gluconate by 0.09

Look for "USP" on the label. The U.S. Pharmacopeia (USP) sets standards for quality that manufacturers must meet. These supplements meet the requirements for quality, purity, and tablet disintegration and will be better absorbed.

Enhance calcium absorption—Various factors may affect calcium absorption:
- *Take supplements with meals*—Although some foods may interfere with calcium absorption, taking a supplement with meals is more convenient. Many older adults also have reduced levels of stomach acid. Eating stimulates acid production and may enhance overall absorption.
- *Add vitamin D*—Be sure to get adequate vitamin D in your diet. Vitamin D helps increase the absorption of calcium. If you are in doubt, take a standard multivitamin, which includes 400 units, the recommended standard.

Minimize side effects—Some forms of calcium may be gas-forming and constipating. To minimize these effects, drink plenty of water, take the supplement with a meal, and take several smaller doses during the day. Try different calcium compounds to find one with fewer side effects.

midday sun 2 or 3 times a week can provide sufficient vitamin D—although not from a winter sun in a northern climate. Older adults or persons with certain diseases often benefit from a supplement. A multivitamin pill with 400 international units of vitamin D provides an adequate amount.

Caffeine

Limiting caffeine is a prudent idea. Too much caffeine can cause extra calcium to be lost in the urine. If an individual has only 2 or 3 cups of coffee a day and is consuming enough calcium in the diet, there should be no problem.

Alcohol

Consuming more than 2 drinks a day can inhibit bone formation and interfere with calcium absorption.

Smoking

Smoking decreases your body's ability to form healthy bones.

Weight-Bearing Exercise

Weight-bearing exercise is any activity done while the bones are supporting the body's weight. It can slow bone loss, strengthen the bones and back, improve posture, and aid in balance, which helps prevent falls. Exercises in which bone sustains repeated impact have added benefit because, for example, the leg bones respond to the impact of the feet striking the ground by slowing bone loss. Remember, it is never too late to begin an exercise program.

Bone-building exercises for prevention of osteoporosis include walking, jogging, running, stair climbing, skiing, and impact-producing sports. Because strong muscles exert more force, and bones respond by becoming stronger, weight lifting (or strength training) is another excellent way to forestall osteoporosis. Of course, anyone with osteoporosis may benefit by consulting with a physician to design a safe exercise program.

Trauma

Avoiding trauma is a given at any stage of life, but even more so for persons with osteoporosis. Wear sturdy, low-heeled shoes with nonslip soles, and check your home for potential obstacles that could cause a fall, such as low tables, loose rugs, or inadequate lighting. The broken bones that come from tumbles and falls can produce serious, even life-threatening, medical complications.

CANCER

Many people have a fear of cancer, perhaps because to them it is always an incurable disease. The facts do not support this idea.

Countless numbers of Americans who are alive today have had cancer and are now considered to be cured. ("Cured" is defined here as being free of any evidence of the disease for 5 years or more.) They may have the same life expectancy as others of the same age and sex who have never had cancer, and they can anticipate leading meaningful and productive lives.

Despite such impressive statistics, cancer remains a serious disease. Annually, cancer is diagnosed in more than a million people (excluding non-melanoma skin cancer) and is one of the most significant causes of mortality in the United States.

There are more than 100 different types of cancer. Some cancers affect just one organ, and others are more generalized. In each of its types, however, cancer is characterized by the uncontrolled growth and spread of abnormal cells (see illustration on page 74).

Why cancer develops in some people who are exposed to potentially cancer-causing agents but not in others is

CANCER TERMS—ANCIENT AND MODERN

Cancer is a disease we have known about since ancient times. The terms used to describe cancer come from ancient languages.

"Cancer" is derived from a Latin word meaning "crab" and was first described in writings attributed to Hippocrates and other Greek physicians (500 B.C. to 200 A.D.). These physicians described various tumors and classified them as either "carcinos" (benign growths that do not spread) or as "carcinomas" or "crab-like" (growths that invade surrounding tissues and cause death). The term "neoplasm" was described by Galen (200 A.D.) as meaning "new growth that is contrary to nature." The word "metastasis" is also a Greek word meaning "to change places." It is used to describe the ability of cancer to migrate to other tissues or organs and to form additional tumors.

not fully understood. But what is known is that many cancers develop slowly. It may be 5 to 40 years after exposure to a cancer-causing agent before there is any evidence of the disease. Cancer of the lung, for example, may not appear until 25 years or more after sustained exposure to tobacco smoke. This long delay between exposure and development of the disease may partly explain why so many people ignore the warnings associated with smoking.

The Nutrition-Cancer Connection

During the past 30 years, research has shown that nutrition plays a significant role in the development of many cancers and that proper food choices might help to reduce the risk of cancer or even prevent it. About a third of the 500,000 cancer deaths that occur each year in the United States can be attributed to dietary factors. The good news is that in addition to engaging in regular exercise and not smoking, people have control over this important factor in cancer development—their food choices.

Any number of dietary factors may be related to the risk of cancer.

The Biology of Cancer

A biomedical revolution is advancing knowledge of the causes of cancer, yielding new and more effective treatments and inspiring greater hope for cancer prevention. This revolution is built on scientific investigation of the basic processes that cause cancer.

The body is a living, growing system that contains billions of individual cells. These cells carry out all of the body's functions, such as metabolism, transportation, excretion, reproduction, and locomotion. The body grows and develops as a result of increases in numbers of new cells and their changes into different types of tissue. New cells are created through the process of cell division (mitosis). Different types of cells are created by a process called cell differentiation, by which they acquire specialized function. Cell division results in the normal pattern of human growth;

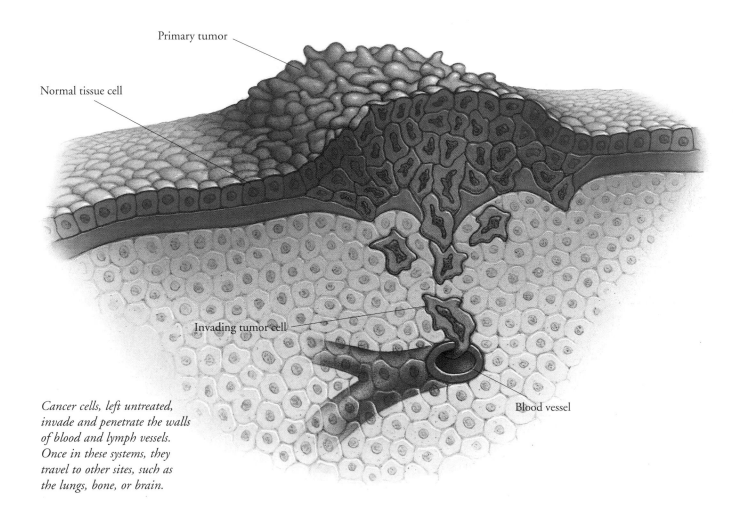

Primary tumor

Normal tissue cell

Invading tumor cell

Blood vessel

Cancer cells, left untreated, invade and penetrate the walls of blood and lymph vessels. Once in these systems, they travel to other sites, such as the lungs, bone, or brain.

cell differentiation makes possible the normal, orderly pattern of growth and development.

Unlike normal cells, cancer cells lack the control mechanisms that stop, or "switch off," growth. They divide without restraint, displacing neighboring normal cells, affecting their normal function and growth, and competing with them for available nutrients. These uncontrolled cells can grow into a mass called a tumor and invade and destroy nearby normal tissue. They also can migrate in a process called "metastasis," spreading via the blood or lymph system to other parts of the body. Not all cells that have rapid or uncontrolled growth are cancerous. Cells may amass as benign tumors, which do not invade or destroy surrounding tissues.

Although science has yet to understand the processes by which all cells grow, divide, communicate, and differentiate, much has been learned about how normal cells are activated or altered into cancerous cells in both inherited and non-inherited forms of cancer.

The Causes of Cancer

Cancer is caused by factors that are external (chemicals, radiation, viruses, and diet) and internal (hormones, immune and metabolic conditions, and inherited [genetic] alterations). Some of these factors are avoidable; others are not. Scientists have identified many of the controllable risk factors that increase the chances of getting cancer. A complex mix of these factors, acting together or in some cascade of events, promotes cancer cell growth.

When the genetic programming of a normal cell is disrupted, its malignant potential is released. Everyone carries this malignant potential within them in normal genes known as proto-oncogenes. Products of these genes perform useful functions, such as regulating cell division and cell differentiation. These functions, however, may be compromised with aging or by exposure to cancer-causing (carcinogenic) agents. When this happens, they may be activated to become oncogenes, coordinating the conversion of normal cells to cancer cells.

Nutrition can influence any of the steps involved in the development of cancer. The development of cancer (carcinogenesis) and its relationship to nutrition is a complex process. Isolating and proving dietary cause-and-effect relationships can be difficult. In addition, studies can be confusing and sometimes show conflicting results. Nevertheless, the potential for nutrition to increase or decrease the risk of various cancers is compelling.

Dietary Guidelines to Reduce the Risk of Cancer

Just as negative dietary and lifestyle choices can significantly increase the risk of cancer, evidence is mounting that appropriate food choices can be powerful tools in reducing risk and even defensive shields in preventing cancer. The American Cancer Society offers these four guidelines to reduce cancer risk:

- Choose most of the foods you eat from plant sources.
- Limit your intake of high-fat foods, particularly from animal sources.
- Be physically active: achieve and maintain a healthful weight (see Chapter 1, page 8).
- Limit consumption of alcoholic beverages, if you drink at all.

The Society's recommendations are consistent with the U.S. Department of Agriculture Food Guide Pyramid and the Dietary Guidelines for Americans (see Chapter 1, pages 8 and 11). Although no diet can guarantee full protection against disease, the American Cancer Society believes that these recommendations offer the best nutrition information currently available to help reduce your risk of cancer. Chapters 4 and 5 provide ideas for planning and preparing healthful meals.

Choose Most of the Foods You Eat From Plant Sources
Choosing foods from plant sources is vital to a healthful diet. Many scientific studies have shown that increased consumption of fruits, vegetables, and whole grains reduces the risk for cancers of the gastrointestinal and respiratory tracts and for lung cancer. This reduction is one of the reasons foods of plant origin form the basis of the Food Guide Pyramid.

Plant foods contain beneficial vitamins, minerals, fibers, and hundreds of other cancer-protective substances. Although more research is needed to understand what specific properties or substances in plant foods may specifically protect against cancer, there are already many candidates under investigation—from vitamins and minerals to fiber and phytochemicals (including carotenoids, flavonoids, terpenes, sterols, indoles, and phenols). Because the positive effects from these components may derive from the whole foods in which they are found, experts recommend food over supplements (see sidebar: Food Versus Supplements, page 76; and see Chapter 2, Phytochemicals—A Food Pharmacy That May Fight Disease, page 35).

Antioxidants found in fruits and vegetables may help to prevent cancer. These are nutrients that seem to offer the body some protection against oxidation—damage done to tissue in the course of normal cellular function which may contribute to the effects of aging and to increased cancer risk. Various antioxidant nutrients—including vitamin C, vitamin E, selenium, and carotenoids —may provide the body with some defense against cancer. Researchers are studying the protective role of antioxidants (see sidebar: Food Sources of Antioxidants, page 32, and see Appendix, Phytochemical Contents of Selected Foods, page 484).

Grains provide vitamins and minerals, such as folate, calcium, and selenium, which may also protect against cancer. Whole grains are preferable to refined grains because they have more fiber and an abundance of certain vitamins and minerals. Beans and legumes are also good sources of nutrients that have cancer-protective qualities. (See page 20 for more information about fiber.)

Although more research is needed to clarify the specific roles of these food components, there is still ample evidence to support eating 5 or more servings of fruits and vegetables a day (especially deep-green and dark-yellow to orange fruits and vegetables, those from the cabbage family, and legumes and soy products) and 6 to 11 servings of grains (with an emphasis on whole grains).

Limit High-Fat Foods, Particularly From Animal Sources

Decreasing the intake of high-fat foods, especially from animal sources, is very important. Studies show that people who eat a high-fat diet have increased rates of cancers of the colon, rectum, prostate, and endometrium (lining of the uterus). Although these relationships exist, it is not clear whether they are due to the total amount of fat in the diet, to a particular kind of fat (saturated, monounsaturated, or polyunsaturated), or to another, unknown, factor.

Because fat, by weight, contains twice the number of calories than protein or carbohydrate, it is difficult to separate the effects of the fat from the effects of its calories. People who eat a high-fat diet are often heavier and tend to eat fewer fruits and vegetables, which also increases cancer risk.

Consumption of red meat, a major source of fat in the American diet, is linked to an increased cancer risk, particularly of the colon and prostate. Scientists are unable to determine whether the connection between red meat and cancer is due to total fat, saturated fat, or other compounds. Meat contains compounds linked to cancer, such as heterocyclic amines, which are produced when it is cooked (see sidebar: Meat—Cooking and Cancer, page 77). This may be a link to colon cancer.

FOOD VERSUS SUPPLEMENTS

Every day there seems to be a headline announcing that some nutrient or food compound has been linked to preventing some sort of cancer. Driving this is the tremendous amount of research exploring specific food components and their role in the cancer process.

DO SUPPLEMENTS OFFER CANCER PROTECTION?

A diet that contains ample amounts of fruits and vegetables is associated with a reduced risk for cancer. There is no evidence that mineral or vitamin supplements are better than obtaining nutrients through whole foods. In fact, studies in which subjects received supplements of beta-carotene showed conflicting and even harmful results. Smokers given supplemental beta-carotene had a higher incidence of lung cancer. Other studies of supplemental beta-carotene have shown neither benefit nor harm.

There are more than 500 known carotenoids. Of these, only a few have been analyzed: alpha-carotene, beta-carotene, gamma-carotene, lycopene, lutein, and zeaxanthin. At this time, it is not known which exerts benefits—or risks.

Experts recommend a diet rich in fruits, grains, and vegetables rather than taking supplements. Plants contain hundreds of cancer-protective substances. It is not known which of the substances may specifically protect against cancer. In addition, it is not known whether these substances work independently or benefit from working together. Plus, relying on supplements rather than eating a variety of foods makes it impossible to benefit from these now unknown food compounds.

MEAT—COOKING AND CANCER

Cooking meat at high temperatures creates chemicals that are not present in uncooked meats. Heterocyclic amines (HCAs) are chemicals linked to cancer that are formed when meat is exposed to high temperatures. Four factors influence HCA formation:

TYPE OF FOOD

HCA is found in cooked muscle meats. Other protein sources (milk, eggs, tofu, and organ meats such as liver) have little or no HCA naturally or when cooked.

TEMPERATURE

Cooking temperature is the most important factor in reducing the formation of HCA. Cooking in an oven at 400° Fahrenheit or less or stewing, boiling, or poaching at 212° Fahrenheit or less forms one-third of HCAs versus meats prepared by frying, broiling, or grilling.

COOKING METHOD

Frying, broiling, and barbecuing produce the largest amounts of HCA—these methods have higher than recommended cooking temperatures. Microwaving meats for 2 minutes helps to decrease HCA. Meats that are microwaved before frying, broiling, or grilling have a 90 percent decrease in HCA content. Marinating meats before cooking also inhibits the production of HCA.

TIME

Meats cooked well done (no pink remaining in center) have more HCA than those cooked medium.

Different fats (such as omega-3 fatty acids and vegetable oil) may have different effects on your risk of cancer. Saturated fat is a particular concern for cancer risk and for coronary artery disease.

How food is prepared is also important. Some cooking methods, such as baking, stewing, boiling, and poaching, are healthier ways than frying, broiling, or grilling.

Limit Consumption of Alcoholic Beverages

People who drink excessive amounts of alcohol have an increased risk of cancer, higher than that of the general population, especially cancer of the larynx, esophagus, stomach, and pancreas.

There is no question that limiting alcohol consumption reduces cancer risk. The risk of cancer begins to increase with an intake of as few as 2 drinks a day. Alcohol along with tobacco use produces a combined cancer risk that is greater than the sum of their individual effects. The risks for cancers of the mouth, throat, esophagus, and larynx are particularly increased.

Research also has noted a connection between alcohol consumption and an increased risk of breast cancer. Although the causes are not known, scientists speculate that alcohol may have a carcinogenic effect, perhaps reflected in its capacity to alter hormone levels.

Drinking too much alcohol also may negatively affect eating habits. The calories in alcohol—with little nutritional value—are perhaps being consumed in place of calories in healthier foods with cancer-protective values. A general rule: men should limit themselves to no more than 2 drinks a day, women to 1. (See page 387 for a further discussion about alcohol.)

In this chapter you will learn:

- *How to plan and select healthful meals whether eating at home or away*
- *How to shop for the healthiest foods*
- *How to read food labels and interpret nutrient and health claims*
- *How changes in technology, such as the use of pesticides, irradiation, additives, genetic engineering, and organic growing methods, affect our food*

PLANNING MEALS

SELECTING HEALTHFUL FOODS
PLUS TWO WEEKS OF MENUS

Whatever your style of eating, several simple planning techniques will help ensure that your food is varied, nutritious, and enjoyable. In this chapter you will learn how to plan and select healthful meals—whether prepared at home or eaten away from home. You also will be introduced to strategies to make your shopping both efficient and effective. Also discussed are some of the new (and not so new) issues in foods that are of current interest. The chapter ends with 2 weeks of healthful menus—just to get you started.

PLAN TO "EAT WELL"

"Eat" and "well" are the key words. To do so, you must be conscious of what you eat and how often you eat. Try to eat at regular times. For most of us, this means eating breakfast, a mid-day meal, and an evening meal. Healthful snacks can be part of your meal plan as well. Going for long periods without eating can affect how you feel and how much you eat. Use the Food Guide Pyramid to help you make healthful selections. Try to be sure that your meals and snacks are rich in plant foods (fruits, vegetables, and grains), because, ounce for ounce, plant-based meals are almost always lower in fat and calories than meat-based meals. Although meat and dairy foods contain many nutrients, they also can be very high in unwanted saturated fat. The key is to avoid high-fat types of meats, make an effort to eat smaller amounts of the lower-fat meats, and focus on low-fat dairy products. Vary your food choices. Try foods that are new to you. Remember that fats and sweets are at the top of the Pyramid, which means enjoy them, but do so only on occasion and in moderation. (See Chapter 1, the Food Guide Pyramid, page 11.)

EAT BREAKFAST

Whether at home or away, start your day with breakfast. "Breaking the fast" provides your body with both nutrients and energy. People who eat breakfast tend to have more energy and, on average, are better able to regulate their appetite during the remainder of the day than their breakfast-skipping counterparts.

Unfortunately, many Americans do not eat breakfast. Some skip breakfast because of their schedule, whereas others do so in a misguided attempt to control weight. However, you can eat a healthful breakfast with the time you have. There are many ways to make what is perhaps the day's most important meal a nutritious, fast, and convenient one.

COLD BREAKFAST CEREALS

With more than 200 brands of breakfast cereal on your grocer's shelves, how do you know what's best? Be sure to read the label.* Choose those that are higher in fiber (more than 3 grams/serving) and lower in calories, fat, and sugar. Ingredients are listed in amounts by weight. Check for hydrogenated fats. If hydrogenated fat is listed near the beginning of the list of ingredients, the product contains trans and saturated fats, which can increase blood cholesterol levels.

BRANDS (1 CUP SERVING)	CALORIES	FAT (GRAMS)	FIBER (GRAMS)	SUGAR (GRAMS)
All-Bran	160	2	20	12
Cracklin' Oat Bran	150	8	8	20
Fiber One	120	2	26	0
Granola	450	17	6	32
Grape-Nuts	400	2	10	14
Raisin Bran	200	1	8	18
Shredded Wheat (1 cup or 2 biscuits)	170	0	5	0
Wheaties	110	1	3	4
Whole Grain Total	150	1	4	7

*Most breakfast cereals are fortified to provide about 25 percent of most daily nutrients. Some cereals have more. Check the label to make your choice. (Use of brand names does not indicate endorsement.)

Build a Better Breakfast

Breakfast is the foundation of a healthful diet. Use the Food Guide Pyramid as a practical resource for planning your breakfasts regardless of whether you choose foods that require preparation or select ones that are ready-to-go. Cereals are a good choice. Simple whole-grain cereals with no added sugar or fat are best (see sidebar: Cold Breakfast Cereals, this page). A breakfast that includes a whole-grain cereal, bread, low-fat milk, and a glass of orange juice is a great starter meal. This breakfast supplies B vitamins, fiber, iron, approximately one-third of the recommended calcium, and 100 percent of the recommended vitamin C for the day. Best of all, it does so in less than 300 calories.

For a change, try a breakfast bagel sandwich. Top a whole-grain bagel with 2 teaspoons of peanut butter and a sliced banana. Add a cup of cold skim milk for a breakfast that is about 400 calories. This breakfast includes foods from most of the food groups, is low in cholesterol, and is a good source of iron, folate, and fiber. For even more variety, top a flour tortilla with 2 ounces of leftover chicken breast and tomato pieces and 1 ounce of low-fat cheese. As a vegetarian option, top with rice and beans. Wrap the tortilla tightly, microwave for a minute or so, and top with salsa. While you are at it, drink a glass of a spicy vegetable juice. Both of these quick-fix breakfasts contribute servings from the vegetable, fruit, and grain groups in just 350 calories. They also give you plenty of vitamins A and C.

Maybe you prefer eggs for breakfast. The current recommendation is to limit your intake of whole eggs to 3 or 4 per week. The reason to limit eggs is that the yolk of a large-sized egg contains about 210 milligrams of cholesterol—more than two-thirds of the daily cholesterol allowance. However, eggs also have many nutrients. People with a low blood cholesterol level probably can safely eat a few more eggs than those who have a high level.

Create an omelet with 1 whole egg plus 2 egg whites, sweet peppers, and onions. Serve with oven-browned potatoes and a slice of whole-grain toast topped lightly with butter or margarine. Or, better yet, top with jam or jelly as a no-fat alternative. Remember to include fruit or juice. This 500-calorie meal—although it contains cholesterol—is a good source of iron and is high in fiber, folate, and vitamin C.

Breakfast Out

Many people are too busy to sit down and eat breakfast at home. The next best bet is to eat breakfast on-the-run,

which at times can pose a nutritional challenge. Fortunately, if you know what you are looking for, a nutritious breakfast can be found almost anywhere food is served. If you are traveling and have time for a "sit-down" breakfast, choose a restaurant that offers a varied menu. If not, try bagel shops, fast-food establishments, the company cafeteria, or even a nearby vending machine. Some may have a "buffet breakfast" that has everything you need, including hot and cold cereals, breads, bagels, fresh fruit and fruit juices, low-fat milk, and yogurt. Others also may offer options such as low-fat burritos, low-fat granola, or low-fat muffins. (See sidebar: Best Bets When Eating Breakfast Out, below.)

WHAT'S FOR LUNCH?

The midday meal is the meal most often eaten away from home. Your options include bringing a lunch from home or purchasing foods in a cafeteria, vending machine, or local delicatessen.

The same rules apply for "eating well" at lunch: plan to eat foods low in saturated fat and high in nutrients. This midday meal helps you meet your goal of 5 fruits and vegetables every day, and it is the perfect time to include whole grains. Lunch helps you to distribute your intake of calories and nutrients evenly throughout the day. It can keep energy levels high and help prevent unplanned snacking on foods that may not be the best choices.

Brown Bagging

Bringing a lunch to work can save you money and help you eat what you really want to eat. To be sure that your lunch tastes good and is good for you, try to incorporate each of the food groups in your lunch. If you bring too little food, you will likely be hungry later in the day. You then increase the chance that you will snack before supper or eat too much at supper.

To keep brown-bag fare interesting and healthful, pack more variety. Think of the options:

Grains—Instead of the usual sandwich bread, try pita (pocket) bread, tortilla wraps, crackers, pretzels, or rice cakes. Or, try salads made with nutritious grains.

Fruits—Include fresh fruits. Every now and then, choose an "exotic" fruit to add interest. Try star fruit, kiwi, papaya, mango, or passion fruit. (See Part II, Fruits, page 153.) Fruit juices can be nutritious and refreshing.

Vegetables—Expand your repertoire from raw carrots and celery sticks to potentially more satisfying vegetable soups and salads. Stuff pocket bread with a variety of cooked vegetables. Use vegetables to make your sandwich more filling: fresh spinach or romaine, and slices of cucumber, tomato, mushrooms, and sweet or chili peppers. A vegetable juice makes a great lunchtime drink.

Dairy—Some days drink milk. Other days eat yogurt or a bit of cheese. Choose the low-fat forms more frequently to be sure your meal is rich in nutrients and not fat and calories.

BEST BETS WHEN EATING BREAKFAST OUT

CHOOSE THESE	CALORIES	FAT (GRAMS)	INSTEAD OF THESE	CALORIES	FAT (GRAMS)
Toast (2 slices)	150	2	Doughnut (1)	250	12
English muffin (1)	135	1	Bran muffin (1 medium)	360	10
Bagel (1 half)	150	1	Croissant (1 small)	230	12
Hard roll (1)	100	1	Sweet roll (1 small)	350	15
Cold cereal, bran type (1/2 cup)	160	2	Granola (1/2 cup)	250	10
4-inch pancake, plain (1)	74	1	Hash browns (1/2 cup)	170	9
Egg substitute (1/4 cup)	30	0	Scrambled egg (1)	100	7

Meat — Meat (which includes poultry, fish, and also beans, legumes, and nuts) is good for you, but remember that "moderation" is the key word. To help take meat off center stage, be sure that you also include plenty of fruits, vegetables, and grains in your lunch. Eat only lean meats. When possible, instead of meat, try to substitute foods such as hummus, lentils, beans, tofu, and nut spreads, which contain large amounts of plant protein (see sidebar: Sandwich Fixings, below).

Vary the temperature of your lunch by including both hot and cold items. A thermos will preserve heat more effectively if you first rinse it with hot water. Likewise, rinsing with cold water will help keep foods cold longer. If your workplace does not have a refrigerator, a thermal lunch bag can help keep your food fresh and safe. A frozen box of juice can help keep your lunch cold, and the juice will thaw by lunchtime.

Going Out for Lunch?

If you know what you are doing, eating lunch out increases your options rather than your temptations. Most fast-food restaurants, eager to please the ever-growing number of people who insist on eating more nutritious foods, now offer "lighter" fare such as salads and chicken. But be careful that you don't choose foods that are healthful in name only. Many foods that at first glance seem to be good choices in fact are loaded with fat and calories.

Salad Savvy

Simple salads are best. To construct a nutritious salad, incorporate lots of fresh vegetables and fruits. Remember that the word "salad" is not synonymous with "healthful." Try to eat salads that are low in calories and fat but high in much-needed nutrients (see sidebar: Salad Fixings, page 83). Many taco salads contain at least 900 calories, more than half of which comes from fat. High-fat meats and cheeses heaped on a chef salad can dominate the vegetables. Chicken and seafood are low in fat; however, this advantage is lost when they are covered with high-fat dressings and oils. Many of today's salad bars look like delicatessens. Pasta salad, potato salad, guacamole, and tortellini are popular items. Depending on how they are made, they too can be high in fat and calories.

Unless used sparingly, some dressings can provide up to 400 calories to your salad. Watch for packaged dressings that contain more than "1" serving; check the label to see

SANDWICH FIXINGS

	Amount	Calories	Fat (grams)
Breads			
Whole wheat	1 slice	70	1
Raisin	1 slice	70	1
Pita, whole wheat	1 (6.5 inch)	170	2
Kaiser roll	1 (3.5 inch)	170	3
Bagel	1 (3.5 inch)	200	1
Fillers			
Onion, lettuce, cucumbers, tomatoes	2-5 pieces	10	0
Chicken or turkey	2 ounces	60	1
Tuna, water pack	2 ounces	66	Trace
Ham	2 ounces	75	2
Roast beef	2 ounces	100	2
Hummus	1/4 cup	105	5
Bologna	2 ounces	180	16
Peanut butter	2 Tblsp	190	16
Salami	2 ounces	220	16
Spreads			
Horseradish	1 Tblsp	6	0
Fat-free mayonnaise	1 Tblsp	10	0
Mustard	1 Tblsp	10	0
Ketchup	1 Tblsp	15	0
Fat-free cream cheese	1 Tblsp	15	0
Reduced-fat margarine	1 Tblsp	50	6
Cream cheese	1 Tblsp	50	5
Margarine	1 Tblsp	100	11
Mayonnaise	1 Tblsp	100	11
Butter	1 Tblsp	110	12

how many servings are in the package. In some instances, the listed ingredients are those contained in a half-ounce serving, even though the package may hold up to 5 servings! A generous-sized ladle can easily drown an otherwise healthful salad with a quarter-cup of dressing (300 calories or more). Instead of using high-fat dressings, try squeezing lemon on your salad, request low-calorie or fat-free dressings, or ask that the dressing be placed on the side.

SALAD FIXINGS

	CALORIES	FAT (GRAMS)	SODIUM (MILLIGRAMS)
Go for the basics			
Mixed greens (2 cups)	20	0	10
Broccoli, green pepper, mushrooms (1/4 cup)	Trace–6	0	0–6
Carrots, shredded (2 tablespoons)	2	0	2
Onion (1 tablespoon)	Trace	0	0
Tomato (1/4)	5	0	2
Extras: proceed with caution			
Bacon bits (1 tablespoon)	40	2	225
Fried noodles (4 tablespoons)	40	2	30
Cottage cheese, fat-free (1/2 cup)	80	0	370
Croutons (2 tablespoons)	30	1	40
Egg, hard boiled (1)	80	6	60
Garbanzo beans (2 tablespoons)	35	1	2
Olives (5 large)	25	3	200
Pasta salad (1/2 cup)	200	15	560
Potato salad (1/2 cup)	180	10	660
Dressings (2 tablespoons)			
Blue cheese, French, ranch, Italian	140–180	14–20	325–440
Blue cheese, French, ranch, Italian: reduced calorie/lite/fat-free	50–100	0–8	250–300
Lemon juice	8	0	0

Tips for salads
- Go for fresh vegetables and fruit
- Go easy on "extras" that add unwanted calories, fat, and sodium
- Put dressings on the side. Even "lite" and "free" dressings add up.

Burgers and Sandwiches

Beware of burgers and sandwiches that are described as "jumbo," "double," or "deluxe." Many contain about 1,000 calories and the majority of your fat allowance for the day. Ask for a regular-sized burger. Stick with lean meat without mayonnaise or cheese. If salt is not a concern, ketchup or mustard adds very few calories. Request extra lettuce or toppings such as tomatoes, cucumbers, onions, or sweet peppers.

SNACK TIME

Carefully chosen, snacks can be healthful and nutritious. Check out the options. Look for "food machines" stocked with fresh fruit, juices, cut-up vegetables or mini-salads, nonfat or low-fat milk, yogurt, cereals, or small bagels. Snack items such as pretzels or baked chips, although high in salt, are relatively low in calories. Negotiate with your coworkers: instead of doughnuts and sweet rolls (which can provide about 250 calories and up to 15 grams of saturated fat), offer to bring fresh fruits, vegetables, mini-sized bagels, or bread with a low-fat cream cheese. Candy bars may seem convenient, but they are almost always high in both fat (mostly saturated) and calories (typically 200 to 300

A variety of vegetables makes for a nutrient-packed low-calorie lunch. Remember to have the dressing served "on-the-side."

SNACKS WITH ABOUT 100 CALORIES

FRUIT GROUP

1 large orange
1 medium apple or pear
1 small banana
1 whole grapefruit
2 medium plums
24 grapes

VEGETABLE GROUP

Up to 2 cups of any cut-up, raw vegetables:
    carrot or celery sticks
    sweet bell pepper strips
    broccoli and cauliflower florets
    summer squash slices
    mushrooms
    radishes
    tomato wedges
    salsa (1 cup)

GRAIN GROUP

One-half English muffin
1 slice whole-grain bread
2 rice or popcorn cakes
2 pretzel rods
3/4 cup cereal (ready-to-eat,
   low-fat, no added sugar)
4 bread sticks
4 cups air-popped popcorn
12 baked tortilla chips
1 ounce (mini-bag) pretzel sticks

MILK GROUP

1/2 cup sugar-free pudding
   made with skim milk
1/2 cup low-fat cottage cheese
1/2 cup frozen yogurt
1 cup fat-free yogurt
1 cup skim milk or 1 percent
2 1-ounce sticks of string cheese
   (part-skim mozzarella)

calories and 10 to 20 grams of fat per bar). In some situations, snacks from home are the best solution (see sidebar: Snacks With About 100 Calories, this page).

Remember that your weight is stable only when the calories you eat equal the calories you burn. If you are heavier than you want to be, just one "extra" portion or snack—if consistently eaten—can keep you from losing weight. Important questions to ask yourself are "Am I really hungry?" "Do I really want a snack?" If you do choose to eat a snack, be sure it is planned, and be sure it is nutritious.

WHAT'S FOR SUPPER?

The evening meal is often the only time of the day when the family sits down together. Unfortunately, a leisurely supper usually gives way to evening demands such as a school meeting, a child's soccer game, or your own "home work." Sometimes the very thought of preparing a nutritious supper seems overwhelming after a hectic day. However, with a little planning, you can still eat well.

Make Time for the Evening Meal

If possible, make time for supper. The evening meal is often the main meal of the day. It therefore can go a long way to ensure that you get all of your daily nutrient requirements. In addition, a satisfying supper decreases your chances of eating an unplanned after-supper snack.

Enticing Evening Meals

Try to incorporate food from each food group into your evening meal. Make an effort to eat more plant-based foods. These nicely complement smaller servings of lean meats and low-fat dairy products. Try cuisine from other parts of the world—particularly those that feature grains, vegetables, and fruits as the centerpiece of the meal. Make your meals interesting by varying the colors, textures, and shapes of foods. You do not have to be a great chef to give supper the visual appeal that everyone wants to eat. Fruits and vegetables add color and all the vitamins, minerals, and fiber your body needs.

Breads come in a broad variety of shapes and textures—from long, thin French baguettes to braided challah bread to the blackest of pumpernickels. The same goes for pasta. In addition to your usual favorites, try whole-wheat spaghetti, farfalle (bow-tie), fusilli (little springs), orzo (rice-shaped), or ziti (large tubes). For more color, try spinach or tomato-

flavored (and colored) pastas. Instead of plain white rice, try basmati, jasmine rice, or nutrient-rich brown rice. Wild rice (not really a "rice" but a seed from a type of grass) adds both crunch and a nutty flavor to a meal.

Instead of red meat, eat fish and poultry more often. Seafood cases are stocked with a wide variety of fish fillets and "steaks" from around the world. In addition to domestic chicken or turkey, poultry choices include wild game birds and small "hens." When you choose a red meat, eat a lean cut and vary the seasonings. Try to have some meatless meals. A supper containing legumes, beans, and other vegetarian fare is high in protein and low in saturated fat.

Top off the evening meal with fresh fruit and a selection of low-fat cheese, a fresh whole-grain cake, frozen low-fat yogurt, ice cream, or sherbet. A fine wine, a cup of café au lait, and a delicious dessert, if part of the day's meal plan, can nicely complement a well-prepared and well-presented meal.

Make It Quick

Evening meals can be quick and yet still contain the freshest of ingredients. Keep in mind that fruits and vegetables are tasty, nutritious, and easy to prepare. Quick but healthful suppers can be made from carefully selected dried, canned, frozen, or bottled foods. Plan to eat leftovers at times when you know you won't have time to cook. Cooking more when you are able, freezing a portion or two, and having it on hand when time is tight all can help you avoid skipping supper or grabbing something that you know you should not eat.

Dining Out

Eating out is an American passion. In fact, Americans spend almost half their food dollars in restaurants. With the wide array of restaurants that are available in most communities, dining out should be a pleasure rather than a threat to your resolve to eat well.

Listed here are five points to remember when you eat out. Think of these as five tips for eating well when eating out.

Choose restaurants that help you achieve your daily 5 or more servings of fruits and vegetables

Choose restaurants that offer a wide selection of fruits and vegetables. Look for fruit and vegetable selections in the appetizer, entrée, and dessert menus which are prepared without added fat. Choose ethnic and vegetarian restaurants that feature grain- and vegetable-based meals.

Lower the fat in four ways

When ordering, consider these four methods to lower the amount of fat in your meals:

Cooking method—If the menu item is described as fried or sautéed, ask that it be baked or broiled. This simple request can save 10 to 30 grams of fat and about 100 to 300 unwanted calories.

"On the side"—Ask for the sauce, condiment, salad dressing, or topping to be put "on the side." If a dish is prepared with a high-fat sauce (such as Alfredo, cream sauce, cheese, or gravy), request that it be placed on the side or ask for a lower-fat alternative such as a marinara sauce. Other high-fat condiments and toppings (such as butter or sour cream) should also be on the side so you can control how much you use.

Serving size—Bigger is not necessarily better, especially when it comes to calories, fat, and your health. Terms such as "mammoth," "deluxe," and "hearty" are your tip-offs. Order smaller portions. Ask for a "take-home" container, or plan to share your meal with your dinner companion so you do not feel obligated to eat it all.

Substitute lower-fat choices—Make sure that your meal has plenty of grains, vegetables, and fruits. Take meat away from center stage by starting with a tossed salad. Ask for extra steamed vegetables. The bread basket generally offers a whole-grain choice. Don't forget to finish the meal with fresh fruit, sherbet, or sorbet and a glass of low-fat milk. Try to avoid the high-fat, calorie-rich desserts that you will later wish you had not eaten.

Limit meat to 3 ounces per serving (or no more than 6 ounces per day)

Health experts agree that Americans eat too much meat (including poultry and fish), especially in light of the fact that our bodies need only a relatively small amount of protein. Large portions of meat almost invariably contribute excess calories, fat, and cholesterol to the meal. A 3-ounce portion is about the size of a deck of cards. Whenever you eat meat, poultry, or fish, try to be sure that you eat a total of only 6 ounces per day (see sidebar: "Cut" Your Portions of Meat, this page).

Limit alcohol to 2 drinks (1 if you are a woman)

Most authorities suggest that alcohol be limited to 2 drinks per day if you are a man, 1 drink per day if you are a woman. All alcohol-containing drinks contain calories (see sidebar: A Drink Defined..., page 87).

Make eating well your number 1 priority

Restaurants are in the business of satisfying their customers. Ask your server to recommend the most healthful foods the kitchen can prepare. Do not rely on "heart-healthy" symbols on the menu. These foods may be low in cholesterol, but they still may be high in total fat, saturated fat, and calories. Feel free to request low-fat foods or that more vegetables and fruits be added to your meal. Ask your server to clarify unfamiliar terms and to answer any of your questions.

Ethnic Cuisine: Your Passport to Healthful Meals

Ethnic restaurants and grocery stores offer a wide variety of culinary alternatives. Often, ethnic cuisines are plant-based, and thus many people assume they are healthful. However, that is not always the case. Some guidelines below, in addition to the suggestions previously mentioned, will help you to savor the exotic flavors while keeping calories, fat, cholesterol, and sodium under control.

Asian cuisine features rice, noodles, and vegetables with little or no meat. However, it may not be as healthful as one believes. For example, fried rice is just what it says—fried. Therefore, white rice or, even better, brown rice would be a healthier option. If you are dining at an Asian restaurant, make sure to request that little oil be used in preparation,

"CUT" YOUR PORTIONS OF MEAT

Note the savings in calories and fat with smaller portions of meat.

	8-OUNCE PORTION		3-OUNCE PORTION	
	CALORIES	FAT (GRAMS)	CALORIES	FAT (GRAMS)
Prime rib of beef	960	83	360	31
Lamb chop	775	66	298	25
Sirloin beef steak	745	59	280	22
Pork chop	454	20	175	8
Roasted chicken breast (without skin)	380	8	145	3
Broiled fish	240	2	90	Trace

and if your portion is large, split it with someone. Stir-fry, grilled meat, or chicken skewers are healthful choices. To decrease sodium, limit foods with soy sauce, salt, and monosodium glutamate (MSG).

Many staples of the Middle Eastern, Indian, and Pakistani cuisines are low-fat, low-sodium items such as pita bread, rice, couscous, and lentils. However, keep in mind that these foods often are combined with large amounts of butter, coconut oil, or palm oil. Request foods without the added fats and that sauces be put on the side. Healthful selections include kabobs or fish that is grilled, rice, couscous, or orzo.

Greek and Mediterranean food is based on pasta, bread, and rice. It is often the sauce that accompanies the grain that adds calories and fat. Sauces to watch out for include pesto and cream-based sauces. Red sauces are usually quite low in calories and fat. If ordering a pizza, remember that the meat and cheese are the culprits for calories and fat. Order a pizza with one meat item at the most, request half the cheese, and load on the vegetables.

Finally, Caribbean, Mexican, and Central and South American cuisine includes whole grains, corn, rice, flour tortillas, beans (pinto, red, black), and salsa. Condiments

A Drink Defined…

Health professionals recommend no more than 2 drinks daily for men, 1 drink for women. A drink is equal to 12 ounces of beer, 5 ounces of wine, 1.5 ounces of 80-proof spirits, or the equal amount of alcohol in other beverages. Alcohol can impair judgment. Never drink and drive.

	CALORIES	ALCOHOL (GRAMS)
Beer—12 ounces		
Beer	150	13
Light beer	100	12
No-alcohol beer	60	trace
Wine—5 ounces		
Wine	100	14
Light wine	80	8
Spritzer (12 ounces)	120	17
Spirits—1.5 ounces		
80 proof	100	14

Remember—mixes such as sweetened sodas, juices, and sugary syrups mean extra calories.

are an issue, so be sure to request less cheese, and have sour cream and guacamole on the side. Grilled fish and chicken are best bets as main courses. Fresh vegetables and fruits accompany most meals.

GROCERY SHOPPING: ANOTHER KEY TO HEALTHFUL MEALS

When it comes to eating at home, your meals can only be as good as the food you have in your kitchen. Stock your kitchen with foods that help you eat well. Be sure you have plenty of fruits, vegetables, and whole grains on hand so you translate your plan for healthful eating into enjoyable and nutritious meals.

Make a List
Make a grocery list. A mental list may work, but a structured plan generally is more effective. A written list helps ensure that you select the ingredients and foods that you want rather than things you select on impulse. A checklist also can save time by avoiding the need to come back to pick up forgotten items.

At the Store
Most Americans make one or more trips to the grocery store every week. Some people consider grocery shopping a form of entertainment, but others regard it as a chore. No matter how you feel about shopping, a few strategies can make your investment of time worthwhile.

Use labels to compare similar foods and to make the healthiest choices.

Shop the perimeter—The freshest foods generally are located along the perimeter of the store. These include fresh fruits, vegetables, breads, dairy products, and meats. Choose whole-grain breads from the bakery, low-fat products from the dairy section, and lean cuts from the meat section. Higher-fat foods often are stocked in the interior of the store. However, grains, legumes, pastas, and canned fruits and vegetables also are frequently located in the interior.

Shop from your list—A list helps ensure that you stick to your plan. Items purchased on impulse generally are more expensive and less nourishing than foods from a well-planned shopping list.

Do not shop when hungry—If you go to the store with a growling stomach, you are apt to make purchases to satisfy your immediate hunger rather than what you will need in the days ahead. You also tend to buy more. Shopping on a full stomach helps you stick to your list and to keep your resolve.

Make bargains count—A bargain is only a bargain when you buy what you want rather than what someone else is trying to sell you. Select bargains that fit your menu, such as in-season fruits and vegetables or bulk quantities of rice, beans, and legumes. Use coupons to reduce the cost of foods that you intend to buy.

Label Smarts
Understanding food labels helps you become a savvy shopper. The U.S. Department of Agriculture has established requirements for food labels in order to make nutrition information accurate, clear, consistent, and useful to consumers. Food labels provide four different types of information: nutrition facts, a list of ingredients, nutrients, and health claims.

Nutrient facts and ingredients
Labels tell you almost everything you need to know about what is contained in the food. The Nutrition Facts panel tells you how many calories and how much fat, cholesterol, protein, and carbohydrate are in a single serving of the food. It also tells you the fiber, vitamin (A and C), and mineral (sodium, potassium, calcium, and iron) contents of the food. This information allows you to compare the nutrients found in similar foods. For example, you can compare two yogurts for their calorie and fat content. Ingredients are listed in descending order by weight, with the first item listed being the predominant one in the food. By reading the label, you can tell which foods have added

saturated fats or ingredients to which you may be intolerant. Use the label to select the best choices—foods that are high in nutrients and low in calories and fat.

To help you make the wisest choices, acquaint yourself with each of the components of the label (see sidebar: Anatomy of a Food Label, below). Some foods, such as fresh fruits, vegetables, and bulk items, do not have nutrition labels. The nutrition information for these foods generally either is on display nearby or is available elsewhere in the store as a handout. When in doubt, ask the store personnel for assistance.

Nutrient and health claims

In addition to the amounts of various nutrients and ingredients found in foods, food labels may carry other types of information. The label may state that the food is a "good

---

ANATOMY OF A FOOD LABEL

At first glance, a food label may look intimidating. But as you become familiar with the format, you'll see how the label can help you compare products for nutritional quality.

SERVING SIZE: Nutrition information is based on consistent and realistic serving sizes that include both household and metric measures. This makes it easier to compare products.

NUTRIENTS: At a minimum, all calories, calories from fat, total fat, saturated fat, cholesterol, sodium, total carbohydrate, dietary fiber, sugars, protein, vitamins A and C, calcium, and iron are listed. Experts believe that too much or too little of these nutrients has the greatest impact on your health.

CALORIES PER GRAM: This shows the calorie content of the energy-producing nutrients.

Nutrition Facts

Serving size 1/2 cup (114 g)
Servings per container 4

Amount per serving	
Calories 90	Calories from fat 30
	Percent Daily Value*
Total fat 3 g	5%
Saturated fat 0 g	0%
Cholesterol 0 mg	0%
Sodium 300 mg	13%
Total carbohydrate 13 g	4%
Dietary fiber 3 g	12%
Sugars 3 g	
Protein 3 g	

Vitamin A	80%	Vitamin C	60%
Calcium	4%	Iron	4%

*Percent Daily Values are based on a 2,000-calorie diet. Your daily values may be higher or lower depending on your calorie needs:

	Calories	2,000	2,500
Total fat	Less than	65 g	80 g
Saturated fat	Less than	20 g	25 g
Cholesterol	Less than	300 mg	300 mg
Sodium	Less than	2,400 mg	2,400 mg
Total carbohydrate		300 g	375 g
Fiber		25 g	30 g

Calories per gram:
Fat 9   Carbohydrate 4   Protein 4

CALORIES FROM FAT: This information underscores the fat content per serving of foods to help you meet the recommendation of no more than 30 percent of calories from fat. Remember, it's your total fat intake over a period of time (such as 24 hours), and not the amount in one food or meal, that's important.

PERCENT DAILY VALUE: The percent daily values show the percentage of a nutrient that is provided by this product serving, based on a 2,000-calorie diet. Use the percent daily value figures to compare products easily and to tell whether a food is high or low in nutrients.

source" of a given nutrient. Nutrient claims help guide you more quickly to certain products. Products also may claim that they benefit certain health conditions. Health claims may suggest that eating a food alters your risk for certain diseases such as heart disease, cancer, or osteoporosis. The government regulates these claims so they are truthful and meet certain criteria. Manufacturers are not required by law to carry nutrient or health claims. (See sidebars: Nutrient Claims, page 92, and Health Claims, page 93.)

FOODS AND ISSUES YOU MAY HAVE WONDERED ABOUT

Farming methods are rapidly evolving. New foods and novel ingredients are continuously being introduced into the marketplace. Pesticides and fertilizers are used to increase yields. Genetic engineering potentially can produce safer, more nutritious, and cheaper forms of foods. Substitutes (such as sugar, fat, and salt substitutes) enable some people to eat foods that would otherwise endanger their health. However, not everyone would agree that these changes are an advance because many have been accompanied by uncertainty and, in some instances, controversy. What about additives? Organic foods? Irradiated foods? What does it mean when a food is enriched or fortified? Read on.

What About Pesticides?

Pesticides and modern pest management practices have helped to ensure that we have a reliable, affordable, varied, nutritious, and safe food supply. Pesticides are chemicals that kill or prevent the growth of weeds (herbicides), bacteria (disinfectants and antibiotics), molds and fungi (fungicides), and harmful insects (insecticides). Some pesticides occur naturally in soil, whereas others are found in compounds isolated from particular plants. Many farmers try to control pests in the most effective, least disruptive manner by practicing what is known as integrated pest management. This approach includes companion planting with plants that contain natural pesticides, crop rotation, use of sterile strains of insects or insect pheromones (to alter reproduction patterns and thereby reduce the insect population), natural insect predators, pest-resistant plant strains, mathematical forecasting techniques, and, when necessary, chemical pesticides.

Should you worry about pesticide residues? The answer is a qualified "no." The upper limit of the amount of pesticide residue permitted on both raw and processed foods has been carefully established and is enforced by several government agencies. These upper limits are far less than the levels of exposure that are considered harmful. However, to reduce your exposure to pesticide residues further, you can take the following steps:

- Carefully select the produce you buy (avoid cuts, holes, or signs of decay).
- Thoroughly wash all produce with water (not soap) to remove surface residues.
- Scrub carrots and potatoes and other root vegetables thoroughly. Wash other fresh fruits and vegetables with a brush. Resist the temptation to peel apples, pears, cucumbers, potatoes, and other produce with edible skin, because peeling removes a valuable source of fiber.
- Remove the outer leaves (and any inner leaves that appear to be damaged) from leafy vegetables such as lettuce and cabbage.
- Eat a variety of foods rather than large amounts of a single food.

What About Food Additives?

Contrary to what many of us think, food additives have been used for centuries (and until the past 100 or so years, without much regard for health considerations). Additives play various roles in foods. Some additives act as preservatives, preventing spoilage, loss of flavor, texture, or nutritive value. These include antioxidants such as vitamin E, BHA (butylated hydroxyanisole), BHT (butylated hydroxytoluene), citric acid, and sulfites; calcium propionate; sodium nitrite (which prevents infection by the dangerous bacteria that causes botulism); and other antimicrobials (germ-fighting additives). Other additives are used as emulsifiers to prevent foods from separating. Emulsifiers include lecithin from soy, egg yolks, or milk and monoglycerides and diglycerides. The leavening agents sodium bicarbonate (baking soda and powder) and yeast allow baked goods to rise. Stabilizers and thickeners create and maintain an even texture and flavor in foods such as ice cream and pudding. Many foods are also fortified or enriched with vitamins, minerals, or proteins or their component amino acids (see page 25).

The government strictly regulates food additives. The safety and effectiveness of each newly proposed additive must be rigorously tested. Some 700 additives belong to a group of substances referred to as GRAS substances

(Generally Recognized As Safe) because extensive testing and use have shown no evidence of adverse effects. Continued approval requires that all additives, including GRAS substances, must be reevaluated regularly with the latest scientific methods. According to law, additives and preservatives must be listed on food labels. People who are sensitive to certain additives such as sulfites or monosodium glutamate should be sure to read labels for additive information.

Organic Foods: Are They Really Better?

If pesticides and food additives pose any potential risk at all, would we be better off eating only "organically grown" foods? What is an organic food? Organic farming methods are those that use only nonsynthetic products (substances that are naturally found in the environment).

Can you be sure that the organic products you buy are really grown and produced organically? In 1998, Congress enacted the Organic Foods Production Act, which regulates production and processing standards for organic foods. This act specifies that foods sold as "organically grown" or "organically processed" must be certified by the U.S. Department of Agriculture. However, organically produced foods may not be 100 percent organic. By law, they must have at least 50 percent of their ingredients produced naturally, and organically processed foods must contain at least 95 percent organically produced ingredients.

Are these foods better? Not necessarily. Nutritionally, organically grown foods may not be significantly different from the same products grown with conventional farming techniques. The nutrient content of a product is determined by many factors, including the composition of the soil, the genetic makeup of the plant, the degree of maturity at harvest, and methods of handling after harvest. The taste of organic products may or may not differ from that of conventionally grown foods. Organic methods tend to be more costly (in terms of both labor and materials), production is lower, and consumer prices tend to be higher. In addition, organic methods are not always safer. Some pesticides reduce the risk of exposure to certain harmful organisms that represent a much greater potential risk than does exposure to the pesticide. In most instances, personal preference rather then proven benefit determines whether you choose to eat an organic food.

What About Irradiated Foods?

Until the 1800s, soaking in salt and natural fermentation were the most commonly used methods for food preservation. This changed with the introduction of canning in the 1800s. Pasteurization began to be used in the late 1800s to kill harmful bacteria, freezing in the early 1900s to extend the shelf life of foods, and freeze drying in the 1960s to preserve foods.

Food is irradiated when it is passed through a beam of radiant energy. Irradiated foods are not radioactive. Irradiation does not diminish the importance of safe food handling (see Chapter 5, Food Safety, page 148), nor does it improve the quality of food. However, irradiation can extend the shelf life, reduce food wastage, and reduce food prices. Most importantly, irradiation can destroy bacteria that cause foodborne illness. Food irradiation is strictly controlled by the Food and

Government regulations require irradiated food at the retail level to be labeled "Treated with Radiation" or "Treated by Irradiation" and to bear this international logo, the radura.

NUTRIENT CLAIMS

TERM	PER SERVING SIZE ON LABEL
Free	Contains no, or only "physiologically or inconsequential" amounts of, fat, saturated fat, cholesterol, sodium, sugars, and calories
Reduced	At least 25 percent lower in the nutrient than the standard product
Lite, light	At least 50 percent less fat 50 percent less sodium; one-third fewer calories (must also contain less than 50 percent of calories from fat)
Low	Low fat: 3 grams or less Low cholesterol: 20 milligrams or less Low saturated fat: 1 gram or less Low sodium: 140 milligrams or less Very low sodium: 35 milligrams or less Low calorie: 40 calories or less
High	20 percent or more of the Daily Value Fiber: 5 grams or more Potassium: 700 milligrams or more Vitamin A: 1,000 IU or more Vitamin C: 12 milligrams or more Folate: 80 micrograms or more Iron: 3.6 milligrams or more
Good source	10 to 19 percent of the Daily Value Fiber: 2.5 to less than 5 grams Potassium: 350 to less than 700 milligrams Vitamin A: 500 to less than 1,000 IU Vitamin C: 6 to less than 12 milligrams Folate: 40 to less than 80 micrograms Iron: 1.8 to less than 3.6 milligrams
Healthy	Food that is low in fat and saturated fat, 480 milligrams or less sodium, and at least 10 percent Daily Value for vitamin A, vitamin C, calcium, iron, protein, or fiber
Lean	Less than 10 grams total fat, 4.5 grams saturated fat, and 95 milligrams cholesterol per 100 grams (about 3 ounces)
Extra lean	Less than 5 grams total fat, 2 grams saturated fat, and 95 milligrams cholesterol per 100 grams (about 3 ounces)

Drug Administration (FDA). All irradiated foods carry a symbol and the phrase "Treated by Irradiation." However, when irradiated foods are used in a restaurant, they are not required to be labeled.

What About Genetically Engineered Foods?

Modern genetic-engineering techniques are a refinement of traditional plant-breeding methods. Genetic engineers first select a desired plant trait. They then isolate and modify the gene(s) responsible for the trait. Finally, they attempt to introduce the altered gene into other plants. In essence, genetic engineering accelerates the natural mixing of genes that normally would occur among various plant species.

Genetic engineering of plants has yielded several benefits. Farmers can protect crops against weeds with genetically engineered biodegradable herbicides that need fewer applications. Genetic engineering also can extend the time before spoilage begins, enabling growers to harvest foods closer to peak freshness. Crops can be genetically engineered to be drought- and temperature-resistant. Genetic engineering can increase crop yields and enable plants to tolerate a wider range of climates. Genetic engineering also can increase the amount and quality of protein in beans and in grains such as rice.

Although there are benefits associated with genetically engineered foods, there also are lingering questions about the long-term effects on the environment and the ability of producers and the government to ensure the safety of the process. Therefore, the debate continues regarding how genetically engineered foods should be developed and regulated.

What Do the Terms "Fortified" and "Enriched" Mean?

Do you know the difference between the terms "fortified" and "enriched"? Both words indicate that nutrients have been added to the food. Fortified means that nutrients were added that were not there originally. Enriched means that nutrients lost during processing are replaced. Fortified and enriched foods have helped eliminate many once-common nutritional deficiency diseases.

Salt was the first commercial food item in the United States to be fortified with additional nutrients. Since 1924, potassium iodide has been added to table salt to help prevent goiter (an enlarged thyroid that may result from iodine deficiency).

HEALTH CLAIMS

The Food and Drug Administration (FDA) strictly regulates health claims. They must be supported by scientific evidence to appear on food labels. The foods must also meet specific nutrient requirements. Below are several examples of FDA-approved health claims.

DISEASE/NUTRIENT OR FOOD	EXAMPLE OF A HEALTH CLAIM
Osteoporosis/calcium	Regular exercise and a healthy diet with enough calcium help teens and young adult white and Asian women maintain good bone health and may reduce their high risk of osteoporosis later in life.
Cancer/fat	Development of cancer depends on many factors. A diet low in total fat may reduce the risk of some cancers.
Cancer/fiber-containing foods	Low-fat diets rich in fiber-containing grain products, fruits, and vegetables may reduce the risk of some types of cancer, a disease associated with many factors.
Cancer/fruits and vegetables	Low-fat diets rich in fruits and vegetables (foods that are low in fat and may contain dietary fiber, vitamin A, or vitamin C) may reduce the risk of some types of cancer, a disease associated with many factors. Broccoli is high in vitamins A and C, and it is a good source of dietary fiber.
Heart disease/fiber-containing foods	Diets low in saturated fat and cholesterol and rich in fruits, vegetables, and grain products which contain some types of dietary fiber, particularly soluble fiber, may reduce the risk of heart disease, a disease associated with many factors.
Heart disease/saturated fat and cholesterol	While many factors affect heart disease, diets low in saturated fat and cholesterol may reduce the risk of this disease.
Hypertension/sodium	Diets low in sodium may reduce the risk of high blood pressure, a disease associated with many factors.
Birth defects/folic acid	Healthful diets with adequate folate may reduce a woman's risk of having a child with a brain or spinal cord defect.

Vitamin D was first added to milk in 1933. Rickets, a deforming bone disease, can result from too little vitamin D. Rickets was very common during the early part of the century when working children were deprived of nutritious diets and sunlight.

Flour, and products made with flour, were first fortified with vitamins and iron in 1940 at the start of World War II. The goal was to build strong, healthy armies by feeding troops well-fortified diets.

Most breakfast cereals are fortified to provide about 25 percent of most daily nutrients. Many of these vitamins may dissolve into the milk that you pour over the cereal, so it is important to drink the milk at the bottom of the bowl. (Plus, milk itself provides its own nutrients.)

Most recently, folic acid has joined the list of fortifiers added to grain-based (bread, cereal, pasta, rice) foods to help meet nutrient needs and to reduce the incidence of birth defects and possibly the risk of heart disease.

TWO WEEKS OF MENUS

The following pages contain 14 days' worth of nutritionally balanced menus. The menus emphasize whole grains, fresh fruits, and vegetables. They are low in fat, saturated fat, and cholesterol, moderate in sodium, and high in taste. In addition, they are a great source of health-enhancing phytonutrients. The actual guidelines for the menus are listed below.

The serving sizes are based on the Food Guide Pyramid (see Chapter 1, page 11). For each day, the breakdown of servings from the Food Guide Pyramid is provided.

Accompanying each menu, there is one recipe. Use these menus as a guide to begin to eat a more nutritiously balanced diet. As needed, make adjustments to suit your needs. If you need additional (or fewer) calories, simply add or take away a serving. If you need to add servings, we suggest adding a whole grain, piece of fresh fruit, or more vegetables.

These menus take the information that has been discussed throughout the book and help you put it on the table. Bon appetit.

MENU GUIDELINES

Calories	1,800-2,000
Fat (less than 30% of calories)	Less than 60 grams
Saturated fat (less than 8% of calories)	Less than 16 grams
Cholesterol	Less than 300 milligrams
Dietary fiber	20-35 grams
Sodium	Less than 2,400 milligrams

FOOD GROUPS	TARGETED NUMBER OF SERVINGS/DAY
Sweets	Sparingly
Fats, oils (emphasis on monounsaturated)	About 3
Meat, poultry, fish, legumes, nuts, seeds (emphasis on poultry, fish, legumes, nuts, and seeds)	2 servings
Dairy (emphasis on low fat)	About 3+
Vegetables (emphasis on fresh)	About 4
Fruits (emphasis on fresh)	About 4
Grains (emphasis on whole)	About 8

Note: A chef and a registered dietitian planned the menus and recipes. The dietitian analyzed the recipes and menus for their nutrient contents.

DAY 1

Nutritional Analysis for Entire Day

	Actual	Goal
Calories	2,000	1,800-2,000
Fat (g)	58	Less than 60
Saturated fat (g)	13	Less than 16
Cholesterol (mg)	80	Less than 300
Fiber (g)	24	20-35
Sodium (mg)	1,800	Less than 2,400

Food Servings

	Actual
Sweets	1
Fats	1
Legumes/nuts	3
Meat, poultry, fish	3 ounces
Milk	3
Vegetables	5
Fruit	3
Grains	7

BREAKFAST
2 whole-wheat pancakes—top with 1/2 cup unsweetened applesauce
1 cup nonfat fruit-flavored yogurt
Coffee—regular or decaffeinated

NOON
Southwest Caesar Salad (see recipe below)
6 wheat crackers
1 fresh orange
1 cup skim milk

EVENING
Broccoli and walnut stir fry (1 tsp sesame oil, 1 cup broccoli florets,
    1/2 cup chopped red or yellow pepper, 1/2 tsp soy sauce, 3 ounces tofu,
    1/3 cup chopped walnuts)
1 cup cooked brown rice
1 whole-grain roll
1 Tblsp honey
Herbal tea

SNACK (ANYTIME)
1 1/2 ounces low-fat cheddar cheese
2 crisp rye wafers
6 ounces pineapple juice

DAY 1 RECIPE

Southwest Caesar Salad
Serves 4

8 cups romaine lettuce
1/2 cup fat-free croutons
2 cups cubed cooked chicken breast
1 can (14 to 16 ounces) low-sodium kidney, black, or pinto beans, drained
1 can (8 ounces) low-sodium whole-kernel corn, drained
1 medium tomato, cut into wedges
1 medium red, yellow, or green bell pepper, thinly sliced
1/2 medium onion, thinly sliced
1/4 cup fat-free Caesar dressing
2 Tblsp Parmesan cheese, grated

Combine romaine lettuce and croutons with chicken, beans, corn, tomato, bell pepper, and onion in a large serving bowl. Pour dressing over salad; toss to coat evenly. Top with Parmesan cheese.

Per serving in this recipe

Calories	220
Fat (g)	3
Saturated fat (g)	1
Cholesterol (mg)	25
Fiber (g)	2
Sodium (mg)	230

DAY 2

BREAKFAST

Herbed scrambled eggs (1 egg, 2 egg whites, 2 tsp chopped chives or 1/4 tsp dill)
2 slices whole-wheat toast
1 tsp soft margarine
1/2 grapefruit
1 cup skim milk
Coffee—regular or decaffeinated

NOON

Tuna salad on whole-wheat bread (3 ounces flaked tuna,* fat-free mayonnaise, lettuce, cucumber, tomato slices, and a few sliced olives as toppings)
1 cup tossed spring greens
2 Tblsp fat-free herbed salad dressing
1 kiwi fruit
1 cup skim milk

EVENING

2 Tacos With Sautéed Vegetables and Smoky Salsa (see recipe below)
1 cup Spanish rice
1 cup tossed green salad
2 Tblsp fat-free red wine vinaigrette dressing
1/4 cantaloupe
Herbal tea

SNACK (ANYTIME)

1 cup low-fat vanilla yogurt
1 sliced banana

*Reduced sodium and water-packed variety recommended.

Nutritional Analysis for Entire Day

	Actual	Goal
Calories	1,850	1,800-2,000
Fat (g)	40	Less than 60
Saturated fat (g)	7	Less than 16
Cholesterol (mg)	270	Less than 300
Fiber (g)	29	20-35
Sodium (mg)	2,375	Less than 2,400

Food Servings

	Actual
Sweets	0
Fats	4
Legumes/nuts	1
Meat, poultry, fish	5 ounces
Milk	3
Vegetables	5
Fruit	4
Grains	9

DAY 2 RECIPE

Soft Tacos With Sautéed Vegetables and Smoky Salsa
Serves 4

1 Tblsp olive oil
1 medium red onion, chopped
1 cup yellow summer squash, sliced
1 cup green zucchini, sliced
3 large garlic cloves, finely chopped
4 medium tomatoes, seeded and chopped
1 jalapeno pepper, seeded and chopped

Kernels from 2 ears of corn (1 cup)
1/2 cup fresh cilantro, chopped
1 cup canned pinto or black beans, rinsed and drained
8 corn tortillas (7-inch diameter)
1/2 cup smoky-flavored salsa

Heat oil in large skillet; add onion and cook until tender. Add squash and zucchini, stir, and continue cooking about 5 minutes. Add garlic, half the tomatoes, and all of the jalapeno pepper. Reduce heat to medium-low and cook until vegetables are tender. Add corn kernels; stir and cook until kernels are tender crisp. Add cilantro, remaining tomatoes, and beans. Stir together and remove from heat. "Warm" the tortillas in a hot, dry skillet. Fill each tortilla with the vegetable mixture. Top with salsa and serve.

Per serving in this recipe

Calories	350
Fat (g)	8
Saturated fat (g)	1
Cholesterol (mg)	10
Fiber (g)	8
Sodium (mg)	450

DAY 3

Nutritional Analysis for Entire Day

	Actual	Goal
Calories	1,810	1,800–2,000
Fat (g)	40	Less than 60
Saturated fat (g)	10	Less than 16
Cholesterol (mg)	120	Less than 300
Fiber (g)	26	20–35
Sodium (mg)	2,225	Less than 2,400

Food Servings

	Actual
Sweets	2
Fats	2
Legumes/nuts	1
Meat, poultry, fish	5 ounces
Milk	3
Vegetables	4
Fruit	4
Grains	8

Per serving in this recipe

Calories	237
Fat (g)	6
Saturated fat (g)	1
Cholesterol (mg)	29
Fiber (g)	6
Sodium (mg)	340

BREAKFAST
1 spiced muffin
2 Tblsp fruit spread
1 banana
1 cup skim milk

NOON
Hamburger on bun (3 ounces extra-lean ground beef, tomato slices, shredded lettuce, whole-grain bun)
1 ounce (individual bag) pretzels
2 cups tossed salad (lettuce, cucumbers, mushrooms, bell peppers)
2 Tblsp fat-free French dressing
15 grapes
Sparkling water

EVENING
Chicken and Couscous Vegetable Salad (see recipe below) on baby greens
1 whole-grain roll
1 Tblsp honey
Moro orange (blood orange)
1 cup skim milk

SNACK (ANYTIME)
4 crispy rye wafers
1 1/2 ounces reduced-fat Swiss cheese

DAY 3 RECIPE

Chicken and Couscous Vegetable Salad
Serves 6

1 1/2 cups low-sodium chicken broth
1 cup uncooked couscous (pearl type)
1 1/2 cups cooked chicken or turkey, chopped
1/2 pound asparagus, cut into 2-inch pieces, cooked and drained

1 1/2 cups chopped green or yellow bell pepper
1 medium tomato, chopped
1/2 cup celery, diced
1/2 cup fat-free Italian dressing
2 Tblsp sliced almonds, toasted

Pour broth into saucepan; bring to a boil. Stir in couscous. Remove from heat; cover, and let stand 5 minutes. Stir with fork.

Stir together couscous, chicken, asparagus, bell pepper, tomato, and celery in a bowl. Add dressing. Stir evenly to coat. Serve at room temperature or chilled. Just before serving, sprinkle with toasted almonds.

DAY 4

BREAKFAST

1 English muffin
2 Tblsp fat-free cream cheese
1 cup fresh strawberries
Coffee—regular or decaffeinated

NOON

Greek salad (2 cups mixed greens, cucumber slices, cherry tomatoes,
    onion slices, 1 1/2 ounces feta cheese, 5 ripe olives, 1/4 cup dates,
    lemon juice, 1 tsp olive oil, vinegar, oregano, to taste)
2 bread sticks
1 fresh pear
1 cup skim milk

EVENING

Quick Pork With Pineapple-Orange-Basil Sauce (see recipe below)
1 cup cooked brown rice with chopped apples
1 cup steamed baby carrots
1 whole-grain roll
1 Tblsp fruit spread
Herbal tea

SNACK (ANYTIME)

1 cup frozen low-fat yogurt
1/2 cup berries

Nutritional Analysis for Entire Day

	Actual	Goal
Calories	1,900	1,800-2,000
Fat (g)	36	Less than 60
Saturated fat (g)	14	Less than 16
Cholesterol (mg)	146	Less than 300
Fiber (g)	30	20-35
Sodium (mg)	2,200	Less than 2,400

Food Servings

	Actual
Sweets	1
Fats	2
Legumes/nuts	0
Meat, poultry, fish	4 ounces
Milk	3
Vegetables	5
Fruit	6
Grains	7

DAY 4 RECIPE

Quick Pork With Pineapple-Orange-Basil Sauce

Serves 4

1 pound pork tenderloin, cut into 1/4-inch
    slices (may also substitute turkey breast)
1 tsp vegetable oil
1 cup pineapple juice
2 garlic cloves, finely chopped

2 tsp cornstarch
1 1/2 tsp dried basil leaves, crushed
1 tsp grated orange peel
2 oranges, sliced into rounds
1 green onion, chopped

In large skillet, cook and stir pork over medium-high heat in hot oil 3 to 5 minutes
or until pork is no longer pink. Remove pork, drain.

Stir together juice, garlic, cornstarch, basil, and orange peel in skillet until blended.
Bring to a boil; cook 2 minutes or until sauce is slightly thickened. Return pork to
skillet; cook 1 minute or until heated through.

Arrange pork slices along with orange slices on dinner plates; sprinkle with green
onion. Top with sauce. Serve immediately.

**Per serving
in this recipe**

Calories	230
Fat (g)	5
Saturated fat (g)	2
Cholesterol (mg)	80
Fiber (g)	Trace
Sodium (mg)	60

DAY 5

Nutritional Analysis for Entire Day

	Actual	Goal
Calories	1,900	1,800-2,000
Fat (g)	45	Less than 60
Saturated fat (g)	6	Less than 16
Cholesterol (mg)	70	Less than 300
Fiber (g)	20	20-35
Sodium (mg)	1,350	Less than 2,400

Food Servings

	Actual
Sweets	1
Fats	2
Legumes/nuts	0
Meat, poultry, fish	3 ounces
Milk	3+
Vegetables	5
Fruit	5
Grains	7

Per serving in this recipe

Calories	395
Fat (g)	5
Saturated fat (g)	1
Cholesterol (mg)	60
Fiber (g)	2
Sodium (mg)	432

BREAKFAST

1 cup old-fashioned oatmeal—topped with 1 Tblsp brown sugar
1 slice whole-wheat toast
1 tsp soft margarine
1 banana
1 cup skim milk
Coffee—regular or decaffeinated

NOON

Southwestern Turkey Pocket Fajita (see recipe below)
1/2 cup fresh vegetables (baby carrots, celery sticks, broccoli florets)
1 peach
1 cup frozen low-fat yogurt
Sparkling water

EVENING

1/4 12-inch pizza (whole-wheat crust, chunky tomato sauce,
    sliced bell pepper, onion, mushrooms, zucchini, 6 ounces mozzarella cheese,
    thyme and basil, to taste)
1 cup tossed greens
2 Tblsp fat-free Italian salad dressing
1 cup mixed fresh fruit
Iced tea

SNACK (ANYTIME)

1 ounce (individual bag) pretzels
6 ounces orange juice

DAY 5 RECIPE

Southwestern Turkey Pocket Fajita
Serves 4

1 cup nonfat plain yogurt
1 tsp grated lime peel
1 can (20 ounces) pineapple tidbits or chunks
12 ounces turkey breast, cut into 1-inch
    cubes (may substitute pork tenderloin)
1 Tblsp lime juice

1 tsp cumin
1/2 cup onion, thinly sliced
1 medium green or red bell pepper, cut
    into 2-inch strips
4 (6-inch) whole-wheat pita breads, cut
    in half crosswise to make pockets

Stir together yogurt and lime peel in a small bowl; set aside.

Drain pineapple; reserve 2 Tblsp juice.

Combine reserved juice, turkey, lime juice, and cumin in non-metallic dish. Cover and marinate 15 minutes in the refrigerator.

In a non-stick skillet, combine the turkey mixture and onion and cook over medium-high heat, stirring until turkey is slightly browned. Add bell pepper and pineapple; cook and stir 2 to 3 more minutes or until the vegetables are tender crisp and the turkey is no longer pink.

BREAKFAST

1 whole-wheat bagel
2 Tblsp peanut butter
1 nectarine
1 cup low-fat yogurt
Coffee—regular or decaffeinated

NOON

Roasted vegetables on herbed focaccia bread (1 cup total of red pepper, eggplant, and zucchini topped with 1 1/2 ounces of part-skim mozzarella cheese)
12 baked tortilla chips
1 apple
1 cup skim milk

EVENING

4-ounce grilled steak topped with 1 cup sautéed mushrooms
1 medium baked potato topped with 1 Tblsp fat-free sour cream and snipped chives
1 cup steamed Brussels sprouts
1 whole-grain dinner roll
1 tsp soft margarine
Poached Pears in Red Wine (see recipe below)
Herbal tea

SNACK (ANYTIME)

2 cups popcorn
6 ounces cranberry juice

DAY 6

Nutritional Analysis for Entire Day

	Actual	Goal
Calories	1,940	1,800-2,000
Fat (g)	45	Less than 60
Saturated fat (g)	14	Less than 16
Cholesterol (mg)	133	Less than 300
Fiber (g)	30	20-35
Sodium (mg)	1,425	Less than 2,400

Food Servings

	Actual
Sweets	1
Fats	2
Legumes/nuts	1
Meat, poultry, fish	4 ounces
Milk	3
Vegetables	5
Fruit	4
Grains	7

DAY 6 RECIPE

Poached Pears in Red Wine
Serves 4

4 whole pears	1/2 cup sugar
1 fresh lemon	1/2 stick cinnamon (1 tsp ground)
1/2 bottle red wine	Pinch of nutmeg

Peel pears, leaving stem. Remove core from the bottom. Lightly rub pears with cut lemon to prevent browning.

In a heavy saucepan, combine wine, sugar, cinnamon, and nutmeg. Add pears and bring to a boil. Turn pears frequently to coat with mixture. Reduce heat and simmer for about 30 minutes or until pears are tender but still firm (a butter knife should go in easily).

Remove from pan and chill in the sauce until serving. Serve on decorative plates and top with sauce.

Per serving in this recipe

Calories	255
Fat (g)	0
Saturated fat (g)	0
Cholesterol (mg)	0
Fiber (g)	4
Sodium (mg)	5

DAY 7

Nutritional Analysis for Entire Day

	Actual	Goal
Calories	2,000	1,800-2,000
Fat (g)	45	Less than 60
Saturated fat (g)	13	Less than 16
Cholesterol (mg)	160	Less than 300
Fiber (g)	33	20-35
Sodium (mg)	2,220	Less than 2,400

Food Servings

	Actual
Sweets	0
Fats	2
Legumes/nuts	1
Meat, poultry, fish	5 ounces
Milk	2
Vegetables	5
Fruit	4
Grains	9

BREAKFAST
1 cup bran cereal—top with 1/2 cup dried mixed fruit
1 cup skim milk
Coffee—regular or decaffeinated

NOON
1 cup split-pea soup*
Sub sandwich (6-inch Italian bread, 2 tsp mustard, chopped lettuce,
    sliced tomato, onion, bell peppers, 2 ounces lean ham, 1 1/2 ounces
    Swiss cheese)
4 soda crackers
1 apple
Sparkling water

EVENING
Grilled Chicken With Tomato and Corn Salsa (see recipe below)
1 cup Spanish rice
1 cup steamed broccoli with lemon zest
1/2 cup sliced mango
Herbal tea

SNACK (ANYTIME)
1 cup raw vegetables (baby carrots, celery sticks, cauliflower florets)
2 Tblsp fat-free ranch dressing (for dip)

*Homemade or reduced sodium variety is recommended.

Per serving in this recipe

Calories	340
Fat (g)	15
Saturated fat (g)	3
Cholesterol (mg)	83
Fiber (g)	2
Sodium (mg)	86

DAY 7 RECIPE

Grilled Chicken With Tomato and Corn Salsa
Serves 4

4 boneless, skinless chicken breasts
1 tsp oil

Kernels from 2 cobs of cooked corn (about 1 cup)
2 tomatoes, peeled, seeded, and chopped
1 shallot, chopped
1/2 cup balsamic vinegar
2 Tblsp olive oil
4 basil leaves, chopped

Brush the chicken breasts with oil. Grill until no longer pink in center.

In the meantime, combine all of the salsa ingredients. Spoon over the grilled chicken and serve immediately.

BREAKFAST

1 cup mixed fresh fruits (melon, berries, banana, pear, cherries)
  —top with 1 cup low-fat yogurt and 1/3 cup toasted almonds
2 slices whole-wheat toast
1 Tblsp honey
Herbal tea

NOON

1 cup vegetable soup*
6 wheat crackers
Tarragon chicken sandwich on whole-wheat bread (2 slices whole-wheat bread,
  1 Tblsp fat-free mayonnaise, lettuce leaf, sliced tomato, alfalfa sprouts,
  2 ounces shaved chicken, tarragon, to taste)
1 apple
1 cup skim milk

EVENING

Grilled Ahi Tuna With Mango-Pineapple Chutney (see recipe below)
1/2 cup small steamed red potatoes in skin
1 cup tossed greens
2 Tblsp fat-free cucumber salad dressing
1 whole-wheat roll
1 Tblsp honey
1 cup sparkling water

SNACK (ANYTIME)

2 graham crackers
1 cup skim milk

*Homemade or reduced-sodium variety is recommended.

DAY 8

Nutritional Analysis for Entire Day

	Actual	Goal
Calories	2,000	1,800-2,000
Fat (g)	60	Less than 60
Saturated fat (g)	9	Less than 16
Cholesterol (mg)	100	Less than 300
Fiber (g)	28	20-35
Sodium (mg)	2,300	Less than 2,400

Food Servings

	Actual
Sweets	2
Fats	2
Legumes/nuts	1
Meat, poultry, fish	5 ounces
Milk	3
Vegetables	4
Fruit	5
Grains	7

DAY 8 RECIPE

Grilled Ahi Tuna With Mango-Pineapple Chutney
Serves 4

4 tsp olive or vegetable oil
2 ounces fresh ginger, peeled, cored,
  and mashed
1 jalapeno pepper, seeded and diced
1 red onion, finely chopped
1/4 cup packed brown sugar
1/4 cup rice wine vinegar
1 cup fresh pineapple, peeled, cored,
  and finely chopped

1 mango, peeled and chopped
1 small bunch fresh cilantro,
  washed and chopped
1/4 tsp salt (use sea salt for different flavor)
4 pieces ahi tuna fillets, about 4 ounces
  each, 1 1/2 inches thick
Cracked pepper, to taste

Chutney: Measure and prepare all ingredients before assembly. In a large sauté pan, over medium heat add 2 tsp of the oil with ginger, jalapeno pepper, and onion. Cook until onions become transparent. Add brown sugar. When sugar is dissolved, add rice wine vinegar, pineapple, and mango. Simmer for about 5 minutes. Turn off heat. When mixture is cool, add the cilantro and season with salt. Cover and refrigerate until serving time.

Tuna: Brush tuna with 2 tsp oil and grill approximately 4 to 5 minutes on each side. Season with cracked pepper to taste.

Per serving
in this recipe

Calories	325
Fat (g)	10
Saturated fat (g)	2
Cholesterol (mg)	43
Fiber (g)	2
Sodium (mg)	310

DAY 9

Nutritional Analysis for Entire Day

	Actual	Goal
Calories	1,900	1,800-2,000
Fat (g)	45	Less than 60
Saturated fat (g)	6	Less than 16
Cholesterol (mg)	95	Less than 300
Fiber (g)	32	20-35
Sodium (mg)	1,200	Less than 2,400

Food Servings

	Actual
Sweets	1
Fats	1
Legumes/nuts	1
Meat, poultry, fish	4 ounces
Milk	3
Vegetables	4
Fruit	5
Grains	7

Per serving in this recipe

Calories	330
Fat (g)	4
Saturated fat (g)	1
Cholesterol (mg)	82
Fiber (g)	5
Sodium (mg)	90

BREAKFAST
French toast (2 slices whole-wheat bread, 2 egg whites, 1/4 cup skim milk, 1/4 tsp vanilla)—top with 1/4 tsp sugar and 1/4 tsp cinnamon
1/2 grapefruit
1 cup skim milk
Coffee—regular or decaffeinated

NOON
Spinach salad (2 cups fresh spinach, a sliced pear, 2 Tblsp red wine vinaigrette, 4 Tblsp sunflower seeds)
12 wheat crackers
1 cup skim milk

EVENING
Turkish Chicken With Spiced Dates (see recipe below)
1 cup cooked couscous
1 cup sautéed root vegetables (such as carrots, beets, potatoes, onions)
Herbal tea

SNACK (ANYTIME)
1/2 cup frozen yogurt—top with 1/2 cup pineapple
4 gingersnaps

DAY 9 RECIPE

Turkish Chicken With Spiced Dates
Serves 4

4 boneless, skinless chicken breasts (may also substitute 1 pound turkey breast slices)
Vegetable cooking spray
1 cup low-sodium chicken broth
1/2 cup onion, chopped

1 box (8 ounces) chopped dates or pitted dates that have been chopped
2 Tblsp apricot or peach fruit spread
1/2 tsp ground cinnamon
Parsley sprigs
4 fresh apricots, cut into wedges

Coat a large skillet with non-stick cooking spray. Cook chicken over medium heat about 5 minutes on each side or until chicken is no longer pink in the center. Remove chicken from skillet; cover and keep warm.

To the skillet, add broth, onion, and dates; bring to boil, stirring occasionally. Reduce heat to low; cook until liquid is reduced by half (about 8 minutes).

Stir apricot spread and cinnamon into sauce until blended; spoon over chicken. Garnish with parsley and apricots. Serve immediately.

DAY 10

BREAKFAST
1 cup wheat flakes—top with 1 peach, sliced
2 slices whole-wheat toast
2 Tblsp honey
1 cup skim milk
Coffee—regular or decaffeinated

NOON
Chicken, fresh pineapple, and black bean salad on spring greens (3 ounces grilled chicken, 1/2 cup pineapple, 1/2 cup black beans, 2 cups greens)
1 ounce (1 individual bag) baked tortilla chips (about 12)
1 cup skim milk

EVENING
1 bowl (2 cups) minestrone soup*
Bruschetta Pomodoro (see recipe below) on 2 slices of Italian or French bread
1/2 cup fresh vegetables (baby carrots, broccoli florets, celery)
2 Tblsp ranch dressing—for dip
1 apple
Herbal tea

SNACK (ANYTIME)
1 cup frozen yogurt
1/2 cup mixed berries

*Homemade or reduced-sodium variety is recommended. See recipe on page 140, Chapter 5.

Nutritional Analysis for Entire Day

	Actual	Goal
Calories	1,850	1,800-2,000
Fat (g)	23	Less than 60
Saturated fat (g)	5	Less than 16
Cholesterol (mg)	75	Less than 300
Fiber (g)	30	20-35
Sodium (mg)	2,350	Less than 2,400

Food Servings

	Actual
Sweets	2
Fats	1
Legumes/nuts	2
Meat, poultry, fish	3 ounces
Milk	3
Vegetables	4
Fruit	4
Grains	7

DAY 10 RECIPE

Bruschetta Pomodoro
Makes about 4 cups

7 to 9 Roma tomatoes, peeled, seeded, diced
3 to 4 garlic cloves, finely chopped
1/2 cup fresh basil, chopped
3 Tblsp virgin olive oil
1/4 tsp salt (use sea salt for different flavor)
Fresh cracked pepper, to taste

Combine all ingredients. Cover and refrigerate. Serve about 2 tablespoons over a slice of Italian or French bread.

Note: This also may be used as a topping over penne pasta.

Per serving (2 Tblsp) in this recipe

Calories	20
Fat (g)	1
Saturated fat (g)	trace
Cholesterol (mg)	0
Fiber (g)	trace
Sodium (mg)	65

DAY 11

Nutritional Analysis for Entire Day

	Actual	Goal
Calories	1,950	1,800-2,000
Fat (g)	44	Less than 60
Saturated fat (g)	8	Less than 16
Cholesterol (mg)	95	Less than 300
Fiber (g)	28	20-35
Sodium (mg)	2,120	Less than 2,400

Food Servings

	Actual
Sweets	1
Fats	3
Legumes/nuts	0
Meat, poultry, fish	3 ounces
Milk	4
Vegetables	5
Fruit	5
Grains	8

Per serving in this recipe

Calories	230
Fat (g)	1
Saturated fat (g)	0
Cholesterol (mg)	0
Fiber (g)	5
Sodium (mg)	55

BREAKFAST

Banana Raspberry Smoothie (see recipe below)
2 slices cinnamon toast (toasted whole-wheat bread, 1 tsp soft margarine, 1/2 tsp sugar, 1/2 tsp cinnamon)
Coffee—regular or decaffeinated

NOON

Lemon pepper chicken on rye sandwich (2 slices rye bread, 1 Tblsp fat-free mayonnaise, shredded lettuce, 3 ounce roasted chicken breast seasoned with lemon pepper)
1 ounce (individual bag) baked potato chips
2 plums
1 cup skim milk

EVENING

Pasta verde (2 cups cooked linguine, 2 tsp olive oil, 2 cups steamed green vegetables—asparagus tips, scallions, snow peas, broccoli florets). Top with 1 Tblsp Parmesan cheese
1 cup romaine lettuce
2 Tblsp fat-free red wine vinaigrette
1 cup (about 10) fresh cherries
Herbal tea

SNACK (ANYTIME)

1 1/2 ounces low-fat cheddar cheese
6 wheat crackers

DAY 11 RECIPE

Banana Raspberry Smoothie
Serves 2

1 1/2 cups pineapple juice
1 cup low-fat vanilla yogurt or frozen low-fat yogurt
1 cup raspberries
2 ripe medium bananas, peeled and cut into chunks

Combine juice, yogurt, raspberries, and bananas in blender. Cover; blend until smooth. Pour into glasses and serve.

ENCYCLOPEDIA
of FOODS

A GUIDE TO HEALTHY NUTRITION

contributions are worthy of special mention. Jennifer K. Nelson, R.D., M. Molly McMahon, M.D., and Robert A. Rizza, M.D., developed the material in Chapters 1 through 5. Kristine A. Kuhnert, R.D., contributed to Chapters 1 through 5 and served as project manager. Judith M. Ashley, Ph.D., R.D., provided the original draft for all of Part II (except the chapters on fruits and vegetables). Sydne J. Newberry, Ph.D., contributed to Chapters 1 and 2 and was the editor and major writer for the fruits and vegetables chapters. Dr. Rizza oversaw the entire process and served as one of the Editors-in-Chief. Other Editors-in-Chief were Vay Liang W. Go, M.D., M. Molly McMahon, M.D., and Gail G. Harrison, Ph.D., R.D.

Several professional staff of Dole Food Company, Inc., contributed substantially to this work. In particular, Lorelei DiSogra, Ed.D., R.D., contributed feedback and advice throughout the process of development; Roberta Wieman provided administrative and moral support at every step of the way; and David A. DeLorenzo provided a grounding in the real world of food production and marketing. Richard Utchell provided oversight of the photography. Donna Skidmore provided expert review of many of the chapters in Part II.

Susan Kaus Eckert, R.D., L.D., provided input to Chapters 3 and 4. Chapter 3 was reviewed by the following Mayo Clinic consultants: Michael D. Jensen, M.D., Sundeep Khosla, M.D., Timothy O'Brien, M.D., Sheldon G. Sheps, M.D., and William F. Young, Jr., M.D. Special thanks go to the following graduate students, faculty, and staff of the University of California Los Angeles (UCLA) for their contributions to the fruits and vegetables section of Part II: Elizabeth Chacko, Ph.D., Nativita M. Dhaiti, M.S., R.D., James Dinh, M.P.H., Roberto Garces, M.P.H., Helanie Hatter, M.P.H., M.A., Yun Kim, M.S., R.D., Leda Nemer, M.P.H., Heiu Ngo, M.P.H., Tuong I. Nguyen, M.P.H., James Pfeiffer, Ph.D., Karen Shih, M.P.H., Judith St. George, Abishek Tewari, Donna Winham, M.P.H., and Osman Galal, M.D., Ph.D. Appreciation is also extended to the research dietitians at UCLA Center for Human Nutrition: Pamela Saltsman, M.P.H., R.D., Shannon Duffy, M.P.H., R.D., Melissa Sherak Resnick, M.P.H., R.D., and Stacy Macris, M.P.H., R.D. Members of the UCLA Nutrition Education Committee provided input and review of sections of the book. Inkham Adams and Jolyn K. Gentemen, students from the University of Nevada Department of Nutrition, assisted with the nutrition tables.

Beverly Parker provided writing expertise during the early stages of book development. Jill Burcum, Anne Christiansen-Bullers, and Mike Dougherty helped with the writing of all chapters in Part II (except fruits and vegetables). In addition, the following staff were involved with the photography: food stylists were Susan Brosious, Sue Brue, Suzanne Finley, Robin Krause, Cindy Syme, and Abigail Wyckoff (also a prop stylist); food stylist assistants were Amy Peterson, Susan Tellen, and Teresa Thell; prop stylists were Michele Joy and Rhonda Watkins; photographers included Kevin Ross Hedden and Mette Nielson; production coordinator was Edward Fruin. Diane M. Knight provided her skills as a computer artist. Photographic separation was completed by Davies Printing Company, Rochester, Minnesota. Executive Chef Patrick Jamon, The Regency Club, Los Angeles, provided culinary expertise. Administration support was provided by Jonathan W. Curtright, Mayo Clinic.

ACKNOWLEDGMENTS

Editorial Staff

Editors-in-Chief	Robert A. Rizza, M.D.
	Vay Liang W. Go, M.D.
	M. Molly McMahon, M.D.
	Gail G. Harrison, Ph.D., R.D.
Associate Editors	Jennifer K. Nelson, R.D.
	Kristine A. Kuhnert, R.D.
Assistant Editor	Sydne J. Newberry, Ph.D.
Editorial Director	LeAnn M. Stee
Art Directors	Karen E. Barrie
	Kathryn K. Shepel
Medical Illustrators	John V. Hagen
	Michael A. King
Editorial Assistant	Sharon L. Wadleigh
Production Consultant	Ronald R. Ward
Photography	Tony Kubat

The vision for this book belongs to David H. Murdock, Chairman and Chief Executive Officer of Dole Food Company, Inc. Mr. Murdock brought his vision and a request for assistance in making it a reality to two of the authors: Robert A. Rizza, M.D., of Mayo Clinic, and Gail G. Harrison, Ph.D., R.D., University of California Los Angeles School of Public Health. They each saw the potential value in this vision and committed themselves to recruit scientific colleagues and technical expertise to bring it to fruition.

The authors all contributed to various stages of the evolution of this volume. Specific

published in the area of nutrition. We decided that Mr. Murdock's goal was achievable, gathered a team of enthusiastic and knowledgeable colleagues, and began to write. From the very beginning, it became obvious that although we all were alleged "experts," none of us knew everything (not surprising) and there was much we could learn from one another. That is when the fun started. We also gained a deep respect for Mr. Murdock, whose unwavering dedication to excellence, without regard for commercial interest, served as an inspiration to us all.

Good food and good nutrition can and should be synonymous. We hope you enjoy and benefit from this book.

R. A. Rizza, M.D.
V. L. W. Go, M.D.

PREFACE

Nutrition is important to all of us. What we eat has a profound effect on our health and our enjoyment of life. Although there is a large amount of valid scientific information dealing with various aspects of nutrition, there is, unfortunately, even more misinformation. The average person thus has difficulty separating fact from fiction.

A team of experts from Mayo Clinic, the University of California Los Angeles, and Dole Food Company, Inc., wrote this book. The team included physicians, nutrition scientists, and clinical nutritionists. The information has been subjected to rigorous peer review not only by the writing group but also by colleagues at our respective institutions who have special expertise in various aspects of the book.

The book seeks to answer three main questions: What am I eating? What should I eat? and Why? The premise of the book is that well-informed people make well-informed decisions. The theme of the book is moderation. The standard is that all recommendations be based on valid scientific evidence. If this is not possible, either because the evidence is not available or it is inconclusive at this time, then the text is so noted and our recommendations are tentative and based on the consensus of nutrition experts. Another premise of the book is that accurate information does not have to be boring. Most of us are curious about what is in the food we eat, where it comes from, and why one food is supposed to be good for us whereas too much of it may be bad.

The book is divided into two parts. Part I provides the reader with an overview of the principles of nutrition, including the basis for the Food Guide Pyramid and for nutrition recommendations, how various nutrients differ, and how our nutrition needs differ as we progress through the different stages of life. Part I also makes suggestions for menu planning, food preparation, and strategies for shopping, food storage, and food safety.

Part II complements Part I by providing information about individual foods and their nutrient content. The sections are organized according to the format of the Food Guide Pyramid. Part II begins with fruits, vegetables, and grains, foods that are at the bottom of the Pyramid and therefore should be the foundation of our food choices. Part II ends with foods that are at the top of the Pyramid and therefore should be eaten sparingly. The range emphasizes the extraordinary choices available to us all. Because of the sheer numbers of foods, those with similar nutrient contents are grouped, whereas those with unique nutrient content are described separately. Nutrient tables also are provided so the reader can gain a greater appreciation of which foods are particularly good sources of vital nutrients.

Writing a book can be both work and fun. In this instance, it was more of the latter. The book began as the vision of Mr. David H. Murdock, Chairman and Chief Executive Officer of Dole Food Company, Inc. Mr. Murdock and his colleagues at Dole have long been advocates of good nutrition. The editors and Mr. Murdock began with a series of conversations as to how the book should be organized and whether such a book would add anything to the large number of books already

TABLE OF CONTENTS

FOREWORD

I believe that knowledge is power. You can put the power of nutrition knowledge to work for you. This is the most important thing you can do to preserve and improve your mind and body. Good health is the key to longevity and provides the foundation that enables you to enjoy life.

The *Encyclopedia of Foods: A Guide to Healthy Nutrition* imparts the knowledge that eating a healthy diet can provide the various nutrients needed to maintain fitness and prevent the many common diseases that affect our health and longevity. Experts at Mayo Clinic and the UCLA Center for Human Nutrition have contributed their knowledge and experience to this book to improve the quality of life through proper nutrition. This book is a 4-year collaborative effort by a large team of experts from the medical profession and the field of nutrition.

It is very difficult to make up one's mind to eat properly and avoid the temptations and health consequences of consuming an excess amount of calories, fats, and refined sugar. Too much of these can detract from healthy living and the enjoyment of life. It is now clear from many studies that what you choose to eat can determine whether you have heart disease, diabetes, or many common forms of cancer. Extensive arrays of books have been written on this subject. With this book we have tried to encapsulate the guidelines for eating foods that are beneficial to the body and that preserve health and longevity.

I, along with most people, have not always been so concerned with health. When I became chairman of Dole Food Company 16 years ago, I truly began to understand the meaning of nutrition and the need for eating a well-balanced diet. A great deal of progress has been made in discovering the benefits to our health provided by fresh fruits and vegetables, whole grains, a healthy diet, and proper exercise and lifestyle. Dole, known as the largest distributor of fresh fruits and vegetables in the world, intends to take a leadership role in disseminating scientific information on the benefits of fruits and vegetables and other foods necessary for promulgating a healthy lifestyle. We intend to publish additional information as it is being developed by institutions throughout the world. We all have the opportunity to instill in our children the knowledge that will enable them to have the healthy life we wish them to enjoy.

The *Encyclopedia of Foods* is a practical guide and personal reference tool of food, nutrition, and health. Many physicians, doctors of philosophy, nutritionists, dietitians, researchers, writers, editors, designers, illustrators, and countless others have worked together to create a comprehensive reference book and present it in an attractive, useful, and friendly fashion.

I personally hope that you will read this book and use it to make the necessary changes in your lifestyle and diet to improve your health and longevity.

David H. Murdock

*Chairman of the Board and
Chief Executive Officer of
Dole Food Company, Inc.*

The *Encyclopedia of Foods: A Guide to Healthy Nutrition* provides practical and easy-to-understand
information on issues relating to good nutrition. This book supplements, but does not replace, the
advice of your personal physician and nutrition advisor, whom you should consult for individual
medical and nutrition issues. The authors of this book and their institutions do not in any case
endorse any company or product.

Academic Press
A Harcourt Science and Technology Company
525 B Street, Suite 1900, San Diego, California 92101-4495, USA
http://www.academicpress.com

Academic Press
Harcourt Place, 32 Jamestown Road, London NW1 7BY, UK
http://www.academicpress.com

Library of Congress Catalog Card Number: 2001093328

International Standard Book Number: 0-12-219803-4

PRINTED IN THE UNITED STATES OF AMERICA
01 02 03 04 05 06 WP 9 8 7 6 5 4 3 2 1

ENCYCLOPEDIA
of FOODS

A GUIDE TO HEALTHY NUTRITION

Prepared by medical and nutrition experts from Mayo Clinic,
University of California Los Angeles, and Dole Food Company, Inc.

Academic Press
San Diego, California

DAY 12

BREAKFAST
2 slices banana bread
1 tsp soft margarine
1/4 cantaloupe
1 cup skim milk

NOON
Bagel sandwich (1 whole-wheat bagel, 2 Tblsp fat-free cream cheese,
     sun-dried tomatoes, cucumbers)
1/2 cup raw vegetables (broccoli florets, celery sticks, green pepper strips)
2 Tblsp fat-free ranch dressing (for dip)
1 apple
1 cup skim milk

EVENING
Grilled Beef Kabobs (see recipe below)
1 cup quinoa with carrots and herbs
1/2 cup steamed spinach with lemon and garlic
2 slices grilled pineapple
Herbal tea

SNACK (ANYTIME)
1 1/2 ounces cheddar cheese
4 crispy rye wafers
6 ounces cranberry juice

Nutritional Analysis for Entire Day

	Actual	Goal
Calories	1,800	1,800-2,000
Fat (g)	24	Less than 60
Saturated fat (g)	7	Less than 16
Cholesterol (mg)	100	Less than 300
Fiber (g)	20	20-35
Sodium (mg)	2,000	Less than 2,400

Food Servings

	Actual
Sweets	0
Fats	3
Legumes/nuts	0
Meat, poultry, fish	3 ounces
Milk	3
Vegetables	6
Fruit	4
Grains	7

DAY 12 RECIPE

Grilled Beef Kabobs
Serves 4

12 ounces top sirloin beef steak
6 Tblsp fat-free Italian dressing
8 cherry tomatoes
2 green bell peppers, cored and seeded and cut into 1-inch chunks
1 onion, cut into 1-inch chunks

Cut meat into 1-inch cubes. Place meat in non-metallic dish, and pour half of the dressing over to coat. Cover; refrigerate for 20 minutes. Turn to coat evenly. Discard marinade. Thread skewers alternately with meat, cherry tomatoes, green pepper, and onions. Grill, basting with remaining fat-free dressing. Serve immediately when cooked.

**Per serving
in this recipe**

Calories	195
Fat (g)	5
Saturated fat (g)	2
Cholesterol (mg)	60
Fiber (g)	2
Sodium (mg)	280

DAY 13

Nutritional Analysis for Entire Day

	Actual	Goal
Calories	1,850	1,800-2,000
Fat (g)	40	Less than 60
Saturated fat (g)	7	Less than 16
Cholesterol (mg)	80	Less than 300
Fiber (g)	22	20-35
Sodium (mg)	2,250	Less than 2,400

Food Servings

	Actual
Sweets	1
Fats	3
Legumes/nuts	1
Meat, poultry, fish	3 ounces
Milk	3
Vegetables	4
Fruit	5
Grains	9

BREAKFAST
1 cup low-fat cottage cheese—top with 1 cup sliced peaches
2 slices toasted oatmeal bread
1 tsp soft margarine
Coffee—regular or decaffeinated

NOON
Rustic Wrap (see recipe below)
1 cup grapes
1 kiwi fruit
6 ounces pineapple juice

EVENING
Pasta with marinara sauce (1 cup linguine, 1 cup marinara sauce*)
    —top with 1 Tblsp Parmesan cheese
1 cup romaine lettuce
2 Tblsp fat-free Caesar dressing
1 sourdough roll—to soak up sauce
1 cup berries
Herbal tea

SNACK (ANYTIME)
4 cups popcorn
1 cup skim milk

*Reduced-sodium variety is recommended. See recipe on page 143, Chapter 5.

Per serving in this recipe

Calories	445
Fat (g)	8
Saturated fat (g)	1
Cholesterol (mg)	60
Fiber (g)	6
Sodium (mg)	140

DAY 13 RECIPE

Rustic Wrap
Serves 4

2 cups mushrooms, sliced
1/2 cup chopped green onions
1 Tblsp orange juice
2 cups cooked wild rice
12 ounces smoked turkey, shredded or chopped
1/2 cup cranberry relish
4 medium-sized tortillas (7-inch diameter)
1/2 cup chopped pecans

In a non-stick skillet, sauté mushrooms and onion over medium heat in 1 Tblsp orange juice. Add cooked wild rice and smoked turkey to warm through. Remove from heat. Spread tortillas with cranberry relish. Top each tortilla with equal amounts of turkey-rice mixture. Top with chopped pecans. Roll up and serve.

DAY 14

BREAKFAST

Breakfast parfait (put in layers: 1/4 cup wheat flakes, 1/4 cup low-fat yogurt, 1/4 cup raisins, 1/4 cup low-fat yogurt, 1/4 cup uncooked rolled oats, 1/4 cup low-fat yogurt, 1/4 cup dried apricots, 1/4 cup low-fat yogurt, sprinkle with nutmeg)

Coffee—regular or decaffeinated

NOON

Mediterranean Pasta Salad (see recipe below) on romaine lettuce leaves

1 crusty seven-grain roll

1 tsp soft margarine

1 cup lemon sherbet (top with 1 cup raspberries)

6 ounces apple juice

EVENING

3 ounces grilled salmon with lemon and dill

1 cup braised baby vegetables (baby carrots, pattypan squash, red potatoes)

1 whole-grain roll

1 tsp soft margarine

1 cup skim milk

SNACK (ANYTIME)

1 cup skim milk

2 medium chocolate chip cookies

Nutritional Analysis for Entire Day

	Actual	Goal
Calories	1,990	1,800-2,000
Fat (g)	40	Less than 60
Saturated fat (g)	7	Less than 16
Cholesterol (mg)	65	Less than 300
Fiber (g)	26	20-35
Sodium (mg)	2,225	Less than 2,400

Food Servings

	Actual
Sweets	1
Fats	2
Legumes/nuts	0
Meat, poultry, fish	5 ounces
Milk	3
Vegetables	4
Fruit	6
Grains	7

DAY 14 RECIPE

Mediterranean Pasta Salad

Serves 4

1 package (8 ounces) refrigerated or frozen cheese tortellini

3 cups broccoli cut into florets (may substitute cauliflower or use both)

1 can (8 ounces) pineapple chunks

2 Tblsp balsamic or red wine vinegar

1 Tblsp olive oil

1/4 pound fresh link turkey sausage, cooked, drained, sliced

1 medium red, yellow, or green bell pepper, cored, seeded, and cut into 1-inch pieces

Prepare tortellini as package directs. Add broccoli during last 2 minutes of cooking. Drain pineapple, reserve 1/4 cup juice. Combine reserved juice, vinegar, and oil in large serving bowl. Drain tortellini and broccoli. Add tortellini, broccoli, sausage, bell pepper, and pineapple to serving bowl. Toss well to coat evenly. Serve at room temperature or chilled.

Per serving in this recipe

Calories	276
Fat (g)	8
Saturated fat (g)	2
Cholesterol (mg)	28
Fiber (g)	2
Sodium (mg)	435

Preparing healthful meals can be fun.
In this chapter you will learn:

- Ways to modify recipes to be lower
  in calories, fat, sugar, and salt and yet
  be nutritious and flavorful
- Ways to create menus that bring
  healthier, plant-based foods — fruits,
  vegetables, and grains — to center stage
- How to prepare, serve, and store food safely

PREPARING
HEALTHFUL MEALS

Much has been learned about how to prepare healthful foods that are enjoyable, convenient to make, and economical. Many of the leading chefs of Europe and the United States have abandoned cooking styles that once depended on fats and oils and are now using healthier cooking methods. This "new" cuisine uses the cook's culinary skill to create delicious meals that bring fruits, vegetables, and grains to center stage.

Simple yet innovative techniques can be used to modify favorite recipes to maximize the nutritious value of the meal without jeopardizing its taste. When you modify an existing recipe, it is generally best to start slowly, making one change at a time. Persistence, willingness to experiment, and a few tried-and-true hints can help you prepare healthful and flavorful meals.

CHANGE IS GOOD

Recently, fat, sugar, and salt have been vilified for the roles they play in increasing the risk of certain diseases such as obesity, diabetes, coronary artery disease, and high blood pressure. However, they are only "bad" when eaten in excess. The key is not to banish them from the kitchen but to use them in moderation.

Fat provides flavor, substance, and a mouth-pleasing creamy texture. Sugar adds sweetness, crispness, tenderness, and color. Salt heightens the flavor of foods and is necessary in baked goods made with yeast.

The art of cooking is to put the proper amounts of these ingredients in each food. Recipe modification is one of the more useful cooking skills. In some instances, modification of the fat, sugar, or salt content actually can make the food tastier, moister, and more satisfying than it was originally.

When Should a Recipe Be Modified?

Sometimes it is difficult to know whether a recipe can be adjusted without sacrificing taste, texture, and appeal. Try modifying a recipe if you answer "yes" to any of the following questions:
• Is the recipe high in fat, sugar, or salt?
• Is this a food I eat frequently?
• Is this a food I eat in large amounts?

Keep in mind that not every recipe needs to be modified. If, for example, a certain high-fat dessert is a family favorite and it is prepared infrequently, there is no need to change it. As long as it is treated as an item from the top of the Food Guide Pyramid (see Chapter 1, page 11)—the occasional food—enjoy it in its familiar form.

Experiment

Because every recipe is different, experimentation is necessary. There are numerous ways to make a recipe healthier. Of course, not every experiment works. It may take several attempts to achieve the desired taste and consistency. Once the modified recipe meets your expectations, file it for future use.

As a start, try these five methods:
- Reduce the amount of fat, sugar, or salt.
- Delete a high-fat ingredient or seasoning.
- Substitute a healthier ingredient.
- Change the method used to prepare the recipe.
- Reduce the amount of meat in the recipe.

Can the Amount of an Ingredient Be Reduced?

Start by reducing the amount one ingredient at a time. In most baked goods, sugar generally can be reduced by one-third to one-half without substantially changing consistency or taste. Because sugar increases moisture, as a rule retain one-fourth cup of sugar, honey, or molasses for every cup of flour in baked goods. To maximize the sweetness of foods, when appropriate, serve the dish warm or at room temperature rather than cold. In addition, there are spices that can enhance sweetness. Some possibilities include cinnamon, cloves, allspice, nutmeg, and vanilla and almond extract or flavoring. Eliminating a cup of sugar in a recipe saves about 800 calories.

Fat also can be reduced by one-third to one-half in baked goods. Use puréed fruit or applesauce to replace the fat in a 1:1 ratio. For example, use one-half cup of oil plus one-half cup of unsweetened applesauce (instead of 1 cup of oil). Eliminating 1 cup of oil or fat saves about 2,000 calories and 225 grams of fat. Another way to decrease fat and cholesterol is to substitute egg whites or egg substitute for a whole egg. For every egg, use 2 egg whites or a quarter cup of egg

---

TIPS FOR USING SPICES AND HERBS

- Conversion: 1 tablespoon fresh herbs = 1 teaspoon dry = 1/4 teaspoon powdered
- Use sparingly: 1/4 teaspoon per pound of meat or pint of sauce. You can always add more.
- When doubling a recipe, add only 50 percent more seasoning.
- Crush or rub dry herbs between your fingers to enhance flavor before adding them to a recipe.
- In long-cooking entrées such as stews, add herbs toward the end of the cooking time.
- In chilled foods such as dips, salads, and dressings, add herbs several hours before serving.
- For maximal freshness, purchase herbs in small quantities and store in airtight containers away from light and heat.

---

substitute. With this replacement, approximately 5 grams of fat, 2 grams of saturated fat, 200 milligrams of cholesterol, and 60 calories are saved.

Reduce but do not totally remove salt because a small amount of salt frequently is required to facilitate the chemical reactions that occur during cooking. Salt is always required with yeast-leavened items. The cooler the food, the saltier it will taste. Try under-salting hot foods that will be chilled before serving. Using a half teaspoon of salt instead of 1 teaspoon in a recipe saves about 1,200 milligrams of sodium. (See sidebar: Tips for Using Spices and Herbs, this page.)

Can an Ingredient Be Omitted?

Determine whether any ingredients can be omitted. Sugar, fat, and salt are likely candidates because in many instances they are used mainly for appearance or by habit. To reduce sugar, omit candy coatings, sugary frostings, and syrups. Nuts, although nutritious, are high in fat and contribute significant calories. Additional condiments that add unwanted fat and calories include coconut, whipped cream, mayonnaise, butter, margarine, and sour cream. Pickles, catsup, olives, and mustard are low in calories. However, because these condiments are high in salt, persons who have high blood pressure or heart disease generally should limit their use.

Can a Substitution Be Made?

Substituting ingredients that are lower in fat, sugar, and salt can make a significant difference in a recipe. For example, use skim milk rather than whole or 2 percent milk. Products made from pureed prunes and apples or mashed bananas often can be used as a replacement for butter, margarine, or oil. These products also can be used in homemade baked goods or box mixes. (See sidebar: Choose These Alternatives to Reduce Fat, Sugar, and Salt, page 127.)

Be cautious when using fat-free spreads (such as fat-free margarine or cream cheese), "artificial" sweeteners, or salt substitutes in cooked foods. Most fat-free spreads contain a significant amount of water. This can change the outcome of the recipe by affecting its leavening or by leaving the food runny. Some sweeteners (such as aspartame) lose their sweetness when exposed to heat. Heat can make some salt substitutes (such as those containing potassium chloride) strong or bitter tasting. For these reasons, these products generally should be restricted to recipes that do not require cooking or are used as condiments when foods are served at the table. In most instances, success depends on patience and a bit of creativity. If one substitution does not yield the desired result, try again. Another substitute or a different amount of the same substitute may work better.

Would Another Cooking Method Be Healthier?

The choice of cooking technique is important. If the usual method of cooking uses fat, try grilling, broiling, braising, or roasting the food instead. Instead of deep-fat frying, try oven baking. French "fries" seasoned with chili powder or oregano are both tasty and low in fat when baked. Although stir "frying" can be a healthful cooking technique, use of generous amounts of oil negates some of the possible benefit. Always measure the oil to be used or, better yet, replace it with wine or a broth that adds flavor but little fat and only a few calories. Cooking food for the proper time (avoiding overcooking) not only makes it taste better but also preserves nutrients.

CHOOSE THESE ALTERNATIVES TO REDUCE FAT, SUGAR, AND SALT

Try these ideas for decreasing fat, sugar, and salt when preparing or eating foods.

FOR FATTY FOODS

Choose:	Instead of:
Two percent, 1 percent, or skim milk; low-fat or fat-free yogurt; low-fat or fat-free sour cream or cheese	Full-fat milk, yogurt, sour cream, or cheese
Lean, trimmed cuts of the loin and round; remove the skin from poultry before eating	Fatty and highly marbled meat
Applesauce or fruit purée in place of half the oil or shortening that is normally used; use the rest of the fat as instructed	Shortening, butter, margarine, or oil in baked goods
Puréed vegetables (carrots or potatoes), mashed potato flakes, or puréed tofu as a thickening agent	Creamed soups and gravy-based stews
Wine, fruit juice, broth, or balsamic vinegar	Oil-based marinades
Vegetable spray	Butter, oil, or margarine to prevent sticking
Two egg whites or egg substitute	A whole egg in a recipe
Roasted garlic; jam, jelly, or honey (although high in sugar, they have half the calories of butter or margarine and no fat)	Butter or margarine on bread or crackers
Salsa, low-fat sour cream with chives, low-fat cottage cheese, yogurt, herbs, or spices on baked potatoes	Butter or sour cream
Fat-free mayonnaise or salad dressing, mustard, cranberry sauce, chutney	Mayonnaise or butter on a sandwich

FOR FOODS WITH A HIGH SUGAR CONTENT

Choose:	Instead of:
Fruit purée, chopped fresh fruit, or applesauce	Sugar, syrup, or honey
Fruit canned in its own juice, fresh fruit	Sweetened fruit

FOR FOODS WITH A HIGH SALT CONTENT

Choose:	Instead of:
Lower-sodium versions	Soups, sauces (barbecue, soy, tartar, cocktail), canned meat or fish, and crackers
Herbs, spices, or marinades	Salt

Many cooking techniques make it possible to prepare more colorful, flavorful, and healthier dishes. These include the following:

Braising—Food is browned, then cooked in a tightly covered pan in a small amount of liquid at low heat over a long period.

Broiling—Food is placed beneath the heat source; basting may be needed.

Grilling—Food is positioned above the heat source; basting may be needed.

Microwaving—This is a quick way to cook food with little added liquid or fat.

Poaching—Food is cooked in a liquid at the simmering point.

Oven roasting—Food is cooked in an uncovered pan by the free circulation of dry air, until the exterior is well browned.

Steaming—Food is placed on a rack in a basket above boiling liquid. Food should not touch the liquid.

Stir frying—Small pieces of food are cooked over high heat and constantly stirred. Use wine, broth, or fruit juice as the liquid instead of the traditional oil.

Marinating food adds flavor and does not have to add fat. Some tips for successful marinating include piercing large cuts of meat, poultry, or fish with a fork to help the marinade permeate the food. Always marinate food in a glass or ceramic dish in the refrigerator. Never place the food in a metal container. Most marinades feature an acid base that may react with metal and change the flavor. Finally, a food safety tip: reserve some of the marinade before you put meat in it. Marinade that has been in contact with raw meat should not be used to baste meat, poultry, or fish during the last 15 minutes of cooking. If you plan to use leftover marinade as a table sauce, it must be boiled for 5 minutes to eliminate bacteria.

There is more that can be done once the food is out of the oven or off the stove. Skim the fat off pan juices, stews, and soups. Instead of topping vegetables with butter or margarine, sprinkle them with lemon juice or herbs. Remove any visible fat (and any skin from poultry) before serving.

No special, expensive equipment is needed to cook healthful foods. A good set of non-stick pans, a skillet, a roasting pan, a baking sheet, measuring cups and spoons, and sharp knives are enough to get you started. Quality, durability, ease of use, and cost should be the primary considerations when outfitting a kitchen.

Can the Amount of Fruits, Grains, and Vegetables in the Recipe Be Increased?

Increasing the amount of vegetables, grains, and fruits in a recipe can both improve taste and increase the nutritional quality of the food. For example, when cooking a soup or stew, use three times as many vegetables (by measure) as meat. Add generous portions of mushrooms, tomatoes, broccoli, and green pepper to pizza. Make pizza even lower in fat and calories by omitting or decreasing the cheese. Alternatively, choose a lower-fat cheese, such as mozzarella (made from skim milk), and use less of it. If possible, eliminate the meat or add only a small amount of lean meat. If you are making your own crust, make it thin and use whole-grain flour. Pizza, if served with a salad and eaten in moderation, can be an enjoyable and nutritious meal.

To get at least 5 servings of fruits and vegetables a day, add them to foods that do not typically include these ingredients. For example, add chopped pieces of fruit or vegetables to rice, add fruit toppings to toast or pancakes, or top meats with chopped vegetables. For each serving of meat (a serving is 2 to 3 ounces of meat), try to eat at least 1 serving each of fruit, vegetables, and grains. When possible, start your meal with a healthful salad. This often helps you decrease the amount of high-calorie food that you eat later in the meal. The more servings of grains, vegetables, or fruit, the better, because these are both filling and high in nutrients (see sidebar: Nutrition Boosters, page 129).

CREATING HEALTHFUL MENUS

Plan menus so that each meal complements what you plan to eat later in the day or what you have eaten earlier in the day. Include plant-based entrées as often as possible. Examples that can result in satisfying meatless meals include pasta with marinara sauce and lots of vegetables, stir-fried vegetables with tofu over rice, or lentil soup with a side dish containing grains, beans, or vegetables. Plant-based entrées can be tasty, filling, and nutritious.

When you choose to eat meat, fish, or poultry, remember that your goal is to eat 6 ounces or less per day. If you ate meat for lunch, appropriately decrease your dinner portion. Avoid red meats that contain a large amount of fat. Instead, emphasize poultry or fish. When you eat red meat, choose a "choice" grade and a cut from the loin or round, because these generally are the leanest types of meat. The skin on poultry holds in moisture and flavor during cooking.

However, the skin is high in fat and calories. Contrary to popular belief, the skin does not need to be removed before cooking. There is minimal fat absorption if the skin is left on. Just make sure to remove the skin before the poultry is eaten.

Many fish are low in fat. Those that are not low in fat generally contain omega-3 fatty acids that may help prevent heart disease. However, remember, all fats are high in calories, so the less fat added during cooking, the better.

Condiments and sauces can add nutrition and enhance flavor. Keep in mind, however, that some are high in fat, sodium, and calories (see Chapter 4, "On the Side," page 86). An example of a high-fat sauce is gravy over mashed potatoes. Instead, try sprinkling mashed potatoes with garlic or other herbs.

When choosing a topping, look for a lower-fat alternative. If none are available, then use less of the original topping. Sliced, chopped, or puréed vegetables can make a nourishing low-fat condiment. Fruits are a delicious complement to almost any meal. They can top meats, enrich salads, or be served for dessert.

A dessert can be a pleasant end to a healthful meal. However, a dessert should not be an "extra." Be sure it is included in your overall meal plan. Make the dessert a bonus by emphasizing fruit, whole grains, and lower-fat items. If you do not have a recipe that emphasizes fruits and whole grains, look for one that can be readily modified. Sorbets and low-fat frozen yogurts or ice creams are good choices. Even cookies, pies, cakes, and chocolates have their place. However, remember, because these desserts generally are high in fat and sugar, they are at the pinnacle of the Food Guide Pyramid. Therefore, they should be the exception rather than the rule. If you plan to eat a dessert, take a small portion. If you are preparing a dessert for a special occasion, make just enough to serve you and your guests. Leftover dessert is a powerful temptation.

With a little thought and planning, you know what foods to emphasize and what foods to limit. You are ready to make a commitment to improve the way you and your family eat. Now it is time to put your plan into action. To help you get started, examples of "makeovers" for breakfast and noon and evening meals are shown on the following pages.

NUTRITION BOOSTERS

There are many ways to help you get more fruits, vegetables, and grains in the foods you eat. Try these ideas.

Instead of:	Substitute:
Syrup on pancakes	Sliced fruit or a purée of a favorite fruit such as apples, berries, pineapple, or peaches
Sugar on breakfast cereal	Fresh fruit or low-fat yogurt

With or in:	Use:
Chicken breast	Kabobs with fruit and vegetables
Traditional or cornbread stuffing	Whole-grain breads, cooked apples, raisins or other dried fruit
Rice	Diced vegetables, dried fruit, or other grains or legumes
Casseroles	Decrease meat and add more vegetables, grains, bran, legumes, or dried fruit
Meat-based stews and soups	Decrease meat, and add more vegetables, legumes, or grains
Pasta	In place of meat sauce, use a meatless tomato sauce and add steamed vegetables

Pizza with less meat and cheese and more vegetable or fruit toppings is enjoyable and nutritious.

BREAKFAST—MAKEOVER 1

Original Meal	Modified Meal
6 ounces apple juice	1 nectarine and 1/2 cup raspberries
1 spiced muffin	**1 spiced muffin**
2 tsp butter	2 tsp marmalade
1 cup 2% milk	1 cup fat-free yogurt
Coffee	Herbal tea
2 tsp cream	
1 tsp sugar	

Meal analysis: 565 calories, 29 g fat, 12 g saturated fat, 77 mg cholesterol, 1 g fiber, 395 mg sodium

Meal analysis: 420 calories, 8 g fat, 1 g saturated fat, 4 mg cholesterol, 7 g fiber, 320 mg sodium

SPICED MUFFINS

Original Recipe	Modified Recipe
2 cups wheat flour	1 3/4 cups wheat flour
1 1/4 cups sugar	3/4 cup sugar
2 tsp baking soda	2 tsp baking soda
1 tsp cinnamon	2 tsp cinnamon
1/2 tsp salt	1/4 tsp salt
3 large eggs	cholesterol-free egg substitute (equivalent to 4 eggs)
1 cup vegetable shortening, melted	1/2 cup canola oil
	1/2 cup unsweetened applesauce
1/2 cup coconut	(Omit coconut)
1 tsp vanilla	2 tsp vanilla
2 cups peeled and chopped apples	2 cups chopped apples (unpeeled)
1/2 cup raisins	1/2 cup raisins
1/2 cup grated carrots	3/4 cup grated carrots
1/2 cup chopped pecans	2 Tblsp chopped pecans

Preheat oven to 350° Fahrenheit. Spray muffin tin with non-stick spray or use paper muffin liners. In a large bowl, mix together the flour, sugar, baking soda, cinnamon, and salt. Mix well. In another bowl, combine the egg substitute, oil, applesauce, and vanilla. Stir in the apples, raisins, and carrots. Add to the flour mixture and stir until just blended. Spoon batter into muffin tins, filling 2/3 full. Sprinkle with chopped pecans and bake for 35 minutes or until springy to the touch. Let cool for 5 minutes, then remove from the pan to a rack and let cool completely. *Note: These freeze well and may be rewarmed before serving.* **Yield: 18 small muffins**

Recipe Analysis (per muffin)

	Original	Modified
Calories	270	175
Fat (g)	15	7
Saturated fat (g)	4	1
Cholesterol (mg)	35	Trace
Fiber (g)	1	2
Sodium (mg)	185	165

BREAKFAST—MAKEOVER 2

Original Meal	Modified Meal
6 ounces pineapple juice	1/2 cup fresh pineapple
1/2 cup granola with 1 cup 2% milk	**1/2 cup reduced-fat granola with 1 cup fat-free plain yogurt**
2 slices white toast	2 slices whole-wheat toast
2 tsp butter	2 tsp jelly
Coffee	Hazelnut-flavored coffee
2 tsp cream	
Meal analysis: 731 calories, 27 g fat, 12 g saturated fat, 42 mg cholesterol, 6 g fiber, 500 mg sodium	*Meal analysis: 560 calories, 7 g fat, 1 g saturated fat, 4 mg cholesterol, 14 g fiber, 500 mg sodium*

GRANOLA WITH RAISINS, APPLES, AND CINNAMON

Original Recipe	Modified Recipe
4 cups old-fashioned oat cereal	3 cups old-fashioned oat cereal
1 cup bran cereal	3/4 cup bran cereal
1 cup slivered almonds	1/4 cup slivered almonds (toasted)
1 cup coconut	3/4 cup dried apple pieces
1 cup raisins	1/2 cup golden raisins
1/3 cup honey	1/4 cup honey
1/4 cup oil	1/4 cup unsweetened applesauce
1 Tblsp vanilla	1 Tblsp vanilla extract
	1 Tblsp cinnamon

Preheat oven to 350° Fahrenheit. Place oat and bran cereals into a large bowl. Toss well. In a small bowl combine the honey, applesauce, vanilla, and cinnamon. Pour over the oat mixture and toss. Don't break clumps apart. Pour onto non-stick baking sheet, spread evenly, and bake for about 25 to 30 minutes, stirring occasionally. Remove when golden brown. Combine almonds, apple pieces, and raisins and stir into hot oat mixture. Cool completely. Store in an airtight container.
Yield: 11 half-cup servings

Recipe Analysis (per half cup)

	Original	Modified
Calories	300	235
Fat (g)	11	4
Saturated fat (g)	3	1
Cholesterol (mg)	0	0
Fiber (g)	5	8
Sodium (mg)	40	38

BREAKFAST—MAKEOVER 3

Original Meal	Modified Meal
6 ounces orange juice	6 ounces orange juice
Southwestern scramble	**Southwestern scramble**
2 slices white toast	1 piece cornbread
2 tsp butter	1 Tblsp honey
1 cup 2% milk	1 cup skim milk
Coffee	Chicory-flavored coffee
2 tsp cream	
1 tsp sugar	

Meal analysis: 845 calories, 47 g fat, 25 g saturated fat, 540 mg cholesterol, 2 g fiber, 1,400 mg sodium

Meal analysis: 460 calories, 11 g fat, 2 g saturated fat, 120 mg cholesterol, 4 g fiber, 600 mg sodium

SOUTHWESTERN SCRAMBLE

Original Recipe	Modified Recipe
1/2 cup diced lean ham	1/2 cup green and red bell pepper, diced
	1/4 cup mushrooms, sliced
4 whole eggs	1 whole egg plus 3 egg whites
2/3 cup cheddar cheese, shredded	2 Tblsp cheddar cheese, shredded
	1/2 cup salsa

Spray a heavy skillet with cooking spray. Place over medium heat and cook vegetables until tender; remove vegetables and keep warm. Combine whole egg and egg whites. Pour into skillet and scramble over medium heat until set. Spoon onto plates. Top with cooked vegetables and sprinkle with cheese. Serve with salsa. **Yield: 2 servings**

Recipe Analysis (per serving)

	Original	Modified
Calories	410	120
Fat (g)	31	6
Saturated fat (g)	16	2
Cholesterol (mg)	500	114
Fiber (g)	trace	1
Sodium (mg)	940	380

NOON MEAL—MAKEOVER 1

Original Meal	Modified Meal
Tuna salad sandwich 1 ounce (individual bag) potato chips 12 ounces cola	**Curried tuna salad with pita triangles** 1/2 cup grapes 1 cup skim milk
Meal analysis: 925 calories, 59 g fat, 12 g saturated fat, 40 mg cholesterol, 3 g fiber, 970 mg sodium	*Meal analysis:* 300 calories, 2 g fat, trace saturated fat, 34 mg cholesterol, 6 g fiber, 415 mg sodium

TUNA SALAD SANDWICH VS CURRIED TUNA SALAD WITH PITA TRIANGLES

Original Recipe (sandwich)	Modified Recipe (salad)
1 can (6 ounces) oil-packed tuna	1 can (6 ounces) water-packed tuna, drained
1/2 cup diced celery	1/2 cup diced celery
	3/4 cup chopped apple
	1/4 cup raisins
	2 Tblsp thinly sliced green onion
1 tsp lemon juice	1 tsp lemon juice
1 cup mayonnaise	1/2 cup fat-free mayonnaise
	1/2 tsp curry powder
	1/4 tsp garlic powder
	dash of cayenne (red) pepper, if desired
4 lettuce leaves	4 lettuce leaves
8 slices bread	2 whole-wheat pita bread rounds (about 6-inch diameter)

In a small bowl, flake the tuna. Add the celery, apple, raisins, and green onion. In a separate bowl, combine the lemon juice, mayonnaise, and spices. Add the tuna mixture and combine. Serve on lettuce leaves along with pita bread that has been cut into triangles. **Yield: 4 servings**

Recipe Analysis (per serving)

	Original	Modified
Calories	625	165
Fat (g)	50	1
Saturated fat (g)	10	trace
Cholesterol (mg)	40	30
Fiber (g)	2	5
Sodium (mg)	800	288

NOON MEAL—MAKEOVER 2

Original Meal	Modified Meal
Spinach salad with bacon and mushrooms 2 bread sticks 1 cup 2% milk	**Citrus spinach salad with honey yogurt dressing** 1 slice sourdough bread 1 Tblsp honey 1 cup skim milk
Meal analysis: 640 calories, 29 g fat, 6 g saturated fat, 25 mg cholesterol, 4 g fiber, 1,460 mg sodium	*Meal analysis:* 425 calories, 3 g fat, 1 g saturated fat, 5 mg cholesterol, 7 g fiber, 610 mg sodium

SPINACH SALAD WITH BACON AND MUSHROOMS
VS
CITRUS SPINACH SALAD WITH HONEY YOGURT DRESSING

Original Recipe	Modified Recipe
6 cups spinach leaves	6 cups spinach leaves
6 strips bacon	1 cup fresh orange segments
1/2 pound sliced mushrooms	1 tart apple, thinly sliced
3 thin slices sourdough (croutons)	1 small red onion, thinly sliced
2 Tblsp sugar	1 cup plain, nonfat yogurt
2 Tblsp cider vinegar	2 Tblsp honey
1/2 cup olive oil	
2 ounces brandy	

Wash and dry spinach, remove stems. Arrange spinach, orange segments, apple, and onion slices onto plates. In a small bowl, combine the nonfat yogurt and honey. Whisk until smooth. Spoon over salads. **Yield: 6 servings**

Recipe Analysis (per serving)

	Original	Modified
Calories	306	110
Fat (g)	22	trace
Saturated fat (g)	3	trace
Cholesterol (mg)	5	trace
Fiber (g)	2	4
Sodium (mg)	243	134

NOON MEAL—MAKEOVER 3

Original Meal	Modified Meal
1 cup minestrone soup	**1 cup minestrone soup**
4 soda crackers	6 multigrain crackers
2 cups tossed salad	1 cup fresh fruit mixed with mint
2 Tblsp Italian dressing	1 cup skim milk
2 chocolate chip cookies	
1 cup 2% milk	

Meal analysis: 815 calories, 47 g fat, 15 g saturated fat, 65 mg cholesterol, 6 g fiber, 2,095 mg sodium

Meal analysis: 400 calories, 8 g fat, 1 g saturated fat, 9 mg cholesterol, 11 g fiber, 534 mg sodium

MINESTRONE SOUP

Original Recipe	Modified Recipe
3 Tblsp olive oil	1 Tblsp olive oil
1/2 cup onion, chopped	1/2 cup onion, chopped
1/2 cup ham, diced	1/3 cup celery, diced
1 carrot, diced	1 carrot, diced
1 clove garlic, minced	1 clove garlic, minced
1 quart chicken broth	1 quart defatted, reduced-sodium chicken broth
1 can (14 ounces) tomatoes, chopped	2 large fresh tomatoes, seeded and chopped
1/2 cup spinach, chopped	1/2 cup spinach, chopped
1/2 cup canned kidney beans	1 can (16 ounces) chickpeas or red
1 cup grated Parmesan cheese	kidney beans, drained and rinsed
	1 small zucchini, diced
	1/2 cup dry small-shell pasta
	2 Tblsp fresh basil, chopped

In a large saucepan over medium heat, cook the onion, celery, and carrot in the olive oil until softened. Add garlic and continue cooking for another minute. Add broth, tomatoes, spinach, chickpeas or kidney beans, and pasta. Bring to a boil over high heat. Reduce heat and simmer for 10 minutes. Add zucchini. Cover and cook for 5 minutes. Stir in basil and serve.
Yield: 4 servings

Recipe Analysis (per serving)

	Original	Modified
Calories	380	190
Fat (g)	22	4
Saturated fat (g)	8	trace
Cholesterol (mg)	33	5
Fiber (g)	3	8
Sodium (mg)	2,500	400

EVENING MEAL—MAKEOVER 1

Original Meal	Modified Meal
Spaghetti with meatballs 2 cups romaine lettuce 2 Tblsp Italian dressing 1 hard roll 1 tsp butter 1 cup ice cream Herbal tea	**Pasta with marinara sauce and grilled vegetables** 2 cups romaine lettuce 2 Tbsp fat-free Italian dressing 1 whole-grain roll (to soak up sauce) Frozen grapes with toasted pecans 4 ounces red wine
Meal analysis: 1,246 calories, 68 g fat, 28 g saturated fat, 165 mg cholesterol, 4 g fiber, 1,580 mg sodium	*Meal analysis:* 355 calories, 15 g fat, 2 g saturated fat, 5 mg cholesterol, 10 g fiber, 830 mg sodium

SPAGHETTI WITH MEATBALLS
VS
PASTA WITH MARINARA SAUCE AND GRILLED VEGETABLES

Original Recipe	Modified Recipe
1 quart ready-made spaghetti sauce	2 Tblsp olive oil
1 pound hamburger	10 large, peeled, diced, fresh tomatoes
1/2 tsp garlic salt	1 tsp salt
1/2 cup onion	1/2 tsp garlic, minced
8-ounce package of spaghetti	2 Tblsp onion, chopped
	1 Tblsp fresh basil, chopped (1 tsp dried)
	1 tsp sugar
	1/2 tsp oregano
	black pepper, to taste
	2 red peppers, sliced into chunks
	1 yellow summer squash, sliced lengthwise
	1 zucchini, sliced lengthwise
	1 sweet onion, sliced into 1/4-inch-wide rounds
	2 bundles fresh garlic, halved
	8-ounce package of whole-wheat spaghetti

Heat oil in a heavy skillet. Add tomatoes, salt, minced garlic, onion, basil, sugar, oregano, and black pepper. Cook slowly, uncovered, for 30 minutes or until sauce is thickened. In the meantime, brush peppers, squashes, onion, and fresh garlic with oil. Place under broiler and cook, turning frequently until browned and tender. Remove to a covered bowl. Keep warm. Cook spaghetti according to package directions. Drain well and portion onto plates. Cover with equal amounts of sauce. Top with equal amounts of vegetables. Serve immediately. **Yield: 4 servings**

Recipe Analysis (per serving)

	Original	Modified
Calories	535	270
Fat (g)	24	6
Saturated fat (g)	9	1
Cholesterol (mg)	66	0
Fiber (g)	1	4
Sodium (mg)	1,045	380

EVENING MEAL—MAKEOVER 2

Original Meal	Modified Meal
Fried chicken	**Balsamic roasted chicken**
1 cup white rice	1 cup roasted vegetables and fruit (new potatoes, onions, pears)
1 cup green beans	2 cups tossed greens
1 dinner roll	2 Tblsp low-fat red wine vinaigrette
1 tsp butter	1 slice crusty Italian or French bread
1 piece apple pie	1 sliced fresh peach sprinkled with nutmeg
1 cup 2% milk	1 cup skim milk
Meal analysis: 1,455 calories, 55 g fat, 17 g saturated fat, 220 mg cholesterol, 7 g fiber, 1,548 mg sodium	*Meal analysis:* 810 calories, 27 g fat, 4 g saturated fat, 195 mg cholesterol, 9 g fiber, 410 mg sodium

FRIED VS BALSAMIC ROASTED CHICKEN

Original Recipe	Modified Recipe
1 4-pound whole chicken, cut into pieces	1 4-pound whole chicken
1/2 cup flour	
1 tsp salt	
1/2 tsp cracked pepper	1 Tblsp fresh rosemary (1 tsp dried)
	1 garlic clove
1/4 cup vegetable oil	1 Tblsp olive oil
	black pepper
	4 sprigs fresh rosemary
	1/4 cup balsamic vinegar
	1/2 tsp brown sugar

Preheat oven to 350° Fahrenheit. Rinse chicken inside and out with cold running water. Dry it with paper towels. Mince together the rosemary leaves and garlic. Loosen the skin from the flesh, then rub the flesh with the oil and then the herb mixture. Sprinkle with black pepper. Put two fresh rosemary sprigs into the cavity of the chicken. Truss the chicken. Place it in a roasting pan and roast for 20 to 25 minutes per pound (about 1 hour and 20 minutes). Baste frequently with pan juices. When browned and juices run clear, transfer the chicken to a serving platter. In a small saucepan, combine the vinegar and the brown sugar. Heat until warm—do not boil. Carve the chicken (remove skin). Top it with the vinegar mixture. Garnish with remaining rosemary. **Yield: 4 servings**

Recipe Analysis (per serving)

	Original	Modified
Calories	603	432
Fat (g)	31	16
Saturated fat (g)	7	3
Cholesterol (mg)	192	192
Fiber (g)	0	0
Sodium (mg)	693	163

EVENING MEAL—MAKEOVER 3

Original Meal	Modified Meal
8-ounce grilled steak	**Steak with steamed vegetables**
1 medium baked potato	**and soba noodles**
2 Tblsp sour cream	1 seven-grain roll
1 cup green peas and onions	1 Tblsp honey
1 dinner roll	1 star fruit, sliced over 1/2 cup sherbet
1 tsp butter	2 fortune cookies
1 piece frosted devils' food cake	Green tea
Coffee—regular or decaffeinated	

Meal analysis: 1,480 calories, 64 g fat, 32 g saturated fat, 230 mg cholesterol, 5 g fiber, 1,203 mg sodium

Meal analysis: 900 calories, 24 g fat, 9 g saturated fat, 70 mg cholesterol, 14 g fiber, 1,040 mg sodium

GRILLED STEAK
VS
STEAK WITH STEAMED VEGETABLES, SOBA NOODLES, AND GINGER SAUCE

Original Recipe
4 8-ounce steaks
8 Tblsp steak sauce

Modified Recipe
1 12-ounce loin steak
1/2 pound soba noodles
2 cups fresh asparagus cut into 1-inch segments
2 cups broccoli florets

Sauce
1/2 cup reduced-sodium soy sauce
1/3 cup rice wine vinegar
1 Tblsp sesame oil
1 1/2-inch piece fresh ginger, peeled, grated
1 tsp sugar
cracked black pepper, to taste

Cook steak —grill, broil, or fry in a non-stick pan until medium rare. Set aside on covered platter to keep warm. Cook soba noodles according to package directions. While noodles are cooking, steam the vegetables until tender crisp. Combine sauce ingredients, heat through. Drain soba noodles, rinse, and redrain. Toss vegetables with the noodles. Place onto plates. Slice steak across grain into thin strips. Arrange on top of vegetables and noodles. Top with sauce. Serve immediately.
Yield: 4 servings

Recipe Analysis (per serving)

	Original	Modified
Calories	790	495
Fat (g)	44	19
Saturated fat (g)	22	8
Cholesterol (mg)	175	58
Fiber (g)	0	58
Sodium (mg)	900	775

FOOD SAFETY

You have learned how to select healthful foods, to modify recipes appropriately, and to make attractive and good-tasting meals. The final step is to ensure that the food you serve is safe to eat.

Bacteria in the Kitchen

Approximately 7 million cases of food poisoning are reported every year in the United States. Many other cases are mistaken for stomach flu or some other infection and therefore are never reported. Food poisoning can be a serious and potentially fatal illness. Fortunately, such severe cases are rare. Bacterial contamination of food can occur if food is handled improperly. Thus, food safety is of paramount importance.

Kitchens are replete with chances for passing along the bacteria (germs) that cause food poisoning. It is the responsibility of the person preparing the meal to make certain that foods and utensils are washed properly. Unclean kitchen utensils can promote food poisoning by growing unwelcome bacteria (see sidebar: Sources of Bacteria, below).

SOURCES OF BACTERIA

BACTERIA SOURCE	SOLUTION
Cutting boards	Keep two on hand—one for meat and the other for produce or breads*
Sponges, dish cloths, and towels	Change and wash frequently
Knives and utensils	Use separate knives and utensils for raw and cooked foods. Wash all utensils with hot water and soap
Countertops	Wash frequently with soap and hot water, particularly after working with meat, poultry, and fish

*A simple cleaning solution that helps to keep bacteria in check is to mix 1 tablespoon of bleach to 1 gallon of water. Generously spray the surface and let it stand for several minutes. Rinse and dry with a clean towel.

Hand Washing

Sometimes in the rush to prepare meals, it is easy to overlook one of the simplest and most important rules in food preparation: wash your hands before handling any food. Bacteria tend to accumulate on your hands, especially around the cuticles and under the fingernails. To actually kill the bacteria, it would take water so hot that it would harm your skin. At least 10 seconds of vigorous rubbing with soap or detergent and warm water is required to rid your hands of germs. You also should wash your hands during meal preparation if they become contaminated by the food you are handling.

Cross-Contamination

Cross-contamination can lead to food poisoning. If uncooked food has been on a plate or cutting board, that plate or cutting board could transfer a potentially infectious agent to any other food that comes in contact with it. Therefore, always use separate utensils, plates, and cutting boards for raw and cooked foods.

Ensuring Food Safety

Food safety begins as soon as you purchase the food. Ideally, perishable foods should be promptly taken home and immediately refrigerated or frozen. However, if you need to make a stop before reaching home, plan to store meat, fish, poultry, and dairy products in a cooler on ice. Always observe the refrigeration recommendations on packaged foods. To decrease the total amount of bacteria found on raw chicken and other poultry, thoroughly rinse, inside and out, under cold water. After a complete rinse, use hot water and soap to wash out the sink. Before freezing meat, poultry, or fish, divide it into the portion size that you will need to prepare one meal. When you need to cool a food that you have cooked, quickly transfer it to a shallow container. Cover it and refrigerate it immediately. Bacteria thrive at temperatures between 40° and 140° Fahrenheit, potentially doubling in number every 20 to 30 minutes. Therefore, the most important food safety rule in the kitchen is: Keep hot foods hot and cold foods cold.

Defrosting Food

Thaw poultry, fish, and meat in the refrigerator. Defrosting at room temperature promotes thawing on the outside while the core remains frozen. The soft outer portion provides a fertile site for bacterial growth. Instead, put frozen food in the refrigerator (which is cool but above freezing)

1 or 2 days before it is to be used. For faster thawing, run cold water over the item or use a microwave for quick defrosting.

Marinade Savvy

Marinate poultry, seafood, and meat in the refrigerator. To play it safe, set some of the marinade aside (to use for basting or as a sauce at the table) before adding it to the raw meat. Avoid using the liquid that the raw meat has been marinating in for basting. If you do, discontinue basting at least 15 minutes before the meat is done so that the marinade can be heated to a high enough temperature to kill any bacteria that may be present. Do not use the leftover marinade as a sauce unless it has not come in contact with the raw meat or you have boiled it for at least 5 minutes.

Cooking Food

Always be sure to cook recipes at the appropriate temperature. Cooking foods to an internal temperature of at least 160° Fahrenheit kills most dangerous bacteria. Uncooked or undercooked meat can harbor pathogens such as the notorious *E. coli* bacteria (see sidebar: Cooking It Safe, this page).

Using Slow Cookers

Using a slow cooker is a popular way of preparing soups, stews, roasts, and other hearty dishes. Because this device cooks at relatively low temperatures—compared with the oven or stovetop—it is vital to exercise safe cooking habits. For example, thaw meat thoroughly and cut it into small pieces. Use recipes that call for plenty of liquid. Bring to a boil quickly and then reduce heat to simmer. Do not overfill the cooker. Be sure to use a thermometer to make certain the temperature stays at 160° Fahrenheit or higher.

SERVING SAFELY

After taking care to prepare and cook food as safely as possible, don't contaminate it while it is being served. Here are a few tips. Avoid letting cooked foods cool on the table. Do not allow foods that contain perishable ingredients (such as raw eggs, homemade sauces, eggnog, or homemade Caesar dressing) to remain at room temperature for longer than a few minutes. Once finished serving, always promptly place cooked or perishable foods in the refrigerator or freezer. At a picnic or party, keep cold foods on ice and hot foods properly heated.

COOKING IT SAFE

TYPE OF FOOD	TEMPERATURE (DEGREES FAHRENHEIT)
Fish and seafood	145
Red meat or pork (including ground)	160
Ground chicken or turkey	165
Poultry—breast	170
Whole poultry and thighs	180
Eggs	Cook until egg white and yolk are firm, not runny

Dishes, serving bowls, or other items made of glazed lead-containing pottery can cause poisoning, particularly in young children. Make sure that the container that is used for cooking and serving is properly glazed (manufacture by a domestic pottery dealer should ensure this). If in doubt, use the pottery for decoration rather than for cooking or serving food.

REFRIGERATING OR FREEZING FOOD

If warm or hot food is headed for storage in the refrigerator or freezer, do not allow it to cool on the countertop. Place warm or hot food into a shallow pan to facilitate cooling and then put it directly into the refrigerator or freezer. If the quantity of food is large, distribute it in two or more containers to enable quicker cooling.

CLEAN IT

When you have finished eating, thoroughly wash pots and pans, utensils, and all kitchen surfaces (counter, stove tops, and sink) with soap and hot water. Let cutting boards and utensils air dry. Wash or replace sponges and dish towels frequently. If you have an automatic dishwasher, it may be helpful to have two sponges so you can wash one with each load of dishes.

THE BOTTOM LINE ON FOOD SAFETY

With a little care, you can minimize the risk that you or others will develop a food-borne illness.

ENCYCLOPEDIA
OF FOODS

PART II

Part I of this book reviewed the relationship of diet to health and provided recommendations for choosing foods and planning diets that contribute to health. The healthiest diets are based on a variety of plant foods—whole grains, vegetables, fruits, legumes, and nuts. Animal products and added fats and oils, sugars, and other sweeteners are best consumed in small quantities. The Food Guide Pyramid reviewed earlier in this book graphically emphasizes the proportions of these foods in the daily diet. Accordingly, we have arranged this section with priority given to grains, fruits, and vegetables—those items that should predominate at every meal and that most people need to consume in greater quantities. Animal products—meat and other high-protein foods and dairy foods—are also discussed. However, these are the foods that should make up relatively smaller parts of our diets.

Part II introduces you to many foods from which you can choose and provides you with knowledge about the nutrients these foods have to offer. In addition, we provide information about the sources of the foods you purchase and eat—the individual plants and animals, how they are processed to the products that appear on store shelves, and some of the history of these foods in our diet.

Before we introduce the foods themselves, we want to explain the arrangement and presentation of food items in these sections. Because this book is written for a North American audience, we have included food products that are available to most North Americans. Within the sections on Fruits and Vegetables, we have listed items by their common names in alphabetical order; when a food has more than one common name, the index should help in locating the item. Where there is a difference between the cultural or common use or perception of an item and its botanical nature, we have listed it according to common usage and mentioned the difference in the text. For example, although cucumbers, eggplant, squash, and tomatoes are botanically fruits, they are listed within the vegetable section, because most American consumers think of them as vegetables.

The nutrient compositions of foods are derived from the current version of the U.S. Department of Agriculture (USDA) nutrient composition database. This database is maintained and updated regularly by USDA laboratories and is the basis of most systems for estimating the nutrient content of foods and diets.

FRUITS

Our earliest ancestors built their diets entirely of vegetables, fruits, seeds, grains, legumes, and nuts. Throughout history, "fruit" has referred to any plant used as a food. More recently, "fruit" has come to mean the edible pulp or fleshy layer around a seed. In the 18th century, the word acquired a botanical definition: the organ derived from the ovary and surrounding the seed. At the same time, culinary custom defined fruits by their sweetness (or the balance of sweet to sour) and by how they are used in the meal, primarily as dessert. Thus, even though eggplant, cucumber, squash, and tomatoes are technically fruits, we call them vegetables.

Until recently, the availability of a fruit during the year depended on its growing season. For example, strawberries appeared in April and May, melons in August and September, whereas some fruits, such as apples and bananas, were available year-round. Today, reliable transportation brings fruit of every type to our markets year-round, although some imported fruits may be more costly during the winter than their domestic counterparts are in the summer when they are in season.

The revised Dietary Guidelines for Americans includes as 1 of its 10 principles the advice to eat a variety of fruits (and vegetables) daily. (For a discussion of the Dietary Guidelines and the 5 a Day program, see Chapter 1, page 8.)

To help you to be better informed and better plan your menus, this section provides information on the origin and nutrient content of many fruits. [(See the Appendix, page 434, for further information about the nutrient content of fruits.)] Fruit is a valuable source of fiber, vitamin C, some of the B vitamins, vitamin A, and other antioxidants and phytonutrients. (See the Appendix, Phytochemical Contents of Selected Foods, page 484).

The tables of nutrient values in the Fruits section are based on serving sizes specified by the U.S. Department of Agriculture Food Guide Pyramid. Nutrient values are rounded (milligrams and micrograms tend to be rounded to one decimal point, grams are rounded to whole numbers). Nutrient claim statements listed beneath the common name of each fruit are based on the serving size specified and the definitions in Chapter 4 (see sidebar: Nutrient Claims, page 92). For example, 1 medium apple is considered a good source of vitamin C, because an apple provides 13 percent of the Daily Value for vitamin C. A food that is high in a particular nutrient provides 20 percent or more of the Daily Value for that nutrient per serving.

ACEROLA

Acerolas are round or oval, cherry-like fruits that range from 2 to 4 inches in diameter. When ripe, the skin turns bright red. The soft, juicy flesh is yellow and has a slightly tart flavor.

Family Malpighiaceae
Scientific name *Malpighia punicifolia* L.,
  Malpighia glabra L.
Common name Barbados cherry, West
  Indian cherry, cereza

♥ **High in vitamin C**

♥ **A good source of vitamin A
  (beta-carotene)**

NUTRIENT COMPOSITION

Acerolas contain the most concentrated source of natural vitamin C of any known fruit, 100 times the vitamin C content of oranges and 10 times that of the guava. Green (unripe) fruits have twice the vitamin C content of ripe fruits. They are also a good source of vitamin A (beta-carotene).

VARIETIES

The Florida Sweet variety, commonly grown in California, yields large, juicy fruits that have a taste similar to apples. Manoa Sweet, a variety developed in Hawaii, has orange-red fruits that are especially sweet. A dwarf variety, which grows to a height of only 2 feet, can tolerate lower temperatures than the other varieties and is suited for container cultivation.

ORIGIN & BOTANICAL FACTS

The acerola is believed to have originated in the Yucatán peninsula of Mexico. Since its discovery, the plant has been introduced throughout the tropical and subtropical regions of the world, but it is still primarily grown in and around the West Indies.

The acerola is a large, bushy shrub that can attain a height of 15 feet. Although the plant grows best in hot tropical lowlands with medium to high rainfall, it is also very drought-tolerant. Acerolas need

protection against frost and winds because their root system is shallow and they can be toppled by high winds. The leaves are covered with hair, are light to dark green, and become glossy when mature. The small, white to pink flowers bloom throughout the year. Because up to 90 percent of the blossoms fall from the plant, only a few of the flowers set fruit. When grown from seed, plants begin to fruit after 2 or 3 years. An 8-year-old tree may yield 30 to 60 pounds of fruit a year.

USES

Because acerolas deteriorate quickly and undergo rapid fermentation once removed from the tree, they should be refrigerated if not used immediately. Unrefrigerated fruits can develop mold within 3 to 5 days. Acerolas can be eaten raw, made into jams and jellies, or puréed into juice. They have been used as a supplemental source of vitamin C, to make baby food, and as an ingredient in ice cream.

SERVING SIZE: *1 cup*

NUTRIENT CONTENT

Energy (kilocalories)	31
Water (%)	91
Dietary fiber (grams)	1
Fat (grams)	0
Carbohydrate (grams)	8
Protein (grams)	0
Minerals (mg)	
Calcium	12
Iron	0
Zinc	0
Manganese	–
Potassium	143
Magnesium	18
Phosphorus	11
Vitamins (mg)	
Vitamin A	75 RE
Vitamin C	1,644
Thiamin	0
Riboflavin	0.1
Niacin	0
Vitamin B$_6$	0
Folate	14 µg
Vitamin E	0

Note: A line (–) indicates that the nutrient value is not available.

APPLE

The apple is a pome, a round fruit that consists of firm, juicy flesh covered by a thin, tough, edible skin and surrounding a cartilaginous, seeded core. The skin color of apples can range from dark green to yellow to bright red, or some combination of these colors. Apples that are just ripe are crisp and juicy, whereas those that are overripe attain an aromatic flavor and a slightly mealy texture.

Family Rosaceae
Scientific name *Malus pumila, Malus sylvestris, Pyrus malus*
Common name apple

♥ **Good source of pectin, a soluble fiber that helps reduce blood cholesterol**

♥ **A good source of vitamin C**

VARIETIES

Thousands of varieties of apples are grown worldwide. As a result, apples are available in a seemingly endless array of colors, crispness, texture, size, sweetness, and aroma. Some of the more popular varieties in the U.S. marketplace are the Red and Golden Delicious, Granny Smith, Fuji, Gala, and Rome Beauty.

ORIGIN & BOTANICAL FACTS

The apple is native to Asia and eastern Europe. The earliest recorded description of apples appears in Greek literature of the 4th century B.C. The first apples cultivated in the New World were grown from seed brought by the Pilgrims. Today, the leading apple-producing nations are Russia, China, the United States, Germany, France, and Italy. In the United States, nearly half the domestic crop is grown in Washington, and New York, California, Michigan, Pennsylvania, North Carolina, and Virginia produce for much of the rest of the domestic market.

Apples can grow virtually anywhere with a moderate climate, although some varieties are better suited to a particular region. Because fruit-bearing seasons vary by variety and region, apples are available all year. Standard-sized trees reach a height and spread of 25 feet and require 5 to 10 years to fruit. Recently, dwarf and semidwarf trees have emerged; these require as few as 2 years to bear fruit.

USES

When selecting apples, choose those with firm flesh and tight skin that is free of bruises, soft spots, and holes. Larger apples tend to be more mealy than small ones. To ripen apples, keep them at room temperature. Apples store well for long periods refrigerated or in a cool, dry place. Sliced apples quickly turn brown on exposure to air; however, this can be prevented by dipping the fruit into acidulated water (dilute lemon juice).

As one of the most popular fruits in the United States, apples are widespread in the American cuisine. They are used in salads, alongside meats, and in pilafs, desserts, preserves, juices, cider, pies, breads, cakes, and alcoholic beverages (such as the liqueur calvados). Dried apples make tasty snacks or additions to breakfast cereal. Characteristics of flavor and texture determine the optimal varieties for each use. Crisp, crunchy, juicy, sweet or sweet-tart apples such as the Granny Smith, Fuji, Gala, or Red Delicious are best for eating. In general, firm-fleshed, tart apples such as the Golden Delicious and Rome Beauty are best for baking whole. Tart or slightly sour varieties are good for pies and applesauce.

NUTRIENT COMPOSITION

A medium-sized fresh apple is a good source of vitamin C; however, most of the vitamin C is lost when the apple is cooked or made into juice.

Apples are a good source of dietary fiber in the form of pectin.

SERVING SIZE: *1 medium (138 g)*

NUTRIENT CONTENT

Energy (kilocalories)	81
Water (%)	84
Dietary fiber (grams)	4
Fat (grams)	0
Carbohydrate (grams)	21
Protein (grams)	0
Minerals (mg)	
Calcium	10
Iron	0
Zinc	0
Manganese	0
Potassium	159
Magnesium	7
Phosphorus	10
Vitamins (mg)	
Vitamin A	7 RE
Vitamin C	8
Thiamin	0
Riboflavin	0.1
Niacin	0
Vitamin B$_6$	0.1
Folate	4 µg
Vitamin E	0

APRICOT

The apricot is a round, fleshy fruit that is closely related to the peach, plum, almond, and cherry. It has a single seed enclosed in a stony shell. The edible, pale-orange skin is smooth and velvety. The flesh is drier than that of most other fruits.

Family Rosaceae
Scientific name *Prunus armeniaca*
Common name apricot

♥ **Good source of vitamin C**

♥ **High in vitamin A (beta-carotene)**

VARIETIES

Approximately 12 varieties of apricots exist, with flesh that varies from yellow to deep orange. Some of the better known varieties are the Blenheim, the Tilton, the Patterson, and the Castlebrite.

ORIGIN & BOTANICAL FACTS

The world's leading producers of apricots are Turkey, Italy, Russia, and Greece. Ninety percent of the U.S. domestic market is supplied by growers in California; Utah and Washington supply the rest. During the off-season, apricots are imported from Chile and New Zealand. Apricot trees grow to about 20 feet in height and spread to a width of 30 feet. The white or pink flowers appear in early spring and give way to fruits in late summer. Because of this early flowering, apricot yield may be limited by late frosts that kill the flowers. The domestic crop is available from mid-May to mid-August, and imports arrive in December and January.

USES

Apricots are best when purchased ripe or slightly underripe and allowed to ripen in a paper bag. Green-tinged fruits will not ripen properly and should be avoided. Ripe apricots can be stored in the refrigerator up to a week, but apricots that are soft and juicy should be eaten within a day or two of purchase. Apricots should be washed just before they are eaten. They are excellent eaten out of hand or used in any recipe that calls for peaches or nectarines. Apricots should not be cooked for an extended time because they tend to lose their flavor rather quickly; poaching is an ideal cooking method. Dried apricots are a convenient, nonperishable snack.

NUTRIENT COMPOSITION

Fresh apricots are high in vitamin A (beta-carotene) and are a good source of vitamin C. (See the Appendix, page 434, for the nutrient content of dried apricots.)

NUTRIENT CONTENT

Energy (kilocalories)	34
Water (%)	86
Dietary fiber (grams)	2
Fat (grams)	0
Carbohydrate (grams)	8
Protein (grams)	1
Minerals (mg)	
Calcium	10
Iron	0
Zinc	0
Manganese	0
Potassium	207
Magnesium	6
Phosphorus	13
Vitamins (mg)	
Vitamin A	183 RE
Vitamin C	7
Thiamin	0
Riboflavin	0
Niacin	0
Vitamin B6	0
Folate	6 µg
Vitamin E	1

AVOCADO

The avocado is a pear-shaped fruit with skin that can be thick or thin, green or purplish black, and smooth or bumpy, depending on the variety. The flesh of the avocado is pale yellow-green and has the consistency of firm butter and a faint nut-like flavor.

Family Lauraceae
Scientific name *Persea americana*
Common name avocado, alligator pear

 Rich source of monounsaturated fat

Good source of fiber

VARIETIES

The two most commonly sold varieties of avocados in the United States are the Hass and Fuerte, both grown in California. The Guatemalan Hass avocado, the most popular variety, has a thick, pebble-textured and purplish skin and usually weighs no more than 12 ounces. The Fuerte avocado, a Guatemalan-Mexican hybrid, has a more pronounced pear shape and is slightly larger than the Hass. It has a shiny, thin, dark-green skin with small, raised, pale spots. Florida-grown varieties, which are Mexican in origin and include the Booth, Waldin, and Lula, are larger, less costly, and more perishable than California avocados. In addition, they contain less fat and fewer calories and lack the rich, creamy flavor of the California varieties.

ORIGIN & BOTANICAL FACTS

The avocado, native to the tropics and subtropics of Central America, was first cultivated in the United States in the mid-1800s in Florida and California. Ninety percent of today's domestic crop of avocados is grown in California. With a harvest of 168,000 tons, the United States is the second-largest grower of avocados in the world, behind Mexico at 718,000 tons.

The avocado tree, a popular shade tree in rural and suburban Hawaii, California, and Florida, is a dense evergreen that may reach a height of 80 feet.

Avocados do not ripen on the tree; ripening is inhibited by hormones produced by the leaves. This delay in ripening is a commercial advantage because the fruit may be left unharvested for long periods (up to 7 months). However, over-ripe avocados may seed internally and become moldy.

USES

Avocados that are unblemished and heavy for their size are best. Ripe avocados yield slightly to finger pressure, but if the finger leaves a dent, the avocado may be over-ripe. Ripening can be hastened by enclosing the fruit in a paper bag and leaving at room temperature. Ripe avocados should be refrigerated and used within 1 to 2 days.

Because cooking destroys the flavor of avocados, it is not recommended. Fresh avocados can be sliced and added to cooked dishes just before serving. They can be diced and mixed into salads, mashed to use in toppings or dips, puréed to use in cold soups and desserts, or julienned to include in sushi rolls. When exposed to air, avocado flesh discolors quickly. Addition of lemon or lime juice to mashed or puréed avocados can delay discoloration. Placing an avocado pit in a bowl of mashed avocados will not prevent discoloration.

NUTRIENT COMPOSITION

Avocados are known for their high fat content; however, most is monounsaturated fat. They are low in saturated fat and are sodium- and cholesterol-free. Avocados are a good source of dietary fiber. They also contain lutein, one of the carotenes that is a phytochemical with antioxidant properties.

NUTRIENT CONTENT

Energy (kilocalories)	324
Water (%)	74
Dietary fiber (grams)	10
Fat (grams)	31
Carbohydrate (grams)	15
Protein (grams)	4
Minerals (mg)	
Calcium	22
Iron	2
Zinc	1
Manganese	–
Potassium	1,204
Magnesium	78
Phosphorus	82
Vitamins (mg)	
Vitamin A	123 RE
Vitamin C	16
Thiamin	0.2
Riboflavin	0.2
Niacin	3.8
Vitamin B$_6$	0.5
Folate	124 µg
Vitamin E	3

Note: A line (–) indicates that the nutrient value is not available.

BANANA

The banana is an elongated, curved, tropical fruit with a smooth outer skin that peels off easily when the fruit is ripe. Bananas are harvested while still green but may be ripened under controlled conditions before being delivered to the grocery store. Yellow bananas are fully ripe when the skin has small flecks of brown. The flesh of the ripe banana has a distinct creamy texture and sweet fragrance.

Family Musaceae
Scientific name *Musa paradisiaca* L.
Common name banana, plantain

♥ **High in vitamin B$_6$**

♥ **A good source of vitamin C, potassium, and fiber**

VARIETIES

The familiar yellow banana sold in the United States is the Cavendish variety, which is 5 to 10 inches in length and available all year. Red bananas from Latin America are slightly wider and are heavier and sweeter than yellow bananas. Their red skin turns purple when ripe. Manzano bananas (also called finger or apple bananas) are short and chubby with a mild, strawberry-apple flavor. They turn fully black when ripe. Plantains (also called green or cooking bananas), thick-skinned bananas that range from green to yellow to brown-black, are a staple food in many parts of the world. When unripe plantains are cooked, they have no banana flavor; however, when cooked ripe, they have a sweet banana taste and a slightly chewy texture.

ORIGIN & BOTANICAL FACTS

Originating in the Malaysian region about 4,000 years ago, the banana was not introduced to the Americas until the Philadelphia Centennial Exhibition of 1876. Today, the banana is the leading fresh fruit sold in the United States and the second leading fruit crop in the world. The United States grows about 4,000 tons of bananas annually and imports a total of 1.6 million tons annually from South America. Worldwide, India is the largest banana grower, followed by Africa, where bananas are mostly kept for local use.

A banana tree is technically not a tree, but rather a tree-like herb that belongs to the grass family. It can attain a height of 10 to 40 feet when fully grown. The banana is actually a berry that has been cultivated to have no seeds. The non-woody banana stalk develops a flowering stem and seven to nine buds that each sustain one cluster (hand) of 10 to 20 bananas (fingers). The stalks are cut after producing the fruit, and new stems grow from buds in the rootstock.

USES

Ripening of green bananas can be hastened by putting the fruit into a paper bag. Ripe bananas can be stored in the refrigerator for up to 2 weeks; although the skin turns dark brown, the fruit remains edible. Unripe bananas should not be refrigerated. Bananas become sweeter as they ripen (as most of the starch converts to sugar) and are most often consumed raw or in desserts such as puddings, pies, and sweet breads. Banana slices should be dipped into acidulated water (dilute lemon juice) to prevent browning. Puréed banana can be added to pancake batter. Because they are rich in tannins, plantains are bitter and must be cooked to be palatable.

NUTRIENT COMPOSITION

Bananas are high in vitamin B$_6$ and are a good source of vitamin C, potassium, and fiber. Red bananas and plantains are good sources of vitamin A. (See the Appendix, page 436, for the nutrient content of plantains.)

SERVING SIZE:
1 medium, raw (118 g)

NUTRIENT CONTENT

Energy (kilocalories)	109
Water (%)	74
Dietary fiber (grams)	3
Fat (grams)	1
Carbohydrate (grams)	28
Protein (grams)	1
Minerals (mg)	
Calcium	7
Iron	0
Zinc	0
Manganese	0
Potassium	467
Magnesium	34
Phosphorus	24
Vitamins (mg)	
Vitamin A	9 RE
Vitamin C	11
Thiamin	0.1
Riboflavin	0.1
Niacin	1
Vitamin B$_6$	0.7
Folate	23 µg
Vitamin E	0

BERRIES

Berry is a general term for fruits that are usually small, rounded, and pulpy with seeds embedded in a juicy flesh. The term is loosely applied to a range of fruits belonging to vastly diverse botanical families. Aside from the more popular berries such as the blackberry, blueberry, cranberry, currant, raspberry, and strawberry, there are a host of less common species, each with its own distinctive shape, color, fragrance, and taste. Berries were a staple in the diets of our hunting-and-gathering ancestors and still play an important role in the culinary traditions of many peoples around the world. American Indians used various types of berries as food, medicine, dyes, and food preservatives. Early American settlers developed a taste for the many varieties growing wild in woods and fields of North America, and they learned to use the berries for food and medicine. Research has shown that several berries have medicinal properties. (Cranberries and blueberries help prevent urinary tract infections.) Most berries contain generous amounts of vitamin C, and some are a good source of fiber because of the skin and seeds.

BLACKBERRY

Family Rosaceae
Scientific name *Rubus fructicosus* (European), *Rubus villosus* (American)
Common name blackberry, bramble berry, dewberry, goutberry

Also called bramble berries because they grow on thorny bushes (brambles), black-berries range from one-half to an inch long when mature and are purplish black. Like raspberries, to which they are related, blackberries are oblong and are made up of small edible seeds that are encased in juicy globules adjoining a fleshy base.

The most common varieties of black-berry are the Cherokee (a sweet variety) and

the Marion (a tart variety). Boysenberries, loganberries, ollalaberries, sylvanberries, and tayberries are hybrids of blackberries and raspberries.

Blackberries are found throughout the temperate zones of the world, growing wild in meadows and at the edge of forests. The bushes flower in spring and bear fruit throughout the summer. Borne in loose clusters on stems that grow from the canes, the berries change from green to red and then to purplish black as they ripen. Blackberry bushes are so vigor-ously invasive that they are considered a weed in some areas.

Plump, deeply colored blackberries are the most delicious to eat, and immature red berries are tart. Blackberries are best used immediately, because they spoil quickly. They can be lightly covered and refrigerated for 1 to 2 days. Blackberries can be eaten fresh; used as a topping for yogurt, ice cream, and pancakes; tossed into a fruit salad; puréed to make a dessert sauce; or made into blackberry pie. About 98 percent of commercially produced berries are processed into jams, fillings, juices, wines, and brandies.

Blackberries are high in vitamin C, are a good source of dietary fiber, and contain ellagic acid, a phytochemical that may help prevent cancer.

BLUEBERRY

Family Ericaceae
Scientific name *Vaccinium myrtillis*
Common name blueberry

Blueberries, a species native to North America, grow in shades varying from light blue to dark purple. Round to oval, the berries have a smooth skin that is somewhat waxy and covered with a powdery silver film

BERRIES

or "bloom." Blueberries were once called star berries because of the star-shaped calyx on the top of each fruit. Cultivated blueberries can be as large as 3/4 inch in diameter, although the "wild" varieties are only 1/4 to 1/2 inch in diameter.

At least 50 species of blueberries, both cultivated and wild, have been identified. The two types of cultivated blueberries are highbrush and rabbiteye. Highbrush blueberries, *V. corymbosum* L., are grown throughout North America, whereas the rabbiteye varieties, *V. ashei* Reade, are better adapted to southern regions of the United States. Lowbush (wild) blueberries, *V. angustifolium* Ait., grow naturally in Maine, Nova Scotia, and Quebec. These plants produce blueberries that are prized for their intense flavor. The lowbush (wild) blueberry varieties grow to about 3 feet in height, whereas the highbush and rabbiteye cultivars can grow to more than 10 feet if not pruned. The desirable flavor, color, and texture of today's cultivars are the result of nearly 100 years of hybridization.

Blueberries have been used as a source of food and folk medicine for thousands of years. Early explorers of North America, such as Lewis and Clark, noted that American Indians smoked the berries to preserve them for winter and pounded the berries with beef to make a jerky called pemmican. Blueberries were also appreciated by the early American settlers as both a food and a medicine.

The blueberry plant is a compact, woody shrub that is related to the bilberry, cranberry, and huckleberry. Blueberries grow in clusters, but because the berries ripen at different times, they must be handpicked to harvest the best of the early fruit. Later, a harvesting machine is used to gently shake each bush so that only the ripe berries fall off. The blueberry season lasts only from mid-April to late September, beginning in the southern states and moving north as the season

progresses. The berries are very perishable and easily damaged by improper handling and extreme temperatures.

Blueberries are one of the most popular berries in the United States, second only to strawberries. They can be eaten dried or fresh as a snack food; added to cereals, salads, yogurt, or ice cream; used as an ingredient in pancakes, muffins, pies, breads, or sauces or as cake topping; or puréed to make jam or jelly. Although the blueberry season is short, berries can be bought in the off-season in frozen, canned, or dried form.

Blueberries are a good source of vitamin C.

Recent research has shown that blueberries may help prevent urinary tract infection by increasing the acidity of urine, which helps destroy bacteria, and by preventing bacteria from colonizing on the bladder walls.

CRANBERRY

Family Ericaceae
Scientific name *Vaccinium macrocarpon, Vaccinium oxycoccus*
Common name cranberry, bounceberry, lingonberry

Cranberries, which are native to North America, are small, smooth-skinned, round berries that are glossy deep red to red-

maroon. About one-third of an inch in diameter and half-inch to an inch long, the cranberry has seeds that are attached to the center of the fruit and are surrounded by a tart white pulp. Also called bounceberries, because they bounce when ripe, cranberries belong to the same family as blueberries and huckleberries; but unlike these fruits, cranberries are too tart to eat raw.

Cranberries are divided into three types. The most common is the large *Vaccinium macrocarpon*, grown for commercial purposes. *Vaccinium oxycoccus*, commonly called the mossberry or small cranberry, is found wild in some areas. *Vaccinium vitis-idaea*, or the lingonberry, grows well in very cold climates and is currently being developed as a crop in several eastern European countries.

Cranberries grow on a flat, woody, evergreen "vine" that thrives in acidic soil. Cranberry vines are planted in peat bogs prepared in a way that allows the plants to be covered with water to protect them from cold damage. The pink or purple cranberry flowers can be self-pollinated, but crop yield is much greater when bees are used to facilitate pollination. The berries are borne on short uprights 6 to 8 inches in length that rise from a dense mass of stems on the soil surface.

Cranberries are extensively cultivated for commercial use in the northern states. Massachusetts is the largest producer, followed by Wisconsin, New Jersey, Washington, and Oregon. Cranberry cultivation is also common throughout Canada. Harvested between Labor Day and Halloween, cranberries enjoy their peak market season from October through December.

The Pilgrims dined on cranberry dishes at the first Thanksgiving in 1621. Once only a traditional holiday food, cranberries are now consumed throughout the year as juice drinks, dried snacks, sauces, and relishes. Because of their sour taste, they must be combined with sweet foods such as sugar or orange juice to make them palatable. Only about 10 percent of the commercial crop is sold fresh; the rest is processed into juice or canned cranberry sauce.

Cranberry juice cocktail is considered effective for preventing or treating urinary tract infections, in part because of its high acidity and its ability to inhibit bacteria from adhering to the lining of the urinary tract.

Fresh cranberries are a good source of vitamin C. In addition, they contain bioflavonoids, plant pigments with antioxidant properties.

CURRANT

Family Saxifragaceae
Scientific name *Ribes rubrum, Ribes vulgare, Ribes petraeum, Ribes sativum, Ribes nigrum, Ribes ussuriense*
Common name currant (red, pink, white, black, and Asian)

Currants are small, spherical berries with thin, translucent skin that can be red, pink, white, or black. They have a soft, juicy pulp that contains several edible seeds. The flavor of currants varies from slightly to exceedingly tart. True currants are not to be confused with the zante currant, a variety of small, dried grape (raisin).

Currants are categorized by their color. Common red currants include the Red Lake, a mild-flavored, bright-red berry, and the Perfection, a medium to large, flavorful variety. The White Imperial, a small, round, white berry that grows on a spreading bush, has the lowest acid content of any currant. The most common pink currant is the Gloire des Sablons, an ancient French variety with pink flesh and colorless skin. The Boskoop is a well-flavored black currant, produced on a vigorous, upright bush. The Willoughby is a mild, black Canadian currant that is hardy to cold and sun and resists mildew.

Currants appear to have originated in northern Europe, northern Africa, Siberia, and in the Western Hemisphere, where they were eaten by American Indians well before their first contact with Europeans. American Indians historically have used currants for both food and medicinal purposes. The Coast Salish Indians of Vancouver Island boiled the fruit and dried it into rectangular cakes for use as a winter food. The Woodlands Cree Indians used currant jam as a condiment for fish, meat, and bread. Before 1550, the English called this fruit ribes, a name of ancient Indo-European origin. Subsequently, the berries came to be called currants, a word derived from the berry's resemblance to the dried Greek raisins that are made from small seedless grapes. English and European colonists in the Americas found currants growing wild in woods and fields and quickly developed a taste for them. Today, currants are commonly cultivated in Europe, Canada, New Zealand, and the United States. Europeans and Canadians seem to prefer black currants, and the less tart red and white varieties are more popular among Americans.

Currant plants are fast-growing, deciduous, perennial shrubs that can reach 5 feet in height and width. Their leaves resemble those of the maple tree in shape, but they are pale green on black currant bushes and dark blue-green on red currant plants. Some varieties are upright, and others spread. The self-fertilizing flowers that give rise to red currants are green, and those that produce black currants are pink. Plants are generally pollinated by insects. The berries, averaging about a fourth of an inch in diameter, hang in clusters from delicate, drooping stems called strigs. Currants prefer cold climates, heavy, moist, enriched soil, and full sun or light shade. Although they can be propagated by seed in the spring or by cuttings in the early fall, bushes grown from seed produce no fruit for 2 to 3 years. Pruning to remove wood that is more than 3 years old encourages the growth of new shoots.

Black currants are harvested selectively as they ripen and before they shrivel and fall from the bush. Red and white currants are pulled by the cluster to avoid damaging the delicate fruit. If the berries are going to be used for jams or jellies, they must be picked before they ripen fully because that is when the fruit pectin levels are highest. Berries grown for eating are allowed to ripen on the bush for several weeks after achieving full color. A mature currant bush can produce up to 4 quarts of fruit each year.

Because of their tartness, currants, particularly black currants, are rarely eaten as fresh fruit. Instead, they are made into jams and jellies or used in pies and sauces. Black currants are sometimes soaked in brandy or made into wine, sometimes mixed with honey and spirits. Black currants are the basis for the French liqueur crème de cassis. An infusion of the young leaves of the black currant shrub makes a drink similar to green tea.

Currants are high in vitamin C. Black currants are a good source of potassium.

BERRIES

ELDERBERRY

Family Caprifoliaceae
Scientific name *Sambucus canadensis,*
    Sambucus coerulea
Common name elderberry

Elderberries are tiny berries that range from purple-red to blue and purple-black.

The elderberry tree is an American version of the common elder tree that is found on European, Asian, and northern African soils. The eastern elderberry *Sambucus canadensis* and the Western *Sambucus coerulea* are two common varieties.

The elder tree, which belongs to the honeysuckle family, has been around for centuries and may date back to the Stone Age. The Egyptians harvested its flowers and extracted their essence to use as medicine and to beautify the skin. In the Middle Ages, it was believed that the elder tree was home to witches and that cutting it down would create trouble by disturbing those residing in the branches. In contrast, the Russians and the English believed that the elder tree warded off evil spirits. Hence, it was considered good luck to plant an elder tree near one's home. The Sicilians believed that sticks of elder wood could kill snakes and drive away thieves.

The plant is an evergreen that lives either as a large shrub, no more than 12 feet in height, or as a small tree, up to 20 feet in height, with hollow stems that support large compounded leaves. Ideal growth conditions include rich, sandy soil and direct sunlight or medium shade. The plant can be found growing wild in meadows or pastures or along roadsides. The plant produces sprays of small, white flowers, up to 6 inches in diameter, that give way to large clusters of berries, 6 to 9 inches wide.

Because of the tartness of the fresh fruit and a toxic alkaloid that is contained in the seeds (which is destroyed by heat), the berries are always cooked before eating. Alternatively, the berries can be added to pies or made into jam or wine.

Elderberries are high in vitamin C, fiber, and bioflavonoids, plant pigments with antioxidant properties.

GOOSEBERRY

Family Saxifragaceae
Scientific name *Ribes hirtellum* (American gooseberry), *Ribes grossularia* (European gooseberry)
Common name gooseberry

Gooseberries are round fruits that vary from white to yellow, green, pink, red, purple, and nearly black. The color of the fruit is most intense in full sunlight.

The fruit consists of a translucent skin tightly surrounding a white pulp that encloses several small seeds. The berries range from a fourth to an inch in diameter.

Most varieties of gooseberry available in the United States are hybrids of the two main species, European and American. The fruits of the European variety are about 1 inch in diameter. The American variety is smaller and rounder and is pink to purplish-red when mature.

The European gooseberry is native to the Caucasus Mountains and northern Africa, and the American variety is native to the northeastern and north central regions of the United States. Gooseberries have been cultivated in Europe since the 15th century. The plants are very resistant to cold temperatures and grow well in cool, temperate climates.

Gooseberry plants are small, deciduous, woody shrubs, about 4 to 5 feet in height, with prominent thorns at the nodes. The fruits are produced along the stems singly or in small groups of two to four. The fruits generally drop from the shrub when they are overripe.

Because of their tartness, gooseberries are usually cooked with sugar and not eaten fresh. This tart but versatile berry can be used by itself or blended with other fruits to make pies, jams, or jellies. Gooseberry sauce prepared from under-ripe berries complements such dishes as roasted goose or duck. Gooseberries are also made into wine or vinegar. For desserts, the larger, thinner-skinned, sweeter types are picked when fully ripe. The European gooseberry is usually preferred to the American type.

Gooseberries are high in vitamin C and are a good source of fiber and bioflavonoids, plant pigments with antioxidant properties.

MULBERRY

Family Moraceae
Scientific name *Morus* species
Common name mulberry

Botanically, the mulberry is not a berry but a collective fruit. After the flowers are pollinated, they and their fleshy bases swell and become succulent and full of juice, like the drupes of a blackberry, which the mulberry resembles in size and shape.

There are three principal species, the names of which refer not to the color of the fruit but to the color of the buds. The black mulberry (*M. nigra*) is native to western Asia and has been grown in Europe and the Middle East since ancient times for its fruits. Large, juicy, and bluish black, the black mulberry is no doubt the most flavorful, with its refreshing combination of sweetness and tartness. The American, or red, mulberry (*M. rubra*), indigenous to the eastern United States, grows wild from Massachusetts to the Gulf Coast. Usually a deep red-purple, the red mulberry is not as tasty as its black cousin. The white mulberry (*M. alba*) is the least tasty of the three, with an unpleasant sweetness that lacks the pleasing tartness of the black mulberry. The plant is native to eastern and central China, where the tree has long been cultivated for its leaves,

which are the essential food for silkworms. The white mulberry became naturalized in Europe, and both the trees and the silkworms were introduced to the United States in early colonial times in an attempt to start a silk industry.

Mulberries can be eaten raw or used to make jams, jellies, sorbet, ice cream, frozen meringue, pudding, and sauces. Slightly unripe, tart berries are best for making pies and tarts. Mulberries also make an interesting wine and are excellent as dried fruits. In medieval England, the berries were puréed to make murrey, which was added to spiced meats or used as a pudding.

Mulberries are high in vitamin C.

RASPBERRY

Family Rosaceae
Scientific name *Rubus idaeus, Rubus strigosus*
Common name raspberry

Raspberries are small aggregate fruits, composed of numerous, small drupelets, each containing a small seed and clustering together around a central core. They range from a half to an inch or more in diameter. When the berry is picked

from the stem, the core remains behind, leaving a hollow cavity in the fruit. Raspberry varieties are distinguished by color. Red berries are the most common and popular, black raspberries are somewhat smaller and less round, and golden berries, which are available only in limited quantities, can vary from yellow to orange, amber, and even white. Raspberries are fragrant and sweet, with a slight tartness. The raspberry is sometimes considered the most intensely flavored of the berry family.

Traces of wild raspberries have been found at prehistoric sites in Asia, and American Indians used wild raspberries medicinally. Red raspberries have been cultivated in Europe for more than 400 years, brought home by Crusaders who found them growing in the Mount Ida region in Turkey. During the 18th century, the cultivation of raspberries improved, and by the 19th century, they were being grown widely throughout Europe and North America. By the 1860s, more than 40 varieties were known. Today, about 90 percent of all domestic raspberries are grown in Oregon, Washington, and California, with some imported from Canada and Chile during the off-peak season.

Raspberries are thorny, perennial bushes that can reach heights of 10 feet. They prefer cool summers, mild winters, and a dry harvest season. Three years is required for the bushes to begin producing the delicate white flowers from which the berries form on erect stalks or canes. Mature berries must be handled carefully because they are fragile and easily damaged. Some are packed in small containers for the fresh market, but the bulk of the harvest is processed into frozen, concentrated, or canned forms.

Raspberries are best eaten within 1 to 2 days of purchase. If possible, they should not be washed, because they absorb water

BERRIES

and become mushy, but they can be rinsed quickly just before serving. Whole berries can be frozen for up to 1 year.

Fresh raspberries make a delicious topping for cereals, pancakes and waffles, yogurt, puddings, cake, and ice cream; a colorful, sweet addition to fruit or green salads; and an excellent snack eaten right out of hand. They can be preserved in brandy or syrup or added to vinegar to make a delicious salad dressing. Raspberries make wonderful tarts, jams, jellies, compotes, wine, and beer and are an elegant addition to champagne and punch. Cooked raspberries, mixed with a touch of lemon or orange juice to enhance their color, make a tasty sauce for chicken and fish dishes.

Raspberries are high in vitamin C and are also a good source of both soluble and insoluble fiber.

STRAWBERRY

Family Rosaceae
Scientific name *Fragaria vesca, Fragaria americana*
Common name strawberry

The sweet, juicy, bright-red strawberry is actually not really a fruit in the botanical sense but a swelling of the plant's stalks that occurs after the flowers are pollinated. The real fruits are the 200 seeds, called achene, that cover the berry's surface. The plant itself is a low-growing perennial that produces horizontal runners, or stolons, that spread out from the base and take root to form new plants.

The hundreds of varieties of strawberries in the United States, which vary in size, color, and taste, are distinguished primarily by their locale. Some California varieties include Chandler, Selva, Seascape, and Camaroso. Florida varieties include the Florida 90, with large, red, flavorful fruit; the Tioga, a large, vigorous plant with medium-quality berries; the Florida Belle, a disease-resistant variety with red, conical fruit; and the Sequoia, with high-quality fruit that tends to be soft when ripe.

Strawberries, which are native to Europe and North and South America, thrive in temperate zones throughout the world and have a history more than 2,000 years old. Wild strawberries, which are smaller but more fragrant and flavorful than cultivated varieties, grew in Italy as early as the 3rd century B.C. American Indians are known to have cultivated strawberries by the 17th century to eat fresh and also dried and added to winter soups. They also used them medicinally, to make dyes, and as preservatives for other food. In the early 18th century, the French developed larger strawberries by crossing two wild varieties. These plants are believed to be the source of the large cultivated strawberries we enjoy today.

Although the source of the name "strawberry" is unknown, it may derive from the practice of placing straw around the plants for protection, from the runners that the plant sends out, or from the Anglo-Saxon verb "to strew," which could have led to names such as streabergen, streberie, straibery, and, finally, the English strawberry.

Strawberries prefer well-drained, moist, sandy soils, warm days, and cool nights. The flowers, usually white but sometimes pink, give rise to berries that ripen about a month after the blossoms form. Most varieties of strawberry continue to bloom and produce fruit throughout the harvest season. The fruit is picked at the peak of its freshness and does not ripen after harvesting. Because strawberries are easily bruised, they are carefully hand-picked, sorted, and packed in the field and then rushed to cooling facilities. They are stored for only 24 hours before being shipped in refrigerated trucks to markets.

In California, where strawberries have been cultivated since the early 1900s, the fruit grows 10 months of the year, from January through November; the peak season falls between April and June. In fact, California produces more than 80 percent of all domestic strawberries, about 1 billion tons per year. In Florida, the second-largest producing state, strawberries are grown in the winter months only, and Oregon cultivates berries mostly for frozen products. Although other states produce strawberries, they usually are available only in the warm summer months for local markets. Some strawberries are also imported from Mexico and New Zealand.

The freshness and flavor of strawberries can be preserved if they are not washed until just before they are to be eaten. Fresh strawberries are most frequently served sliced over small shortcakes, topped with whipped cream; used as a garnish for appetizer and cheese platters; or added to fresh fruit tarts. Whole, long-stemmed strawberries dipped in chocolate make an elegant dessert. Strawberries are also added to rhubarb pies and made into preserves. Mixed in a blender with low-fat milk or yogurt, honey, and other fruits, they make a refreshing, nutritious shake.

Strawberries are high in vitamin C.

SERVING SIZE: *1/2 cup*

NUTRIENT CONTENT

	Black-berry	Blue-berry	Cran-berry	Currant (red)	Elder-berry	Goose-berry	Mul-berry	Rasp-berry	Straw-berry
Energy (kilocalories)	37	41	23	31	53	33	30	30	22
Water (%)	86	85	87	84	80	88	88	87	92
Dietary fiber (grams)	4	2	2	2	5	3	1	4	2
Fat (grams)	0	0	0	0	0	0	0	0	0
Carbohydrate (grams)	9	10	6	8	13	8	7	7	5
Protein (grams)	1	0	0	1	0	1	1	1	0
Minerals (mg)									
Calcium	23	4	3	18	28	19	27	14	10
Iron	0	0	0	1	1	0	1	0	0
Zinc	0	0	0	0	0	0	0	0	0
Manganese	1	0	0	0	–	0	–	1	0
Potassium	141	65	34	154	203	149	136	93	120
Magnesium	14	4	2	7	4	8	13	11	7
Phosphorus	15	7	4	25	28	20	27	7	14
Vitamins (mg)									
Vitamin A (RE)	12	7	2	7	44	22	2	8	2
Vitamin C	15	9	6	23	26	21	25	15	41
Thiamin	0	0	0	0	0.1	0	0	0	0
Riboflavin	0	0	0	0	0	0	0.1	0.1	0
Niacin	0	0	0	0	0	0	0	1	0
Vitamin B$_6$	0	0	0	0	0.2	0.1	0	0	0
Folate (µg)	24	5	1	4	4	5	4	16	13
Vitamin E	1	1	0	0	1	0	0	0	0

Note: A line (–) indicates that the nutrient value is not available.

BREADFRUIT

Breadfruit is a large oblong or round fruit, 8 to 10 inches in diameter and up to 10 pounds in weight, with a thin, bumpy skin that turns green-brown to yellow as the fruit ripens. The meat is cream-colored, mealy, and starchy in texture, and it is blandly sweet, similar to the potato. Thus, it is not eaten as a fruit but as a high-carbohydrate vegetable. Mature breadfruit is dark, dull, greenish brown, with stains on the surface from the milky sap that is exuded by the fruit.

Family Moraceae (fig or mulberry)
Scientific name *Artocarpus communis, Artocarpus altilis*
Common name breadfruit

♥ **High in vitamin C and dietary fiber**

♥ **A good source of potassium**

VARIETIES

On the island of Maui in Hawaii, almost 100 varieties of breadfruit, called "ulu," are grown at Kahanu Gardens of the National Tropical Botanical Garden.

ORIGIN & BOTANICAL FACTS

Native to the Pacific, particularly Polynesia and southeast Asia, the beautiful, smooth-barked breadfruit tree grows to about 60 feet tall, with dark-green, palmate leaves up to 3 feet long. Breadfruit was very important in the lives of early Polynesian people, who carried it with them in their canoes and planted trees wherever they settled throughout the Pacific Islands. In Hawaiian tradition, breadfruit is a symbol of creation and of the creator's generosity and love. Today, however, the largest producers of breadfruit are the Caribbean Islands.

Each breadfruit actually is composed of thousands of small fruit growing together around a core. Breadfruit is generally picked while it is firm and before it ripens, becomes overly sweet, and falls to the ground.

Breadfruit grows in hot, wet, tropical lowlands, tolerating a variety of well-drained soils. The fruit is propagated from shoots that develop from the tree's roots, or from root cuttings themselves. The tree produces an extensive root system, so it must be planted where it will have room to grow. It does not transplant easily. Trees bear fruit 5 to 7 years after the shoots are planted, and generally two crops of fruit mature during the year, once between April and June, and once between October and January.

Breadfruit must be harvested by hand, by climbing the tree and cutting or snapping off the stem close to the branch. If knocked from the tree, bruises will cause rapid softening. Because individual breadfruits do not develop at the same rate, each tree must be harvested several times during the season.

USES

Breadfruit that is slightly soft with a yellow to tan rind and no bruises should be chosen. The fruit can be stored up to 10 days if wrapped in plastic and placed in a cool area. Like squash or potatoes, breadfruit can be peeled and boiled, steamed, baked, grilled, stir-fried, or made into a salad resembling potato salad. It also can be preserved through fermentation. In Hawaii it is sometimes pounded into a paste called "ulu poi." (Hawaiian poi usually is made from taro root.) Despite its name, it is not used to make bread. In the Pacific, the sap and wood of the breadfruit plant have various non-culinary uses. Breadfruit is sold fresh in some ethnic markets or specialty stores, or it is sometimes available canned.

NUTRIENT COMPOSITION

Breadfruit is high in vitamin C and fiber and is a good source of potassium.

SERVING SIZE:
1/4 small (96 g)

NUTRIENT CONTENT

Energy (kilocalories)	99
Water (%)	70
Dietary fiber (grams)	5
Fat (grams)	0
Carbohydrate (grams)	26
Protein (grams)	1
Minerals (mg)	
Calcium	16
Iron	1
Zinc	0
Manganese	0
Potassium	470
Magnesium	24
Phosphorus	29
Vitamins (mg)	
Vitamin A	4 RE
Vitamin C	28
Thiamin	0.1
Riboflavin	0
Niacin	1
Vitamin B6	0.1
Folate	13 µg
Vitamin E	1

CALAMONDIN

Calamondin, also called "acid orange," is a citrus fruit resembling a miniature orange. It is a slightly oblong fruit about 1 to 1.5 inches in diameter. The edible peel is smooth and tender and varies in color, ranging from yellowish green when premature to deep orange when ripe. The flesh is juicy and orange and forms a segmented crown around a small semi-hollow axis. Calamondin contains a small number of seeds with green cotyledons. The fruit is extremely sour but can also be very bitter if picked before maturity.

Family Rutaceae
Scientific name *Citrofortunella mitis*
Common name calamondin

♥ **No nutritional information is available**

VARIETIES

Calamondin is one of several hundred subspecies of the genus *Citrus*. It belongs to the family that includes lemons, limes, and kumquats. Cross-breeders believe that it may be a hybrid of lime and mandarin. Others think it resembles a cross between the kumquat and the tangerine. It is a close relative to the "kalamansi," also known as "musk lime," which is used extensively in southeast Asian cuisine.

ORIGIN & BOTANICAL FACTS

Calamondin is a native of the Philippines but has its origin in China. The fruit was brought to Florida from Panama via Chile in the late 1800s. It is cultivated in Florida and in California and is mostly recognized for its ornamental value. According to ancient Chinese beliefs, a flourishing calamondin tree will bring good luck to the household. Unlike many of its cousins in the citrus family, it is able to withstand mild cold temperatures. However, it thrives best in filtered sunlight and acidic soil at temperatures ranging between 60° and 85° Fahrenheit. Excess moisture may damage its roots. The dwarf tree produces very decorative, fragrant white flowers about an inch in diameter, and it is valued as an ornamental houseplant whose beauty lasts through the year. Its golden fruits can take up to 12 months to mature and ripen.

USES

Aside from its use as a garnish, calamondin is appreciated for its distinctive flavor. The entire fruit, except for the seeds, can be consumed. The fruit is best used within a week of harvesting when it is still green. Once it reaches deep yellow, it must be kept refrigerated to retain its crispness and aroma. The fruit can be kept refrigerated up to 2 weeks. This tiny fruit releases a highly acidic (almost caustic in taste) juice that works wonderfully as a flavor enhancer in a variety of dishes, from fish to noodles, soups, sauces, and desserts. It also is used to make preserves. Calamondin juice serves as a base in many beverages.

CARAMBOLA

The carambola is an oval to elliptical fruit with a thin, shiny, waxy surface and a greenish yellow skin. Its length ranges from 2 to 6 inches with four to six prominent vertical lobes (cells) that result in star-shaped slices when cut crosswise. The flesh is light to dark yellow, crunchy, juicy, and translucent. The flavor resembles a blend of the flavors of many fruits. Up to 12 small, thin, edible seeds are contained in each fruit, enclosed by a thin gelatinous pocket.

Family Oxalidaceae
Scientific name *Averrhoa carambola*
Common name starfruit, carambola, star apple

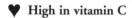

♥ **High in vitamin C**

♥ **Good source of vitamin A (beta-carotene) and fiber**

VARIETIES

Two types of carambola are available: the tart varieties and the sweet. The leading commercial variety, the Arkin, is sweet and has a bright-yellow to yellow-orange skin and flesh. Another common variety is the Golden Star, a fruit that is slightly larger than the Arkin and mildly tart. Other varieties include the Fwang Tung, Hoku, Kaiang, Maha, Sri Kembanqan, Wheeler, Thayer, and Newcombe.

ORIGIN & BOTANICAL FACTS

The carambola has been cultivated in southeast Asia for many centuries and is thought to have originated in what is now Sri Lanka or in Malaysia. The carambola was introduced into Florida around 1887, and later Hawaii. Currently, the major suppliers worldwide include Taiwan, Malaysia, Guyana, India, the Philippines, Australia, and Israel.

The carambola tree is a slow-growing, short evergreen (25 to 30 feet high and 20 to 25 feet wide) that can be single-trunked or multitrunked. Carambola

leaves are compound structures composed of smaller ovoid to oblong leaflets. The leaves are spirally arranged on the branch and are sensitive to light and sudden movements (they fold up during the night or when the tree is abruptly shaken). Although classified as a subtropical plant, the tree can tolerate short periods of frost with little damage.

If picked before ripening, green carambola fruit eventually turns yellow. However, the fruit is sweetest if allowed to ripen on the tree.

USES

Carambolas are easily damaged, and it is best to choose fruits that are firm and shiny. The fruit can be refrigerated in a moderately humid area for about 3 weeks without damage or loss in fruit quality. When transferred to room temperature, fruits that have been picked before fully ripe will turn yellow. The sweet variety is generally eaten fresh, either whole or sliced. Juiced, preserved, dried, and canned versions also are available. The tart variety is used for making jams. Before the fruit is served, the darker edge of the cells (or ridges) should be removed and the fruit sliced crosswise. The star-shaped sections are often used as garnishes for light summer entrées. Rubbing a very small amount of salt onto the exposed flesh will prevent the darkening that is caused by exposure to the air. Other uses for the fruit include pickling, adding it to salsa and salads, puréeing it for chutney, grilling it on skewers with seafood or chicken, using it as a garnish, and adding it to puddings, tarts, stews, and curries. In Hawaii, carambola juice is mixed with gelatin, sugar, lemon juice, and boiling water to make sherbet.

NUTRIENT COMPOSITION

Carambolas are high in vitamin C and are a good source of vitamin A (beta-carotene). The fruit is also a good source of dietary fiber.

SERVING SIZE:
1, raw (127 g)

NUTRIENT CONTENT

Energy (kilocalories)	42
Water (%)	91
Dietary fiber (grams)	3
Fat (grams)	0
Carbohydrate (grams)	10
Protein (grams)	1
Minerals (mg)	
Calcium	5
Iron	0
Zinc	0
Manganese	0
Potassium	207
Magnesium	11
Phosphorus	20
Vitamins (mg)	
Vitamin A	62 RE
Vitamin C	27
Thiamin	0
Riboflavin	0
Niacin	1
Vitamin B6	0.1
Folate	18 µg
Vitamin E	0

CHERIMOYA

The cherimoya is a large compound fruit, about 4 to 8 inches long and weighing up to 6 pounds, with a conical or heart shape. Its relatively thin skin may be smooth with fingerprint-like markings or covered with scale-like overlapping lobes. The fruit can be green or bronze, turning almost black as it ripens. The fragrant, juicy white flesh is strewn with black, almond-shaped seeds, has the texture of firm custard, and has a flavor resembling a mixture of pineapple, papaya, and banana.

Family Annonaceae
Scientific name *Annona cherimola*
Common name cherimoya, custard apple

♥ **A good source of vitamin C**

♥ **Provides some dietary fiber**

VARIETIES

Of the more than 50 varieties of cherimoya, most were developed in California. The Bays, from Ventura, California, is a medium-sized fruit with a lemony flavor, and the Booth, which tastes like papaya, is one of the hardiest.

ORIGIN & BOTANICAL FACTS

As with other members of the Annonaceae family (such as atemoya, soursop, and sweetsop), the cherimoya is believed to have originated in the inter-Andean valleys of Ecuador, Colombia, and Peru. The seeds were brought to California in 1871 and planted in the area of Carpinteria, south of Santa Barbara. Today, cherimoyas are grown in many parts of the tropical and subtropical world, including El Salvador, Mexico, Malaysia, the Philippines, and Vietnam. California is the only North American producer of the cherimoya, and the fruit is not exported to other states.

The cherimoya tree is a dense, fast-growing, subtropical or mild-temperate evergreen that can grow to 30 feet tall if not pruned. The large, dark-green leaves have velvety undersides and prominent veins. Cherimoya trees can grow in a wide range of soil types but seem to grow best in well-drained, medium soil of moderate fertility. They do not flourish in hot, humid climates, but prefer sunny exposure, light coastal air, and cool nights. The trees can tolerate a light frost and require some chilling to produce well.

Cherimoyas generally are propagated by seed or grafting. A tree grown from seed will produce fruit after 5 or 6 years, but grafted trees will produce fruit in 3 to 4 years. The greenish brown flowers of the cherimoya tree open first as female flowers for 36 hours, and later as male flowers. However, they usually are hand-pollinated. The fruits are clipped from the tree while they are still firm, because they usually crack open and decay if left to ripen on the tree.

USES

Because the pulp of the cherimoya is the only edible portion, the peel and seed must be removed before eating. Unripe fruits can be ripened at room temperature. Ripe fruits tend to ferment quickly and should be stored in the refrigerator for no more than 1 to 2 days. Care should be used when handling the fruits, because cherimoyas are very fragile. The fruit is best served chilled. The ripe fruit is cut in half or quartered and the flesh spooned out, cubed, or sliced and added to fruit salads. The pulp also can be puréed and used as a topping for puddings and frozen desserts or made into refreshing sorbets, ice creams, or milk shakes. The fruit itself also can be served frozen.

NUTRIENT COMPOSITION

The cherimoya is a good source of vitamin C and provides some dietary fiber.

SERVING SIZE: 1/8 (68 g)

NUTRIENT CONTENT

Energy (kilocalories)	64
Water (%)	74
Dietary fiber (grams)	2
Fat (grams)	0
Carbohydrate (grams)	16
Protein (grams)	1
Minerals (mg)	
Calcium	16
Iron	0
Zinc	–
Manganese	–
Potassium	–
Magnesium	–
Phosphorus	27
Vitamins (mg)	
Vitamin A	1 RE
Vitamin C	6
Thiamin	0.1
Riboflavin	0.1
Niacin	1
Vitamin B_6	–
Folate	–
Vitamin E	–

Note: A line (–) indicates that the nutrient value is not available.

CHERRY

Cherry fruits are round with a depression at the stem. They are a fourth to an inch in diameter and have a smooth, thin skin that adheres to the fleshy pulp. The color of the skin, as well as the pulp, can range from yellow to red to near black, depending on the variety. Each fruit has a hard seed at its center.

Family Rosaceae
Scientific name *Prunus avium*
Common name cherry, sweet cherry

♥ **Sour cherries are a good source of vitamin C and vitamin A (carotene)**

♥ **Contain terpenes, phytochemicals that may help prevent cancer**

VARIETIES

Cherries are categorized as "sweet" or "sour" according to their flavor. Bing and Lambert are popular dark-red, sweet cherries. Rainier and Royal Ann are sweet varieties that are golden with a slight touch of red. Sour cherries are smaller, softer, and more globular, and the best-selling varieties are Early Richmond, Montmorency, and Morello.

ORIGIN & BOTANICAL FACTS

Named after the Turkish town of Cesarus where they were first cultivated, cherries are believed to have originated in northeastern Asia. They were mentioned by Theophratus, a Greek philosopher and naturalist, in the *History of Plants*, written in 400 B.C. Currently, the United States produces about 90,000 tons of cherries annually, with Washington, Oregon, Idaho, and Utah producing 70 percent of the nation's crop. Worldwide, Europe is the leading producer.

Cherries are related to other deciduous flowering fruit trees such as the peach. Until recently, cherry trees were difficult to grow in a home garden because of their large spread and height: a cherry tree can reach 40 feet in height. This problem has been eliminated by the development of new self-fertilizing hybrids that reach no more than 6 to 8 feet in height. Cherry trees provide a spectacular display of white or pink blossoms in spring, and some varieties are grown purely for their ornamental value.

Domestically grown cherries are available only from late May through early August. After August, cherries that appear in the market often have been kept in cold storage. In addition, small quantities are imported from Chile and New Zealand during the off-season.

USES

When selecting cherries, choose those that are firm, bright, and shiny. Soft or shriveled fruits with darkened stems are a sign of old age or poor storage conditions. After purchase, cherries should be covered and refrigerated if not used immediately, because they tend to absorb odors. Fresh cherries can be stored in the refrigerator for up to 1 week or frozen for up to 1 year.

Sweet cherries are usually eaten fresh. They can be used to top ice cream, yogurt, or pancakes and waffles, or they can be tossed into a fruit salad. Pitted sour cherries are used as a pie filling or made into delicious compotes and jams. Candied cherries are an important ingredient in baked items such as fruitcake and Black Forest cake. Dried cherries are also available for snacks or to be added to desserts or baked goods.

NUTRIENT COMPOSITION

Sour cherries are higher in vitamin C and vitamin A (carotene) than the sweet varieties. They also contain terpenes, phytochemicals that may help prevent cancer. (See the Appendix, page 434, for the nutrient content of sour cherries.)

SERVING SIZE:
1/2 cup sweet cherries (73 g)

NUTRIENT CONTENT

Energy (kilocalories)	52
Water (%)	81
Dietary fiber (grams)	2
Fat (grams)	1
Carbohydrate (grams)	12
Protein (grams)	1
Minerals (mg)	
Calcium	11
Iron	0
Zinc	0
Manganese	0
Potassium	162
Magnesium	8
Phosphorus	14
Vitamins (mg)	
Vitamin A	15 RE
Vitamin C	5
Thiamin	0
Riboflavin	0
Niacin	0
Vitamin B$_6$	0
Folate	3 µg
Vitamin E	0

COCONUT

The coconut is the fruit of the coconut palm. Roughly oval, the fruit is up to 15 inches long and 12 inches wide. Each coconut has several layers: a smooth outer covering; a fibrous husk; a hard, brittle, dark-brown, hairy shell with three indented "eyes" at one end; a thin brown skin; the edible fleshy white coconut meat inside this skin; and the clear coconut "milk" at the center. The unripe coconut is usually green, although some varieties have a yellowish covering.

Family Arecaceae or Palmaceae
Scientific name *Cocos nucifera*
Common name coconut

♥ **High in saturated fat**

VARIETIES

There are several types of coconut palm, varying from genetically engineered dwarf varieties to the familiar tall varieties, which attain heights of 80 to 100 feet.

ORIGIN & BOTANICAL FACTS

The coconut palm is found throughout the tropics, although experts believe it is a native of the West Pacific and Indian Ocean islands. It is cultivated in the hot, wet lowlands of South and Central America, India, and Hawaii and throughout the Pacific Islands. Because this palm tolerates brackish soils and salt spray, it is typically found along tropical, sandy shorelines. In the United States, the coconut palm is found in Hawaii, the southern tip of Florida, Puerto Rico, and the Virgin Islands.

The coconut palm is tall and slender, with a cluster of leaves at the top of a slightly curved trunk. The tree has a swollen base and a strong, flexible, ringed trunk. The yellowish green, pinnate, compound leaves that form the crown are 15 to 17 feet in length, made up of lanceolate leaflets that can reach lengths of 3 feet. The tree typically begins to bear fruit when it is about 7 years old. The fruits are produced in clusters near the base of the leaf fronds at the rate of about 50 per year. Thus, during its lifetime of 70 to 100 years, the coconut palm produces thousands of fruits. Fresh coconuts are available year-round, with the peak season from October through December. Coconuts that are available for sale in the United States almost always have the two outer layers removed. Upon ripening, the flesh of the coconut transforms from a translucent yellow gel to a firm, white meat.

USES

When selecting coconuts, choose those that are free from cracks and heavy for their size and sound full of liquid when shaken. The "eyes" should be dry and clean. Unopened coconuts can be stored at room temperature up to 6 months. The coconut is opened by piercing two of the eyes. The thin, slightly sweet coconut water inside the nut can be mixed with lemon or lime juice and used as a beverage. Chunks of ripe coconut meat can be grated or chopped and eaten directly or substituted for dried, pack-aged coconut in recipes. Grated fresh coconut can be refrigerated tightly sealed up to 4 days or frozen up to 6 months. Coconut milk and cream are made by heating water and shredded fresh or desiccated coconut. Both coconut milk and cream are used in cooking and in preparing drinks. Dried coconut meat, called copra, is pressed to extract coconut oil.

NUTRIENT COMPOSITION

Coconut meat provides some fiber but is high in fat, a substantial amount of which is saturated fat. Coconut oil has the dubious distinction of being one of the most highly saturated of all plant-based oils and is best consumed in limited amounts.

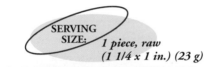

SERVING SIZE: *1 piece, raw (1 1/4 x 1 in.) (23 g)*

NUTRIENT CONTENT

Energy (kilocalories)	80
Water (%)	47
Dietary fiber (grams)	2
Fat (grams)	8
Carbohydrate (grams)	3
Protein (grams)	1
Minerals (mg)	
Calcium	3
Iron	1
Zinc	0
Manganese	0
Potassium	80
Magnesium	7
Phosphorus	25
Vitamins (mg)	
Vitamin A	0 RE
Vitamin C	1
Thiamin	0
Riboflavin	0
Niacin	0
Vitamin B6	0
Folate	6 µg
Vitamin E	0

DATE

The date is a small, oblong fruit of the date palm tree that grows in clusters of as many as 200. The mature date is approximately 2 inches long and 1 inch wide and has a somewhat wrinkled skin and a single, central pit. Dates can be yellow to orange, red, green, or brown.

Family Palmaceae
Scientific name *Phoenix dactylifera* L.
Common name date

♥ **A good source of fiber**

VARIETIES

Fresh dates are classified as "soft," "semisoft," and "dry," depending on their moisture content. The most common type is "semisoft," a well-known example of which is the large, flavorful Medjool from Morocco. Other "semisoft" varieties are the firm-fleshed, amber Deglet Noor and the small, golden Zahidi. The Barhi, Khadrawy, and Halawy are "soft" dates. "Dry" varieties contain relatively little moisture when ripe. Thus, the term "dry" does not mean "dehydrated" or "dried."

ORIGIN & BOTANICAL FACTS

Dates originated somewhere in the desert area that stretches from India to North Africa. Cultivation seems to have begun at least 8,000 years ago, when settlement began along the Jordan River and around the Dead Sea. Archaeological evidence indicates that cultivation of dates was well established by 3000 B.C. in what is now Iraq.

The northern coastal region of the Middle East was originally called Phoenicia, a name that may be the source of the early Greek term for the date, phoenix. The word "date" appears to have been derived from the Greek *daktylos*, which is related to part of the

fruit's scientific name, *dactylifera*. *Dactylifera* means "the finger-bearer," an apt description of the date palm, with its brown bunches of finger-like fruit.

Dates were first brought to the Americas in the 18th century by Spanish missionaries, who planted date palms around their missions. Some of these original trees still stand in southern California and in Mexico. Today, although the Middle East supplies three-fourths of the world's dates, much of the American demand is supplied by dates from California and Arizona. Seventy-five percent of California dates are of the Deglet Noor variety, but some Medjools are grown along the Colorado River.

The date palm grows to about 100 feet in height. The tree itself can thrive in almost any warm climate, but fruit production requires a hot, dry environment with an underground water supply. Humidity prevents the fruit from setting, and temperatures below 70° Fahrenheit prevent ripening.

USES

The dates most often available in stores are either fresh or partly dehydrated. These may be difficult to distinguish,

because fresh dates are rather wrinkled, and both types are usually packaged in cellophane. Covered and refrigerated, both types should keep indefinitely.

Fresh dates can be eaten as a snack or chopped and added to dry and cooked cereal, yogurt, puddings, breads and muffins, cookies, and ice cream. Middle Eastern recipes include dates in stews, poultry stuffing, and pilafs.

NUTRIENT COMPOSITION

One serving of dates provides minimal amounts of vitamins and minerals, but dates are a good source of dietary fiber.

SERVING SIZE: *5, dried (42 g)*

NUTRIENT CONTENT

Energy (kilocalories)	114
Water (%)	22
Dietary fiber (grams)	3
Fat (grams)	0
Carbohydrate (grams)	31
Protein (grams)	1
Minerals (mg)	
Calcium	13
Iron	0
Zinc	0
Manganese	0
Potassium	271
Magnesium	15
Phosphorus	17
Vitamins (mg)	
Vitamin A	2 RE
Vitamin C	0
Thiamin	0
Riboflavin	0
Niacin	1
Vitamin B$_6$	0.1
Folate	5 µg
Vitamin E	0

DURIAN

The durian varies from olive-green to yellow to brown and can be pendulous, round, or oblong. The fruit ranges in length from 20 to 35 cm and can weigh up to 10 pounds. The semihard shell of the durian is covered with short, pointed spines that make the fruit difficult to open. The hard shell protects the fruit from bruises and damage when the ripe fruit drops from the tree. The fruit itself is a capsule that divides into five lobes, or segments, when ripe. Each segment contains one or more brown seeds that are covered with a thick, creamy, strong-smelling pulp, the edible part of the fruit.

Family Bombacaceae
Scientific name *Durio zibethinus* Murr
Common name durian

 **A good source of vitamin C**

VARIETIES

In Malaysia, more than 100 durian varieties have been developed, and they are identified only by number. The better varieties of fruit have a thick, well-flavored pulp with a creamy custard-like consistency. The pulp varies from deep cream, yellow, and orange to a violet-swirled yellow. This swirled variety is noted for its flavor, which alternates between bitter and sweet.

ORIGIN & BOTANICAL FACTS

Commercial production of durian is concentrated in its native Thailand, Malaysia, and Indonesia. Thailand is by far the largest producer of durian, followed by Malaysia and Indonesia. However, Malaysia is the largest exporter of fresh durian. Other southeast Asian countries such as the Philippines also produce durian but on a smaller scale and mostly for domestic markets. Fresh durians are usually shipped to nearby countries such as Singapore,

Hong Kong, and Taiwan. Nearly all of the small quantity of frozen durian exported from Thailand is shipped to the United States, Australia, and Canada.

The durian tree can reach a height of about 125 feet and may bear fruit twice a year. The crop is heaviest between June and August. The fruits take 3 months to develop. The yield increases with the age of the tree, beginning with 10 to 40 fruits during the 1st year, increasing to about 100 fruits during the 6th year, and commonly reaching a yield of 200 after the 10th year. Ripe fruits are usually allowed to fall and are collected daily. The fruits also may be harvested directly from the tree, a common practice in Thailand. Harvested fruits taste better and have a shelf life of 9 to 11 days, compared with 2 to 5 days when the fruit is allowed to drop from the tree.

USES

Durian can be stored at room temperature 2 to 5 days. The ripeness of durian may be indicated by the emission of a strong, but not sour, smell when a knife is inserted into the center of the fruit; however, an inserted knife that comes out sticky is the best indication that the fruit is ripe. Durian is generally eaten fresh or made into desserts such as milk shakes, ice cream, or custard. In Indonesia, fermented durian is wrapped in palm leaves and served as a popular side dish called "tempoya." The fruit also is mixed with rice and sugar to make a dessert called "lempog." In addition, durian seeds can be roasted or cut into slices and fried in spiced coconut oil. They are then eaten with rice or mixed with sugar to make a sweet.

NUTRIENT COMPOSITION

The durian is a good source of vitamin C.

SERVING SIZE: *1/4 cup, raw (61 g)*

NUTRIENT CONTENT

Energy (kilocalories)	89
Water (%)	65
Dietary fiber (grams)	2
Fat (grams)	3
Carbohydrate (grams)	16
Protein (grams)	0
Minerals (mg)	
Calcium	4
Iron	0
Zinc	0
Manganese	0
Potassium	265
Magnesium	18
Phosphorus	23
Vitamins (mg)	
Vitamin A	3 RE
Vitamin C	12
Thiamin	0
Riboflavin	0
Niacin	0
Vitamin B6	0
Folate	–
Vitamin E	–

Note: A line (–) indicates that the nutrient value is not available.

FEIJOA

The feijoa, also called pineapple guava, is an oval fruit that grows up to about 3 inches in length. It has a thin, waxy, blue-green to olive skin that at times has a red or orange blush. The flesh is creamy white and somewhat granular and surrounds a translucent, jelly-like center that encloses 20 to 40 tiny, edible, oblong seeds. Feijoa has a fairly complex flavor that is often compared to that of pineapple but also contains hints of strawberry, guava, quince, and mint.

Family Myrtaceae
Scientific name *Feijoa sellowiana* O.
Common name feijoa, pineapple guava

♥ **High in vitamin C**

♥ **A good source of folate**

VARIETIES

Most varieties of feijoa cultivated today originated in Australia, New Zealand, or California. The Choiceana from Australia is a small to medium-sized fruit with a smooth skin and pleasant flavor. Selected from Choiceana seedlings, the Mammoth is a larger variety from New Zealand with thick, wrinkled skin. The most widely cultivated variety in California is the Coolidge, a small to medium-sized fruit with wrinkled skin and mild flavor.

ORIGIN & BOTANICAL FACTS

The feijoa is native to South America, specifically the cool subtropical and tropical highland areas of southern Brazil, Uruguay, Paraguay, and Argentina. The plant was introduced to California in the 1890s, and even though the feijoa is not in great demand commercially, 1,000 acres in California are dedicated to its cultivation. Some fruit is imported to the United States from New Zealand.

The feijoa is a slow-growing evergreen shrub that can be retrained to be a small tree or pruned to form a dense hedge or screen. Unpruned, it can reach 15 to 20 feet in height and in width. Its thick, oval leaves are green on top and silvery underneath, a feature that makes it an attractive plant when ruffled by a breeze. The flowers of the feijoa, formed singly or in clusters, have white petals with bristly, scarlet stamens.

Feijoas thrive in a variety of soils, but they do best in well-drained, non-saline soil. They prefer cool winters and moderate summers; the fruit is less flavorful in warm climates. Propagation is most successful by cuttings and by layering and grafting. Trees propagated from seed do not produce fruit until they are 3 to 5 years old, and the fruit may be inferior in quality.

Feijoas mature 4 1/2 to 7 months after the flowers bloom, depending on the climate. Fruit that is picked when it is still firm will ripen at room temperature, but feijoas are most flavorful when allowed to remain on the tree until they are ready to drop. Harvesting is accomplished by shaking the tree and letting the fruit fall onto a tarpaulin to prevent bruising.

USES

Feijoas should be firm and unblemished. They should be eaten within 3 to 4 days of purchase or refrigerated up to a month. Feijoas should be peeled before eating, because the skin is bitter. Immersing the peeled fruit in water and fresh lemon juice keeps it from turning brown. Feijoas usually are eaten fresh as desserts or used as garnishes or in fruit salads. They can be stewed or baked in puddings, pies, and pastries or made into jellies and preserves.

NUTRIENT COMPOSITION

Feijoas are high in vitamin C and are a good source of folate.

SERVING SIZE:
3, raw (150 g)

NUTRIENT CONTENT

Energy (kilocalories)	74
Water (%)	87
Dietary fiber (grams)	–
Fat (grams)	1
Carbohydrate (grams)	16
Protein (grams)	2
Minerals (mg)	
Calcium	26
Iron	0
Zinc	0
Manganese	0
Potassium	233
Magnesium	14
Phosphorus	30
Vitamins (mg)	
Vitamin A	0 RE
Vitamin C	30
Thiamin	0
Riboflavin	0
Niacin	0
Vitamin B$_6$	0.1
Folate	57 µg
Vitamin E	–

Note: A line (–) indicates that the nutrient value is not available.

FIG

The fig is a pleasantly sweet fruit that consists of a soft flesh pursed around a large number of tiny edible seeds. It can be eaten whole, peeled or unpeeled. Under certain circumstances, the natural sugars crystallize on the surface of the fruit, making the fruit sweeter. Figs are small, about 1 to 3 inches long. The shape varies from plain round or oval to gourd shaped, and the color ranges from brown to purple-black to almost white.

Family Moraceae
Scientific name *Ficus carica*
Common name fig

♥ **Raw figs are a good source of vitamin B₆ and are high in fiber**

♥ **Dried figs are high in fiber**

VARIETIES

Hundreds of varieties of figs exist throughout the world. Although they vary in shape and color, all have the same fleshy, gelatinous pulp. The most well-known varieties are the greenish Adriatic fig, which has a white flesh; the Smyrna, a familiar pear-shaped Turkish purple-brown fig; the Kadota; the Celeste; the Magnolia or Brunswick; and the Mission. Most domestic figs are grown in the Fresno area of California. These varieties include the Calimyrna (a Californian version of the Smyrna), the Mission, the Adriatic, and the Kadota.

ORIGIN & BOTANICAL FACTS

The fig is believed to be as old as humankind. In the Bible, fig leaves served as the first clothing for Adam and Eve in the Garden of Eden. Fig remnants have been found in excavation sites dating to 5000 B.C., and among the ancient Greeks, Romans, Egyptians, and Muslims, the fig had a symbolic and spiritual significance.

The fig's origin has been traced to western Asia and to Egypt, Greece, and Italy. Today, figs are found in all warm, dry climates, especially sunny areas of the Mediterranean. Rainy seasons are not favorable to the fruit's development. Excess moisture can split the skin and accelerate decay. Figs were brought to the Americas by Spanish conquistadors in the 16th century. The fruits arrived in California through Catholic missions and were planted in areas around San Diego and Sonoma.

The fig is a broad, irregular, picturesque deciduous tree that generally reaches 10 to 30 feet in height but can sometimes reach 50 feet. The leaves are large, bright-green, and hairy on both sides. Fig trees are valued for their shade.

Fig tree blossoms do not appear on the branches. Instead, the flower grows inside the fruit, which is actually a flower that is inverted into itself. The seeds are actually underdeveloped, unfertilized ovaries of the real fruit which impart the resin-like flavor associated with figs. Because figs will not continue to ripen after harvest, they must be allowed to ripen fully on the tree. Because fresh figs are delicate, highly perishable, and very sensitive to cold, 90 percent of all harvested figs are dried. Dried figs are available year-round, but the peak season for fresh figs lasts from June to October.

USES

Fresh figs should be plump and fairly soft but free of bruises. Figs are quite perishable and should be refrigerated no more than 7 days. Delicious as a snack, figs also can be diced and added to salads and other dishes or used for pie fillings and preserves. Figs also add sweetness and moisture to baked goods.

NUTRIENT COMPOSITION

Raw figs are a good source of vitamin B₆ and are high in dietary fiber. (See the Appendix, page 434, for the nutrient content of dried figs.)

SERVING SIZE:
3 medium, raw (150 g)

NUTRIENT CONTENT

Energy (kilocalories)	111
Water (%)	79
Dietary fiber (grams)	5
Fat (grams)	0
Carbohydrate (grams)	29
Protein (grams)	1
Minerals (mg)	
Calcium	53
Iron	1
Zinc	0
Manganese	0
Potassium	348
Magnesium	26
Phosphorus	21
Vitamins (mg)	
Vitamin A	21 RE
Vitamin C	3
Thiamin	0.1
Riboflavin	0.1
Niacin	1
Vitamin B₆	0.2
Folate	9 µg
Vitamin E	1

GRAPEFRUIT

The grapefruit, one of the largest members of the citrus family, measures up to 5 or 6 inches in diameter. It is a plump, imperfectly round fruit with thick, glossy skin that varies from yellow to pink-tinged yellow. Like all citrus fruits, the flesh of the grapefruit is segmented and each segment is tightly wrapped in a semiopaque, thin, fibrous membrane, the albedo. The segments are arranged spherically around a solid axis. The juicy flesh has a refreshing tart taste.

Family Rutaceae
Scientific name *Citrus paradisi*
Common name grapefruit

♥ **High in vitamin C**

♥ **Contains antioxidants that may help prevent certain forms of cancer**

VARIETIES

The varieties of grapefruit are categorized by the colors of their flesh, which range from white to bright pink or red. The white grapefruit has pale-yellow skin and flesh, whereas the pink or red grapefruit has rose to bright-pink flesh and pink-tinged yellow skin. The flavor varies from a biting, bitter tang to honey-sweet; the white is the more bitter. The most common variety of white grapefruit is the White Marsh, and the most popular pigmented varieties are the Flame, the Rio Red, and the Star Ruby. Some less familiar varieties are the Duncan and the Golden.

ORIGIN & BOTANICAL FACTS

Citrus fruits have been part of the human diet since the Stone Age, but the origins of the grapefruit are a mystery. Some evidence suggests that grapefruit may have originated in China 4,000 years ago and its seeds spread worldwide by insects. Others believe the grapefruit may be a descendent of the pomelo, dropped on Jamaican land by seagulls traveling from the island of Barbados, where the fruit was brought by a captain who worked for one of the East Indian trading companies. Disagreement even exists about the origins of the name "grapefruit." One theory holds that it was so named because the growing fruits resemble a cluster of grapes.

The grapefruit tree is a large evergreen with dark, glossy, green leaves. The grapefruit prefers warmer climates and therefore thrives best in the southern states. Today, Florida, Texas, and California supply 90 percent of the world's grapefruit. Because the fruits ripen at different times in different areas, the fruits are available year-round.

USES

Fresh grapefruit may be left at room temperature in a well-ventilated area for up to a week or kept up to 6 to 8 weeks in the crisper of a refrigerator. Exposure to ethylene gas from other ripening fruits may accelerate decay.

Fresh grapefruit halves are refreshing at breakfast, for a snack, or as a first course before dinner. Grapefruit that has been lightly sprinkled with sugar and broiled makes a pleasant, old-fashioned dessert.

Grapefruit sections can be added to fruit or vegetable salads, paired with avocado, or served as a complement to seafood salad. Grapefruit skin can be candied or used to make marmalade.

NUTRIENT COMPOSITION

Grapefruit is high in vitamin C. The pink and red varieties contain vitamin A (beta-carotene) and lycopene, an antioxidant that may help prevent cancer.

Grapefruit contains a chemical that can alter intestinal absorption of some medications and lead to higher than normal blood levels of some drugs and potential problems. Individuals who take prescription medications and who frequently drink grapefruit juice or eat grapefruit should notify their health care practitioners.

SERVING SIZE:
1/2 medium, raw (128 g

NUTRIENT CONTENT

Energy (kilocalories)	41
Water (%)	91
Dietary fiber (grams)	1
Fat (grams)	0
Carbohydrate (grams)	10
Protein (grams)	1
Minerals (mg)	
Calcium	15
Iron	0
Zinc	0
Manganese	0
Potassium	178
Magnesium	10
Phosphorus	10
Vitamins (mg)	
Vitamin A	15 RE
Vitamin C	44
Thiamin	0
Riboflavin	0
Niacin	0
Vitamin B₆	0.1
Folate	13 µg
Vitamin E	0

GRAPES

More grapes are grown than any other fruit in the world. These popular berries are produced in thousands of varieties, growing in clusters on climbing vines and low shrubs throughout most of the world's temperate zones. Grapes have juicy, sweet flesh and smooth skins that range from pale yellowish green to purplish black.

Family Vitaceae
Scientific name *Vitis* species
Common name grapes

♥ **Contain phytochemicals that may reduce heart disease**

VARIETIES

The thousands of varieties of grapes can be divided into two basic types: European (*Vitis vinifera*) and American (*Vitis labrusca*). Both are grown in the United States, but the European varieties are the more popular. Most American grapes (such as the Concord) are slip-skin types, meaning that the skins slide off easily, whereas the skins of most European grapes cling tightly to the flesh. Grapes are classified by whether they have seeds or are seedless. They also can be classified by their uses, such as for the making of wine (such as cabernet), for commercial foods (such as concord grapes for jelly), or for eating at the table (such as Thompson).

ORIGIN & BOTANICAL FACTS

Grapes are among the oldest cultivated fruits. Fossil evidence indicates that grapes were consumed, and possibly cultivated, as early as 8,000 years ago near what is now northern Iran, between the Black and Caspian seas.

In precolonial America, native grapes (*Vitis girdiana*) grew wild along the banks of rivers and streams, but these grapes were very sour. Spanish missionaries traveling north from Mexico in the late 18th century are believed to have brought the cultivation of European grapes to California.

Today, California produces about 97 percent of all domestic grapes.

Grapes can grow in almost any climate, but they thrive in temperate regions with average annual temperatures above 50° Fahrenheit. Although modern farm machinery is used, some aspects of grape growing, or "viticulture," are still done by hand. Grapevines generally are propagated from grafts and cuttings rather than from seed. Five years is required for a young grapevine to reach optimal production. The woody vines must be staked to support the weight of the fruit. Like most fruit, grapes develop sugar as they ripen, but they do not get sweeter after they are picked. Domestic grapes are available from May through January or March, and imported grapes fill the gap during late winter and spring.

USES

When selecting grapes, it is best to choose those with a powdery-looking coating called "bloom." Green grapes should have a slight gold cast, and dark grapes should be uniform in color. Grapes can be refrigerated in a perforated plastic bag for up to 3 days. Table grapes such as Thompson seedless are served fresh or frozen. Concord grapes are made into preserves, jams, jellies, and juices. Others are dried into raisins and currants or crushed to make juice and wine, depending on variety. Red and purple wine or grape juice is made by including the skins in the processing of the grapes, whereas the skins are removed to make white wine and juice.

NUTRIENT COMPOSITION

Some varieties of grapes are good to excellent sources of vitamin C, whereas others are not. Moderate consumption of red wine, which contains the phytochemical resveratrol, along with a heart-healthy diet may contribute to the prevention of heart disease. (See the Appendix, page 436, for the nutrient content of raisins.)

SERVING SIZE: *European type, raw, 1/2 cup (18 fruits) (80 g)*

NUTRIENT CONTENT

Energy (kilocalories)	57
Water (%)	71
Dietary fiber (grams)	1
Fat (grams)	0
Carbohydrate (grams)	14
Protein (grams)	1
Minerals (mg)	
Calcium	9
Iron	0
Zinc	0
Manganese	0
Potassium	148
Magnesium	5
Phosphorus	10
Vitamins (mg)	
Vitamin A	6 RE
Vitamin C	9
Thiamin	0.1
Riboflavin	0
Niacin	0
Vitamin B$_6$	0.1
Folate	3 µg
Vitamin E	0

GUAVA

Guavas are usually round or oval and approximately 2 to 4 inches in diameter. Embedded in the center of the pulp are numerous (100 to 500) tiny, peach-colored, round edible seeds. The seeds encircle a pulp that is softer, sweeter, and less granular than the outer part of the fruit. The thin skin, green and tart when unripe, can take on shades of yellow, white, pink, or light green when ripe and edible.

Family Myrtaceae
Scientific name *Psidium guajava*
Common name guava, guyava

♥ **High in vitamin C and fiber**

♥ **A good source of vitamin A
   (beta-carotene)**

VARIETIES

Guavas differ greatly in flavor, and the pulp can vary from white to pink, yellow, or red depending on the variety. The varieties found most often in U.S. markets are the common, lemon, and strawberry guava. The juice varieties usually have deep-pink flesh and hard, inedible seeds.

ORIGIN & BOTANICAL FACTS

The guava is believed to have originated in an area extending from southern Mexico through parts of Central America. Today, the guava is grown throughout the tropics and subtropics and is an important fruit in many parts of the world, including Mexico, India, and southeast Asia. Domestically, guavas are grown in Hawaii, Florida, and parts of southern coastal California.

The evergreen guava tree grows to a height of about 35 feet with spreading branches. The leaves are long, leathery, and aromatic when crushed. The fruit, technically a berry, generally matures 90

to 120 days after flowering. Although it can survive outside subtropical areas, the guava prefers warm, frost-free climates. Fruits grown in cooler climates tend to be inferior in flavor.

USES

The softest, yellowest guavas, free of blemishes, are best for purchase. They can be ripened at room temperature and refrigerated in a perforated plastic bag. Mature but green guavas can be kept refrigerated for several weeks and will ripen at room temperature in 1 to 5 days. The ripening process can be accelerated by placing the fruit in a paper bag. Ripe fruit that has changed color should be eaten within a couple of days because it will bruise easily and rot quickly. The just-ripened fruit is crisper in taste than the fully ripe fruit. Guavas can be frozen for extended periods of storage. The flesh of the guava can be eaten with a spoon or peeled and sliced. Puréed guava is used as a marinade or a dessert sauce or to make smoothies or sorbet. Commercially, guava is often made into juice.

NUTRIENT COMPOSITION

Guavas are a good source of vitamin A (beta-carotene) and are rich in vitamin C, although much of the vitamin C is in the rind of the fruit. Guavas are also high in dietary fiber and contain lycopene, a carotenoid with antioxidant properties.

SERVING SIZE:
1, without seed (90 g)

NUTRIENT CONTENT

Energy (kilocalories)	46
Water (%)	78
Dietary fiber (grams)	5
Fat (grams)	1
Carbohydrate (grams)	11
Protein (grams)	1
Minerals (mg)	
Calcium	18
Iron	0
Zinc	0
Manganese	0
Potassium	256
Magnesium	9
Phosphorus	23
Vitamins (mg)	
Vitamin A	71 RE
Vitamin C	165
Thiamin	0
Riboflavin	0
Niacin	1
Vitamin B$_6$	0.1
Folate	13 µg
Vitamin E	1

JACKFRUIT

The jackfruit is the largest tree-borne fruit in the world, reaching 80 pounds in weight, up to 36 inches in length, and 20 inches in diameter. This oval fruit has a pale-green to dark-yellow rind when ripe and is covered with short, sharp, hexagonal, fleshy spines. The interior consists of large, soft, yellow bulbs that taste like banana. The flesh encloses hundreds of smooth, oval, light-brown seeds.

Family Moraceae
Scientific name *Artocarpus heterophyllus*
Common name jackfruit, jakfruit

♥ **A good source of vitamin C**

♥ **Provides a moderate amount of vitamin A (beta-carotene)**

VARIETIES

A relative of breadfruit, jackfruit comes in two main varieties. One variety has a fibrous, soft, sweet flesh with a texture similar to that of raw oysters. The other, more commercially important, variety is crisp and almost crunchy with a flavor that is not quite as sweet. This latter variety is more palatable to western tastes.

ORIGIN & BOTANICAL FACTS

Believed to be indigenous to the rain forests of India, the jackfruit has spread to other parts of India, southeast Asia, the East Indies, the Philippines, central and eastern Africa, Brazil, and Suriname. Although adapted to humid tropical and near-tropical climates where it can reach the size of a large eastern oak, the mature jackfruit can withstand bouts of frost, unlike its cousin, the breadfruit.

Jackfruits mature 3 to 8 months after flowering, as indicated by a change in fruit color from light green to yellow-brown. After ripening, the fruits turn brown and spoil very quickly.

USES

Throughout Asia, unripe jackfruit is often boiled, fried, or roasted. The ripe fruit, which emits a pleasant smell and has a sweet taste, is usually eaten fresh as a dessert, or fermented and distilled to produce a liquor. Jackfruit also is preserved by drying or canning. Jackfruit seeds are roasted or boiled and eaten like chestnuts or, in India, used in curries.

NUTRIENT COMPOSITION

The jackfruit is a good source of vitamin C. One serving also provides a moderate amount of vitamin A (beta-carotene).

SERVING SIZE:
1/2 cup (83 g)

NUTRIENT CONTENT

Energy (kilocalories)	78
Water (%)	121
Dietary fiber (grams)	1
Fat (grams)	0
Carbohydrate (grams)	20
Protein (grams)	1
Minerals (mg)	
Calcium	28
Iron	0
Zinc	0
Manganese	0
Potassium	250
Magnesium	31
Phosphorus	30
Vitamins (mg)	
Vitamin A	25 RE
Vitamin C	6
Thiamin	0
Riboflavin	0.1
Niacin	0
Vitamin B$_6$	0.1
Folate	12 µg
Vitamin E	0

JUJUBE

The jujube may be round or oblong and about the size of an olive or a date, depending on the variety. As the fruit ripens, maroon spots begin to appear on the thin, shiny, green skin until the entire fruit is reddish brown or almost black. Shortly after turning color, the crunchy fruit begins to soften and wrinkle. The yellow or green flesh surrounds a single hard stone that contains two seeds. Although not particularly juicy, the flesh is sweet, especially when the fruit has changed color.

Family Rhamnaceae
Scientific name *Ziziphus jujuba*
Common name jujube, Chinese jujube, Chinese date, red date, Tsao

 Fresh jujube is high in vitamin C

VARIETIES

Of the more than 400 jujube types, Li and Lang are the two most commonly available. Li, an early ripening variety, yields round fruits that are best picked and eaten while still green. Lang produces pear-shaped fruits that are most flavorful when left to brown and dry on the tree.

ORIGIN & BOTANICAL FACTS

The jujube is native to China, where it has been cultivated for more than 4,000 years. Jujube plants were brought to Europe around the year 1 A.D. and subsequently became widely cultivated throughout the Mediterranean region. From Europe, the jujube was introduced to the United States in the early 19th century. Although most of the jujube supply in the United States is imported from China, some is grown on the West Coast.

Growing up to 40 feet in height, the deciduous jujube tree is graceful and ornamental with small, shiny green leaves and drooping, zigzag-shaped, thorned branches. The tiny, somewhat fragrant, flowers are produced in large numbers, but only a small number set fruit. Although capable of withstanding a wide range of tropical and subtropical climates, the tree nevertheless requires summer sun and heat to maximize fruit production.

USES

Jujubes can be used fresh, dried, canned, or preserved. Fresh jujubes should be firm and free of blemishes. Ripe jujubes should be refrigerated in a perforated plastic bag. Dried jujubes should be heavy and wrinkled and are usually soaked before being used. They can be candied; added to cakes and other desserts, soups, stews, or stuffings; or substituted in recipes that call for raisins or dates. Poached jujubes can be added to fruit compotes.

A candy called "jujube," which is made from jujube paste, is available in the United States. Jujubes also can be pressed to make juice or fermented to make an alcoholic beverage.

NUTRIENT COMPOSITION

One serving of raw jujube is high in vitamin C. (See the Appendix, page 436, for the nutrient content of dried jujube.)

SERVING SIZE:
Jujube, fresh (100 g)

NUTRIENT CONTENT

Energy (kilocalories)	79
Water (%)	78
Dietary fiber (grams)	–
Fat (grams)	0
Carbohydrate (grams)	20
Protein (grams)	1
Minerals (mg)	
Calcium	21
Iron	0
Zinc	0
Manganese	0
Potassium	250
Magnesium	10
Phosphorus	23
Vitamins (mg)	
Vitamin A	4 RE
Vitamin C	69
Thiamin	0
Riboflavin	0
Niacin	1
Vitamin B$_6$	0.1
Folate	–
Vitamin E	–

Note: A line (–) indicates that the nutrient value is not available.

KIWI

The kiwi is a small fruit (approximately the size and shape of a large hen's egg) with a brown, hairy skin. Its flesh is bright green, with tiny, black, edible seeds arranged in circular rows. The fruit has a mild, sweet flavor, which has variously been described as resembling citrus, melon, and strawberry, with a hint of pineapple.

Family Actinidiaceae
Scientific name *Actinidia deliciosa*
Common name kiwi, kiwi fruit, Chinese gooseberry

♥ **High in vitamin C**

♥ **A good source of fiber**

VARIETIES

The most common variety of kiwi grown commercially is the Hayward, a domestic variety with little cold tolerance but comparatively large size, full flavor, and excellent keeping quality.

The *Actinidia arguta* and *Actinidia kolomikta* varieties are more winter hardy than *Actinidia deliciosa*. However, despite sweeter taste and superior hardiness, these varieties have not been commercially successful because of smaller size, softer consistency, and shorter shelf life.

ORIGIN & BOTANICAL FACTS

Kiwi originated in China's Yangtze River Valley, where its vines grow wild on trees and bushes (thus its original English name of "Chinese gooseberry"). Introduced to New Zealand in the early 1900s, the fruit got its common name from its resemblance to the small, brown, fuzzy-looking native bird. The fruit was introduced to the United Kingdom, Europe, and the United States about the same time. Widespread planting began in the 1960s in California, where kiwi is now a major commercial crop. Kiwi also is supplied by China and South Africa.

Kiwi grows on woody, deciduous vines with large, thick leaves. Strong trellising is necessary to support the size and weight of the plant when it is heavy with fruit. The plants are dioecious, meaning that male and female flowers develop on different plants. Thus, both male and female plants are needed for pollination. The male plant does not produce fruit but is sometimes used as a landscape decoration because of its attractive flowers.

Kiwi is propagated by seeds, cuttings, and grafting. Plants grown from cuttings or grafting take 1 year to produce fruit, whereas vines propagated from seeds need more time to mature and will produce fruit only after 2 to 3 years. The plants require a long, frost-free growing season of about 220 days for fruit ripening. In California, the vines leaf in mid to late March and flower in May. Although the fruit may achieve full size in midsummer, it is not sufficiently ripe for picking until late October or early November. If temperatures fall below 29° Fahrenheit between leafing and harvesting, the leaves, blossoms, and fruit will be damaged.

USES

When selecting ripe kiwi, look for those that are plump and slightly soft. Unripe kiwi can be ripened in 2 to 3 days by placing in a paper bag with a ripe apple and leaving at room temperature. Kiwi can be stored in the refrigerator in a plastic bag for up to 2 weeks. Kiwi can be peeled and eaten fresh, cooked, frozen, or canned. Its juice can be consumed alone or in combination with other beverages. Kiwi also contains enzymes that are similar to papain, an enzyme in the juice of unripe papayas which digests protein and can be used as a meat tenderizer.

NUTRIENT COMPOSITION

Kiwi is high in vitamin C and a good source of fiber.

SERVING SIZE:
1 large (91 g)

NUTRIENT CONTENT

Energy (kilocalories)	56
Water (%)	42
Dietary fiber (grams)	3
Fat (grams)	0
Carbohydrate (grams)	14
Protein (grams)	1
Minerals (mg)	
Calcium	24
Iron	0
Zinc	–
Manganese	–
Potassium	302
Magnesium	27
Phosphorus	36
Vitamins (mg)	
Vitamin A	16 RE
Vitamin C	89
Thiamin	0
Riboflavin	0
Niacin	0
Vitamin B$_6$	0.1
Folate	35 µg
Vitamin E	–

Note: A line (–) indicates that the nutrient value is not available.

KUMQUAT

The kumquat is a small fruit, about 1 to 2 inches in diameter, round or oval, that resembles a small orange in flavor and appearance. Its name is derived from the Chinese word "kam kwat," meaning gold orange. Its thin, bright-orange skin has a sweet, spicy taste, and its slightly dry flesh, which contains numerous small white seeds, is quite tart. The kumquat has a distinctive flavor that is both sweet and sour.

Family Rutaceae
Scientific name *Fortunella japonica, Fortunella margarita*
Common name kumquat

♥ **High in vitamin C and dietary fiber**

VARIETIES

The two major varieties of kumquat are the oval *Fortunella margarita* and the round *Fortunella japonica*. A common oval type is the Negami kumquat, a hardy variety grown in the United States.

ORIGIN & BOTANICAL FACTS

Kumquats are native to China, but they also have been cultivated in Japan, southeast Asia, and Java for centuries. Today the kumquat tree remains a sacred symbol of the Chinese lunar New Year. The fruit signifies gold and good fortune. The cultivation of kumquats has spread to Australia, Israel, Spain, and the Americas. In the United States, California and Florida are the leading producers.

Kumquats are resistant to cold but grow best in mild, temperate climates. The tree is a small, shrub-like evergreen, usually from 6 to 12 feet high, thornless, with glossy, dark-green leaves and white flowers that resemble orange blossoms. The tree also can thrive in a pot, but it may

be sensitive to overwatering. The fruit ripens in the fall. One tree can produce as much as 40 pounds of fruit annually.

Although kumquats are not classified botanically as citrus fruits, they are closely related and can hybridize well with citrus. Recently, they have been crossed with limes and oranges to create limequats and orangequats. The calamondin, another small, orange-like fruit used in Philippine cooking, may be a cross between the kumquat and the mandarin orange.

USES

Kumquats that are plump, not shriveled, should be chosen. Kumquats are delicious eaten fresh and whole. Because the skin is also edible, the fruit should be washed before eating. The bright-orange fruits, fresh, candied, or preserved in syrup or brandy, also make an attractive decoration for cakes and other desserts. Kumquats soaked for several months in a mixture of vodka and honey are used as a garnish or snack. Cooked kumquats can be made into jams, preserves, and marmalades; used as garnishes for green salads and main courses; or substituted for oranges in sauces for meat and poultry.

Kumquats also can be pickled and made into relish. Kumquat trees often are used as ornamental plants.

NUTRIENT COMPOSITION

Kumquats are high in vitamin C and dietary fiber.

SERVING SIZE:
4, raw (76 g)

NUTRIENT CONTENT

Energy (kilocalories)	48
Water (%)	82
Dietary fiber (grams)	5
Fat (grams)	0
Carbohydrate (grams)	12
Protein (grams)	1
Minerals (mg)	
Calcium	33
Iron	0
Zinc	0
Manganese	0
Potassium	148
Magnesium	10
Phosphorus	14
Vitamins (mg)	
Vitamin A	23 RE
Vitamin C	28
Thiamin	0.1
Riboflavin	0.1
Niacin	0
Vitamin B6	0
Folate	12 µg
Vitamin E	0

LEMON

The lemon is a small, oval, bright-yellow citrus fruit that bulges at the blossom end. The flesh is tart and acidic and is not usually eaten out of hand. Lemons are available year-round, but production is slightly higher in the spring and summer.

Family Rutaceae
Scientific name *Citrus limonia*
Common name lemon

♥ **High in vitamin C and fiber**

♥ **Contains bioflavonoids (antioxidants) that may help prevent cancer**

VARIETIES

Lemons can be acid or sweet, but only acidic lemons are grown commercially. The two most common varieties of commercially grown lemons are the large Eureka, which has a pitted skin and few seeds, and the Lisbon, which is smaller and has a smooth skin and no seeds. Sweet lemon trees are used almost exclusively by home gardeners as ornamental plants.

ORIGIN & BOTANICAL FACTS

Lemons originated in southeast Asia, between south China and India. They may have been grown in the Mediterranean region as early as the 1st or 2nd century, because they appear in Roman artwork of the period. From there, they were brought to the rest of Europe about the time of the Crusades. Christopher Columbus brought lemon seeds to the Americas, and by the 17th century, lemons and other citrus fruits were well established in what is now Florida. Throughout the 1800s, however, Florida lemon groves were repeatedly destroyed by frost. California lemon cultivation began during the Gold Rush to alleviate the shortages of fresh fruits and vegetables that led to scurvy, a disease caused by vitamin C deficiency. Today, California is the primary source of lemons in the United States, and Arizona ranks second. Other countries with significant commercial lemon crops are Italy, Mexico, Spain, Brazil, Argentina, Iran, Turkey, India, and Egypt.

Lemon trees are tropical plants and can grow only in frost-free regions. They can be standard or dwarf size, and like other citrus trees, they have large, dark-green, evergreen leaves and produce very fragrant white flowers. Although they bloom most abundantly in the spring, they also may flower at other times of the year, depending on the climate. Only about 2 percent of the blossoms produce fruit, but that number still can bring a large harvest. Lemon and other citrus trees can live and continue to bear fruit for as long as 100 years.

USES

When selecting lemons, choose those that are heavy for their size and bright yellow. Lemons can be kept up to 2 weeks in plastic bags in the refrigerator.

Although lemons are too tart and acidic to eat as fresh fruit, they are among the most versatile and widely used fruits. The juice and grated peel are used to flavor a wide variety of foods and beverages. Spread on the surface of cut fruits (such as apples) and vegetables (such as potatoes), lemon juice prevents browning that results from oxidation. Frozen lemon juice, but not the processed type (labeled as "reconstituted"), is an acceptable substitute for fresh juice.

NUTRIENT COMPOSITION

Lemons are high in vitamin C and fiber and contain bioflavonoids (antioxidants) that may help prevent cancer.

SERVING SIZE: *1, raw without seeds (108 g)*

NUTRIENT CONTENT

Energy (kilocalories)	22
Water (%)	87
Dietary fiber (grams)	5
Fat (grams)	0
Carbohydrate (grams)	12
Protein (grams)	1
Minerals (mg)	
Calcium	66
Iron	1
Zinc	0
Manganese	–
Potassium	157
Magnesium	13
Phosphorus	16
Vitamins (mg)	
Vitamin A	3 RE
Vitamin C	83
Thiamin	0.1
Riboflavin	0
Niacin	0
Vitamin B$_6$	0.1
Folate	–
Vitamin E	–

Note: A line (–) indicates that the nutrient value is not available.

LIME

The lime is a small citrus fruit with thin, smooth, dark-green skin, measuring about 1 to 1.5 inches in diameter. The pulp is pale green and is divided into 10 to 12 segments. The fruit has an aromatic taste but is too tart for eating out of hand. Its primary use is to flavor other foods.

Family Rutaceae
Scientific name *Citrus aurantifolia*
Common name lime

♥ **High in vitamin C**

♥ **Contains antioxidants that promote health**

VARIETIES

The many varieties of lime are nearly identical in shape and appearance, but their degree of acidity ranges from nearly neutral to extremely tart. Among the high-acidity varieties are the small Mexican or Key lime, which has a sweet-tart taste. The Tahitian lime comes in two strains, the Persian and Bears, all of unknown origin and nearly seedless. These limes are grown commercially in California and the coastal areas of Florida. The Rangpur lime is highly acidic and is somewhat different in appearance from the others, with its pale-yellow peel, orange-red pulp, and green cotyledons. It is very seedy but has ornamental value. The Palestine "sweet" lime, less acidic than the others, also grows in south Florida.

ORIGIN & BOTANICAL FACTS

Limes may have originated in Asia, in the vicinity of India, Burma, and Malaysia. The silhouette of the lime can be observed in 2nd- to 3rd-century Roman art, and limes appear to have been popular in Europe around the time of the Crusades. Limes probably were brought to the New World along with other citrus species by Columbus. The Key lime was brought to the Americas by the Spaniards and cultivated in Mexico, the West Indies, some Central American countries, and the Florida Keys. The lime became popular in the West as a preventive and treatment for scurvy among British sailors. For the same reason, its popularity rose further in the United States during the California Gold Rush of 1849 and the construction of the transcontinental railroad. Four decades later, lime production ceased after a damaging freeze in the 1890s but underwent a resurgence after World War I.

The lime tree is small and crooked with thorny branches. Like its cousins in the citrus family, the tree produces small white flowers that later become the fruits. In the United States, lime trees grow best in the southern states. Southern Florida is the source of more than 85 percent of North American limes. Limes are available throughout the year, but the supply is most plentiful from May to October.

USES

When choosing limes, select those that are brightly colored and smooth-skinned. Uncut limes can be refrigerated in a plastic bag for up to 10 days. Cut limes can be tightly wrapped in plastic and refrigerated up to 5 days.

Lime juice is an excellent meat tenderizer and flavor enhancer, and it is well known as an ingredient in the mixed drink known as the margarita. Lime also has many nonculinary uses, including the manufacture of perfumes, suntan products, and cattlefeed (lime seeds are believed by some farmers to keep cattle's coats shiny and to prevent the appearance of parasites such as ticks).

NUTRIENT COMPOSITION

Lime juice is high in vitamin C and contains some antioxidants that promote health.

SERVING SIZE: *1 (2" diameter), raw (67 g)*

NUTRIENT CONTENT

Energy (kilocalories)	20
Water (%)	88
Dietary fiber (grams)	2
Fat (grams)	0
Carbohydrate (grams)	7
Protein (grams)	0
Minerals (mg)	
Calcium	22
Iron	0
Zinc	0
Manganese	0
Potassium	68
Magnesium	4
Phosphorus	12
Vitamins (mg)	
Vitamin A	1 RE
Vitamin C	19
Thiamin	0
Riboflavin	0
Niacin	0
Vitamin B₆	0
Folate	5 µg
Vitamin E	0

LONGAN

The longan is a small fruit similar to the lychee. The fruit, which develops in drooping clusters, is about 1 inch in diameter and has a smooth, yellow-brown skin. Inside is a single black seed surrounded by white, translucent flesh that has a sweet, slightly musky flavor.

Family Sapindaceae
Scientific name *Nephelium longana*
Common name longan

♥ **High in vitamin C**

NUTRIENT COMPOSITION

Longans are high in vitamin C. Fresh longans are significantly higher in vitamin C than the dried form. (See the Appendix, page 436, for the nutrient content of dried longan.)

VARIETIES

The most popular varieties of longan are the Blackball, cultivated in China, the E Bure, E Dol, and E Haw from Thailand, the Shek Kip from Hong Kong, and the Kohala, which was developed in Hawaii.

ORIGIN & BOTANICAL FACTS

The longan is native to southern China, where it remains a popular fruit. In 1903, the United States Department of Agriculture introduced the Chinese varieties of the longan to Florida, and cultivation was brought to Bermuda, Cuba, Puerto Rico, and Hawaii. Today the longan is cultivated throughout southeast Asia, in Central America, and in the United States, where the leading producers are Hawaii and Florida.

The longan tree is a tropical to subtropical evergreen that can reach heights of up to 35 feet and widths to 45 feet. Although it prefers warm weather, the mature tree can tolerate brief exposure to temperatures slightly below freezing. The tree's large leaves, up to a foot long, create a dense, dark green foliage. The tree blooms once a year with small, greenish yellow flowers. The fruit develops in large, drooping clusters over about a 4-month period. Even in the best growing conditions, fruit yield can be erratic.

The tree is propagated easily from seed, but because the tree must be 6 to 9 years old before bearing fruit (and even then the quality is not predictable), commercial propagation is usually accomplished by air layering or grafting.

USES

The longan is particularly popular in China and southeast Asia, where it is eaten fresh, dried, and canned. The fruit also stores well when frozen. Because it is similar to the lychee, the longan can be used as a substitute in a variety of recipes. In addition to providing delicious fruit, the longan tree is an attractive addition to the garden, furnishing significant shade because of the length and density of its foliage.

SERVING SIZE: *About 10, raw (32 g)*

NUTRIENT CONTENT

Energy (kilocalories)	19
Water (%)	83
Dietary fiber (grams)	0
Fat (grams)	0
Carbohydrate (grams)	5
Protein (grams)	0
Minerals (mg)	
Calcium	0
Iron	0
Zinc	0
Manganese	0
Potassium	85
Magnesium	3
Phosphorus	7
Vitamins (mg)	
Vitamin A	–
Vitamin C	27
Thiamin	0
Riboflavin	0
Niacin	0
Vitamin B$_6$	–
Folate	–
Vitamin E	–

Note: A line (–) indicates that the nutrient value is not available.

LOQUAT

The loquat is a small pear-shaped fruit that grows to about 3 inches in length and has thin yellow skin that is sometimes covered with a fine down. When the fruit is ripe, the skin peels easily from the flesh. The flesh is juicy, translucent white to orange, and slightly tart, although immature fruits can be quite sour. Each fruit contains about three to five large, smooth, dark-brown seeds.

Family Rosaceae
Scientific name *Eriobotrya japonica*
Common name loquat, May apple,
Japanese medlar, Japanese plum

♥ **High in vitamin A**

VARIETIES

Loquats are available in two varieties: orange-fleshed and white-fleshed. Orange-fleshed varieties include Gold Nugget, Strawberry, and Tanaka; white-fleshed types include Advance, Champagne, and Vista White. Gold Nugget fruits have a flavor similar to that of an apricot, whereas Strawberry fruits have a flavor similar to that of strawberries. Tanaka varieties bear long-lasting, very large, firm, orange fruits with an aromatic, sweet flavor. The translucent white-fleshed Advance fruits are juicy and pleasantly flavored. Vista White fruits have pure white flesh and a high sugar content.

ORIGIN & BOTANICAL FACTS

Although the loquat is indigenous to southeastern China, the Japanese have cultivated the plant for more than 1,000 years and have considerably improved and popularized the fruit. Loquats were introduced to Europe in the late 18th century, where they were grown initially for purely ornamental purposes. It is believed that the plants were introduced to Hawaii by the Chinese.

The loquat tree is a large evergreen that belongs to the same family as the apple, peach, and plum and can grow up to 30 feet in height. Easy to grow, the plant is often used as an ornamental plant because its long, boldly textured, dark-green leaves add a tropical look to the garden. Small white flowers with a sweet fragrance bloom in fall or early winter, and the fruits appear in clusters in early spring.

The loquat has adapted to subtropical and mild-temperate climates, but the tree will not bear fruit if the weather is too cool or excessively hot and humid. The white-fleshed varieties are better adapted to cool coastal areas than are the orange-fleshed types. Today, loquats are grown in China, Japan, India, Central and South America, the Mediterranean, the Middle East, and the United States, where the leading producers are California and Florida. Worldwide, Japan is the leading producer, followed by Israel and Brazil. Because fresh loquats bruise and perish easily, they usually are found only in the regions where they are grown. Consequently, they are not as popular or commercially successful as some other fruits.

USES

Loquats are available in fresh, dried, and canned forms in Asian markets. Fresh fruits can be stored at room temperature or, if very ripe, can be refrigerated in a plastic bag. With or without the skin, loquats are refreshing as a snack. Their tangy flavor livens up poultry dishes. They can be added to fruit salads or pies, made into jams and jellies, candied, or made into a liqueur.

NUTRIENT COMPOSITION

Loquats are high in vitamin A.

SERVING SIZE: *1/2 cup (75 g)*

NUTRIENT CONTENT

Energy (kilocalories)	35
Water (%)	87
Dietary fiber (grams)	1
Fat (grams)	0
Carbohydrate (grams)	9
Protein (grams)	0
Minerals (mg)	
Calcium	12
Iron	0
Zinc	0
Manganese	0
Potassium	198
Magnesium	10
Phosphorus	20
Vitamins (mg)	
Vitamin A	114 RE
Vitamin C	1
Thiamin	0
Riboflavin	0
Niacin	0
Vitamin B$_6$	0.1
Folate	10 µg
Vitamin E	1

LYCHEE

The lychee is a small spherical fruit, 1 to 2 inches in diameter, with a rough, inedible, bright-red shell. Inside the shell, the creamy translucent flesh surrounds a single dark, shiny seed. The texture is smooth, chewy, and sweet.

Family Sapindaceae
Scientific name *Litchi chinensis* Sonn.
Common name lychee, litchi

♥ **High in vitamin C**

NUTRIENT COMPOSITION

Lychees are high in vitamin C. (See the Appendix, page 436, for the nutrient content of dried lychee.)

VARIETIES

Of the nearly 75 varieties of lychee commonly grown today, the two most common types in the United States are the Brewster and the Mauritius, introduced from China and South Africa, respectively.

ORIGIN & BOTANICAL FACTS

Lychees originated in southern China, where the fruit is considered a symbol of love.

The lychee tree is a long-lived evergreen that reaches heights of up to 40 feet. Its leaves are pale green with tinges of pink when young, and they turn dark green and leathery when mature. In spring, large sprays of yellowish green flowers cover the trees. For the best flavor, fruits should ripen on the trees approximately 60 to 90 days. The tree requires moist, well-drained soil and a climate that is cool and dry for several months preceding flowering and hot and humid for the rest of the year. Most of the world production of lychee is concentrated in Asia, with Taiwan being the leading exporter. Australia, Israel, Mexico, and the United States also produce lychees. American production is concentrated in Florida, Hawaii, and California.

USES

Although lychees are usually eaten fresh in tropical countries, the canned versions are more often found in U.S. markets. Fresh lychees should be brightly colored and full (not shriveled) with shells that are intact and free of blemishes and with the stem still attached. The fruits can be placed in a plastic bag and stored in the refrigerator for 2 to 3 weeks or in the freezer for up to 6 months. Lychees can be eaten on their own, sprinkled with lemon or lime juice, or combined with berries and other fruit in a salad. In Hawaii, lychees are often stuffed with low-fat cream cheese, topped with crushed nuts, and served as an appetizer. Used in cooking, the sweet, aromatic flavor of the fruit complements entrées made with ham, chicken, fish, or beef.

SERVING SIZE:
10, raw (96 g)

NUTRIENT CONTENT

Energy (kilocalories)	63
Water (%)	81
Dietary fiber (grams)	1
Fat (grams)	0
Carbohydrate (grams)	16
Protein (grams)	1
Minerals (mg)	
Calcium	5
Iron	0
Zinc	0
Manganese	0
Potassium	164
Magnesium	10
Phosphorus	30
Vitamins (mg)	
Vitamin A	0 RE
Vitamin C	69
Thiamin	0
Riboflavin	0.1
Niacin	1
Vitamin B6	0.1
Folate	13 µg
Vitamin E	1

MANGO

The mango is an oval fruit with a smooth, inedible skin that varies from green to yellow to red. The fruit ranges from 2 to 9 inches long. The yellow to orange flesh of a mango is soft and very juicy. The flesh encloses one large, fibrous seed. When ripe, the fruit exudes a rich smell and the flavor is both sweet and sour.

Family Anacardiaceae
Scientific name *Mangifera indica* L.
Common name mango, mangot, manga

♥ **High in vitamin A (beta-carotene) and vitamin C**

VARIETIES

Mangoes are available in two main types, the Indian and the Indochinese (sometimes referred to as the Philippine). Between these two types, more than 100 different varieties are grown worldwide.

ORIGIN & BOTANICAL FACTS

Mangoes are indigenous to southeast Asia and India. Around the 5th century B.C., they were brought from India to other parts of tropical Asia, from where their cultivation spread to other parts of the world. The Portuguese may have introduced the fruit to the New World when they brought seeds and seedlings to Brazil. From there, the mango found its way into Florida in the late 18th century.

The mango belongs to the same family as the cashew and pistachio; it is a medium-sized to large evergreen tree. Classified as a drupe (a fruit with a single seed), most popular commercial varieties of the fruit have been cultivated to be less fibrous and more flavorful than their predecessors.

Mangoes are as popular in the tropics as the apple is in the United States. Most U.S. imports come from Mexico, with smaller numbers from Haiti, Brazil, and Peru. Puerto Rico produces most of the U.S. crop, and Florida and California produce the rest.

Worldwide, India is the leading producer and consumer of the fruit. Mexico is second to India in production and is the leading exporter of mangoes today. Pakistan, Indonesia, Thailand, and Brazil are also major producers.

USES

Mangoes are picked for shipping while still firm and green. The ripe fruit yields to slight pressure and has an intense flowery fragrance. Partially ripe mangoes will ripen at room temperature in about 3 to 5 days. Ripe fruit will keep for 2 to 3 days in the refrigerator.

Sliced or cubed, mango is often combined with papayas, bananas, and coconut to make a tropical fruit salad. Mangoes can be used to top waffles or pancakes and can be blended with yogurt and ice to make smoothies. Puréed mango can be used to make marinade for grilling meats or a dessert sauce.

Mangoes are generally consumed fresh, but canned and preserved versions and juices of the fruit also are available.

NUTRIENT COMPOSITION

Mangoes are high in vitamin A (beta-carotene) and vitamin C.

SERVING SIZE:
1/2, raw (104 g)

NUTRIENT CONTENT

Energy (kilocalories)	67
Water (%)	81
Dietary fiber (grams)	2
Fat (grams)	0
Carbohydrate (grams)	18
Protein (grams)	1
Minerals (mg)	
Calcium	10
Iron	0
Zinc	0
Manganese	–
Potassium	161
Magnesium	9
Phosphorus	11
Vitamins (mg)	
Vitamin A	403 RE
Vitamin C	29
Thiamin	0.1
Riboflavin	0.1
Niacin	1
Vitamin B$_6$	0.1
Folate	14 µg
Vitamin E	1

Note: A line (–) indicates that the nutrient value is not available.

MELONS

Melons, sweet-flavored members of the family that includes squash, cucumber, and gourds, come in an array of shapes, colors, sizes, and textures. They range in diameter from 3 inches to more than 3 feet. The skin may be white, green, yellow, orange, tan, or even black and has a surface texture that is smooth, ribbed, grooved, or netted. Inside the thick rind, the flesh may be pink, red, orange, yellow, green, or white and usually contains numerous seeds.

Family Cucurbitaceae
Scientific name *Cucumis melo* (melon), *Cucumis melo L. indorus* (honeydew melon), *Cucumis melo* var. *reticulatus* (cantaloupe), *Citrullus lanatus* (watermelon)
Common name melon, honeydew, cantaloupe (muskmelon), watermelon

ORIGIN & BOTANICAL FACTS

Melons are believed to have originated in Africa, Persia, and India. Egyptian hieroglyphics that date to 2400 B.C. provide evidence that melons have been cultivated and enjoyed for thousands of years. Melons were introduced by the Spanish Moors to most of Europe and were later introduced to the Americas in the late 15th century.

The cantaloupe is thought to be named either for Cantaloup, a village in southern France, or Cantaluppi, a papal summer residence near Rome, Italy. The true cantaloupe is a European melon that is not exported to the United States, and American "cantaloupes" are actually a type of muskmelon.

VARIETIES

Sweet melons are generally divided into two broad categories, dessert melons and watermelon. Dessert melons are further subdivided into smooth (or winter) melons, a group that includes the honeydew and casaba and the lesser known canary, Crenshaw, and Santa Claus; netted melons, including cantaloupe (also known as muskmelon) and Persian melon; and the much less familiar tropical melons, including Haogen and Galia. Popular melons include cantaloupe, honeydew, casaba, and watermelon.

Honeydew melons (*Cucumis melo L. indorus*) weigh 4 to 8 pounds and are characterized by a slightly oval shape and smooth, creamy-yellow rind. Two types of honeydew are available, those with green flesh and those with orange flesh. The orange-fleshed varieties are similar to cantaloupe in flavor and texture. Casaba melon (*Cucumis melo* var.) is globular with a pointed stem end and usually weighs 4 to 7 pounds. Casaba

rind is chartreuse-yellow with longitudinal wrinkles, and the flesh is smooth, pale green, and subtly sweet.

The netted melons are generally oval and range from 5 to 8 inches in diameter. When ripe, the fruits have a raised "netting" on a smooth, grayish beige skin. The juicy, fragrant flesh is pale to bright orange and contains numerous white seeds.

Watermelons (*Citrullus lanatus*) average 15 to 35 pounds and may be round or oval. The rind may be two-toned green or gray-green and variegated or striped, and the sweet, juicy flesh (usually red, but also occasionally orange, yellow, or white) contains rows of shiny, black seeds, although some newer varieties are seedless. Over 200 varieties of watermelon exist, with some 50 varieties grown in the United States. Varieties tend to be localized to specific regions. Smaller varieties are referred to as Icebox or apartment-size melons.

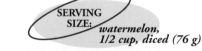

SERVING SIZE: *watermelon, 1/2 cup, diced (76 g)*

NUTRIENT CONTENT

Energy (kilocalories)	24
Water (%)	92
Dietary fiber (grams)	0
Fat (grams)	0
Carbohydrate (grams)	5
Protein (grams)	0
Minerals (mg)	
Calcium	6
Iron	0
Zinc	0
Manganese	0
Potassium	88
Magnesium	8
Phosphorus	7
Vitamins (mg)	
Vitamin A	28 RE
Vitamin C	7
Thiamin	0
Riboflavin	0
Niacin	0
Vitamin B$_6$	0.1
Folate	2 µg
Vitamin E	0

The watermelon appears to be native to Africa, where in ancient times the fruit was valued as a source of portable water. Like the other melons, cultivation spread to India, China, and Egypt, where 5,000-year-old pictures of watermelon adorn ancient tombs. By the 1600s, watermelons were cultivated in England, Spain, and beyond. Watermelons also may be native to North America, because early French explorers found American Indians cultivating the plants in the Mississippi Valley. Thomas Jefferson grew watermelons at Monticello, and during the Civil War the Confederate army boiled down watermelons to produce sugar and molasses.

Melons grow on annual vine plants that creep along the ground, attaining lengths of 6 to 10 feet. Their leaves form a canopy over the fruits. Although melons can thrive in many kinds of soil, the highest yields and the best melons are produced with fertile, well-drained, slightly acidic sandy or silt loam. A long, frost-free season with ample sunlight, warm temperatures, and low humidity is ideal for melons. With such climatic requirements, it should not be surprising that in the United States, all but a few varieties are grown only in the south. Florida, Texas, Georgia, and California are the leading domestic producers.

Melons are ready to harvest about 70 to 120 days after seeds are sown. The ripeness of honeydew melons is determined solely by rind color. A change in color from predominantly green to predominantly white indicates the melons are ready for harvest. The maturity of cantaloupe is indicated by a thick, raised netting on the surface. In contrast, the ripeness of watermelons is difficult to judge; however, mature melons tend to have a hollow ring, the spot on the melon that touches the ground turns from white to yellow, and the leaves closest to the fruit dry and turn brown.

USES

Unripe melons can be ripened in a paper bag at room temperature. Because some melons readily absorb the odor of other foods even when uncut, they should be wrapped with plastic if kept in the refrigerator for more than a day. Melons keep well in the refrigerator up to 1 week.

Most melons can be used interchangeably in a variety of ways. They can be sliced and, if desired, the flavor can be enhanced with lemon or lime juice. The flesh can be cubed or scooped out and mixed with other fruits to make a salad. Puréed melons also make toppings for ice cream and can be used as a base for cold soups. Watermelon is always eaten fresh or puréed to make a refreshing drink. The rind is pickled to make a condiment, and in Asia the roasted seeds are eaten as a snack similar to sunflower seeds.

NUTRIENT COMPOSITION

Honeydew melon is high in vitamin C and is a good source of potassium. Watermelon is high in vitamin C. Cantaloupe is high in vitamin C and is a good source of vitamin B_6 and vitamin A. Watermelon is a source of lycopene, an antioxidant that may help protect against cancer.

SERVING SIZE: *1/8 honeydew melon (125 g)*

NUTRIENT CONTENT

Energy (kilocalories)	44
Water (%)	90
Dietary fiber (grams)	1
Fat (grams)	0
Carbohydrate (grams)	11
Protein (grams)	1
Minerals (mg)	
Calcium	8
Iron	0
Zinc	0
Manganese	0
Potassium	339
Magnesium	9
Phosphorus	13
Vitamins (mg)	
Vitamin A	5 RE
Vitamin C	31
Thiamin	0.1
Riboflavin	0
Niacin	1
Vitamin B_6	0.1
Folate	8 µg
Vitamin E	0

SERVING SIZE: *1/4 medium cantaloupe (138 g)*

NUTRIENT CONTENT

Energy (kilocalories)	48
Water (%)	90
Dietary fiber (grams)	1
Fat (grams)	0
Carbohydrate (grams)	12
Protein (grams)	1
Minerals (mg)	
Calcium	15
Iron	0
Zinc	0
Manganese	–
Potassium	426
Magnesium	15
Phosphorus	23
Vitamins (mg)	
Vitamin A	444 RE
Vitamin C	58
Thiamin	0.1
Riboflavin	0.1
Niacin	1
Vitamin B_6	0.2
Folate	23 µg
Vitamin E	0

Note: A line (–) indicates that the nutrient value is not available.

NECTARINE

Nectarines, often called "peaches without the fuzz," are generally the same size, shape, and color as their counterparts. However, nectarines tend to be sweeter, and because their flesh is firmer than that of peaches, nectarines are less juicy. The skin of a ripe nectarine is a brilliant, golden yellow with generous blushes of red. Because they contain a pit, or "stone," nectarines are classified as drupes.

Family Rosaceae
Scientific name *Prunus persica* var. *nectarina*
Common name nectarine

♥ **High in vitamin A (beta-carotene)**

♥ **A good source of vitamin C**

♥ **Contains phytochemicals that promote health**

VARIETIES

Of the more than 150 varieties of nectarines, the most popular are Fantasia, Summer Grand, Royal Giant, and May Grand. Early nectarine varieties were small and white-fleshed, and the skins were uniformly green, red, or yellow. Today's modern cross-breeding techniques (in which nectarine varieties are cross-bred with one another and with peaches) have yielded larger, more peach-like nectarines with a gold and crimson skin and yellow flesh.

ORIGIN & BOTANICAL FACTS

The nectarine is indigenous to Asia, from where it made its way to Europe and finally to the Americas. The name "nectarine" is thought to be a derivative of the name of the Greek god Nektar, and the juice of the fruit has been referred to as the "drink of the gods." The nectarine is a member of the rose (Rosaceae) family. About 98 percent of the domestic crop is grown in California, where cultivation began just over 130 years ago. These nectarines are available throughout the summer, reaching their peak in July and August. Smaller quantities are imported from South America or the Middle East in winter and early spring.

USES

Slightly underripe nectarines can be ripened at room temperature in a paper bag. Ripe fruit should be refrigerated and used within 5 days. Nectarines can be eaten out of hand or used in salads, in a variety of fresh and cooked desserts, and as a garnish for many hot and cold dishes. At breakfast, they can be eaten sliced and topped with yogurt and crunchy cereal, or used as a topping for waffles, pancakes, or French toast. Because cooking softens the fruit and enhances its sweetness, nectarines are sometimes used in recipes that require baking, grilling, broiling, or poaching. Nectarines can be substituted in any dish that calls for peaches or apricots.

Baked nectarine halves can be served with baked chicken or ham, and they make a delicious dessert.

NUTRIENT COMPOSITION

Nectarines are high in vitamin A (beta-carotene) and are a good source of vitamin C. The pit of the nectarine contains amygdalin, a compound that is converted to cyanide in the stomach. Although not likely, swallowing an occasional pit accidentally is not harmful.

SERVING SIZE: *1 medium, raw (136 g)*

NUTRIENT CONTENT

Energy (kilocalories)	67
Water (%)	86
Dietary fiber (grams)	2
Fat (grams)	1
Carbohydrate (grams)	16
Protein (grams)	1
Minerals (mg)	
Calcium	7
Iron	0
Zinc	0
Manganese	0
Potassium	288
Magnesium	11
Phosphorus	22
Vitamins (mg)	
Vitamin A	101 RE
Vitamin C	7
Thiamin	0
Riboflavin	0.1
Niacin	1
Vitamin B$_6$	0
Folate	5 μg
Vitamin E	1

OLIVES

Although considered more of a condiment, the olive is an oblong fruit, slightly pointed at one end, one-half to an inch in length. The edible skin is thin and smooth in unprocessed fruits but can be smooth or wrinkled in processed fruits and varies from the yellow-green of unripe olives to dull green, red, yellow, tan, rosy brown, and black, depending on variety, ripeness, and method of processing. The skin covers flesh of the same color, enclosing a hard pit. The flavor of fresh olives is always bitter, but the final flavor depends on both variety and method of processing.

Family Oleaceae
Scientific name *Olea europaea*
Common name olive

♥ **A source of monounsaturated fat**

VARIETIES

Olive varieties are usually divided by use into table olives and those that are pressed for oil. The oil varieties significantly outnumber the table varieties.

Table olive varieties grown in Greece include the slender, oval, purple to black Kalamata, the dark-green, cracked naphlion, and the reddish Royal (Greek black and oil-cured olives are the result of alternative methods of processing). Italy grows the brownish black Gaeta, the tartly flavored Sicilian Green, the tan Calabrese, the firm black Lugano, and the piquant brown-black Liguria. France produces the tiny brown Nicoise. Table olives grown in the United States include the Manzanilla, Sevillano, and Mission.

ORIGIN & BOTANICAL FACTS

The olive is one of the oldest known cultivated fruits; cultivation of what had been wild olive trees predates recorded history and probably began on the Greek peninsula of Attica. In the early 18th century, Spanish explorers brought olive cuttings to Peru, from where they were carried to Mexico by Franciscan monks, who later brought them to California.

The cultivated olive tree is a long-lived evergreen that requires a mild climate with warm summers and relatively cold winters. Its relative drought resistance has enabled it to thrive in the Mediterranean climate and in California. Olive trees bloom in May, producing delicate cream-colored flowers, and the fruit is harvested from early autumn to winter. The Mediterranean countries account for 95 percent of the world's olive cultivation.

USES

Fresh olives contain tannins that render them inedible; hence, all olives are processed or cured. Table olives are processed by four methods. With the Spanish method, unripe olives are fermented in brine for up to 7 months. The brine-soaking method is used on Italian and Greek olives such as Kalamatas, often added to Greek salads. Brief soaking (1 to 2 weeks) produces crunchy olives, and prolonged soaking (a month or more) results in chewier, sweeter olives. A third method, typically used in Greece, involves packing and aging olives in salt or oil, which produces olives with shriveled skins and flesh. The fourth method, developed and practiced solely in the United States, is the rapid soaking of ripe olives in lye, followed by boiling in iron (for color preservation) and canning, to produce the familiar bland, soft black olive.

Unopened cans or jars of olives can be stored at room temperature for up to 2 years. Loose olives and opened cans should be loosely covered with plastic wrap and refrigerated for no more than 2 weeks.

NUTRIENT COMPOSITION

Although olives are among the fruits with the highest fat content, it is mostly monounsaturated. (See the Appendix, page 436, for other nutrients found in olives.)

SERVING SIZE: *10 large ripe olives, 1/3 cup (44 g)*

NUTRIENT CONTENT

Energy (kilocalories)	51
Water (%)	80
Dietary fiber (grams)	1
Fat (grams)	5
Carbohydrate (grams)	3
Protein (grams)	0
Minerals (mg)	
Calcium	39
Iron	1
Zinc	0
Manganese	0
Potassium	4
Magnesium	2
Phosphorus	1
Sodium	380
Vitamins (mg)	
Vitamin A	18 RE
Vitamin C	0
Thiamin	0
Riboflavin	0
Niacin	0
Vitamin B$_6$	0
Folate	0 µg
Vitamin E	1

ORANGE

The orange is a reddish yellow, round fruit of the citrus family with a rich, juicy pulp that varies in flavor from very sweet to sour. The pulp of the orange is a segmented ball, each segment wrapped tightly in a thin semi-opaque membrane called the albedo. The flesh is encased in a sturdy, glossy skin composed of two layers. The outer layer, called the zest, has a pungent but pleasant fragrance, and the inner layer, called the pith, is white, spongy, and bitter. Although some varieties are seedless, most have seeds.

Family Rutaceae
Scientific name *Citrus aurantium* L.,
  Citrus sinensis L.
Common name orange, sweet orange,
  sour orange

♥ **High in vitamin C**

♥ **A good source of folate and fiber**

♥ **Contains antioxidants that promote health**

good guide to quality because some oranges are artificially colored to preserve shelf life and to enhance appeal and marketability.

The fruit is a great snack, although most Americans consume oranges in the form of juice. Oranges are a versatile cooking ingredient. The skin is used commercially in candy and is the base for various liqueurs and cordials.

NUTRIENT COMPOSITION

Oranges are high in vitamin C and are a good source of folate. Oranges (but not their juice) are good sources of fiber. Both contain antioxidants that promote health.

VARIETIES

There are two common types of oranges: the sweet orange, which is the more common, and the sour orange. The sour orange has a thick skin and is used predominantly in making marmalades and liqueurs. The sweet varieties are prized both for eating and for their juice. The two most common varieties of sweet orange are the navel and the Valencia. The navel orange has a thick, easy-to-peel skin, is seedless, and has a mild flavor. Valencia oranges are more commonly known as juice oranges because of their abundant juice content and thinner skin, which makes them easy to squeeze. Other sweet oranges include the blood orange, with its red pulp, and the Jaffa, imported from Israel.

ORIGIN & BOTANICAL FACTS

The name orange, "naranga" in Sanskrit, comes from the Tamil "naru" and means

"fragrant." The orange is a native of Southeast Asia. The seeds and seedlings of this golden fruit were brought to the New World by European conquerors around 1520. By the 1820s, the orange was a flourishing crop in Florida. Oranges survived the severe freeze during the winter of 1894-1895 to become the most popular fruit in the United States after apples and bananas.

Oranges grow best in areas that have a subtropical to semitropical climate. The orange tree is a lush evergreen that thrives in warm climates and can simultaneously produce flowers, fruit, and foliage. For this reason, it is nicknamed the "fertility tree." When in full bloom, the tree has a fragrant smell.

USES

Oranges may keep up to 7 days in the refrigerator or in cool room temperatures in ventilated areas. Skin color is not a

SERVING SIZE: *1 medium, raw (131 g)*

NUTRIENT CONTENT

Energy (kilocalories)	62
Water (%)	87
Dietary fiber (grams)	3
Fat (grams)	0
Carbohydrate (grams)	15
Protein (grams)	1
Minerals (mg)	
Calcium	52
Iron	0
Zinc	0
Manganese	0
Potassium	237
Magnesium	13
Phosphorus	18
Vitamins (mg)	
Vitamin A	28 RE
Vitamin C	70
Thiamin	0.1
Riboflavin	0.1
Niacin	0
Vitamin B$_6$	0.1
Folate	40 µg
Vitamin E	0

PAPAYA

Papayas are round to oval fruits that have a smooth, thin skin. When ripe, papayas have yellow skin and firm, sweet flesh that ranges from yellow-orange to salmon pink. Numerous tiny black seeds are clustered in the center of the fruit.

Family Caricaceae
Scientific name *Carica papaya*
Common name papaya, tree melon

♥ **High in vitamin C**

VARIETIES

Of the two types of papayas, Hawaiian and Mexican, the Hawaiian is the smaller, pear-shaped type that is generally found in supermarkets. A whole Hawaiian papaya weighs about a pound. The Mexican papaya is usually larger and more elongated, often weighing up to 10 pounds. This variety usually has darker flesh and, although the flavor is less intense than that of its Hawaiian counterpart, it is still quite juicy and delicious. Mexican varieties include Mexican Yellow and Mexican Red, named for the flesh color. The most common Hawaiian variety is called the Solo.

ORIGIN & BOTANICAL FACTS

Although the papaya is native to southern Mexico and Central America, it is now grown in every tropical and subtropical country. Brazil is the leading producer of papayas in the world, dominating exports to Europe. Mexico is the largest supplier of papayas to the United States and Canada. The United States crop is concentrated in Hawaii and is used to supply papayas to Japan and Canada.

Technically an herb, the papaya tree can grow to 10 or 12 feet in height. Mexican varieties are usually taller. A

thin, cylindrical, non-woody trunk is topped off by spiraling leaves that contain five to nine main segments each. All parts of the plant contain large amounts of latex. The fruits, which hang from short, thick peduncles at the base of the leaves, are usually harvested at color break (when a streak of yellow appears in the green) and should still exhibit some green in the supermarket.

Papaya plants exist in one of three sex types: male, female, and hermaphrodite. Male plants have tubular flowers but bear no fruit. Female plants have round flowers and bear round fruits. Hermaphrodite plants have characteristics of both male and female flowers and produce the pear-shaped fruit that is preferred by consumers. Some plants produce flowers of more than one sex type, depending on climatic factors. High temperatures seem to favor male flowers. Papaya seeds can remain viable for years if storage conditions are dry and cool.

USES

Papayas ripen in 3 to 5 days at room temperature, and ripe fruit can be stored in the refrigerator for a week. Because cold

temperature permanently halts the ripening process, unripe fruits should not be refrigerated.

Papaya is usually eaten raw. Its cool, bland flavor complements spicy foods. Green papayas contain latex and should not be eaten raw, but they can be cooked and used in salsa or added to stews and soups. Papayas cannot be used in gelatin desserts because an enzyme in the fruit prevents the gelatin from solidifying.

NUTRIENT COMPOSITION

Papayas are high in vitamin C and contain beta-cryptoxanthin, a phytochemical that promotes health.

SERVING SIZE:
1/4 medium (76 g)

NUTRIENT CONTENT

Energy (kilocalories)	30
Water (%)	89
Dietary fiber (grams)	2
Fat (grams)	0
Carbohydrate (grams)	8
Protein (grams)	0.5
Minerals (mg)	
Calcium	18
Iron	0
Zinc	0
Manganese	0
Potassium	196
Magnesium	8
Phosphorus	4
Vitamins (mg)	
Vitamin A	22 RE
Vitamin C	47
Thiamin	0
Riboflavin	0
Niacin	1
Vitamin B$_6$	0
Folate	29 µg
Vitamin E	1

PASSION FRUIT

Shaped like an egg, the passion fruit ranges from 2 to 8 inches long. As the fruit ripens, the inedible leathery skin, which can be purple or yellow, darkens, wrinkles, and becomes brittle. The yellow pulp has a jelly-like consistency and contains many edible grape-sized, flesh-covered, black seeds that are somewhat like those of a pomegranate. The pulp's flavor is sweet-tart and lemony, and it is highly fragrant.

Family Passifloraceae
Scientific name *Passiflora edulis*
Common name passion fruit, granadilla

♥ **High in vitamin C**

♥ **A good source of vitamin A (beta-carotene)**

♥ **High in fiber**

VARIETIES

With more than 400 varieties, about 30 of which are edible, passion fruits are divided into three main categories: purple passion fruit (*Passiflora edulis* Sims), yellow passion fruit (*P. edulis f. flavicarpa* Deg.), and giant granadilla (*P. quadrangularis* L.). The purple varieties bear dark purple to black fruits that are about 2 inches long. The yellow varieties bear slightly longer (about 2 1/2 inches) deep-yellow fruits. Less commonly found are fruits of the giant granadilla varieties, which can reach 8 inches in length.

ORIGIN & BOTANICAL FACTS

Passion fruit is native to the South American tropics. Spanish missionaries, upon discovering the plant in South America, are said to have given the fruit its name, because its flowers resembled instruments of the Passion and crucifixion of Christ, such as the crown of thorns, hammers, and nails.

Being a tropical to subtropical fruit, the passion fruit is best grown in frost-free climates. In addition to South America, New Zealand, Africa, the West Indies, Malaysia, and the United States also grow passion fruit. California and Florida account for the majority of domestic production of the purple varieties, whereas Hawaii produces mainly the yellow varieties. As a commercial item, fresh passion fruit is currently considered a specialty, low-volume item in the United States. The fruit is more commonly used as an ingredient in commercial food and drink products.

The plant is a vigorous, climbing vine that can grow 15 to 20 feet a year once established. Each fruit develops from a single fragrant flower that is 2 to 3 inches wide with green and white petals. The fruits quickly turn from green to purple (or yellow) when ripe and fall to the ground within a few days. They can be picked from the vine when ripe or harvested off the ground.

USES

When ripe, passion fruits are heavy with wrinkled skin. Fruits that are heavy and firm should be chosen. Unripe fruits can be left at room temperature to ripen. Ripe fruits can be refrigerated up to 1 week.

The fruit's pulp can be eaten plain or spooned over ice cream, cakes, and other desserts. Passion fruit also makes delicious jams and jellies, to which the seeds add a crunchy texture. The fruit also can be pressed to extract a highly fragrant juice that adds a pleasant flavor to beverages such as iced tea, punch, and cocktails.

NUTRIENT COMPOSITION

Passion fruit is high in vitamin C and dietary fiber if the seeds are consumed along with the pulp. In addition, it is a good source of vitamin A (beta-carotene).

SERVING SIZE: *4 fruits, raw (72 g) (pulp only)*

NUTRIENT CONTENT

Energy (kilocalories)	68
Water (%)	73
Dietary fiber (grams)	8
Fat (grams)	0
Carbohydrate (grams)	16
Protein (grams)	2
Minerals (mg)	
Calcium	8
Iron	0
Zinc	0
Manganese	–
Potassium	252
Magnesium	20
Phosphorus	48
Vitamins (mg)	
Vitamin A	52 RE
Vitamin C	20
Thiamin	0
Riboflavin	0
Niacin	1
Vitamin B6	0
Folate	12 µg
Vitamin E	1

Note: A line (–) indicates that the nutrient value is not available.

PEACH

Peaches are round to oblong with a slight tip. Because of the hard seed, or "stone," at their core, they are known as a "stone fruit," or drupe. The fuzzy skin of peaches is the only characteristic that distinguishes them in appearance from the smooth-skinned nectarine. Ripe peaches can assume a range of colors from creamy-white to light-pink, yellow, orange, and red. The flesh also can range from a pinkish white to an intense yellow-gold. The firmness and juiciness of a peach depend largely on variety and on the degree of ripeness.

Family Rosaceae
Scientific name *Prunus persica*
Common name peach

 A good source of vitamin A (beta-carotene) and vitamin C

VARIETIES

Peaches are generally classified into one of two categories: "freestone" or "clingstone," although some are also considered "semi-freestone." Freestone peaches, the ones more commonly available, are those whose pits are easily removed, whereas the pit of clingstones is enmeshed within the flesh. Both freestone and clingstone peaches have numerous varieties that differ in skin color, flesh color, firmness, and juiciness. Two of the most popular varieties of yellow-fleshed freestone peaches are Elegant Lady and O-Henry. Other varieties include the Hale, Rio Oso Gem, and Elberta.

ORIGIN & BOTANICAL FACTS

A native of China, where they have been grown for more than 2,500 years, peaches were once revered as a symbol of longevity and immortality. The fruit made its way to Europe by way of Persia. Spanish explorers brought the plant to the New World, where Spanish missionaries planted the trees in California. Since the early 1800s, peaches have been grown commercially in the United States, which now produces one-fourth of the world's market crop. Other major producers of peaches include Italy, Greece, and China. Georgia was once the largest producer of peaches in the United States, earning it the nickname "Peach State." Today, the fruit is grown in more than 30 states, and California is the largest producer. Peaches are related to other deciduous flowering fruit trees, including plum, cherry, apricot, and almond. Although originally grown only in moderate climates, the many new varieties make it possible for peaches to be grown throughout much of the United States.

Standard trees may grow as high as 30 feet and can live up to 40 years. Some dwarf varieties may reach no more than 3 feet. Beautiful flowers, ranging from pale pink to red, appear in the spring and give way to fruits that usually ripen in midsummer. An 8- to 10-year-old tree can produce up to 6 bushels of fruit annually.

USES

Peaches that are slightly soft to firm when pressed and are free of blemishes or soft spots should be chosen. To hasten the ripening process, underripe peaches can be left in a loosely closed paper bag at room temperature for 2 to 3 days. Once ripe, they can be kept at room temperature for about 3 to 4 days or slightly longer in the refrigerator.

For cooking purposes, the skin of a peach can be easily peeled by blanching for 30 seconds.

NUTRIENT COMPOSITION

Peaches are a good source of vitamin A (beta-carotene) and vitamin C.

SERVING SIZE: *1 medium peeled, raw (98 g)*

NUTRIENT CONTENT

Energy (kilocalories)	42
Water (%)	88
Dietary fiber (grams)	2
Fat (grams)	0
Carbohydrate (grams)	11
Protein (grams)	1
Minerals (mg)	
Calcium	5
Iron	0
Zinc	0
Manganese	0
Potassium	193
Magnesium	7
Phosphorus	12
Vitamins (mg)	
Vitamin A	53 RE
Vitamin C	6
Thiamin	0
Riboflavin	0
Niacin	1
Vitamin B6	0
Folate	3 µg
Vitamin E	1

PEAR

Pears are bell-shaped fruits, wide and round at the bottom, narrowing toward the stem; however, some varieties are nearly round. Pears range in size from less than 1 inch to 3 inches in diameter. They have a smooth, thin skin that may be green, yellow, brown, or red when ripe. The juicy, sweet flesh is usually white with a tinge of yellow. The flesh of some pears is sandy in texture. Enclosed within the flesh is a cartilaginous core that contains as many as 10 seeds.

Family Rosaceae
Scientific name *Pyrus communis* (common or European pear), *Pyrus pyrifolia* (Asian pear)
Common name pear, common pear, European pear, Asian pear, sand pear

♥ **A good source of dietary fiber and vitamin C**

VARIETIES

Hundreds of pear varieties, varying in shape, size, color, texture, flavor, aroma, and time of ripening, grow throughout the world. In the United States, the four main varieties are the Bartlett, Anjou, Bosc, and Comice. The Bartlett, the most popular summer pear, is the principal variety used for canning and the only variety sold dried. Bartletts are large, juicy, fragrant, and sweet and turn from dark-green to golden-yellow when ripe. A red-skinned strain called the Red Bartlett is also available. Blander in taste are Anjou pears, which are oval with smooth yellow-green skin and a creamy flesh. The Anjou is the most abundant winter pear. Bosc pears have dull, reddish brown skin and very firm flesh. Reputedly the most flavorful and sweetest variety is the Comice, which is squat in shape with dull-green skin. A variety that is growing in popularity is the Asian pear (also known as the Oriental pear, Chinese pear, or Japanese pear). Asian pears are crunchier than the common pears, round, and golden brown to yellow-green.

ORIGIN & BOTANICAL FACTS

Native to the northern regions of central Asia, pears have been cultivated for more than 3,000 years. Pear trees were introduced to North America by the early colonists, who brought cuttings from European stock. The largest producers of pears today are China, Italy, Russia, and the United States. California, Oregon, and Washington account for 98 percent of the United States pear crop. Pears are closely related to apples in that both are pome fruits (fruits with a distinct seed-containing core) and members of the rose family. However, pear trees tend to be more upright than apple trees (commercially grown trees are usually pruned to about 20 feet high). Ideal growing conditions require a combination of warm days, cool nights, rich volcanic soil, and ample water. Between 100 and 170 days are required from bloom to harvest. Like bananas and avocados, pears are usually picked before they are fully ripe, because they do not ripen well on the tree.

USES

Firm pears, such as the Bosc, are best for baking and poaching. Because pears are picked while still green, they should be ripened at room temperature until the stem end yields slightly to pressure. Once ripe, pears should be refrigerated. Sliced pears should be sprinkled with lemon juice to prevent browning.

NUTRIENT COMPOSITION

Raw pears (with the skin on) are a good source of vitamin C and dietary fiber.

SERVING SIZE: *1 medium, raw, with skin (166 g)*

NUTRIENT CONTENT

Energy (kilocalories)	98
Water (%)	84
Dietary fiber (grams)	4
Fat (grams)	1
Carbohydrate (grams)	25
Protein (grams)	1
Minerals (mg)	
Calcium	18
Iron	0
Zinc	0
Manganese	0
Potassium	208
Magnesium	10
Phosphorus	18
Vitamins (mg)	
Vitamin A	3 RE
Vitamin C	7
Thiamin	0
Riboflavin	0.1
Niacin	0
Vitamin B$_6$	–
Folate	12 µg
Vitamin E	1

Note: A line (–) indicates that the nutrient value is not available.

PERSIMMON

The persimmon, sometimes called the "apple of the Orient," is a spherical or acorn-like, smooth-skinned fruit that ranges from a light yellow-orange to a brilliant orange-red. Persimmons vary from 1 to more than 3 inches in diameter. Except for the seeds, the entire fruit is edible.

Family Ebenaceae
Scientific name *Diospyros kaki* (Oriental persimmon), *Diospyros virginiana* (native persimmon)
Common name persimmon, Oriental persimmon, Japanese persimmon, kaki

♥ **High in vitamin A (carotenes)**
♥ **A good source of vitamin C and fiber**

VARIETIES

Persimmons are divided into two types, based on flavor and texture. The astringent type is inedible until it ripens and becomes soft, and the non-astringent type can be eaten while it is underripe and crisp. The astringent varieties are harvested while still firm and allowed to ripen fully and soften. Astringent persimmons can ripen off the tree when stored at room temperature. Non-astringent varieties are harvested when they are fully colored and ripe. Persimmons are also divided into two classes by their origin: Japanese and American. Although Japanese persimmons exist in both astringent and non-astringent varieties, they tend to be less astringent than American. The most common variety of persimmon available in the United States is the Hachiya, an astringent Japanese persimmon that is large and acorn-shaped. A popular non-astringent Japanese variety is the Fuyu, which is smaller than the Hachiya and shaped like a tomato.

ORIGIN & BOTANICAL FACTS

The Oriental persimmon, cultivated for centuries in China and later brought to Japan and Korea, was introduced to California in the 1870s. The U.S. Department of Agriculture imported persimmon trees to Florida and Georgia. In contrast, native persimmons have flourished over much of what is now the continental United States for centuries. American Indians dried the fruit to eat throughout the winter.

The Oriental persimmon is a droop-leafed deciduous tree that can attain a height and width of about 25 feet. Because the tree is relatively cold-sensitive, it is grown only in the Deep South. The native persimmon can reach heights of 30 to 40 feet and is more tolerant of poor soils and cold than the Oriental type.

Oriental persimmon trees are self-pollinating and also can produce seedless fruits from unfertilized flowers. The inconspicuous flowers are cream-colored or pink. The fruit is in season from October to February. Native persimmons are not self-pollinating. Their flowers range from white to yellow and appear in May. The fruit appears in September, but it does not ripen until the weather cools. Oriental and native varieties cannot cross-pollinate.

USES

Unripe fruit will ripen in a few days at room temperature. Ripe fruit can be refrigerated 2 to 3 days. Freezing astringent persimmons and thawing them the next day also may help remove some of the astringency.

Persimmons can be added to cakes, cookies, rolls, and breads. Persimmon pulp is used to make preserves, beer, and brandy. Puréed persimmon can be used as a sauce for poultry or dessert.

NUTRIENT COMPOSITION

Japanese persimmons are an excellent source of vitamin A (carotenes) and are a good source of vitamin C and fiber.

SERVING SIZE: *1/2 persimmon, raw (84 g)*

NUTRIENT CONTENT

Energy (kilocalories)	59
Water (%)	80
Dietary fiber (grams)	3
Fat (grams)	0
Carbohydrate (grams)	16
Protein (grams)	0
Minerals (mg)	
Calcium	7
Iron	0
Zinc	0
Manganese	0
Potassium	135
Magnesium	8
Phosphorus	14
Vitamins (mg)	
Vitamin A	182 RE
Vitamin C	6
Thiamin	0
Riboflavin	0
Niacin	0
Vitamin B6	0.1
Folate	6 µg
Vitamin E	0

PINEAPPLE

The pineapple is a cylindrical fruit that is approximately 4 to 8 inches in diameter and can reach a length of 12 inches, weighing up to 10 pounds. It has a waxy, tough rind covering a juicy flesh that surrounds a fibrous core. The flesh and core range from nearly white to yellow, and the flavor is a combination of apples, strawberries, and peaches.

Family Bromeliaceae
Scientific name *Ananas comosus*
Common name pineapple

♥ **High in vitamin C**

VARIETIES

Three varieties of fresh pineapple are available in the United States. The popular Smooth Cayenne, from Hawaii, weighs 3 to 5 pounds, and its flesh ranges from pale yellow to yellow. The Red Spanish is nearly square and has a tougher shell that makes it well suited to shipping. Its flesh is pale yellow and has a pleasant aroma. Weighing up to 10 pounds, the Sugar Loaf is the largest of the three varieties. Its white flesh lacks the woodiness often found in the core of other varieties.

ORIGIN & BOTANICAL FACTS

The pineapple is indigenous to southern Brazil and Paraguay. Columbus encountered the pineapple on his 1493 journey to the Caribbean and took it back to Europe, from where it spread to many other parts of the world on ships that carried it as protection against scurvy, a disease caused by a deficiency of vitamin C. The name is derived from piña, a name given by the Spanish, who thought that the fruit resembled a pinecone.

Pineapples do not grow on trees. They grow on a plant that is technically an herb, with large, waxy, pointed leaves. Each plant bears one fruit in the center, and each pineapple is actually the result of the fusion of many individual fruits. Unlike most other fruits (with the exception of some melons), pineapples do not have a reserve of starch that converts to sugar after harvest. Instead, the starch is stored in the stem of the plant and enters the fruit as sugar just before it ripens completely. As a result, the fruit will not become any sweeter after harvest, so growers must allow the pineapple to ripen on the plant to maximize the sugar and juice content of the fruit.

USES

When selecting a pineapple, choose one with fresh, green leaves and no obvious soft or brown spots, especially at the base. When ripe, the rind can be dark green, yellow, or reddish yellow; however, most pineapples on the market are already ripe, regardless of their color, and should be refrigerated in a plastic bag after purchase. Freshly cut pineapple may be kept, sealed airtight, in the refrigerator for up to a week.

Pineapples are consumed fresh and canned and as juice. Fresh pineapple cannot be added to gelatin, yogurt, or cottage cheese because the fruit contains a digestive enzyme called bromelain that can break down the protein in milk, meat, and gelatin and makes these foods watery. However, bromelain is degraded by heat, so canned or boiled pineapple can be used instead. Because of this enzyme activity, fresh pineapples are often used in marinades to tenderize meats and poultry, although meat that is allowed to sit in the pineapple marinade for too long can turn mushy.

NUTRIENT COMPOSITION

Pineapple is high in vitamin C and contains phytochemicals that promote health.

SERVING SIZE: *1/2 cup, diced, raw (78 g)*

NUTRIENT CONTENT

Energy (kilocalories)	38
Water (%)	87
Dietary fiber (grams)	1
Fat (grams)	0
Carbohydrate (grams)	10
Protein (grams)	0
Minerals (mg)	
Calcium	5
Iron	0
Zinc	0
Manganese	1
Potassium	89
Magnesium	11
Phosphorus	5
Vitamins (mg)	
Vitamin A	2 RE
Vitamin C	12
Thiamin	0.1
Riboflavin	0
Niacin	0
Vitamin B$_6$	0.1
Folate	8 µg
Vitamin E	0

PLUM

The plum is a drupe, a fruit with a single pit that is related to the peach, nectarine, and apricot. However, whereas only two or three varieties of those exist, plums are available in a wide variety of shapes, sizes, and colors. Plums grow in clusters and have smooth, richly colored skins. The thousands of varieties identified worldwide range from 1 to 3 inches in diameter, in flavor from sweet to tart, and in skin color from yellow to green, red, purple, and indigo blue.

Family Rosaceae
Scientific name *Prunus domestica, Prunus salicina*
Common name plum

♥ **Plums are a good source of vitamin C**

♥ **Prunes (dried plums) are a good source of vitamin A (carotenes) and fiber and contain isatin, a natural laxative**

VARIETIES

Of the more than 1,000 varieties of plums in Europe and 140 in North America, about 20 dominate the commercial supply of plums in the United States, most of which are Japanese or European varieties. The Japanese types have juicy yellow or reddish flesh and skin colors that range from crimson to black-red. The Santa Rosa and Red Beaut are two of the more popular Japanese varieties. European plums, or *Prunus domestica*, are smaller, denser, and less juicy than their Japanese counterparts. Their skin color is always blue or purple, and their pits are usually freestone, which means that they separate easily from the flesh. Among the better-known varieties are Italian, President, Empress, Stanley, and Tragedy. In the United States, the bulk of European plums are grown in the Pacific Northwest, but some varieties are successfully cultivated in the eastern states. With its firmer flesh and higher sugar and acid contents, the European variety is best suited for prunes (also called dried plums). The most common variety of plum used for prunes is the California French, also known as d'Agen. A few varieties of prune plums

are sold fresh and are called fresh prunes or purple plums.

ORIGIN & BOTANICAL FACTS

Although plums are native to several temperate regions around the world, including North America, early colonists brought European varieties with them that supplanted native American plums. In the late 19th century, dozens of varieties from Europe and Asia were cultivated in the United States, primarily in California. One of the most influential plum breeders was the famed horticulturist Luther Burbank, who in 1907 developed the Santa Rosa variety, which now accounts for about a third of the total domestic crop. The California French plum is a descendant of the first prune plums brought to California from France by Louis Pellier in the 1850s. The domestic plum season extends from May to October, beginning with the Japanese varieties and ending with the European types. Today, about 70 percent of the world's prune supply and nearly 100 percent of domestic prunes come from California.

USES

The majority of plums are eaten fresh. Plums are a nutritious, low-calorie food that can be eaten out of hand or added to fruit salads, baked goods, compotes, and meat dishes. Plums also can be made into jams, purées, or sauces. A famous food prepared from plums is the Chinese plum sauce, also known as duck sauce. Puréed prunes make a good substitute for butter and other fat in baked goods. Prunes are also made into juice.

NUTRIENT COMPOSITION

Plums are a good source of vitamin C. Prunes are a good source of vitamin A (carotenes) and fiber. (See the Appendix, page 436, for the nutrient content of prunes [dried plums].)

SERVING SIZE: *1 medium, raw (66 g)*

NUTRIENT CONTENT

Energy (kilocalories)	36
Water (%)	78
Dietary fiber (grams)	1
Fat (grams)	0
Carbohydrate (grams)	9
Protein (grams)	1
Minerals (mg)	
Calcium	3
Iron	0
Zinc	0
Manganese	0
Potassium	114
Magnesium	5
Phosphorus	7
Vitamins (mg)	
Vitamin A	21 RE
Vitamin C	6
Thiamin	0
Riboflavin	0.1
Niacin	0
Vitamin B6	0.1
Folate	1 µg
Vitamin E	0

POMEGRANATE

The pomegranate is a round fruit the size of a large orange with a protruding crown and smooth, leathery skin that can range from red to yellowish pink. Each fruit contains hundreds of ruby-colored seeds that are individually encased in a translucent, red, juicy pulp that is sweet to tart. The seeds are packed into compartments that are separated by cream-colored, bitter-tasting membranes. Both the seeds and pulp are edible.

Family Punicaceae
Scientific name *Punica granatum*
Common name pomegranate, grenadier, granada, Chinese apple

♥ **A good source of potassium and vitamins C and B₆**

VARIETIES

There are three kinds of pomegranates: one that is very sour and two that are sweet. The sour type is used in place of unripe grapes to make juice, and the sweet types are eaten as a dessert. In addition, some nonfruiting varieties are grown purely for their double flowers, and a dwarf variety has been developed that grows only 2 to 3 feet tall and makes a decorative container plant.

ORIGIN & BOTANICAL FACTS

The pomegranate is a fruit with a colorful history. The name is derived from the Old French terms "pome," for "apple," and "grenate," for "many-seeded." Native to southeastern Europe and Asia, pomegranates have long been celebrated in art and literature. The seeds have been a Hebrew symbol of fertility since biblical times, and the fruit once formed part of the decoration on the pillars of King Solomon's temple. In the 16th century, Spanish missionaries brought the plant to the New World. Currently, the pome-

granate is a crop of minor commercial importance.

The pomegranate plant is a dense, deciduous shrub that can grow up to 12 feet in height. Crimson flowers are borne on slender, somewhat thorny branches that have glossy, dark-green leaves about an inch long. The plant grows best in subtropical climates but can tolerate subfreezing temperatures. If grown from seed, plants begin to fruit after 3 to 4 years.

USES

Pomegranates are available in the United States only from October to December. Fruits should be heavy for their size and plump, as if bursting, with a slightly soft crown and shiny skin. The fruit can be refrigerated for up to 2 months or stored in a cool, dark place for about a month.

The pomegranate is rather labor-intensive to eat. After the skin has been peeled, the seeds can be removed individually, or the fruit can be cut in half and the seeds scooped away from the membrane with a spoon. The seeds also can be used as a garnish for desserts and in salads or pressed to make a refreshing drink. Used as a spice in northern India, dried pomegranate can

be substituted for raisins in cakes. Grenadine, a light syrup made from pomegranates, is used as a flavoring in cocktails, soft drinks, and confections. Pomegranate molasses is a popular ingredient in Mediterranean and Middle Eastern cooking. Aside from the fruit's culinary uses, crushed pomegranate flowers produce a brilliant red dye. The bark is used in tanning and is the source of the yellow hue of Moroccan leather.

NUTRIENT COMPOSITION

Pomegranates are a good source of potassium and vitamins C and B₆.

SERVING SIZE:
1 fruit (154 g)

NUTRIENT CONTENT

Energy (kilocalories)	105
Water (%)	81
Dietary fiber (grams)	1
Fat (grams)	0
Carbohydrate (grams)	26
Protein (grams)	1
Minerals (mg)	
Calcium	5
Iron	0
Zinc	0
Manganese	–
Potassium	399
Magnesium	5
Phosphorus	12
Vitamins (mg)	
Vitamin A	0 RE
Vitamin C	9
Thiamin	0
Riboflavin	0
Niacin	1
Vitamin B₆	0.2
Folate	9 µg
Pantothenic acid	1
Vitamin E	1

Note: A line (–) indicates that the nutrient value is not available.

PRICKLY PEAR

Two to 4 inches long and shaped like an egg, the prickly pear has a coarse, thick skin that can be yellow, orange, pink, magenta, or red, depending on the variety. The inedible skin is dotted with tubercles that have small, almost invisible spines capable of pricking the skin. The prickly pear's flesh is mildly sweet, juicy, and fragrant and contains numerous edible, small, crunchy seeds. Like the skin, the flesh can range from yellow to dark red.

Family Cactaceae
Scientific name *Opuntia ficus-indica*
Common name prickly pear, Indian fig, nopal, nopalitos, Sharon's fruit

♥ **High in vitamin C and magnesium**

♥ **A good source of fiber**

VARIETIES

The genus *Opuntia* contains as many as 1,000 species, most of which bear edible fruits. Commonly cultivated as a source of food are varieties of the species *Opuntia ficus-indica*. Other species are planted for purely ornamental purposes.

ORIGIN & BOTANICAL FACTS

Species of the genus *Opuntia* are believed to have originated in central Mexico and the Caribbean. Since pre-Columbian days, American Indians have collected the ripe fruits and tender stems, or "pads," for use as a food source and the older pads for livestock feed. Spanish explorers introduced the plants to Spain, from where they were brought to North Africa by the Moors. Today, the plant is grown worldwide in areas with a moderate climate. Prickly pear is the national fruit of Israel, where it is called Sharon's fruit.

The prickly pear plant is a perennial of the cactus family that prefers a hot, dry environment and, like other cacti, can withstand long periods of drought.

Growing up to 15 feet high, the plant has no real leaves; the segmented, flat, oval-shaped pads serve as both leaves and water-storage organs. These pads are covered with sharp spines. In midsummer, brilliant flowers bloom along the edges of the pads, from which fleshy fruits develop. The plants are easily propagated by detaching the pads and planting them in soil. Roots form quickly, and new plants soon become established. Some prickly pears harbor an interesting parasite, the cochineal, a red insect less than an eighth-inch long that is the source of a brilliant red dye. Along with the cactus that harbors the insect, the technique of isolating the dye was brought back to Europe by the conquistadors.

USES

Although commercially sold prickly pears have already had their spines removed, caution should still be used when handling the fruit. If the spines have not been removed, they should be scraped off carefully with a knife or rubbed off with a towel. The skin should be peeled before consumption. Immature fruits can be left at room temperature to ripen. When ripe, the fruits yield when gently pressed. Ripe fruits can be stored in a perforated plastic bag in the refrigerator up to 2 days.

Prickly pears are refreshing when eaten with a sprinkle of lime or lemon juice. They can be diced and used to top ice cream, sorbet, yogurt, and various desserts, or they can be puréed to make marmalade and dessert sauces. The pads (nopales), which are served as a vegetable, can be cut into pieces, steamed or stewed, and added to omelets, salads, and soups.

NUTRIENT COMPOSITION

Prickly pears are high in magnesium and vitamin C and are a good source of fiber.

SERVING SIZE:
1 fruit, raw (103 g)

NUTRIENT CONTENT

Energy (kilocalories)	42
Water (%)	88
Dietary fiber (grams)	4
Fat (grams)	1
Carbohydrate (grams)	10
Protein (grams)	1
Minerals (mg)	
Calcium	58
Iron	0
Zinc	0
Manganese	–
Potassium	227
Magnesium	88
Phosphorus	25
Vitamins (mg)	
Vitamin A	5 RE
Vitamin C	14
Thiamin	0
Riboflavin	0.1
Niacin	0
Vitamin B6	0.1
Folate	6 µg
Vitamin E	0

Note: A line (–) indicates that the nutrient value is not available.

PUMMELO

The pummelo is a pear-shaped citrus fruit that comes in a variety of sizes and colors and is believed to be the ancestor of the grapefruit. Normally between 4 and 7 inches in diameter (about the size of a cantaloupe), the pummelo can grow to the size of a large watermelon and weigh up to 20 pounds. The fruit is covered by a soft, easily peeled rind that may be half an inch or more in thickness and ranges from yellow to pink. Thick membranes separate the inner segments of this fruit. The flesh of the pummelo also varies from a light yellow to a dark pink. Although tart, the pummelo is sweeter (but firmer and less juicy) than grapefruit.

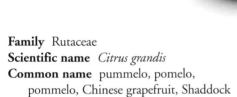

Family Rutaceae
Scientific name *Citrus grandis*
Common name pummelo, pomelo, pommelo, Chinese grapefruit, Shaddock

♥ **High in vitamin C**

VARIETIES

Among the common varieties of pummelo available in the United States are the Chandler, Ichang, Red Shaddock, Reinking, Tresca, and Webber.

ORIGIN & BOTANICAL FACTS

A native of Malaysia, the pummelo is a popular fruit in east, southeast, and south Asia. It is believed to have been introduced to the West Indies by an English sea captain named Shaddock, by whose name this fruit is sometimes called. Pummelo trees are strictly tropical and grow only in frost-free regions. They achieve heights of 15 to 30 feet, and the crown of the tree is round. Like other citrus trees, they bear fruit for many decades. Although the pummelo is not grown commercially in the United States, imported fruits are available at the market from November through March.

USES

Pummelos can be used in the same way as grapefruits. By themselves, they can be eaten as a breakfast fruit or a refreshing snack, or they can be added to fruit salads. The thick rind and pith should be peeled before use and the fruit sectioned like a grapefruit or pulled into pieces. The skin and white pith of the pummelo are candied to make a traditional Chinese treat. The fruit can be stored at room temperature for up to a week or up to 2 weeks in the refrigerator.

NUTRIENT COMPOSITION

Like most citrus fruits, the pummelo is high in vitamin C.

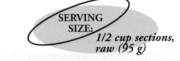
SERVING SIZE: *1/2 cup sections, raw (95 g)*

NUTRIENT CONTENT

Energy (kilocalories)	36
Water (%)	89
Dietary fiber (grams)	1
Fat (grams)	0
Carbohydrate (grams)	9
Protein (grams)	1
Minerals (mg)	
Calcium	4
Iron	0
Zinc	0
Manganese	0
Potassium	205
Magnesium	6
Phosphorus	16
Vitamins (mg)	
Vitamin A	0 RE
Vitamin C	58
Thiamin	0
Riboflavin	0
Niacin	0
Vitamin B6	0
Folate	–
Vitamin E	–

Note: A line (–) indicates that the nutrient value is not available.

QUINCE

The quince is a member of the same family as apples and pears. The mature fruit ranges in size and shape from that of a small plum to that of a large pear, depending on variety. In cool, temperate climates, the quince's rough, woolly rind develops a golden color when ripe. Its firm, white flesh has a strong fragrance, but the fruit is hard and sour and is generally inedible unless cooked.

Family Rosaceae
Scientific name *Cydonia oblonga*
Common name quince

♥ **High in vitamin C**

VARIETIES

The two most common varieties of American quince are the pineapple quince and the perfumed quince. The pineapple quince is round and has a yellow skin and white flesh that is somewhat dry. The flavor is similar to that of pineapple. The perfumed quince is the shape of a small football and has a tart flesh. Great Britain produces the Portugal, the apple-shaped, and the pear-shaped quince. The Japanese quince has a slightly more acidic flavor. Several small varieties are often used for bonsai plants.

ORIGIN & BOTANICAL FACTS

Known throughout Asia and the Mediterranean region for about 4,000 years, the quince originated somewhere in the Middle East (possibly Iran), where it still grows wild. The ancient Greeks cultivated a common variety of quince but grafted onto it a better variety from Cydon, a town in Crete, from which the word "quince" is derived.

The ancient Romans believed the quince had medicinal and mystical powers because it had been held sacred by the goddess Venus. The quince became a symbol of love and happiness, a symbolism that lasted into the Middle Ages. Quince was eaten at weddings, shared by brides and grooms as a token of their love. Medieval English manuscripts contain recipes mentioning "char de Quynce," the old name for quince marmalade. In fact, the word "marmalade" is derived from the Portuguese word for quince, "marmelo." Today the quince is cultivated throughout the Mediterranean, in South America, and in the United States, where California is the leading producer.

Quince grows as a many-branched deciduous shrub or small tree, no more than 10 to 12 feet tall, and produces large, fragrant white, pink, or red flowers before the leaves appear. The plants are propagated by seeds, shoots, cuttings, or layering. The flowering quince is popularly grown as an ornamental plant.

USES

Quince is available only in the autumn. Unripe fruits can be ripened at room temperature. Ripe quince, which is fragrant, can be kept in a perforated plastic bag in the refrigerator up to 2 weeks. Because of the quince's dry, hard texture and its astringent flavor, it is better consumed cooked than raw. Before being cooked, quince must be peeled, cored (the seeds contain a cyanide compound), and placed in a mixture of water and lemon juice to prevent discoloration. When cooked, the hard pulp of the fruit softens, turns pink, and takes on the texture of a pear. The flavor becomes more mellow and sweeter.

NUTRIENT COMPOSITION

Quince is high in vitamin C.

SERVING SIZE: *1 fruit, raw (92 g)*

NUTRIENT CONTENT

Energy (kilocalories)	52
Water (%)	84
Dietary fiber (grams)	2
Fat (grams)	0
Carbohydrate (grams)	14
Protein (grams)	0
Minerals (mg)	
Calcium	10
Iron	1
Zinc	0
Manganese	–
Potassium	181
Magnesium	7
Phosphorus	16
Vitamins (mg)	
Vitamin A	4 RE
Vitamin C	14
Thiamin	0
Riboflavin	0
Niacin	0
Vitamin B₆	0
Folate	3 µg
Vitamin E	1

Note: A line (–) indicates that the nutrient value is not available.

RAMBUTAN

The rambutan is a rubbery red fruit about the size and shape of a golf ball. Short, flexible, curved spines give the fruit its name, which means "hairy" in Malay. Underneath the spiny shell is a sweet, juicy, translucent flesh similar to that of the lychee fruit. Another similarity to lychees is the one shiny seed in the center of the flesh of the rambutan.

Family Sapindaceae
Scientific name *Nephelium lappaceum*
Common name rambutan

NUTRIENT COMPOSITION

The fresh fruit has not been analyzed for nutrient content. Canned rambutan provides a small amount of vitamin C.

VARIETIES

Several varieties of rambutan are available at tropical markets. They vary in shape from round to slightly ellipsoid and in color from green to yellow, orange, and red. The Thai green rambutan has a thinner rind and a more delicate flavor than the others.

ORIGIN & BOTANICAL FACTS

Rambutan is indigenous to Malaysia and Indonesia and is distributed throughout the tropical regions of southeast Asia. Internationally, Malaysia, Thailand, and Indonesia are the leaders in rambutan export, and Singapore consumes more than 60 percent of the fruit. Because of the short shelf life of the fruit, the market for fresh rambutan is concentrated in Asia. Domestically, rambutans are grown in Hawaii; however, restrictions on imports to the continental United States limit the market for fresh Hawaiian rambutans.

The rambutan tree, which reaches heights of 8 to 15 feet, is slightly shorter than the lychee tree but bears considerable resemblance to its well-known cousin. The rambutan flowers in terminal clusters that give rise to "bouquets" of fruits. These flowers are used decoratively in floral arrangements. However, unlike the lychee tree, the rambutan is strictly tropical and requires well-irrigated soil to flourish.

USES

The numerous spines of rambutans provide a large surface area for dehydration. To prevent this moisture loss (which results in darkening of the color), the fruit should be refrigerated in sealed plastic bags. Under these conditions, rambutans can maintain their bright color for up to 12 days. To eat the fruit, it is necessary to cut around the middle of the spiny shell with a knife and peel the shell away to reveal the pale, juicy flesh. The seeds should not be ingested. Canned rambutans, which retain the flavor and texture of the fresh fruit, are available in specialty food stores.

SERVING SIZE: *1/2 cup canned in syrup, drained (75 g)*

NUTRIENT CONTENT

Energy (kilocalories)	62
Water (%)	78
Dietary fiber (grams)	1
Fat (grams)	0
Carbohydrate (grams)	16
Protein (grams)	0
Minerals (mg)	
Calcium	17
Iron	0
Zinc	0.1
Manganese	0.5
Potassium	32
Magnesium	5
Phosphorus	7
Vitamins (mg)	
Vitamin A	0 RE
Vitamin C	4
Thiamin	0
Riboflavin	0
Niacin	1
Vitamin B$_6$	0
Folate	6 µg
Vitamin E	–

Note: A line (–) indicates that the nutrient value is not available.

RHUBARB

Although rhubarb is botanically a vegetable, it is used as a fruit, sometimes even referred to as "pie plant" because of its frequent use as pie filling. Except for its pink color, rhubarb is similar in appearance to celery. The acidity and intensity of flavor vary, and young stalks are more tender than older stalks. The roots and leaves of rhubarb are not eaten because they contain significant amounts of oxalic acid and are highly poisonous.

Family Polygonaceae
Scientific name *Rheum officinale, Rheum palmatum, Rheum rhaponticum*
Common name rhubarb, rheum

♥ **Contains some vitamin C**

VARIETIES

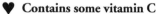

Rhubarb is available in two main types, each of which includes many species and dozens of varieties. Hothouse-grown rhubarb has pink or light red stalks and yellow leaves, and field-grown rhubarb has dark-red stalks and green leaves. The hothouse variety has a milder flavor and is less stringy.

ORIGIN & BOTANICAL FACTS

Much of the history of cultivation of rhubarb is related to its use as a medicinal plant. The earliest recorded use of rhubarb for medicinal purposes appeared in 2700 B.C. Marco Polo was the first to introduce rhubarb to Europe. Although rhubarb stalks were eaten in eastern Turkey as early as the 13th century, it was not until the 18th century that Europeans began to use rhubarb as a food. By 1830, rhubarb had become a popular winter vegetable in the London markets. In the late 18th or early 19th century, seeds and rootstock of rhubarb were brought to New England, where it was cultivated and began to appear in Massachusetts produce markets by the 1820s.

Rhubarb is a large, leafy perennial that can grow to 10 feet. It sends out thick, brown, branching roots. The field-grown variety grows to only 3 feet and has red roots. Rhubarb is a heat-intolerant, cool-season crop that is grown in fields and hothouses throughout Washington, Oregon, and Michigan.

USES

Rhubarb is available frozen, canned, or fresh, sold as loose stalks or bagged. Most cooks prefer to use the fresh stalks. The leaves should be cut off before storing the stalks in plastic bags. They will stay fresh in the refrigerator for about a week.

Rhubarb is too tart to eat raw. Instead, the stalks are sliced thinly or diced, baked or stewed, and then sweetened. Because rhubarb becomes slightly sweeter when cooked, sweeteners should be added after cooking. Cooked rhubarb may be sweetened with sugar, honey, maple syrup, orange or pineapple juice, or berry preserves. Combining rhubarb with sweet fruits such as strawberries decreases the amount of sweetener needed and hence the number of calories. Cooking causes rhubarb to turn brown, but this can be prevented by the addition of a cooked beet to the pot. Aluminum or cast iron saucepans should not be used to cook rhubarb because the acid in the vegetable will blacken the pot and the rhubarb.

NUTRIENT COMPOSITION

Raw rhubarb contains some vitamin C.

SERVING SIZE:
1/2 cup, raw (61 g)

NUTRIENT CONTENT

Energy (kilocalories)	13
Water (%)	94
Dietary fiber (grams)	1
Fat (grams)	0
Carbohydrate (grams)	3
Protein (grams)	1
Minerals (mg)	
Calcium	52
Iron	0
Zinc	0
Manganese	0
Potassium	176
Magnesium	7
Phosphorus	9
Vitamins (mg)	
Vitamin A	6 RE
Vitamin C	5
Thiamin	0
Riboflavin	0
Niacin	0
Vitamin B_6	0
Folate	4 µg
Vitamin E	0

SAPODILLA

The sapodilla is nearly round and about 2 to 4 inches in diameter. Its thin, brownish skin is easy to peel when the fruit is ripe. The translucent flesh has a sweet flavor reminiscent of honey and apricots and a "melt in the mouth" texture. The center of the fruit contains about 3 to 12 hard, shiny black seeds.

Family Sapotaceae
Scientific name *Manilkara zapota*
Common name sapodilla

♥ **High in vitamin C and fiber**

VARIETIES

The extensive cultivation of the sapodilla in India has resulted in numerous varieties. Brown Sugar produces fragrant, juicy fruits whose flesh is pale brown and richly sweet. The flesh of the Prolific variety is light pinkish tan, mildly fragrant, smooth-textured, and sweet. Russel bears large fruits that are rich and sweet, but it is not a prolific producer. A new selection, Tikal, yields fruits that have an excellent flavor but are smaller.

ORIGIN & BOTANICAL FACTS

The sapodilla plant is believed to have originated in the Yucatán peninsula of Mexico, northern Belize, and northeast Guatemala. The plant was highly prized by the Aztecs, who called the fruit "tzapotl," from which the Spanish derived the name sapodilla. The plant is now grown in almost all the tropical and subtropical regions of Africa, Asia, the East Indies, and the Americas. The main producers are the Central American countries, Australia, India, Indonesia, and, in the United States, California and Florida.

Equally at home in humid and relatively dry environments, the sapodilla tree is a slow-growing evergreen that can reach up to 100 feet in height. The tree bears small, ball-shaped white flowers borne on slender stalks at the leaf bases. A resinous sap called "chicle" was once collected from sapodilla tree trunks for making chewing gum. This practice has largely been replaced by the use of synthetic ingredients. The mature sapodilla tree can yield from 2,000 to 3,000 fruits in a single year. Because the fruits are easily perishable and fragile, they do not ship well and therefore are relatively unknown outside their areas of origin.

USES

Because the high tannin and latex contents of unripe fruits make them astringent and unpalatable, sapodillas should be eaten only when ripe. Unripe sapodilla fruits should be left to ripen at room temperature and refrigerated after ripening. The sapodilla is best eaten raw and chilled, by cutting in half and spooning the pulp out of the skin. It can be added to salads or desserts such as ice cream and sorbet. In Malaysia, the fruit is stewed with lime juice or fried with ginger. In India, it is eaten as a dried fruit.

NUTRIENT COMPOSITION

Sapodilla is high in vitamin C and in dietary fiber.

SERVING SIZE:
1 fruit, raw (170 g)

NUTRIENT CONTENT

Energy (kilocalories)	141
Water (%)	78
Dietary fiber (grams)	9
Fat (grams)	2
Carbohydrate (grams)	34
Protein (grams)	1
Minerals (mg)	
Calcium	36
Iron	1
Zinc	0
Manganese	–
Potassium	328
Magnesium	20
Phosphorus	20
Vitamins (mg)	
Vitamin A	10 RE
Vitamin C	25
Thiamin	0
Riboflavin	0
Niacin	0
Vitamin B$_6$	0.1
Folate	24 µg
Vitamin E	0

Note: A line (–) indicates that the nutrient value is not available.

TAMARIND

The tamarind is a brown, flat, irregularly curved pod about 3 to 8 inches long with a sour, fruity taste. The pod may have as many as 12 large, flat, glossy seeds embedded in a brown, edible pulp. As the pod matures, it fills out somewhat and the juicy, acerbic pulp turns brown or reddish brown. The sweet, tart taste of the pulp is the result of its high content of both acid and sugar; however, the pulp becomes extremely sour when dried. The shells become brittle and crack readily when the fruit is fully ripe.

Family Leguminosae
Scientific name *Tamarindus indica*
Common name tamarind, Indian date

VARIETIES

The size and flavor of tamarinds are determined by their variety. Indian varieties have long pods with 6 to 12 seeds, and the West Indian and American varieties have shorter pods containing only 3 to 6 seeds. Sweeter pulp is found in selected varieties such as the Makham Waan from Thailand and the Manila Sweet from the United States Department of Agriculture's subtropical horticulture research unit in Miami.

ORIGIN & BOTANICAL FACTS

The tamarind, also known as "Indian date," is one of the few fruits native to Africa that is enjoyed in the cuisines of many other continents. In China, it is called Asam koh; in Vietnam, it is called Me; in France, Tamarin; in Cambodia, Ampil khui or tum; in Thailand, Mak kham; and in Italy and Spain, Tamarindo. Although the tamarind is native to tropical Africa and grows wild throughout the Sudan, the fruit has been cultivated in India for centuries. During the 16th century, the fruit was brought to the Americas, and it is now widely grown in

Mexico and Central and South America. Belize, Brazil, Guatemala, and India are the major commercial producers of tamarind worldwide. The tree is a slow-growing, long-lived evergreen with supple branches and bright-green leaves that appear in pairs, 1 to 2-1/2 inches in length, and fold up at night. Under favorable conditions, the tree may grow up to 80 feet tall and 20 to 35 feet wide. In severe drought, the leaves often drop off the tree. A young tree bears fruit within 4 years and continues to fruit for up to 60 years. Tamarind fruits may be left on the tree for up to 6 months after maturity without loss of moisture.

USES

Tamarind is available in Indian and Asian markets as a fresh fruit, as a concentrated pulp with seeds, as a paste, as whole pods dried into "bricks," and as a powder. Tamarind has a variety of uses in cooking. The immature fruit can be roasted and served as a "vegetable," or it can be used to season rice, fish, or meat. Ripe tamarind is eaten fresh or made into sauces, chutneys, or curry dishes. It is also one of the many ingredients in Worcestershire sauce. Tamarind pulp concentrate is often used as

a flavoring in East Indian and Middle Eastern dishes, in much the same way lemon juice is used in Western cuisine. Tamarind's sweet-sour flavor combines well with the spicy flavor of chili in the Thai and Vietnamese cuisines, where unripe pods are used in soups and stews. In Indian cooking, tamarind is used as a seasoning in lentil and bean dishes and in the dish called "vindaloo." Tamarind syrup, which can be found in Dutch, Indonesian, and East Indian markets, is used to flavor soft drinks.

NUTRIENT COMPOSITION

In the amounts customarily eaten, tamarind is not a significant source of nutrients.

SERVING SIZE:
10 fruits (20 g)

NUTRIENT CONTENT

Energy (kilocalories)	48
Water (%)	31
Dietary fiber (grams)	1
Fat (grams)	0
Carbohydrate (grams)	13
Protein (grams)	1
Minerals (mg)	
Calcium	15
Iron	1
Zinc	0
Manganese	–
Potassium	126
Magnesium	18
Phosphorus	23
Vitamins (mg)	
Vitamin A	1 RE
Vitamin C	1
Thiamin	0.1
Riboflavin	0
Niacin	0
Vitamin B6	0
Folate	3 µg
Vitamin E	0

Note: A line (–) indicates that the nutrient value is not available.

TANGERINE

The tangerine is a citrus fruit, usually round and about 2 1/2 inches in diameter, smaller than the orange. Its rough, fragrant rind is generally orange or red-orange, thin, and loose on the fruit, so that it peels very easily. Inside, numerous fibers loosely hold the 8 to 15 easily separated segments, or carpels, that contain the juice sacs and white seeds. The center is hollow. Although the terms "tangerine" and "mandarin orange" are sometimes used interchangeably, tangerines actually are a subgroup of the mandarin.

Family Rutaceae
Scientific name *Citrus reticulata*
Common name tangerine

♥ **High in vitamin C**

♥ **A good source of vitamin A (carotenes)**

is a refreshing thirst quencher, either alone or combined in a blender with other fresh fruits. Tangerines can be substituted for oranges in various dishes. Meat, fish, and poultry can be marinated in tangerine juice before grilling, and tangerine juice poured over freshly sliced fruit helps keep the fruit from turning brown and adds a distinctive flavor. Tangerines also make an excellent marmalade.

NUTRIENT COMPOSITION

Although tangerines have about 43 percent less vitamin C than oranges, they are still an excellent source for this vitamin. Tangerines also contain more vitamin A (carotenes).

VARIETIES

The most popular variety of tangerine in the United States is the Dancy, a very sweet fruit with a red-orange color and a mellow flavor. Honey tangerines, with their slightly green-tinged peel, are true to their name. They have a high sugar content and a rich taste. The Fallglo is a large tangerine with dark-orange rind and flesh. Two smaller sized varieties are the Clementine, also called the Algerian tangerine, and the Sunburst, which has a thin skin and deep-orange flesh.

were crossed with other citrus fruits, producing numerous hybrids such as the tangelo (a cross between a tangerine and a grapefruit) and the tangor (a hybrid of a tangerine and an orange).

The tangerine tree is an evergreen that grows to a height of about 10 feet. The five-petaled tangerine blossoms are white and fragrant. It takes 6 to 10 months from the time the blooms appear until the fruit is ready for harvest. Tangerines grow year-round in warmer climates, and they are a traditional Christmas or New Year treat in some parts of the world.

SERVING SIZE:	1 medium, raw (84 g)

NUTRIENT CONTENT

Energy (kilocalories)	37
Water (%)	88
Dietary fiber (grams)	2
Fat (grams)	0
Carbohydrate (grams)	9
Protein (grams)	1
Minerals (mg)	
Calcium	12
Iron	0
Zinc	0
Manganese	0
Potassium	132
Magnesium	10
Phosphorus	8
Vitamins (mg)	
Vitamin A	77 RE
Vitamin C	26
Thiamin	0.1
Riboflavin	0
Niacin	0
Vitamin B6	0.1
Folate	17 µg
Vitamin E	0

ORIGIN & BOTANICAL FACTS

Tangerines are native to China, but today they are grown all over the world. In the United States, the leading producers are California, Arizona, and Florida.

The name "tangerine" is derived from the ancient, walled Moorish town of Tangier in northern Morocco, where the fruit grows in abundance. As the cultivation of tangerines was carried around the globe, the original mandarin oranges

USES

Tangerines are always picked when they are ripe, so they are ready for immediate consumption. Fruits should be heavy for their size and free of bruises. Color is not a reliable indicator of the quality of tangerines. The fruit is most often eaten as a snack or dessert or used in green salad or fruit salad. Tangerine slices also make an attractive garnish for cakes and other desserts. Freshly squeezed tangerine juice

VEGETABLES

Vegetables and other foods of plant origin were the primary source of sustenance for early humans. The plant foods that we call vegetables came under cultivation later than the grains and legumes. Less protein- and carbohydrate-dense than the grains and legumes, vegetables have always served more as accompaniments or accessory ingredients than as staples. Some fragrant, highly flavored vegetables, such as scallions, garlic, ginger, parsley, basil, oregano, fenugreek, and dill, are really used as herbs and spices in sparing amounts as flavorings for other foods. Only a few of these "vegetables" are described in this section. Other herbs and spices are discussed on pages 363 to 375.

Some plant foods that we consider vegetables are, botanically, fruits (for example, avocado, squash, cucumber, olives, tomatoes, and eggplant), that is, edible flesh surrounding seeds. Nevertheless, the term "vegetable" has come to denote plant foods eaten as side dishes or used in the preparation of any part of the meal except dessert. Vegetables tend to be less sweet than foods considered fruits. In addition, some foods that are served as vegetables are really grains (corn), legumes (green beans, lima beans, peas), or fungi (mushrooms, truffles). So what is a vegetable, really? A vegetable is essentially any edible part of the plant (leaves, roots or tubers, and stalks) except, in most cases, the fruit. Multiple parts of some plants are eaten as separate vegetables with very different nutrient contents. The most common example is beet roots and their greens.

Vegetables contribute significant amounts of vitamins, minerals, soluble and insoluble fiber, and other phytonutrients to our diets. The Dietary Guidelines for Americans advise us to eat a wide variety of vegetables and fruits every day, because the nutrient content varies considerably from one to another. (For a discussion of the Dietary Guidelines and the 5 a Day program, see Chapter 1, page 8.) With the exception of olives and avocados, which are really fruits, few vegetables, by themselves, provide significant amounts of fat. The fat provided by olives and avocados is high in monounsaturated fatty acids, which may help prevent heart disease (see Chapters 2 and 3).

To assist you in menu planning, the following section provides information regarding the nutrient content of many vegetables.

The tables of nutrient values are based on serving sizes specified by the U.S. Department of Agriculture Food Guide Pyramid. Nutrient values are rounded (milligrams and micrograms tend to be rounded to one decimal point, grams are rounded to whole numbers). Nutrient claim statements listed beneath the common name of each vegetable are based on the serving size specified and the definitions in Chapter 4 (see sidebar: Nutrient Claims, page 92). For example, asparagus is considered a good source of vitamin C, because it provides about 18 percent of the Daily Value for vitamin C. A food that is high in a particular nutrient provides 20 percent or more of the Daily Value for that nutrient per serving.

See the Appendix, Nutrients in Foods, page 434, and Phytochemical Contents of Selected Foods, page 484, for a more complete listing of nutrient and phytonutrient contents of selected vegetables.

AMARANTH

Amaranth is cultivated as both a vegetable and a cereal grain. The upright herbaceous plant usually reaches 6 to 7 feet in height but may grow to 13 feet in favorable environments. The foliage varies in shape and color, although the leaves of most varieties are large, broad, and dark green with deep purplish veins. The flowers are small, green, and clover-like. The leaves from most varieties are edible and are delicious when cooked. The tiny grain is very nutritious. (See Grains, page 272, for a discussion of the grain.)

Family   Amaranthaceae
Scientific name   *Amaranthus dubius*
Common name   amaranth

♥ **Cooked amaranth leaves are high in vitamin A, vitamin C, and potassium**

♥ **Amaranth leaves are a good source of calcium**

NUTRIENT COMPOSITION

Amaranth is a nutrient-dense food. One-half cup of cooked leaves is high in vitamin A and vitamin C and is a good source of calcium.

SERVING SIZE: *1/2 cup leaves, cooked (66 g)*

VARIETIES

The genus *Amaranthus* includes a number of amaranth species. The common types include tampala, hon-moi-toi, bush greens, pigweed, Chinese spinach, and wild amaranth, some of which are edible and some not. *Amaranth gangeticus*, one of the edible varieties, is available in the United States as the green-leafed tampala. The red-leafed amaranth, known as *Amaranth tricolor* L., is also available, as is another familiar type called Joseph's Coat.

ORIGIN & BOTANICAL FACTS

Reports of the existence of amaranth date back to the 2nd century A.D. The grain was a staple of the Aztec diet and was used in religious ceremonies until its cultivation was outlawed by the Spanish conquerors. Amaranth was brought to Asia after the 15th century and was cultivated in China and India. Today, China and Central America are the world's leading suppliers.

Amaranths are hot weather plants that thrive best in well-fertilized, well-irrigated raised beds in sunny areas. Amaranth greens are harvested 4 to 6 weeks after the planting season. They are transported to the market packed in ice and are sold in bunches. Once considered a weed in the United States, amaranth is now consumed as a vegetable green.

USES

Young amaranth leaves are preferred for cooking. The leaves wilt easily and have a very short shelf life. They must be refrigerated or kept in ice water to retain their freshness and crispness. Amaranth leaves can be boiled, steamed, or stir-fried as a side dish to accompany meats, fish, or other vegetables. As a grain, amaranth is mostly ground into flour and used to make breads, pasta, pastries, and cereals. The flour is also available commercially.

NUTRIENT CONTENT

Energy (kilocalories)	14
Water (%)	92
Dietary fiber (grams)	-
Fat (grams)	0
Carbohydrate (grams)	3
Protein (grams)	1
Minerals (mg)	
Calcium	138
Iron	2
Zinc	1
Manganese	1
Potassium	423
Magnesium	36
Phosphorus	48
Vitamins (mg)	
Vitamin A	183 RE
Vitamin C	27
Thiamin	0
Riboflavin	0.1
Niacin	0
Vitamin B$_6$	0.1
Folate	37 μg
Vitamin E	–

Note: A line (–) indicates that the nutrient value is not available.

ARTICHOKE

The artichoke plant is a member of the thistle or sunflower family. The cones, or spherically shaped buds, are enclosed by overlapping outer scales (bracts) and are edible at the base. At the center of the bud is an inedible thistle (choke). The edible "heart" of an artichoke is the round, tender, firm base of the bud that is revealed after pulling off the petals. Commercially sold artichoke hearts are the tender central portions of small artichokes that have almost no choke. Their flavor is delicately nutty and slightly bitter.

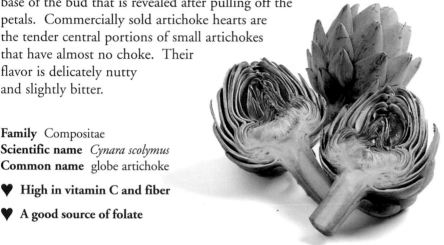

Family Compositae
Scientific name *Cynara scolymus*
Common name globe artichoke

♥ **High in vitamin C and fiber**

♥ **A good source of folate**

VARIETIES

The most popular variety of artichoke is the Green Globe, which is usually solid green. Other varieties, rarely seen in the U.S. marketplace, include the Violetta and the Purple Roscoff, which have hints of purple either on the scales or the choke itself.

ORIGIN & BOTANICAL FACTS

The artichoke is native to the eastern Mediterranean region. Its use was documented by the ancient Greeks and Romans thousands of years ago. Today, most of the European supply of artichokes is still grown by the countries surrounding the Mediterranean Sea. Artichokes were introduced to the United States in the 19th century by European immigrants and soon found their way to the midcoast region of California, where most of the domestic crop is cultivated today.

An artichoke is an immature bud that, if left to bloom, boasts a bright-purple, thistle-like flower that can be found at outdoor farmers' markets and can be dried for use in flower arrangements. The size of an artichoke bud is determined by the stalk on which it grows and is not indicative of quality. Thick stalks, which are usually concentrated around the center of the plant, produce large artichokes, and the thinner side stalks produce smaller artichokes.

USES

Artichokes picked in the fall or winter months may have bronze-tipped leaves or a slightly gray tint, which is a sign of exposure to frost. However, this should not affect the flavor of the artichoke. Squeezing the artichoke slightly should elicit a squeak if the leaves are still plump and crisp. Although the artichoke looks tough and hardy, it should be kept in the refrigerator for no more than 4 or 5 days. A sprinkle of water in a plastic bag will help maintain the moisture of the artichoke, but it should not be trimmed, cut, or washed before storing. Artichokes are most often boiled or steamed and can be eaten hot or cold. Each petal is pulled off and the base is dipped into melted butter or lemon juice. Only the tender portion at the base of the petal is edible. Underneath the rough outer petals, the thinner rose-colored petals can be bitten off or removed to find the choke. After removing the choke, the heart can be eaten whole. The hearts also can be added to pasta sauces or green salads and used to top pizzas. Whole steamed artichokes can be filled with well-seasoned stuffing and served as is or baked.

NUTRIENT COMPOSITION

The artichoke is high in vitamin C and dietary fiber and is a good source of folate.

SERVING SIZE:
1 artichoke (120 g)

NUTRIENT CONTENT

Energy (kilocalories)	60
Water (%)	84
Dietary fiber (grams)	6
Fat (grams)	0
Carbohydrate (grams)	13
Protein (grams)	4
Minerals (mg)	
Calcium	54
Iron	2
Zinc	1
Manganese	0
Potassium	425
Magnesium	72
Phosphorus	103
Vitamins (mg)	
Vitamin A	22 RE
Vitamin C	12
Thiamin	0.1
Riboflavin	0.1
Niacin	1
Vitamin B6	0.1
Folate	61 µg
Vitamin E	0

ARUGULA

Arugula is an annual that grows 8 to 24 inches high. It has green, deeply cut, compound leaves that are edible and have spicy, pungent flavor resembling horseradish. The vegetab called *roquette* (the French word for rocket), but the term "arugula" is now becoming more common.

Family Cruciferae
Scientific name *Eruca vesicaria sativa*
Common name arugula, Italian cress, rocket, roquette, tira, white pepper, garden rocket

♥ **Provides some vitamin A**

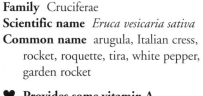

VARIETIES

The arugula can be divided into smooth-edged or serrated-leafed varieties. Some arugula varieties are wild, although most are cultivated. The flavor varies depending on variety.

ORIGIN & BOTANICAL FACTS

Arugula belongs to the Cruciferae family and is a close relative of the mustard. Ancient Egyptians and Romans considered arugula leaves in salads to be an aphrodisiac. It is a minor crop in the southeastern United States, grown to a limited extent commercially and in home vegetable gardens. Seeds often are listed in seed catalogs as "roquette" under the category of herbs. A cool season vegetable best grown in Florida during the fall, winter, and spring, it matures from seed in 2 to 3 months. Periods of very warm temperatures cause it to bolt (go to seed) rather quickly. Few pests attack the arugula.

USES

The freshest, crispest leaves free of brown spots should be chosen. Arugula should be used as soon as possible after purchasing. If necessary, after removing any wilted leaves, the remaining unwashed leaves can be refrigerated for no more than 2 days in a plastic bag. The zesty leaves can be used raw in salads by themselves or tossed with other greens. Arugula also can be added to soups or lightly cooked and served as a side dish. The arugula is widely consumed in the Middle East as a garnish on meats and sandwiches. Tiny arugula blossoms also can be added to salads.

NUTRIENT COMPOSITION

Arugula provides some vitamin A.

SERVING SIZE:
1 cup, raw (20 g)

NUTRIENT CONTENT

Energy (kilocalories)	5
Water (%)	92
Dietary fiber (grams)	0
Fat (grams)	0
Carbohydrate (grams)	1
Protein (grams)	1
Minerals (mg)	
Calcium	32
Iron	0
Zinc	0
Manganese	0
Potassium	74
Magnesium	9
Phosphorus	10
Vitamins (mg)	
Vitamin A	47 RE
Vitamin C	3
Thiamin	0
Riboflavin	0
Niacin	0
Vitamin B$_6$	0
Folate	19 µg
Vitamin E	0

ASPARAGUS

Asparagus is a member of the lily family and is related to onions, leeks, and garlic. It is cultivated for its edible young shoots, which are long and unbranched with compact, pointed tips made of tiny leaves.

Family Liliaceae
Scientific name *Asparagus officinalis*
Common name asparagus

♥ **High in folate**

♥ **A good source of vitamin C**

♥ **Contains glutathione, an antioxidant that promotes health**

VARIETIES

Two basic varieties of asparagus, white and green, are cultivated. The green variety is the only one grown on a commercial scale in the United States, whereas the white is preferred in Europe. White asparagus is produced by banking soil against the plant to keep out sunlight, which otherwise would turn the stalks green. Also available, although not common, is a violet variety, with pinkish purple shoots and tips.

ORIGIN & BOTANICAL FACTS

Asparagus was first cultivated in Greece about 2,500 years ago. In fact, the name asparagus is Greek for "stalk" or "shoot." The ancient Greeks believed that asparagus had medicinal qualities and could cure toothaches and bee stings. The cultivation of asparagus was adopted by the Romans, who carried it throughout Europe and Great Britain. From there, its popularity spread to the rest of the world. Traditionally, asparagus was a Northern Hemisphere crop, but today it is cultivated worldwide. The United States is the world's largest supplier of asparagus, with most cultivation concentrated in California.

The asparagus plant is a perennial but requires three seasons to mature. In its first season, a crown forms with 6 inches of root. In its second season, the crown develops into a fern. Asparagus can be harvested in its third season, but the plant does not reach its prime until 6 to 8 years of age. At peak age, an asparagus field can yield up to 2 tons per acre. Because its growing season is short and it must be harvested by hand, asparagus can be expensive. Asparagus appears in American markets as early as February, when the first California crops are harvested, but the peak season in the West is from late April to late May and, elsewhere in the United States, from May through July. Throughout the rest of the year, fresh asparagus may be available from Mexico and South America.

USES

Asparagus stalks of similar width with tightly closed tips should be selected. Young asparagus is thinner and generally more tender. Fresh asparagus should be stored in the refrigerator with the cut ends immersed in water and should be used within a day or two.

Fresh asparagus is best steamed or microwaved until just crisp-tender. Steaming should be done quickly, with the spears in an upright position to heat the stalks evenly. The spears also can be roasted briefly in the oven with a little olive oil. Cooked asparagus is best served immediately and simply, without rich sauces. Asparagus spears also can be cut into diagonal pieces and stir-fried.

Asparagus is also available canned or frozen. Frozen spears are closer to fresh spears in flavor and nutrition. The canned variety is less nutritious.

NUTRIENT COMPOSITION

Asparagus is a good source of vitamin C and is an excellent source of folate. It also contains glutathione, an antioxidant that promotes health.

NUTRIENT CONTENT

	4 spears, raw (64 g)	6 spears, cooked (1/2 cup) (72 g)
Energy (kilocalories)	14	22
Water (%)	92	92
Dietary fiber (grams)	1	1
Fat (grams)	0	0
Carbohydrate (grams)	3	4
Protein (grams)	1	2
Minerals (mg)		
Calcium	13	18
Iron	1	1
Zinc	0	0
Manganese	0	0
Potassium	175	144
Magnesium	11	9
Phosphorus	36	49
Vitamins (mg)		
Vitamin A	37 RE	49 RE
Vitamin C	8	10
Thiamin	0.1	0.1
Riboflavin	0.1	0.1
Niacin	1	1
Vitamin B$_6$	0.1	0
Folate	82 µg	131 µg
Vitamin E	1	0

BAMBOO

Although the bamboo is often thought to be a tree, it is actually a type of evergreen perennial grass that is woody when mature but whose young shoots are edible. The mature stalks are characterized by green internodes ribbed with cream-colored, brown-speckled sheaths and hanging leaves up to 8 inches long and 3/4 inch wide. Fresh bamboo shoots are light yellow or brown, purple at the root end, and white at the stalk end. The cooked young shoots are crisp, fragrant, and mild in flavor.

Family Graminaceae
Scientific name *Phyllostachys* species, *Bambusa* species
Common name bamboo

♥ **A source of potassium**

VARIETIES

The dozens of varieties of bamboo can be classified in several ways. All bamboo can be divided into those that grow uncontrollably (the invasive type) and those that tend to clump. Bamboo also can be divided into those that are cold-hardy and those that are tropical and sub-tropical. Finally, within each of the above categories, the numerous species can be classified by their mature size (giant, large, medium, and dwarf). Bamboo grown in the United States is almost exclusively the cold-hardy *Phyllostachys* species, most of which are invasive.

ORIGIN & BOTANICAL FACTS

Bamboo is native to China, Japan, southeast Asia, India, Africa, South America, and parts of Mexico. Although bamboo shoots have been an important vegetable in Asian diets for thousands of years, Asian-grown bamboo is mostly consumed locally, with only small quantities processed for export. Bamboo is still a rare vegetable in Western countries, used exclusively in Oriental dishes. However, Europe and the United States are beginning to develop bamboo crops.

Bamboo grows by sending out new rhizomes (underground, horizontal stems) from which new shoots emerge. The constant appearance of new shoots and leaves gives the plant its evergreen appearance. The nutrients made by the leaves are stored in the rhizomes and then converted into the following year's new growth. Because large crops occur in alternate years, growers maintain plants of various ages. To keep the shoots white, soil is sometimes piled against new growth areas to prevent them from developing chlorophyll.

Nations that export bamboo shoots may harvest cultivated plantations or native forests. Unfortunately, lack of regulation has allowed excessive harvesting, which has led to a decline in some native forests. In contrast, because many hardy bamboo species spread uncontrollably, U.S. home gardeners who want to grow bamboo should construct an underground barrier wall to prevent its spread.

USES

The tenderest shoots are those about 6 inches or less in height. Fresh shoots should be stored in cold water for no more than 2 days, or wrapped tightly in plastic and refrigerated up to a week. The shoots also can be blanched and frozen for up to a year. Fresh shoots should be boiled in one or two changes of plain or slightly salted water until tender, then husked and sliced lengthwise. Canned or frozen bamboo shoots should be rinsed, heated, and served as is or stir-fried with meats and other vegetables. The tender parts also can be used in salads.

NUTRIENT COMPOSITION

Cooked bamboo shoots are a source of potassium.

SERVING SIZE: *1/2 cup, cooked (60 g)*

NUTRIENT CONTENT

Energy (kilocalories)	7
Water (%)	95
Dietary fiber (grams)	1
Fat (grams)	0
Carbohydrate (grams)	1
Protein (grams)	1
Minerals (mg)	
Calcium	7
Iron	0
Zinc	0
Manganese	0
Potassium	320
Magnesium	2
Phosphorus	12
Vitamins (mg)	
Vitamin A	0 RE
Vitamin C	0
Thiamin	0
Riboflavin	0
Niacin	0
Vitamin B$_6$	0.1
Folate	1 µg
Vitamin E	–

Note: A line (–) indicates that the nutrient value is not available.

BEETS

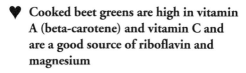

Beets are round, firm root vegetables with edible, leafy green tops. Although the most common root color is bright red, colors may vary from deep red to yellow or white, and one variety displays concentric rings of red and white. When cooked, they have a tender-crisp texture and a sweet flavor.

Family Chenopodiaceae
Scientific name *Beta vulgaris*
Common name beet

♥ **Beets are a good source of folate**

♥ **Cooked beet greens are high in vitamin A (beta-carotene) and vitamin C and are a good source of riboflavin and magnesium**

VARIETIES

The garden or table beet is the type most commonly grown for human consumption. Three common varieties are the Lutz salad leaf beet, the Detroit dark red beet, and the Chioggia beet, an Italian favorite with alternate red and white concentric rings. Another type of beet is the spinach or leaf beet, grown not for its root but for its leaves, which are better known as Swiss chard. A third type of beet, the sugar beet, is not grown as a vegetable. This beet contains twice the sugar of table beets and provides about a third of the world's sugar supply. This type of beet also is used as fodder.

ORIGIN & BOTANICAL FACTS

Modern varieties of beets are derived from the sea beet, an inedible plant that grows wild along the coasts of Europe, North Africa, and Asia. The garden beet has been cultivated for thousands of years. In ancient Greece, beets were so highly valued that, according to myth, a beet was offered on a silver platter to Apollo at Delphi.

Today, beets are grown in many regions of the world. The leading beet-producing regions of the United States are California, Colorado, New Jersey, Ohio, and Texas. The beet is a cool-weather biennial that is cultivated as an annual. Beets are grown from seeds sown in early spring and are ready to harvest 60 to 80 days after planting. Beets are not harmed by frost, but hot weather can toughen the roots. Thus, in regions with hotter summers, they are planted in early fall for winter and spring harvest. Consequently, fresh beets are available all year.

USES

When buying beets with the leaves attached, those with the youngest, freshest looking leaves should be selected. Otherwise, avoid beets that are dried, cracked, or shriveled. Large beets may be tough, and small ones are the most tender and flavorful. Leaves should be crisp and should be rinsed well before using. Beets should be stored separately from the leaves in perforated plastic bags in the refrigerator vegetable crisper.

Beets can be grated and eaten raw in salads, boiled, steamed, stewed, baked, sautéed, or pickled. To preserve their color and nutrients, it is best not to peel beets before cooking. They should be scrubbed gently and at least a half inch of stem should be left on. Beets also keep their color better if an acid ingredient such as vinegar or lemon juice is added during cooking. Canned beets are available, but fresh beets are crisper and more flavorful. Beets are used to make the traditional Russian soup borscht, which is colored red by the beet juice.

NUTRIENT COMPOSITION

Cooked beets are a good source of folate. Cooked beet greens are high in vitamin A (beta-carotene) and vitamin C. They are also a good source of riboflavin and magnesium.

NUTRIENT CONTENT

	1/2 cup sliced beets, cooked (85 g)	1/2 cup greens, cooked (72 g)
Energy (kilocalories)	37	19
Water (%)	87	89
Dietary fiber (grams)	2	2
Fat (grams)	0	0
Carbohydrate (grams)	8	4
Protein (grams)	1	2
Minerals (mg)		
Calcium	14	82
Iron	1	1
Zinc	0	0
Manganese	0	0
Potassium	259	655
Magnesium	20	49
Phosphorus	32	30
Vitamins (mg)		
Vitamin A	3 RE	367 RE
Vitamin C	3	18
Thiamin	0	0.1
Riboflavin	0	0.2
Niacin	0	0
Vitamin B$_6$	0	0.1
Folate	68 µg	10 µg
Vitamin E	0	0

BITTER MELON (BALSAM PEAR)

About 4 to 10 inches long, bitter melons are shaped like cucumbers and have wrinkled, bumpy skin. The vegetable's name is derived from its distinctive bitter taste, the result of a high quinine content. The bitterness increases as the melon matures; young, green melons have a delicate, sour flavor, whereas older (yellow) ones are very bitter and acrid. When fully mature, the melon's rind dries and splits lengthwise into three sections, revealing the bright-red arils that enclose the seeds. Bitter melons are normally eaten as immature fruits, but some people prefer the bitter-tasting, more mature fruits.

Family Cucurbitaceae
Scientific name *Momordica charantia*
Common name bitter melon, balsam pear, bitter cucumber, bitter gourd

♥ **High in vitamin C**

VARIETIES

The bitter melon is a variety of squash. A closely related variety, the balsam apple (*Momordica balsamita*), bears fruits similar to bitter melons except that they are egg-shaped and smaller, with smoother skin. Balsam apples, which have a taste similar to that of the bitter melon, are also cultivated, harvested, and prepared like the bitter melon.

ORIGIN & BOTANICAL FACTS

Bitter melons originated in tropical India and have been cultivated for centuries throughout Asia.

The bitter melon is an annual that grows in tropical and subtropical areas. Reaching up to 30 feet in length, the plant grows as a vine with tendrils that attach to plants or other objects for support. Although highly popular as a food crop in India, China, and southeast Asia, bitter melons have been introduced only recently as a food item in U.S. markets. In this country, bitter melons are often grown on trellises and fences as decorative plants.

USES

Bitter melons are available fresh from April through September in Asian markets and also are sold canned or dried. They can be refrigerated in a plastic bag for up to a week. When buying bitter melons, choose green ones if a less bitter taste is desired or yellow ones for a stronger, more bitter flavor. Before it is cooked, the fruit should be cut lengthwise to remove the seeds and the surrounding white fibers. The skin can be either left intact or removed. Bitter melons are always cooked before eating. In India, bitter melons are combined with potatoes or lentils and seasoned with cumin and turmeric. In China, they are steamed or used as an ingredient in soup. They can be thinly sliced and stir-fried with eggs, meats, or other vegetables. Stuffed with meat, shrimp, wood ears, and thin rice noodles, bitter melons can be braised in a light broth to make a bitter-sweet soup. The young leaves of the plant can be boiled and stir-fried like greens or used fresh in salads.

NUTRIENT COMPOSITION

Bitter melons are high in vitamin C. They also contain many phytochemicals, including elasterol, lutein, and lycopene. The leafy tips are a good source of vitamin A.

SERVING SIZE: *pods, 1/2 cup, cooked (62 g)*

NUTRIENT CONTENT

Energy (kilocalories)	12
Water (%)	94
Dietary fiber (grams)	1
Fat (grams)	0
Carbohydrate (grams)	3
Protein (grams)	1
Minerals (mg)	
Calcium	6
Iron	0
Zinc	0
Manganese	0
Potassium	198
Magnesium	10
Phosphorus	22
Vitamins (mg)	
Vitamin A	7 RE
Vitamin C	20
Thiamin	0
Riboflavin	0
Niacin	0
Vitamin B₆	0
Folate	32 µg
Vitamin E	0

BROCCOLI

The broccoli plant is a dark-green vegetable with a firm stalk and branching arms that end in florets. The name comes from the Latin word *brachium*, meaning "arm" or "branch," or the Italian word *broccolo*, for "cabbage sprout." The edible portions are the florets and 6 to 8 inches of the supporting stem. Broccoli is closely related to cauliflower, cabbage, and Brussels sprouts.

Family Cruciferae
Scientific name *Brassica oleracea*
Common name broccoli

♥ **High in vitamin A (beta-carotene) and vitamin C**

♥ **A cruciferous vegetable that contains phytochemicals that may help prevent cancer**

VARIETIES

The most common type of broccoli in the United States today is the sprouting, or Italian, green broccoli. The light-green stalks are topped by umbrella-shaped clusters of dark-green florets. This variety is also called the Calabrese, named after the Italian province in which it was first grown. Broccoli rabe, a distinct but related type, has smaller florets and a stronger flavor.

ORIGIN & BOTANICAL FACTS

Broccoli dates back to the time of the Roman Empire, when it was cultivated from wild cabbage native to coastal Europe. It was brought to the United States in the early 1900s by Italian immigrants to northern California. Currently, 90 percent of the domestic commercial market is supplied by California producers. Although it is not a popular vegetable worldwide, broccoli began gaining popularity in the 1970s, when consumption per person increased from about a half pound per year to the current 4 1/2 pounds. Today, broccoli ranks 11th among leading U.S. vegetable crops.

The broccoli plant is an upright annual, able to reach a height of 3 feet, with large spreading leaves. Usually grown from seed, broccoli is harvested 80 to 120 days after planting. The consumed portion of broccoli is actually a group of buds that are almost ready to flower. Overmature broccoli is tough and woody because the plant sugar is converted to lignin, a type of fiber that is not softened by cooking.

USES

Broccoli with the tiniest buds and the darkest blue-green color should be selected. Avoid those with a yellowish cast. Broccoli should be stored unwashed in an open bag in the refrigerator, because excess moisture encourages the growth of mold. Before use, broccoli should be rinsed thoroughly under cold running water to remove any dirt. Broccoli can be consumed raw as an appetizer with dip or in salads, or it can be cooked in a variety of ways. Well-cooked broccoli should be tender enough to yield to a fork, yet remain crisp and bright. Because the florets tend to cook faster than the stalks, stalks should be split to expose more surface area, which ensures even cooking. The florets also may be cut from the stalks and added after the stalks have been cooking for 2 to 3 minutes. Broccoli can be boiled, steamed, microwaved, stir-fried, or puréed and added to soups.

NUTRIENT COMPOSITION

Broccoli is high in vitamin A (beta-carotene) and vitamin C. The vitamin A and various phytochemicals, such as isothiocyanates, indoles, and bioflavonoids, in broccoli may help prevent cancer.

NUTRIENT CONTENT

	½ cup, raw (44 g)	½ cup, cooked (about 2 spears) (78 g)
Energy (kilocalories)	12	22
Water (%)	91	91
Dietary fiber (grams)	1	2
Fat (grams)	0	0
Carbohydrate (grams)	2	4
Protein (grams)	1	2
Minerals (mg)		
Calcium	21	36
Iron	0	1
Zinc	0	0
Manganese	0	0
Potassium	143	228
Magnesium	11	19
Phosphorus	29	46
Vitamins (mg)		
Vitamin A	68 RE	108 RE
Vitamin C	41	58
Thiamin	0	0
Riboflavin	0.1	0.1
Niacin	0	0
Vitamin B$_6$	0.1	0.1
Folate	31 µg	39 µg
Vitamin E	1	1

BRUSSELS SPROUTS

Brussels sprouts look like miniature dark-green cabbages and are, in fact, related to the cabbage. The sprouts range from 1 to 1 1/2 inches in diameter. As many as a hundred of these ball-like sprouts may grow in bunches from a single, long plant stalk that is usually between 2 and 4 feet in height. Brussels sprouts are similar to the cabbage in flavor but are milder and have a denser texture.

Family Cruciferae
Scientific name *Brassica oleracea* var. *gemmifera*
Common name Brussels sprouts

♥ **High in vitamin C**

♥ **A good source of folate and vitamin A (beta-carotene)**

♥ **A cruciferous vegetable that contains phytochemicals that may help prevent cancer**

VARIETIES

Among the common varieties of Brussels sprouts are the Noisette and Bedford Fillbasket. The Rubine is a red-leafed variety; the Mallard, Captain Marvel, Prince Marvel, Montgomery, and Jade Cross are all hybrids. The Early Half Tal is another variety.

ORIGIN & BOTANICAL FACTS

Brussels sprouts, named after the capital of Belgium, are one of the few vegetables that originated in northern Europe. They were first cultivated in Belgium in the 16th century, introduced to France and England in the 19th century, and probably brought to North America by French settlers, who grew them in Louisiana. In the United States, they are grown primarily along the east and west coasts where summer daytime temperatures average 65 degrees or less. Brussels sprouts are grown from seed, and the first sprouts are ready to pick about 4 months after the seeds are sown. The plant continues to produce sprouts for approximately 6 weeks. Brussels sprouts are very resistant to cold, and the tastiest sprouts are often those that mature after the first fall frost. They are usually available throughout the year. The peak season is from late August through March. California is the major supplier of Brussels sprouts in the United States.

USES

Brussels sprouts are usually selected on the basis of size and appearance. Small, compact, fresh sprouts that are bright green will have the freshest flavor and the crispiest texture. They may be stored in a loosely closed plastic bag in the refrigerator for up to 5 days. Any wilted or yellow outer leaves should be removed and the stems of the sprouts trimmed, although not flush with the bottoms, before cooking. Cutting an "X" in the base of the sprouts helps the heat penetrate the solid core and allows the sprouts to cook evenly. Brussels sprouts can be cooked in a variety of ways, although care must be taken to avoid overcooking, which turns the stems mushy. Sprouts may be boiled, braised, steamed, or microwaved and can be seasoned with mustard, dill, caraway, poppy seeds, or sage leaves. Brussels sprouts are a good accompaniment to strong-flavored meats and cheeses.

NUTRIENT COMPOSITION

Brussels sprouts are high in vitamin C and are a good source of folate and vitamin A (beta-carotene). They are cruciferous vegetables and contain phytochemicals that may help prevent cancer.

SERVING SIZE: *1/2 cup, cooked (about 4 medium) (78 g)*

NUTRIENT CONTENT

Energy (kilocalories)	30
Water (%)	87
Dietary fiber (grams)	2
Fat (grams)	0
Carbohydrate (grams)	7
Protein (grams)	2
Minerals (mg)	
Calcium	28
Iron	1
Zinc	0
Manganese	0
Potassium	247
Magnesium	16
Phosphorus	44
Vitamins (mg)	
Vitamin A	56 RE
Vitamin C	48
Thiamin	0.1
Riboflavin	0.1
Niacin	0
Vitamin B6	0.1
Folate	47 µg
Vitamin E	1

CABBAGE

Cabbage is a leafy vegetable that grows in heads close to the ground. The leaves may be loosely or tightly compacted and range from pale-green to dark purple-red, depending on the variety.

Family Cruciferae
Scientific name *Brassica oleracea* L.
Common name cabbage

♥ **High in vitamin C**

♥ **A cruciferous vegetable that contains phytochemicals called indoles that may help prevent cancer**

VARIETIES

Of the hundreds of types of cabbage, three are grown and sold in the United States: green, red, and savoy. Green cabbage has smooth, green outer leaves and pale interior leaves. The three most commonly grown varieties of green cabbage are Danish, with very compact, round or oval heads, produced for sale in the late fall; Domestic, with looser heads of curled leaves; and Pointed, grown primarily in the Southwest for the spring market, with small, conical heads and smooth leaves. Red cabbage has dark-red to purple leaves with white veins. Red cabbage has a tougher texture and a flavor that is similar to but slightly sweeter than that of the green variety. Savoy cabbage has pale, yellow-green, crinkled leaves forming a less compact, more oblong head. Its flavor tends to be milder than that of red or green cabbage.

ORIGIN & BOTANICAL FACTS

The oldest accounts of cultivated cabbage appear in Greek literature and date from about 600 B.C. However, the cabbage eaten by the early Greeks and Romans appears to have been a loose-leaved, non-heading type. Modern compact-headed varieties with overlapping leaves were developed by northern European farmers during the Middle Ages. Because this type thrived through cold winters, it became almost as much a staple in the European diet as potatoes and corn. Cabbage is an inexpensive vegetable that is easy to grow and stores well. It is particularly popular in Germany, Austria, Poland, and Russia. In the United States, the primary regions of cultivation are California, Florida, Georgia, New York, and Texas.

Cabbage is propagated from seed sown first in a seedbed and then transplanted after 1 to 2 months. Tall varieties must be staked to prevent damage from wind or heavy rain. Cabbage is a relatively slow-growing crop. Some varieties take up to 200 days to mature. Other vegetables that develop more quickly, such as lettuce or green beans, may be sown between rows of cabbage plants.

USES

Uncut cabbage can be stored for months in perforated vegetable bags in the refrigerator crisper.

Raw cabbage can be shredded for salads and cole slaw. Cooked cabbage has a strong flavor and mushy consistency when overcooked, but it can be prepared so that its mild taste and crisp texture are retained. Cabbage can be microwaved, steamed, stir-fried, or added to soups and stews. Individual cabbage leaves can be separated and used to wrap a variety of stuffings, such as meats and rice or other grains. Seasonings that work well with both raw and cooked cabbage include caraway, dill, mustard, and curry.

NUTRIENT COMPOSITION

Cabbage is high in vitamin C. As a cruciferous vegetable, it contains significant amounts of nitrogen compounds called indoles, which are phytochemicals that may help prevent some types of cancer.

NUTRIENT CONTENT

	1 cup shredded, raw (70 g)	½ cup shredded, boiled (75 g)
Energy (kilocalories)	18	17
Water (%)	92	94
Dietary fiber (grams)	2	2
Fat (grams)	0	0
Carbohydrate (grams)	4	3
Protein (grams)	1	1
Minerals (mg)		
Calcium	33	23
Iron	0	0
Zinc	0	0
Manganese	0	0
Potassium	172	73
Magnesium	11	6
Phosphorus	16	11
Vitamins (mg)		
Vitamin A	9 RE	10 RE
Vitamin C	23	15
Thiamin	0	0
Riboflavin	0	0
Niacin	0	0
Vitamin B$_6$	0.1	0.1
Folate	30 µg	15 µg
Vitamin E	0	0

CARROT

The carrot plant is a member of the parsley family, characterized by light, feathery leaves. Other members of this family include fennel, dill, and celery. The edible root of the plant is usually orange and shaped like a long cylindrical cone. A fibrous channel or core runs the length of the vegetable; usually, the smaller the core, the younger and sweeter the vegetable.

Family Umbelliferae
Scientific name *Daucus carota*
Common name carrot

♥ **High in vitamin A (carotenes)**

♥ **Good source of fiber**

♥ **Contains phytochemicals that may help prevent cancer and heart disease**

VARIETIES

Many varieties of carrots are grown throughout the world. Colors range from white to yellow to crimson. A carrot may be as short as 3 to 6 inches and as long as several feet. However, most carrots on the U.S. market today are orange and 7 to 9 inches long. Mini-peeled carrots are cut from the smaller, sweeter "caropak" carrots, which have been grown tightly together especially for this purpose. Despite packaging and labeling claims, mini-peeled carrots are not baby carrots. True baby carrots are carrots harvested earlier than usual and do, in fact, look like miniature carrots. They are often sold with their green tops still on them in specialty food stores.

ORIGIN & BOTANICAL FACTS

The first carrots, which were white, purple, and yellow, were cultivated in Afghanistan and then brought to the Mediterranean area. Today's orange carrots descend from Dutch-bred carrots and have been grown in the United States since colonial times. Domestically, California produces about 60 percent of the United States crop, 25 percent of which goes into the production of mini-peeled carrots.

USES

Carrots should be firm and brightly colored from top to bottom. Near the leafy crown of the root, there may be a greenish tinge, but dark or black coloring is an indication of age. Keeping carrots refrigerated in moisture-retaining packaging will preserve them for up to a month. The green leaves should be twisted off before storage, because they wilt quickly and draw moisture from the carrots. Fruits that produce ethylene gas as they ripen, such as apples or pears, should not be stored in the same bag with carrots.

With the exception of beets, carrots contain more sugar than any other vegetable. They are a satisfying snack when eaten raw and are a tasty addition to a variety of mixed dishes. Grated raw carrots may be added to fruit or vegetable salads, mixed with peanut butter as a sandwich filling, or used in baking cakes, muffins, or breads. Cooked carrots enhance the flavor of casseroles, soups, and stews. Puréed carrots may be used in cookies, puddings, and soufflés.

NUTRIENT COMPOSITION

A medium-sized raw carrot is an excellent source of beta-carotene, which is converted into vitamin A. Carrots are a relatively good source of fiber. In addition to beta-carotene, carrots contain two other carotenoids: alpha-carotene and lutein. The carotenoids, which are responsible for the bright-orange color of carrots, have antioxidant properties and may help prevent cancer and heart disease. Lutein also has been looked at for its role in protecting the eye from free-radical damage and maintaining vision. Cooking carrots makes them more digestible and appears to increase the amount of vitamin A available for use in the body. However, the vitamin A content of fresh or frozen carrots is twice that of canned versions.

NUTRIENT CONTENT

	1 medium, raw (61 g)	½ cup, cooked (78 g)
Energy (kilocalories)	26	35
Water (%)	88	87
Dietary fiber (grams)	2	3
Fat (grams)	0	0
Carbohydrate (grams)	6	8
Protein (grams)	1	1
Minerals (mg)		
Calcium	16	24
Iron	0	0
Zinc	0	0
Manganese	0	1
Potassium	197	177
Magnesium	9	10
Phosphorus	27	23
Vitamins (mg)		
Vitamin A	1,716 RE	1,915 RE
Vitamin C	6	2
Thiamin	0.1	0
Riboflavin	0	0
Niacin	1	0
Vitamin B$_6$	0.1	0.2
Folate	9 µg	11 µg
Vitamin E	0	0

CASSAVA

The cassava is a root 2 to 3 inches in diameter and 6 to 12 inches long, covered with a coarse, inedible brown skin. To help preserve the root, the skin is often coated with a shiny film of wax. The flesh of the tuber is white with thin veins running through it and is potato-like in texture.

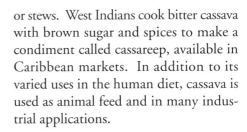

Family Euphorbiaceae
Scientific names *Manihot esculenta* Crantz
   (Manihot ultissima Phol [sweet]*;*
   Manihot aipi Phol [bitter])*
Common name yuca, tapioca, manioc, apple

VARIETIES

Until recently, the many varieties of cassava were divided into two main categories: bitter and sweet. Because the bitter root contains substances that are converted to toxic cyanide compounds when the root is cut, it must not be eaten raw; cooking destroys these substances. Although sweet cassava is believed to have low amounts of these potential toxins, taste is not a reliable predictor of toxin content, and experts recommend that all cassava be cooked.

ORIGIN & BOTANICAL FACTS

The cassava originated in Brazil, Paraguay, and the Caribbean Islands, from where it was introduced to Africa and the Far East. Africa is now the leading producer, and cassava is an important dietary staple throughout the continent. In the United States, cassava is grown in Florida and is imported from Mexico, Central America, South America, and the Antilles.

Cassava is propagated from stem cuttings. Ideal growing conditions include temperatures between 77° and 86° Fahrenheit (the plants cease to grow if temperatures fall below 50° Fahrenheit). Most cassava roots are harvested by hand, although Brazil has developed mechanical harvesters. Because the roots are extremely sensitive to physical damage, harvesting must be done carefully.

To increase the short shelf life of the cassava further, the leaves are removed 2 weeks before harvest. In addition to dipping the roots into wax, storing the newly harvested roots in plastic bags extends the shelf life by 3 to 4 weeks.

USES

Cassava should be refrigerated no more than 4 days. The peeled cassava can be boiled and mashed, baked, or sliced and fried, identical to the cooking of potatoes. Alternatively, the peeled root can be grated and the starch extracted to make breads, crackers, pasta, and tapioca pearls (a commercial product used to make pudding). In Africa, the roots are fermented in water, after which they are made into an alcoholic beverage; sundried for storage; or grated, formed into a dough, and cooked alone or in soups or stews. West Indians cook bitter cassava with brown sugar and spices to make a condiment called cassareep, available in Caribbean markets. In addition to its varied uses in the human diet, cassava is used as animal feed and in many industrial applications.

NUTRIENT COMPOSITION

Cassava is composed mostly of carbohydrate and is a major source of calories in Third-World countries.

SERVING SIZE:
1/4 cup, raw (51 g)

NUTRIENT CONTENT

Energy (kilocalories)	83
Dietary fiber (grams)	1
Fat (grams)	0
Carbohydrate (grams)	20
Protein (grams)	1
Minerals (mg)	
Calcium	8
Iron	0
Zinc	0
Manganese	0
Potassium	140
Magnesium	11
Phosphorus	14
Vitamins (mg)	
Vitamin A	1 RE
Vitamin C	11
Thiamin	0
Riboflavin	0
Niacin	0
Vitamin B$_6$	0
Folate	14 µg
Vitamin E	0

CAULIFLOWER

As their names imply, cauliflower and broccoflower are actually flowers. The part of the plant that is eaten is the head of underdeveloped, tender flower stems and buds. While growing, the head is surrounded by heavy green leaves that protect it from sunlight and discoloration. Many of the leaves are trimmed off during preparation for shipment and sale. Cauliflower has a strong odor when cooked and a rich, cabbage-like flavor. Broccoflower is a hybrid of broccoli and cauliflower.

Family Cruciferae
Scientific name *Brassica oleracea* L.
  (*botrytis*)
Common name cauliflower

♥ **High in vitamin C**

♥ **A cruciferous vegetable that contains phytochemicals that may help prevent cancer**

VARIETIES

Cauliflower falls into three types. The most commonly grown and sold is the white cauliflower, which has creamy curds and bright-green leaves. The green variety is actually the hybrid broccoflower developed about 10 years ago, which has bright lime-green curds. Less commonly known is the purple-headed cauliflower.

ORIGIN & BOTANICAL FACTS

Cauliflower is native to the Mediterranean region and Asia Minor, where it was cultivated more than 2,000 years ago. By the 16th century, its cultivation had spread throughout western Europe. In the United States, cauliflower did not become an important vegetable until the early part of the 20th century. Today it is grown in numerous states. California and New York are the leading producers, and it is also grown in Arizona, Michigan, Oregon, Florida, Washington, and Texas.

Cauliflower requires cool tempera-

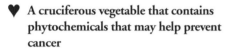

tures and rich, fertile soil with good moisture. It is usually planted as an annual, but milder climates can support winter varieties, so it is available year-round, with peak seasons in spring and fall. Cauliflower is propagated from seed, usually sown first in beds and then transplanted to the field after 4 or 5 weeks. The flower bud forms on a single stalk and is surrounded by large, heavy green leaves that protect it from the discoloring effect of sunlight. Heads are ready for harvest between 80 and 110 days after transplantation. In the field, many of the leaves are trimmed off, and the heads, which bruise easily, are packed gently for shipping, usually in plastic wrap that has been perforated to allow the escape of carbon dioxide, which can discolor the head and cause an unpleasant taste.

USES

A cauliflower head should be unbruised, firm, and uniformly cream-colored without a trace of black, and it should be heavy for its size. It can be stored in perforated plastic vegetable bags in the refrigerator crisper for several days.

The raw florets are tasty in salads, as a crunchy appetizer with dips, or pickled. Cauliflower can be boiled, steamed, microwaved, or baked. It is a flavorful addition to soups and stews or to other vegetables such as carrots, tomatoes, peas, bell pepper, or broccoli. Cauliflower also works well on its own, seasoned with nutmeg, dillweed, chives, or almonds. Cauliflower is available fresh and frozen, but the processing required for freezing destroys some nutrients and can turn the florets watery in flavor and appearance.

NUTRIENT COMPOSITION

Cauliflower is high in vitamin C. As a cruciferous vegetable, cauliflower contains phytochemicals that may help prevent cancer.

NUTRIENT CONTENT

	½ cup, raw (50 g)	½ cup, boiled (62 g)
Energy (kilocalories)	13	14
Water (%)	92	93
Dietary fiber (grams)	1	2
Fat (grams)	0	0
Carbohydrate (grams)	3	3
Protein (grams)	1	1
Minerals (mg)		
Calcium	11	10
Iron	0	0
Zinc	0	0
Manganese	0	0
Potassium	152	88
Magnesium	8	6
Phosphorus	22	20
Vitamins (mg)		
Vitamin A	1 RE	1 RE
Vitamin C	23	27
Thiamin	0	0
Riboflavin	0	0
Niacin	0	0
Vitamin B$_6$	0.1	0.1
Folate	29 µg	27 µg
Vitamin E	0	0

CELERIAC

Closely related to celery, celeriac (also called celery root or celery knob) is a knobby, bulb-shaped root about the size of a baseball. It has a rough brown skin and rootlets. It has a crisp texture and a nutty flavor that resembles that of strong celery or parsley.

Family Umbelliferae
Scientific name *Apium graveolens*
Common name celeriac

♥ **Raw celeriac is a good source of vitamin C and provides some potassium and phosphorus**

VARIETIES

Celeriac is available in three varieties: Iram, a medium-sized, globe-shaped root with few side shoots; Marble Ball, a round, white root; and Tellus, a quick-growing, round root with brownish red leaf stems.

ORIGIN & BOTANICAL FACTS

A native of the Mediterranean region, celeriac is a popular vegetable in Europe, particularly in France and Italy, but it is less well known in the United States.

Celeriac is propagated from seed, usually sown in pots or greenhouses in early spring, and then transplanted to the field in May. It requires a fertile soil that is rich in organic matter. For a large corm (underground stem base) to develop, a long growing season and plenty of water are required. Thus, celeriac thrives in moist, temperate climates. In midsummer, the outer leaves are removed, and the plant is mulched or fertilized to assist growth and moisture retention. Lateral shoots also are removed to create a single

growing point. Celeriac is hardier than celery and keeps well in winter if stored in a cool place. Thus, it is available year-round, with a peak season lasting from November through April.

USES

Small to medium-sized celeriac roots that are heavy for their size and free of cuts, bruises, and soft spots are the best. The roots should be stored with their stalks intact in perforated plastic bags in the refrigerator crisper.

Celeriac can be consumed either raw or cooked. Regardless of how it is to be used, the root must be rinsed well and peeled. After it is peeled, the pieces should be dropped into a bowl of acidulated water (water to which a few drops of lemon juice or vinegar have been added) to prevent the discoloration that occurs from exposure to the air. The raw root can be cut into sticks for dipping, or it can be grated or julienned for salads. Celeriac can be boiled, braised, baked, or steamed until it is tender. It can be

cooked whole and then peeled, diced, or puréed and added to soups, stews, and stir-fries or eaten alone with a bit of butter or margarine and fresh herbs. In Europe, celeriac is often added to mashed potatoes and served with butter or cream sauce. Like celery, the celeriac root as well as the stalks are often used as a seasoning.

NUTRIENT COMPOSITION

Raw celeriac is a good source of vitamin C and provides some potassium and phosphorus.

NUTRIENT CONTENT

	½ cup, raw (78 g)	½ cup, cooked (77 g)
Energy (kilocalories)	33	21
Water (%)	88	92
Dietary fiber (grams)	1	1
Fat (grams)	0	0
Carbohydrate (grams)	7	5
Protein (grams)	1	1
Minerals (mg)		
Calcium	34	20
Iron	0	0
Zinc	0	0
Manganese	0	0
Potassium	234	134
Magnesium	16	9
Phosphorus	90	51
Vitamins (mg)		
Vitamin A	0 RE	0 RE
Vitamin C	6	3
Thiamin	0	0
Riboflavin	0	0
Niacin	1	0
Vitamin B6	0.1	0.1
Folate	6 µg	3 µg
Vitamin E	0	–

Note: A line (–) indicates that the nutrient value is not available.

CELERY

Celery is a vegetable that is enjoyed for its crisp texture and distinctive flavor. A bunch of celery is actually a single stalk consisting of separate ribs, and the most tender, inner ribs are called the hearts. The crispness of celery comes from the rigidity of its cell walls and its high water content. In fact, celery is mostly water, which makes it low in calories and an ideal snack food.

Family Umbelliferae
Scientific name *Apium graveolens* L.
Common name celery

♥ **A good source of vitamin C**

VARIETIES

Although celery is available in many colors, most of the celery grown in the United States belongs to the green varieties, which range in shade from pale to dark and are referred to as Pascal. One common green variety is the American Green, also known as the Tall Utah or Greensnap, which does not require blanching (the banking of soil against the plant to keep it from turning dark green upon exposure to sunlight). Other varieties of celery include the Giant Pink, with pink or red stems and dark-green leaves; the Golden Self-Blanching, with pale, golden-yellow leaves and golden stems; and the Ivory Tower, a fast-maturing, self-blanching variety with pale leaves.

ORIGIN & BOTANICAL FACTS

Wild celery is a biennial or annual herb native to southern Europe, Asia, and Africa, growing in marshes along the muddy banks of tidal rivers or in other saltwater areas. Although it resembles domestic celery, it is smaller, with a stronger, more pungent odor and flavor. Before the familiar milder, thick-stalked forms were cultivated, celery probably was used solely as a seasoning and medicinal herb. Leafy cultivated varieties may date back 2,000 years or more, but stalk celery seems to have been grown first in Italy in the 16th century. In the 1690s, John Evelyn, an English diarist, described celery as a new vegetable. Today, celery is grown in Great Britain, India, the United States, and Canada.

Celery requires a moist, rich soil for good growth. It is especially successful in low-lying, alkaline areas such as the eastern regions of Florida and Great Britain. Because of its high water content, celery requires large amounts of moisture; otherwise, the stalks become stringy and tough. Celery is propagated from seeds so tiny that it takes more than a million of them to add up to a pound. Most commercially grown celery is planted in March or April, in greenhouses or seedbeds with controlled watering.

About 2 months after sowing, when the seedlings are 4 to 6 inches tall, they are transplanted to fields. Varieties that require blanching usually are planted in trenches to facilitate the banking of the soil against the plants. Three to 4 months after field planting, celery is ready for harvesting.

USES

Celery that is light in color and shiny has the best flavor. Celery should be stored by the bunch in perforated plastic vegetable bags in the refrigerator crisper.

Celery is a versatile vegetable. Raw celery adds crunch to chicken, seafood, egg, potato, and green salads. It is an excellent snack food or appetizer. Celery also can be microwaved, stir-fried, braised, or steamed to serve as a main vegetable, but it usually is combined with other vegetables or is included in stuffings for poultry and fish. Celery even has been made into a uniquely flavored soft drink, Dr. Brown's Cel-Ray tonic.

NUTRIENT COMPOSITION

Celery is a good source of vitamin C.

SERVING SIZE: 2 stalks (80 g)

NUTRIENT CONTENT

Energy (kilocalories)	13
Water (%)	95
Dietary fiber (grams)	1
Fat (grams)	0
Carbohydrate (grams)	3
Protein (grams)	1
Minerals (mg)	
Calcium	32
Iron	0
Zinc	0
Manganese	0
Potassium	230
Magnesium	9
Phosphorus	20
Vitamins (mg)	
Vitamin A	10 RE
Vitamin C	6
Thiamin	0
Riboflavin	0
Niacin	0
Vitamin B₆	0.1
Folate	22 µg
Vitamin E	0

CHAYOTE

The chayote, a tropical member of the cucumber and squash family, is actually a fruit. It resembles a summer squash or avocado in shape and appearance but has deep, lengthwise ridges and a single, flat, nut-like seed. The fruit can be variable in size, color, texture, and flavor. The skin of the fruit can be smooth, deeply fissured, or even wrinkled and prickly. Colors range from light-green to almost white. The opaque flesh has a cucumber-like texture and varies in color. Once heated, the flesh becomes somewhat translucent. Except for the seed, the entire fruit is edible.

Family Cucurbitaceae
Scientific name *Sechium edule*
Common name chayote

♥ **A good source of vitamin C**

VARIETIES

Although definite strains of chayote-producing plants exist, distinctive varieties of the fruit are yet to be identified. However, the fruit is identified by a variety of names. In many places, it is recognized as mango squash, chocho, christophine, and choke. In Louisiana, it is called mirliton, and in Florida it is called a vegetable pear.

ORIGIN & BOTANICAL FACTS

Chayote is native to Mexico, Central America, and the West Indies. The fruit is believed to have been cultivated by the Aztecs and Mayans long before Columbus arrived. The name chayote is derived from the Mayan word "chayotli." The plant is now grown in South America, North Africa, and in subtropical parts of southern Florida. The fruit grows abundantly from a fast-growing tropical climbing vine that may reach up to 100 feet in a single season and is covered with large, heart-shaped, lobed leaves that measure 4 to 6 inches. Ideal growing conditions include full sunlight, high moisture levels, and rich, well-drained soil. Under the proper conditions, some plants can produce up to 100 fruits in a single season. The fruits must be harvested young or they will become tough. Chayote is available year-round, but the peak season is late summer through early fall.

USES

A firm, unblemished, clear-green chayote is the best choice. The chayote keeps up to 1 month stored uncovered in a cool, dry, dark place, or it can be stored in a perforated plastic bag in a refrigerator vegetable crisper for up to a week.

The chayote is most easily prepared by peeling the fruit and microwaving or steaming it for a few minutes. Although very young fruit can be prepared with the skin left on, more mature fruit should be peeled under running water to prevent being irritated by the sticky sap under the skin. Cooked chayote can be seasoned to taste and eaten as is; sliced or diced and added to other dishes such as salads in place of cucumbers; or prepared like french fries. Chayote halves that are stuffed and baked make a filling main dish. The chayote can be substituted for many other fruits and vegetables in recipes. A fully mature fruit may be used in place of potatoes in soups and purées. The grated fruit also is useful as a substitute for carrots and zucchini in breads and pastries. Cooked, mashed, and seasoned with sweet spices, it resembles applesauce and can be served as a light snack or dessert. The leaves and stems of the chayote plant are used as a low-cost animal feed and can also be spun into cord.

NUTRITIONAL FACTS

Chayote is a good source of vitamin C.

SERVING SIZE: *1/2 cup pieces, cooked (80 g)*

NUTRIENT CONTENT

Energy (kilocalories)	17
Water (%)	93
Dietary fiber (grams)	2
Fat (grams)	0
Carbohydrate (grams)	4
Protein (grams)	1
Minerals (mg)	
Calcium	10
Iron	0
Zinc	0.2
Manganese	0
Potassium	138
Magnesium	10
Phosphorus	23
Vitamins (mg)	
Vitamin A	4 RE
Vitamin C	6
Thiamin	0
Riboflavin	0
Niacin	0
Vitamin B$_6$	0.1
Folate	14 µg
Vitamin E	–

Note: A line (–) indicates that the nutrient value is not available.

CHICORY

Chicory is a perennial that forms long, stick-like stems and ragged, widely spaced bunches of leaves, sometimes in tight heads or in loose formations. The outer leaves may be green, white, or red, depending on variety, and have a strong, slightly bitter taste. The inner leaves are usually paler in color and milder in flavor.

Family Compositae
Scientific name *Chichorium intybus*
Common name chicory, Belgian endive, radicchio

♥ **Leaves are high in folate, vitamin A, vitamin C, potassium, and fiber**

♥ **A good source of calcium, magnesium, riboflavin, and vitamin B₆**

VARIETIES

The two basic types of chicory, forcing and nonforcing, are distinguished by their method of cultivation. The forcing chicories are initially sown outdoors, but because exposure to light tends to create a bitter taste, the plants are transferred to a dark area (a process called blanching) for the latter half of their growth. As a result, forcing chicories have a milder flavor. The most common forcing varieties are the Witloof chicory, sometimes called Belgian endive, and the red-leaf radicchio, an Italian chicory that is becoming increasingly popular in the United States.

Nonforcing chicories do not require blanching. These varieties are grown like lettuce, without forcing. The Italian radichetta, more common in North America than in Europe, has narrow leaves that grow on wide stalks and are cooked like asparagus.

ORIGIN & BOTANICAL FACTS

Chicory is native to Europe and western Asia. Evidence suggests that it was grown in ancient Egypt, where, along with endive and escarole, it was believed to have been one of the bitter herbs consumed during the Jewish Passover. Works by Horace, Aristophanes, and Pliny attest to the use of chicory by the early Greeks and Romans. Later, it was brought to North America, where it now grows wild and in cultivated form.

Chicory can be grown in a variety of soil types. Seeds generally are sown directly into open ground. Because most types of chicory thrive in cool temperatures, planting is done in early spring or late fall. Chicory plants have shallow roots, so frequent irrigation is necessary. The plants grow with a scruffy appearance and with multiple stick-like stems that are 2 to 3 feet tall. Bright, almost iridescent, blue flowers appear on the stems in the second year.

Forcing varieties are dug up in late fall, the leaves and roots are cut back, and the plants are laid horizontally in pots of moist peat in a dark, warm place. After 3 or 4 weeks, the heads are ready to be cut. Nonforcing varieties of chicory are planted in the spring and picked like lettuce in the fall. They can be used immediately or stored in a cool place for later use. Leading domestic producers of chicory, particularly the newly popular radicchio, are California and New Jersey. Some chicory is also grown in Mexico and Italy.

USES

Chicory, Belgian endive, and radicchio should be selected and stored in a manner similar to arugula and lettuce. Chicory leaves most often are used raw in salads. The roots of some varieties of chicory are roasted and ground to make a coffee substitute or flavoring popular in Louisiana.

NUTRIENT COMPOSITION

Chicory is high in folate, vitamin A, vitamin C, potassium, and fiber. It is also a good source of calcium, magnesium, riboflavin, and vitamin B₆.

SERVING SIZE:
1 cup greens, raw (180 g)

NUTRIENT CONTENT

Energy (kilocalories)	41
Water (%)	92
Dietary fiber (grams)	7
Fat (grams)	1
Carbohydrate (grams)	8
Protein (grams)	3
Minerals (mg)	
Calcium	180
Iron	2
Zinc	1
Manganese	1
Potassium	756
Magnesium	54
Phosphorus	85
Vitamins (mg)	
Vitamin A	720 RE
Vitamin C	43
Thiamin	0.1
Riboflavin	0.2
Niacin	1
Vitamin B₆	0.2
Folate	197 µg
Vitamin E	4

CHINESE CABBAGE (BOK CHOY & NAPA)

Bok choy and napa are two varieties of Chinese cabbage, a member of the same family as broccoli and Brussels sprouts. Resembling a cross between celery and Swiss chard, bok choy has white, celery-like stalks with dark-green, long, rounded leaves. Napa cabbage is similar in shape and size to romaine lettuce and has white, crisp stalks.

Family Cruciferae
Scientific name *Brassica campestris* L.
Common name bok choy, napa cabbage

♥ **Raw bok choy is high in vitamin A and vitamin C**

♥ **Raw bok choy is a good source of folate**

♥ **Napa cabbage is a good source of zinc**

VARIETIES

As many as 33 varieties of Chinese cabbage exist, each with a different name. The two most common varieties are bok choy (var. *chinensis*) and napa cabbage (var. *pekinensis*). Bok choy also is known as pak-choi, qing cai, taisai, chongee, and Chinese mustard cabbage. Baby bok choy is a variety that grows to a fraction of the size of regular bok choy and is consumed whole. Among its other names, napa cabbage is sometimes called Chinese cabbage, which adds to the confusion among varieties.

ORIGIN & BOTANICAL FACTS

Native to China and eastern Asia, Chinese cabbages are annual plants that grow best in cool, moist environments. Both bok choy and napa cabbage have been cultivated in China for thousands of years and are popular in that country and in Korea and Japan. Introduced to the United States by Chinese immigrants in the late 19th century, Chinese cabbage is now grown in California, New Jersey, Hawaii, and Florida.

USES

Both bok choy and napa are available throughout the year. Bok choy should have bright, white stalks and fresh green leaves and should show no signs of wilting. Napa heads should be tightly closed and have unblemished leaves. Uncut, unwashed cabbage can be refrigerated in a plastic bag for up to 3 days. Mild-flavored and versatile, both bok choy and napa cabbage can be prepared in the same ways as regular cabbage. They can be used raw in salads or steamed, boiled, braised, stuffed, or stir-fried. Cooking softens the flavor of the leaves and sweetens the flavor of the stalks. Before cooking, the stalks must be sliced crosswise or on the diagonal, and the leaves cut into thick shreds. When stir-frying bok choy, the stems should be cooked a few minutes before adding the more tender leaves. Bok choy and napa cabbage are delicious cooked alone or with meat, poultry, and other vegetables. A mild-flavored soup can be prepared by adding the leaves and stalks to either a chicken or miso broth with scallions and cubes of chicken or tofu. In Korea, kimchee, a spicy dish made from pickled Chinese cabbage, is served at most meals.

NUTRIENT COMPOSITION

Raw bok choy is high in vitamin A and vitamin C. One serving of cooked napa cabbage is a good source of zinc.

NUTRIENT CONTENT

	1 cup bok choy, raw, shredded (70 g)	½ cup napa, cooked (55 g)
Energy (kilocalories)	9	7
Water (%)	95	96
Dietary fiber (grams)	1	–
Fat (grams)	0	0
Carbohydrate (grams)	2	1
Protein (grams)	1	1
Minerals (mg)		
Calcium	74	16
Iron	1	0
Zinc	0	2
Manganese	0	0
Potassium	176	47
Magnesium	13	4
Phosphorus	26	10
Vitamins (mg)		
Vitamin A	210 RE	5 RE
Vitamin C	32	2
Thiamin	0	0
Riboflavin	0	0
Niacin	0	0
Vitamin B$_6$	0.1	0
Folate	46 µg	23 µg
Vitamin E	0	–

Note: A line (–) indicates that the nutrient value is not available.

COLLARDS

Collards are plain-leafed, nonheading members of the cabbage family, closely related to kale. In flavor, they resemble a cross between cabbage and kale and are considered one of the milder greens.

Family Cruciferae
Scientific name *Brassica oleracea* var. *acephala*
Common name collards

♥ **High in vitamin A (beta-carotene), vitamin C, and folate**

♥ **A good source of fiber and calcium**

♥ **A cruciferous vegetable that contains phytochemicals that may help prevent cancer**

VARIETIES

Collard varieties include the Plant Vates, Carolina Improved Heading (or Morris), Georgia Southern, Blue Max, and Heavi Crop.

ORIGIN & BOTANICAL FACTS

Collards, one of the oldest members of the cabbage family, are similar to the wild, nonheading forms of cabbage that were among the first foods eaten by prehistoric people. They are native to the eastern Mediterranean region and Asia Minor and were cultivated by the ancient Greeks and Romans. Collards were introduced to Britain and France around 400 B.C. by either the Romans or the Celts. Although collards were first mentioned in the American colonies in 1669, they may have been present here before that time.

Collards are a cool-season crop that thrives in temperate climates. They grow well in warm weather but can tolerate cold temperatures in late fall, and their flavor is enhanced by light autumn frost. Because they can survive cold tempera-

tures, the cultivation of collards has spread north from the southeastern United States, where they have long been a popular vegetable. Collard seeds can be sown directly into fields, or they can be planted in protected beds and the seedlings transplanted 6 or 8 weeks later into the fields. The plants mature in about 60 days. Collards can grow in a variety of soils and are tolerant of poor soil.

Collard greens can be harvested in these ways: the tender, young leaves can be removed from mature plants (which encourages new growth), or the entire plant can be cut when it is very young, half-grown, or fully mature. Maximal yield occurs when the leaves are removed from the bottom of the plant before they age. The peak season for collards is January through April, but they generally are available in markets year-round.

USES

Crisp bunches of intact leaves with no yellowing are best. Collards can be stored in perforated plastic bags in the refrigerator crisper.

All green parts of the collard plant are edible. The southern style of cooking

fresh collards is to boil them with a chunk of bacon or salt pork, but they can be prepared similarly to cabbage, spinach, or other greens. They can be steamed or microwaved and added to soups and stews or casseroles. Cooked collards make a tasty salad served chilled with olive oil and lemon juice. Collard leaves also are available frozen and canned.

NUTRIENT COMPOSITION

One-half cup of cooked collards is a good source of fiber and calcium and is high in vitamin A (beta-carotene), vitamin C, and folate. As a cruciferous vegetable, collards contain phytochemicals that may help prevent cancer.

SERVING SIZE:
1/2 cup, boiled (95 g)

NUTRIENT CONTENT

Energy (kilocalories)	25
Water (%)	92
Dietary fiber (grams)	3
Fat (grams)	0
Carbohydrate (grams)	5
Protein (grams)	2
Minerals (mg)	
Calcium	113
Iron	0
Zinc	0
Manganese	1
Potassium	247
Magnesium	16
Phosphorus	25
Vitamins (mg)	
Vitamin A	297 RE
Vitamin C	17
Thiamin	0
Riboflavin	0.1
Niacin	1
Vitamin B6	0.1
Folate	88 µg
Vitamin E	1

CORN

Because corn is a member of the grass family, it is not strictly a vegetable but a grain (see Grains, page 269). However, one type, sweet corn, is prepared and served as a fresh vegetable. The seeds, or kernels, which are the edible part of the plant, form in spikelets on a woody axis called an ear. They are covered with a green husk. Ears of corn vary in size, and the kernels range from white to yellow, orange, red, brown, blue, purple, and black, although sweet corn is always white or butter-yellow.

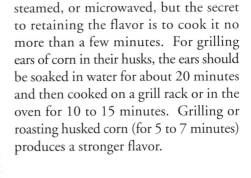

Family Gramineae (Poaceae)
Scientific name *Zea mays* L.
Common names corn, sweet corn

♥ **Moderately good source of fiber, vitamin C, and folate**

VARIETY

Of the many types of corn, sweet corn is the only variety that is eaten fresh as a vegetable. Sweet corn is available in several varieties divided by kernel color. Some sweet corn is pure yellow; some, like Silver Queen, is white; and some, like Butter and Sugar, is bicolored, that is, it has both yellow and white kernels. Popping corn, baby corn, and the white corn known as hominy are different types of eating corn.

ORIGIN & BOTANICAL FACTS

Corn is native to the Americas, probably having originated in Mexico or Guatemala, where historians believe it evolved from a wild grain called teosinte that still grows in the Mexican highlands. The corn plant, which can attain heights of 20 feet, has a hard, jointed stalk. Male flowers develop in the tassel at the top of the stalk, and the female flower is a cluster at the joint of the stalk. Corn is pollinated by the wind. The corn silk hanging from the husk of each ear is the pollen receptor; each thread must receive a grain of pollen for the kernels to develop. Corn grows best during long, hot summers. It requires rich soil and regular fertilizing, particularly with nitrogen. Because the flowers are wind-pollinated, plants should be spaced closely together. The ears do not ripen well in cold weather, so the seeds must be planted in plenty of time for the kernels to develop before the first autumn frost. Although each ear of corn produces many kernels, the plant has no natural mechanism for dispersing its seeds. To ensure that it will be tender and succulent for eating, sweet corn is picked before it reaches maturity.

USES

Sweet corn is a popular vegetable in the United States. The average American eats about 25 pounds of corn every year, most of it frozen or canned, but in summer, corn is preferred fresh on the cob. Because the sugars in the kernels of sweet corn begin to convert to starch as soon as the ear is picked, corn should be eaten as soon after harvest as possible.

While still on the cob and after the husks are removed, corn can be boiled, steamed, or microwaved, but the secret to retaining the flavor is to cook it no more than a few minutes. For grilling ears of corn in their husks, the ears should be soaked in water for about 20 minutes and then cooked on a grill rack or in the oven for 10 to 15 minutes. Grilling or roasting husked corn (for 5 to 7 minutes) produces a stronger flavor.

NUTRIENT COMPOSITION

Corn provides some fiber, vitamin C, and folate.

NUTRIENT CONTENT

	½ cup (about 1 ear) yellow corn, boiled (82 g)	1 cup popcorn, air-popped (8 g)
Energy (kilocalories)	89	31
Water (%)	70	4
Dietary fiber (grams)	2	1
Fat (grams)	1	0
Carbohydrate (grams)	21	6
Protein (grams)	3	1
Minerals (mg)		
Calcium	2	1
Iron	1	0
Zinc	0	0
Manganese	0	0
Potassium	204	24
Magnesium	26	10
Phosphorus	84	24
Vitamins (mg)		
Vitamin A	18 RE	2 RE
Vitamin C	5	0
Thiamin	0.2	0
Riboflavin	0.1	0
Niacin	1	0
Vitamin B₆	0	0
Folate	38 μg	2 μg
Vitamin E	0	0

CRESS (WATERCRESS)

Cress is a member of the mustard family, a cruciferous vegetable whose small, dark-green leaves add a slight crunch and a tangy, peppery flavor to dishes.

Family Cruciferae
Scientific name *Lepidium sativum*
(garden cress)
Common name cress, garden cress, watercress

♥ **High in vitamin C**

♥ **A good source of folate and vitamin A**

♥ **A cruciferous vegetable that contains phytochemicals that may help prevent cancer**

VARIETIES

The most common variety of cress is watercress, which grows in flooded soil beds and has small, heart-shaped leaves and a slightly bitter taste. Watercress is difficult to grow in home gardens. An easily grown, hardy alternative is winter cress, with dark-green, strongly flavored leaves that form rosettes. Other varieties include broad-leaved cress, with oval leaves; peppergrass, or curly cress, with an attractive, ornamental appearance; extra curled, a compact plant with short stalks and fine leaves; and garden cress, a tall, cool-season annual.

ORIGIN & BOTANICAL FACTS

Cress grows wild in many parts of the world, including Asia, the Middle East, Europe, North America, and New Zealand. Watercress is native to Europe, and garden cress originated in Persia, later spreading to India, Syria, Greece, and Egypt. Watercress has been cultivated since ancient Roman times. Commercial cultivation was first recorded in Germany in 1750 and later in Great Britain in 1808. Today, watercress is grown in Great Britain and in the United States, mainly in California, Florida, and Virginia. Winter cress is cultivated as a substitute for watercress when flowing water is not available for planting. It is produced mostly in the United States, favored by growers because of its hardiness.

Cress is propagated from seeds or stem cuttings and generally prefers the cool growing conditions of early spring and late fall. The growth of watercress requires a special environment with flooded soil beds containing absolutely pure water, because any water-borne contaminants could be deposited on the plant. The leafy stems are generally harvested about 180 days after planting. Requirements for other varieties vary. Garden cress, which can grow up to 18 inches tall, prefers sun or light shade and well-drained soil. Seeds can be sown at intervals from early spring through the summer in order to furnish a constant supply of young leaves. The whole plants are harvested about 60 days after planting. Hardy winter cress is not planted until July or August and can be harvested from late fall through the winter, until the plants begin to flower in spring. The leaves of most cresses are picked when they are 3 to 5 inches long, or the entire plant can be harvested before seed stalks form.

USES

The raw, young leaves of the cress plant are sold in bunches. When selecting watercress, choose crisp leaves with deep color, and avoid those with yellow leaves. Watercress can be refrigerated up to 5 days in a plastic bag or with the stem ends in a glass of water and the leaves covered with a plastic bag. The leaves and trimmed stems may be added to salads, sandwiches, and soups and used as a garnish for a variety of dishes.

NUTRIENT COMPOSITION

Watercress is high in vitamin A and vitamin C. As a cruciferous vegetable, it contains phytochemicals that may help prevent cancer.

SERVING SIZE:
1 cup, raw (34 g)

NUTRIENT CONTENT

Energy (kilocalories)	4
Water (%)	89
Dietary fiber (grams)	1
Fat (grams)	0
Carbohydrate (grams)	0
Protein (grams)	1
Minerals (mg)	
Calcium	41
Iron	0
Zinc	0
Manganese	0
Potassium	112
Magnesium	7
Phosphorus	20
Vitamins (mg)	
Vitamin A	159 RE
Vitamin C	15
Thiamin	0
Riboflavin	0
Niacin	0
Vitamin B₆	0
Folate	40 μg
Vitamin E	0

CUCUMBER

The cucumber is a member of the same family as gourds, melons, and squash. Although the cucumber is botanically a fruit, it is more commonly thought of and used as a vegetable. It is usually oblong and 1 to 8 inches long. It has glossy, dark-green skin and tapering ends. Its interior is generally pale green to white, with rows of tender, edible seeds down the center. Cucumbers are largely water. They are moist but crisp, and their flavor is sweet and mild.

Family Cucurbitaceae
Scientific name *Cucumis sativus*
Common name cucumber

VARIETIES

Cucumbers grow in a variety of shapes and sizes, from tiny gherkins to greenhouse types that are up to 20 inches long. All have a similar flavor. Those grown primarily for eating fresh are called slicing varieties and include both field-grown and greenhouse cucumbers. The greenhouse varieties, sometimes called English cucumbers, tend to be longer and narrower, milder, and seedless. One type of English cucumber is sometimes referred to as "burpless." Other varieties are cultivated for pickling. These are usually smaller than slicing cucumbers and have bumpy, lighter-colored skins. The smallest is the gherkin, which is only 1 or 2 inches long. One of the pickling varieties, the Kirby, is often sold fresh and is enjoyed for its thin skin, crispness, and very small seeds.

ORIGIN & BOTANICAL FACTS

The cucumber is believed to have originated in wild form in the mountains of northern India, where a similar wild species still grows. It was also in India that the cucumber was first cultivated, about 3,000 years ago. From there it was brought to Greece and then to Western Europe. Columbus transported the cucumber to the Americas, where it eventually was cultivated by American Indians and European colonists in eastern North America and as far north as Canada. Today, the leading producers of cucumbers in the United States are Florida, North Carolina, Texas, Georgia, and South Carolina.

Cucumbers require warm temperatures and should not be planted until all danger of frost has passed. Distinct male and female flowers develop on one cucumber plant, and pollen is carried by insects; 10 to 20 bee visits per flower per day are required to produce long, straight fruit. In contrast, greenhouse cucumbers are not pollinated, so they form without seeds. Both field and greenhouse types are picked as soon as they are of edible size so that the plants will continue to produce flowers and fruit.

USES

Whole cucumbers can be refrigerated in a crisper up to 1 week, tightly wrapped in plastic.

With its high water content, the cucumber is especially refreshing in warm weather, chilled and eaten fresh, pickled and eaten alone, or added to green salads or sandwiches. In the Mediterranean region, cucumber often is grated into yogurt, to which spices and raisins or nuts are added to make a cooling condiment for spicy dishes.

NUTRIENT COMPOSITION

Cucumbers are composed mostly of water and contain only small amounts of nutrients.

SERVING SIZE: *1/2 cup, sliced (52 g)*

NUTRIENT CONTENT

Energy (kilocalories)	7
Water (%)	96
Dietary fiber (grams)	0
Fat (grams)	0
Carbohydrate (grams)	1
Protein (grams)	0
Minerals (mg)	
Calcium	7
Iron	0
Zinc	0
Manganese	0
Potassium	75
Magnesium	6
Phosphorus	10
Vitamins (mg)	
Vitamin A	21 RE
Vitamin C	3
Thiamin	0
Riboflavin	0
Niacin	0
Vitamin B$_6$	0
Folate	7 µg
Vitamin E	0

EGGPLANT

Although often thought of as a vegetable, the eggplant is botanically a fruit. It is a member of the nightshade family, which includes the tomato, potato, and pepper. In addition to the purple eggplant, there are many other types of eggplant, varying from white to green-yellow and purple-black. Eggplants also vary in shape and may be oblong, round, tear-dropped, or lobed. Eggplants range in length from 2 to 12 inches.

Family Solanaceae
Scientific name *Solanum melogena esculentum*
Common name eggplant, aubergine

VARIETIES

The eggplant is available in many varieties. In the United States, the most common eggplant is the large, cylindrical or pear-shaped variety with a smooth, glossy, dark-purple skin. Another popular variety is the Japanese or Asian eggplant, which ranges from solid purple to striated shades and has tender, slightly sweet flesh. The Italian, or baby, eggplant looks like a miniature version of the larger common variety. The egg-shaped White Egg has tougher skin and firmer, smoother flesh.

ORIGIN & BOTANICAL FACTS

First cultivated more than 4,000 years ago, the eggplant is believed to be native to India, from where it was subsequently brought to China. In the Middle Ages, Arab traders brought it to Spain and northern Africa. By the 18th century, both the French and the Italians cultivated eggplant, which they called aubergine. Thomas Jefferson introduced the eggplant to the United States. However, it was not until the 20th century that Americans began to use the eggplant as a food. Previously, it was used as a table decoration. Today, the eggplant is most popular in the cuisines of southern Italy and the Middle East.

The eggplant is a frost-intolerant perennial grown as an annual. It will sustain damage if the temperature falls below 65° Fahrenheit. Eggplant is available throughout the year. The peak season is from July to October. Florida and North Carolina produce half the domestic crop. In the winter months, California and Mexico are also major suppliers.

USES

Eggplant is very perishable and should be stored in a cool, dry place. It can be refrigerated in a plastic bag for 3 to 4 days. Because it contains a heat-sensitive toxin that can induce diarrhea and vomiting, eggplant must be cooked before eating. To prevent the discoloration that occurs when the flesh is exposed to air, eggplant should be left intact until just before cooking. Peeling is recommended for older eggplant, because the skin toughens with age. Because a carbon steel blade will blacken the flesh, a stainless steel knife is preferred for cutting eggplant.

Eggplant can be stuffed and baked, broiled, roasted, fried, stir-fried, or stewed. Because they are very porous, eggplants soak up oil easily during frying. Oil absorption can be minimized by salting to draw out the moisture and compact the flesh. This process also eliminates the natural bitter taste. Using a nonstick pan also can help cut down on fat absorption.

NUTRIENT COMPOSITION

Eggplant is composed mostly of water and contains only small amounts of nutrients.

SERVING SIZE: *1/2 cup cubes, cooked (50 g)*

NUTRIENT CONTENT

Energy (kilocalories)	14
Water (%)	92
Dietary fiber (grams)	1
Fat (grams)	0
Carbohydrate (grams)	3
Protein (grams)	0
Minerals (mg)	
Calcium	3
Iron	0
Zinc	0
Manganese	0
Potassium	123
Magnesium	6
Phosphorus	11
Vitamins (mg)	
Vitamin A	3 RE
Vitamin C	1
Thiamin	0
Riboflavin	0
Niacin	0
Vitamin B6	0
Folate	7 µg
Vitamin E	0

FENNEL

Fennel is an aromatic herb similar in appearance to dill. It has pale yellowish-green, hollow stems and bright-green, feathery leaves. The clear yellow flowers of fennel produce seed structures that resemble umbrellas. Both the root and the leaves have a mild licorice flavor. The seeds have a pungent, aromatic scent.

Family Umbelliferae
Scientific name *Foeniculum vulgare (dulce)*
Common name fennel, sweet anise

VARIETY

The two basic types of fennel, common fennel and Florence fennel, bear a close resemblance to one another but are used differently. The shoots, leaves, and seeds (called "fruit") of the common fennel are used primarily as flavoring agents for food. Several varieties of common fennel have seeds that differ in length, width, and taste. These include the sweet variety, also known as French or Roman fennel, characterized by long, yellowish green fruit with a sweet flavor; Indian fennel, which is brownish, smaller, and less rounded; and the pale-green Persian and Japanese varieties, which are the smallest and have a stronger anise flavor and odor.

Florence fennel, also called finocchio, is somewhat smaller than common fennel and is grown mainly for its broad, bulbous leaf base, which is eaten as a vegetable.

ORIGIN & BOTANICAL FACTS

Fennel is native to the Mediterranean region and was well known to the ancient Greeks and Romans. The Romans enjoyed the young shoots as both a flavoring and, according to their belief, an aid to controlling obesity. The Greeks called it "marathon," a name derived from "maraino," meaning "to grow thin."

From the Mediterranean, fennel was carried east to India and also north to Europe and England, especially to Roman colonies. Spanish settlers are believed to have brought fennel to the Western Hemisphere more than 200 years ago. In 1824, the American consul at Liverpool gave Thomas Jefferson fennel seeds for his garden at Monticello.

Fennel is a long-lived plant that thrives almost anywhere. Fennel plants usually grow to 3 or 4 feet but have been known to reach 7 feet. Young plants form a bulbous, thick root the first year and flower the following summer. By mid-summer, the clusters of small yellow blossoms begin to droop with the weight of the heavy seeds. Leaves can be picked at any time, and seeds should be harvested when they begin to turn brown. A single plant produces about 1/4 cup of seeds and 1 cup of leaves.

USES

Fennel can be refrigerated unwashed in a plastic bag up to 1 week.

All parts of the fennel plant are edible. The mildly licorice-flavored leaves are used as a seasoning for fish. They also can be chopped for salads, dressings, dips, and cream sauces. The seeds have a more pungent flavor and are used either whole or ground as an ingredient in curries, pies, breads, and sausages and in a variety of soups and stews. The bulb can be sliced and eaten raw in salads, cooked in stews, added to pasta dishes, sautéed in oil, or baked and served with grated cheese and breadcrumbs.

NUTRIENT COMPOSITION

Fennel is not a significant source of nutrients.

SERVING SIZE: *fennel bulb, 1/2 cup, sliced, raw (87 g)*

NUTRIENT CONTENT

Energy (kilocalories)	13
Water (%)	90
Dietary fiber (grams)	1
Fat (grams)	0
Carbohydrate (grams)	3
Protein (grams)	0
Minerals (mg)	
Calcium	21
Iron	0
Zinc	0
Manganese	0
Potassium	180
Magnesium	7
Phosphorus	22
Vitamins (mg)	
Vitamin A	6 RE
Vitamin C	5
Thiamin	0
Riboflavin	0
Niacin	0
Vitamin B6	0
Folate	12 µg
Vitamin E	–

Note: A line (–) indicates that the nutrient value is not available.

GARLIC

Garlic is a member of the *Allium* genus, as are the onion, leek, and scallion. Covered in a loose, thin outer skin, the garlic bulb consists of small sections called cloves that are individually wrapped by a more tight-fitting, paper-like sheath.

Family Amaryllidaceae
Scientific name *Allium sativum* L.
Common name garlic, stinking rose

♥ **Contains phytochemicals that may promote health**

VARIETIES

Some 300 varieties of garlic are grown around the world. In the United States only two main types, "early" and "late," are grown. About 90 percent of the garlic is grown in California. The early variety is harvested in mid-summer, and the late variety is harvested a few weeks later. The late variety is slightly denser and has a longer shelf life than the early variety. Other, rare varieties of garlic are the Chileno and Elephant garlic, which is actually a form of leek and has a milder flavor.

ORIGIN & BOTANICAL FACTS

Garlic is native to central Asia, where it has been cultivated for more than 5,000 years. Garlic was known to the Egyptians as early as 3200 B.C. When taking solemn oaths, the ancient Egyptians swore on garlic in much the same way people swear on the Bible today. Today, garlic is among the leading vegetable crops of the world. Some 2.3 million metric tons are produced worldwide annually. Some of the leading garlic-producing countries are China, South Korea, India, Spain, the United States, Thailand, Egypt, Turkey, and Brazil.

Garlic is available year-round as a result of staggered harvests and a long shelf life. The California harvest begins in June, and garlic is shipped to markets from July through December. When the California supply is depleted, it is replaced by imported garlic from Mexico and South America.

USES

Garlic's strength varies with the season and variety, a factor to keep in mind when cooking with garlic. Because garlic that has sprouted is less pungent than younger garlic, sprouting should be prevented by keeping the bulbs in a cool, dark place. Garlic is the basic flavoring in most Chinese dishes and in much of the cooking of southern and central Europe. Garlic is potent when raw, milder when sautéed, and sweetly delicate when boiled or baked, because heat destroys some of the flavor- and odor-producing compounds. However, when garlic is sautéed, care must be taken not to burn the garlic, because it will turn bitter. Slow baking produces garlic that is sweet and nutty with a buttery consistency. Baked garlic can be spread on bread to make an appetizer. Rubbing a salad bowl with a cut garlic clove before adding the ingredients will give the salad a mild and fresh garlic flavor. Garlic juice also can be used to make salad dressing. Slivers of garlic can be inserted into slits made in roast beef, veal, or lamb before cooking. In addition, whole garlic can be baked or roasted with meat or poultry.

NUTRIENT COMPOSITION

Garlic contains the phytochemicals allicin, ajoene, saponins, and phenolic compounds that may have antioxidant and immune-promoting functions.

SERVING SIZE:
1 tsp., raw (3 g)

NUTRIENT CONTENT

Energy (kilocalories)	4
Water (%)	59
Dietary fiber (grams)	0
Fat (grams)	0
Carbohydrate (grams)	1
Protein (grams)	0
Minerals (mg)	
Calcium	5
Iron	0
Zinc	0
Manganese	0
Potassium	11
Magnesium	1
Phosphorus	4
Vitamins (mg)	
Vitamin A	0 RE
Vitamin C	1
Thiamin	0
Riboflavin	0
Niacin	0
Vitamin B$_6$	0
Folate	0 µg
Vitamin E	0

GINGER

Ginger is a tropical Asian herb grown for its pungent and spicy aromatic roots. Gingerroot is peppery and slightly sweet. Its light-brown skin covers a firm flesh that ranges from greenish yellow to ivory.

Family Zingiberaceae
Scientific name *Zingiber officinale*
Common name Jamaican ginger, African ginger, Cochin or Asian ginger

VARIETIES

Several hundred varieties of ginger exist. In addition, fresh gingerroot may be young or mature. Spring ginger, as young ginger is sometimes called, has a pale, thin skin that does not require peeling. Young ginger is delicate and milder than mature ginger.

ORIGIN & BOTANICAL FACTS

Ginger is believed to be native to South China or India, where it has been cultivated since ancient times. The earliest recorded mention of ginger appears in Chinese writings. According to the Pen Tsao Ching (*Classics of Herbs*), written by Shen Nung around 3000 B.C., ginger "eliminates body odor and puts a person in touch with the spiritual realm." In ancient India, ginger was believed to cleanse the body spiritually. Ginger also was used to preserve food and treat digestive problems. As in India, the ancient Greeks used ginger for digestive problems by eating ginger wrapped in bread after large meals. Eventually, ginger was added to the bread dough, and the product became known as gingerbread. The Romans also used ginger as a digestive aid. Arab traders introduced ginger to the Mediterranean area, and in the 16th century, Francisco de Mendoza of Spain brought it to the West Indies. In England and Colonial America, ginger was made into ginger beer, a popular home remedy for diarrhea, nausea, and vomiting and a precursor to today's ginger ale.

Ginger thrives in the tropics and in warmer regions of the temperate zone. Currently, the herb is grown in several regions of West Africa and the West Indies, and in India and China. The plant reaches maturity in the late summer when the foliage begins to turn yellow. However, the root can be harvested at any stage simply by digging it up. The finest quality ginger comes from Jamaica, where production is most abundant. In the United States, ginger is grown in Florida, Hawaii, and along the east coast of Texas.

USES

Ginger is a popular ingredient in Asian cooking, for which it has been used for centuries in both its fresh and dried forms. Fresh ginger can be shredded, grated, finely minced, or sliced and used in curries and stir-fried dishes. When buying fresh ginger, choose roots that have a firm, smooth skin with a fresh, spicy smell. Fresh unpeeled ginger can be tightly wrapped in a paper towel and plastic wrap or placed in a sealed plastic bag and refrigerated up to 2 weeks or frozen for 6 months. Powdered, dried ginger, which has a more spicy, intense flavor, is used for making gingerbread, gingersnaps, and other spice cookies. Ginger also is available in crystallized or candied form, preserved, and pickled. Dried powdered ginger should not be substituted for fresh or crystallized ginger in recipes, because it will not provide the same flavor.

NUTRIENT COMPOSITION

Ginger is not a significant source of nutrients.

SERVING SIZE: *5 slices ginger, raw (11 g)*

NUTRIENT CONTENT

Energy (kilocalories)	8
Water (%)	82
Dietary fiber (grams)	0
Fat (grams)	0
Carbohydrate (grams)	2
Protein (grams)	0
Minerals (mg)	
Calcium	2
Iron	0
Zinc	0
Manganese	0
Potassium	46
Magnesium	5
Phosphorus	3
Vitamins (mg)	
Vitamin A	0 RE
Vitamin C	1
Thiamin	0
Riboflavin	0
Niacin	0
Vitamin B$_6$	0
Folate	1 µg
Vitamin E	0

HORSERADISH

Horseradish is a root crop belonging to the botanical family that includes cabbage, mustard, and radish. Only the long, fleshy roots of this vegetable are used, because the leaves contain a slightly poisonous compound and have no culinary value. Rough and cream-colored, the parsnip-like root can grow to 20 inches in length and 1 to 3 inches in diameter. A sulfur-containing compound known as allyl isothiocyanate is responsible for the root's strong, pungent odor and hot, biting flavor, which is reminiscent of mustard and results from a chemical reaction that occurs only when the root is bruised or cut.

Family Cruciferae
Scientific name *Armoracia rusticana*
Common name horseradish, mountain radish, great raifort, red cole

ORIGIN & BOTANICAL FACTS

The horseradish is native to Eastern Europe and has been used as an herb since ancient times. It was grown in Greece more than 3,000 years ago. Mentioned in the Bible, it is one of the bitter herbs served during the Jewish Passover festival. In medieval Europe, the root was believed to be a cure-all. Initially used only for its medicinal properties, by the 1600s horseradish had become a common condiment for fish and meat in Europe. The word "horse" in horseradish, often used to imply coarseness, as in horse-mint or horse chestnut, is used similarly in this case to distinguish the plant from the edible radish (*Raphanus sativus*).

The perennial horseradish plant grows worldwide and is often found growing wild along roadsides throughout Europe and North America. From the long main root, stems sprout that grow to about 3 feet in height and give rise to large, jagged, wavy leaves. Once established, horseradish can tolerate any amount of neglect and can easily become a weed. Most of the fresh horseradish sold in the United States is grown in California.

USES

Roots that are firm and free of blemishes should be selected. The root can be refrigerated in a plastic bag for up to a week. Usually grated and used raw, the root must be washed, scrubbed, and peeled before grating by hand or with a food processor. Vinegar or lemon juice can be added to the grated horseradish to retard the enzyme process that produces the distinctive bite. For a mild sauce, 2 to 3 tablespoons of lemon juice or vinegar can be added to a cup of horseradish along with a half teaspoon of salt immediately after grating. For a hot sauce, the grated horseradish should be allowed to stand a few minutes before the lemon juice or vinegar is added. Because heat causes the root to release a pungent smell, horseradish should never be cooked. Grated horseradish is used as a condiment on fish, beef, chicken, and sausages. It is usually combined with oil and vinegar or with cream to make sauces for beef, smoked fish, or asparagus. Horseradish is the ingredient that provides the fresh, pungent flavor to seafood cocktail sauce. Blending horseradish with yogurt or applesauce makes a traditional Austrian accompaniment to meat. Preserved, grated horseradish is available bottled in vinegar or in beet juice, which gives it a reddish hue. Horseradish also is available in a dried form that must be reconstituted with water before using. Wasabi, a pungent green condiment sometimes referred to as "Japanese horseradish," is traditionally made from the root of a semiaquatic Asian plant, *Wasabia japonica*, from the same family of cruciferous vegetables. However, some inexpensive commercial wasabi powder and paste may contain domestic horseradish instead of wasabi.

NUTRIENT COMPOSITION

One serving of prepared horseradish provides small amounts of nutrients.

SERVING SIZE:
1 tsp., prepared (5 g)

NUTRIENT CONTENT

Energy (kilocalories)	2
Water (%)	85
Dietary fiber (grams)	0
Fat (grams)	0
Carbohydrate (grams)	1
Protein (grams)	0
Minerals (mg)	
Calcium	3
Iron	0
Zinc	0
Manganese	0
Potassium	12
Magnesium	1
Phosphorus	2
Vitamins (mg)	
Vitamin A	0 RE
Vitamin C	1
Thiamin	0
Riboflavin	0
Niacin	0
Vitamin B$_6$	0
Folate	3 µg
Vitamin E	0

JERUSALEM ARTICHOKE

Paradoxically, the Jerusalem artichoke is neither an artichoke nor from Jerusalem. This vegetable is the thick, brown-skinned root of a variety of sunflower. The mature tubers or roots, most of which are 3 to 4 inches long and about half as thick, resemble small, lumpy Irish potatoes. The white flesh of the Jerusalem artichoke is usually described as nutty, sweet, and crunchy.

Family Compositae
Scientific name *Helianthus tuberosus*
Common name Jerusalem artichoke, sunflower artichoke, sunchoke, topinambour

♥ **A good source of thiamin and iron**

of ways, including baking, boiling, mashing, or frying. It can also be prepared in combination with other vegetables. Added to soups, Jerusalem artichokes impart a sweet, nutty flavor.

NUTRIENT COMPOSITION

The Jerusalem artichoke is a good source of thiamin and iron.

VARIETIES

Jerusalem artichokes exist in both red- and white-skinned varieties. Smooth Garnet and Brazilian Red are red-skinned varieties. White-skinned varieties such as the New White Mammoth and Brazilian White have a clean, white skin and are also more rounded than the red-tinted ones. Other varieties include the Golden Nugget, which has carrot-like tubers; Stampede, a quick-maturing variety with large tubers; and Dwarf Sunray, which is a small variety. Modern varieties produce less knobby tubers that are easier to peel.

ORIGIN & BOTANICAL FACTS

One of the few vegetables that is native to the North American plains, the Jerusalem artichoke is indigenous to the lake regions of Canada. In the United States, it grows as far south as Arkansas and Georgia. Reportedly cultivated by American Indians before the 16th century, the tuber prefers to grow in damp places with good soil. It is a hardy perennial that tolerates frost and can be left in the ground all year. The plants have stiff stems that may grow to a height of 10 feet, and some varieties produce small sunflowers late in the summer. Several tubers are produced at the base of each flower stalk.

USES

Jerusalem artichokes may be stored in a sealed plastic bag in the refrigerator for up to a week. After that, they lose moisture and become withered. Jerusalem artichokes can be eaten raw in salads or served with a dip. Although they should be washed thoroughly, they need not be peeled. If peeled, the vegetable should be immersed in acidulated water (dilute lemon juice) to prevent discoloration. The tuber also can be cooked in a variety

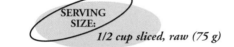

SERVING SIZE:
1/2 cup sliced, raw (75 g)

NUTRIENT CONTENT

Energy (kilocalories)	57
Water (%)	78
Dietary fiber (grams)	1
Fat (grams)	0
Carbohydrate (grams)	13
Protein (grams)	2
Minerals (mg)	
Calcium	11
Iron	3
Zinc	0
Manganese	0
Potassium	322
Magnesium	13
Phosphorus	59
Vitamins (mg)	
Vitamin A	2 RE
Vitamin C	3
Thiamin	0.2
Riboflavin	0
Niacin	1
Vitamin B$_6$	0.1
Folate	10 µg
Vitamin E	0

JICAMA

Jicama (pronounced hi-ca-ma) is a large, edible, tuberous root that can weigh up to 50 pounds. It has a slightly squat shape, a thin brown skin, and ivory flesh that is crunchy (resembling a raw potato), very juicy, and slightly sweet.

Family Leguminosae
Scientific name *Pachyrhizus erosus*
Common name Xiquima, Mexican turnip, Mexican potato, yam bean

♥ **High in vitamin C**

♥ **A good source of fiber**

VARIETIES

The two most popular cultivated forms of jicama are jicama de agua and jicama de leche. The former, which produces a translucent juice, is usually preferred, while jicama de leche, which has elongated roots and a milky juice, is less familiar.

ORIGIN & BOTANICAL FACTS

A legume native to Central America and Mexico, jicama is a perennial vine that grows to a length of 20 feet or more and has compound leaves with pointed edges. The vine bears beautiful sprays of mauve, white, or blue flowers and poisonous seeds in pods 6 to 8 inches long. Because the tuber requires a very long and warm growing season, most of the jicama available in U.S. supermarkets is imported from Mexico and South America, although the plant is also grown in parts of east and southeast Asia.

USES

Jicama is ready to be eaten at any stage of growth. Tubers should be firm, smooth-skinned, heavy, and free of bruises, wrinkles, and cracks. Jicama can be stored in the refrigerator for up to 2 weeks. If it is stored for longer periods, its starch content converts to sugar. Ideally the tuber should be stored in a cool, dry place, because too much moisture can cause mold to form. The peeled tuber can be eaten raw or cooked. The sliced or julienned root retains its crispness even after sautéing or stir-frying and can be added to soups or stews. Raw jicama can be served as part of a vegetable platter or added to salads. It serves as a substitute for water chestnut.

NUTRIENT COMPOSITION

The jicama is high in vitamin C and is a good source of fiber.

SERVING SIZE: *1/2 cup, sliced (60 g)*

NUTRIENT CONTENT

Energy (kilocalories)	23
Water (%)	90
Dietary fiber (grams)	3
Fat (grams)	0
Carbohydrate (grams)	5
Protein (grams)	0
Minerals (mg)	
Calcium	7
Iron	0
Zinc	0
Manganese	0
Potassium	90
Magnesium	7
Phosphorus	11
Vitamins (mg)	
Vitamin A	1 RE
Vitamin C	12
Thiamin	0
Riboflavin	0
Niacin	0
Vitamin B6	0.1
Folate	7 µg
Vitamin E	0

KELP

Kelp is a type of seaweed that grows up to 200 feet in length in the waters off Japan, Europe, and North America. Like other seaweed, it is a large form of algae. Fresh seaweed and other sea vegetables are similar in texture to some dry-land greens, but they have a strong, salty, seawater flavor.

Family Laminariaceae
Scientific name *Laminaria*
Common name kelp

VARIETIES

The thousands of varieties of algae grow in a broad spectrum of colors, shapes, and sizes, ranging from the small freshwater algae to the long-stemmed kelp that flourishes in the oceans. Also called seaweed, agar-agar, carrageenan, and dulse, algae are generally classified by their color, which varies from brown to red, green, or blue-green. Kelp is one of the brown algae. Some of the varieties of kelp available in Japan include Wakame, Arame, Kombu, and Hijiki.

ORIGIN & BOTANICAL FACTS

The word "seaweed" generally refers to the large red or brown varieties of algae. Brown algae grow in cold waters, and red algae thrive in tropical seas. Algae compose two-thirds of the plant material on earth and are among the very few plants that have not changed for centuries. In ancient times, sailors harvested the kelp beds that thrived off the coasts of England and France and burned the plants for fuel. They also wrapped fish in the fronds and baked them. Unlike the Japanese, Europeans have never made significant use of kelp as a food. However, 18th-century

European physicians noted that enlarged thyroid gland (goiter) rarely developed in people who lived along the coast. In 18th-century England, a physician successfully used charred kelp to treat a patient who had goiter. It was not until the 19th century, however, that scientists discovered that goiter is caused by iodine deficiency and that kelp is rich in iodine. For several decades after this discovery, Europeans and Americans harvested kelp from undersea rocks to use as a source of iodine.

USES

Today, seaweed is a staple in the diets of some people. In Japan, seaweed constitutes approximately 25 percent of the diet. Kombu and Wakame, two popular types of kelp, are used extensively. Kombu is used to make a tasty broth, and Wakame is used as an ingredient in soups and stir-fries. Sheets of Wakame are used to prepare sushi. Powdered kelp can serve as a salt substitute, helping to flavor soups, salads, and tomato juice. In the United States, dried sheets of seaweed can be found in Asian groceries and in some supermarkets.

NUTRIENT COMPOSITION

The nutritional value of seaweed depends on the type, but most provide calcium, iodine, folate, and magnesium. Dried kelp contains so much iodine that consumption of large quantities can be harmful. Some varieties, such as Kombu and Wakame, are also high in sodium.

SERVING SIZE: *1/8 cup, raw (10 g)*

NUTRIENT CONTENT

Energy (kilocalories)	4
Water (%)	82
Dietary fiber (grams)	0
Fat (grams)	0
Carbohydrate (grams)	1
Protein (grams)	0
Minerals (mg)	
Calcium	17
Iron	0
Zinc	0
Manganese	0
Potassium	9
Magnesium	12
Phosphorus	4
Sodium	23
Vitamins (mg)	
Vitamin A	1 RE
Vitamin C	0
Thiamin	0
Riboflavin	0
Niacin	0
Vitamin B$_6$	0
Folate	18 µg
Vitamin E	0

KOHLRABI

The kohlrabi is a member of the cabbage family, grown for the swollen, globe-shaped portion of the stem, which rests on the surface of the ground. The best kohlrabi bulbs are between 2 and 3 inches in diameter. The vegetable can be white, purple, or green and has a creamy-white interior that is somewhat sweet and similar in texture to a turnip.

Family Cruciferae
Scientific name *Brassica oleracea* var. *caulorapo*
Common name kohlrabi, stem turnip, colinabo, cabbage turnip

♥ **Kohlrabi is high in vitamin C**

♥ **Raw kohlrabi is a good source of fiber**

VARIETIES

Kohlrabi varieties are distinguished by color. Among the popular varieties are the Grand Duke, Kolibri F1, Purple Danube, Purple Vienna, and White Vienna.

ORIGIN & BOTANICAL FACTS

The kohlrabi, which literally means "cabbage turnip," is descended from both the wild cabbage and the wild turnip. Although the origin of this vegetable is uncertain, a vegetable answering to the same description was mentioned by the Roman botanist Pliny in the 1st century A.D. No such vegetable was again described until after the Middle Ages. Today, the kohlrabi is mainly eaten in central and eastern Europe.

Unlike other cruciferous vegetables, the part of kohlrabi that is eaten is a swollen part of the stem. Kohlrabi has so declined in popularity in the last century that it can be difficult to purchase. The reason for this decline is not known but may be its increased toughness with larger sizes.

Kohlrabi bulbs should be harvested when they reach no more than 2 to 3 inches in diameter. Kohlrabi cultivation in the United States is quite limited.

USES

Kohlrabi bulbs that are plum-size or smaller, firm, and unblemished and have leaves that are still attached are best. After the leaf stems are removed, the bulbs of kohlrabi can be stored in the refrigerator for several weeks (longer if placed in sealed plastic bags). The kohlrabi bulb may be eaten raw or cooked. Although small, tender bulbs generally do not require peeling, the skin of medium and large ones should be removed before use. The crisp flesh can be served raw in salads or made into a relish. The bulb also can be cubed, sliced, or julienned and steamed until tender. The practice in many central European countries is to hollow out the vegetable and stuff it with meat or other vegetables before baking or steaming.

Kohlrabi leaves are similar to collard greens or kale in flavor. The succulent, tender leaves of young kohlrabi plants can be cooked like spinach or mustard greens.

NUTRIENT COMPOSITION

Kohlrabi is high in vitamin C. Raw kohlrabi is a good source of fiber. Both raw and cooked forms also contain antioxidants and bioflavonoids.

NUTRIENT CONTENT

	½ cup, raw (70 g)	½ cup, cooked (83 g)
Energy (kilocalories)	19	24
Water (%)	91	90
Dietary fiber (grams)	3	1
Fat (grams)	0	0
Carbohydrate (grams)	4	6
Protein (grams)	1	1
Minerals (mg)		
Calcium	17	21
Iron	0	0
Zinc	–	0
Manganese	0	0
Potassium	245	281
Magnesium	32	16
Phosphorus	13	37
Vitamins (mg)		
Vitamin A	3 RE	3 RE
Vitamin C	43	45
Thiamin	–	0
Riboflavin	–	0
Niacin	0	0
Vitamin B_6	0	0.1
Folate	11 µg	10 µg
Vitamin E	0	1

Note: A line (–) indicates that the nutrient value is not available.

LEEK

The leek is related to garlic, scallions, and onions and resembles a large scallion. Unlike the onion, however, it does not form a real bulb but grows as a thick, fleshy stalk with flattened leaves. The leaves are green to blue-green or purple and wrap tightly around each other like rolled paper. The white leaf base—the part that is most commonly eaten—has a flavor and fragrance similar to but milder than onions.

Family Amaryllidaceae
Scientific name *Allium ampeloprasum*
Common name leek

VARIETIES

The many varieties of leeks differ mostly in the color of their leaves, their general cold hardiness, and the degree of bulbing at the stem base. Those with blue or purple leaves tend to be the hardiest. An old French variety called Bleu Solaise, which is known to be resistant to cold, has blue-green leaves.

ORIGIN & BOTANICAL FACTS

Leeks are native to southern Europe and the Mediterranean region, where they still grow wild. Both wild and cultivated leeks have been consumed for thousands of years. In ancient Rome, Emperor Nero consumed large quantities of leeks in the belief that they would improve his singing voice. In the 6th century, Wales adopted the leek as its national symbol in the belief that leeks worn on the helmets of Welsh soldiers, to distinguish them from enemy troops, helped them achieve victory. Today, France, Belgium, and the Netherlands lead the world in leek production. Although not as popular in the United States, the leek is cultivated in California, Michigan, New Jersey, and Virginia. Leeks can be grown from seed or transplants, but transplanting is the preferred method, for which plants are begun in containers between December and April. Leeks prefer a cool to moderate climate, rich but well-drained soil, and uniform watering. The base is blanched by tilling the soil up around each plant when it is about the size of a pencil. Blanching makes the edible portion longer and whiter. Leek plants produce flowering stems more than 6 feet in height, with white, pink, or dark-red flowers. On most leek varieties, numerous bulbils, or secondary bulbs, form around the base of the plant, and these can be used to start new plants.

USES

Before being used, leeks should be washed thoroughly to remove any soil or grit trapped between the leaves, and the rootlets and leaf ends should be trimmed off. Like onions and garlic, leeks are used primarily to add flavor to a variety of dishes. Raw leeks can be sliced thin and added to salads of all types. Sliced or puréed, they add zest to quiches, stews, casseroles, mixed vegetable dishes, and soups. They are one of the key ingredients in French vichyssoise, a classic cold potato and leek soup. They also may be baked, braised in broth or wine, broiled, sautéed, or microwaved. Leeks should be cooked only until barely tender.

NUTRIENT COMPOSITION

Leeks contain only small amounts of nutrients.

NUTRIENT CONTENT

	½ cup chopped, raw (45 g)	½ cup chopped, cooked (52 g)
Energy (kilocalories)	27	16
Water (%)	83	91
Dietary fiber (grams)	1	1
Fat (grams)	0	0
Carbohydrate (grams)	6	4
Protein (grams)	1	0
Minerals (mg)		
Calcium	26	16
Iron	1	1
Zinc	0	0
Manganese	0	0
Potassium	80	45
Magnesium	12	7
Phosphorus	16	9
Vitamins (mg)		
Vitamin A	4 RE	3 RE
Vitamin C	5	2
Thiamin	0	0
Riboflavin	0	0
Niacin	0	0
Vitamin B6	0.1	0.1
Folate	29 µg	13 µg
Vitamin E	0	–

Note: A line (–) indicates that the nutrient value is not available.

LETTUCE

Lettuce is a salad green that grows in forms ranging from tightly compacted heads to loose leaves, depending on the variety. The leaves are pale to dark green or green with red edges, crisp, and mild to pungent.

Family Asteraceae
Scientific name *Lactuca sativa*
Common name lettuce

♥ **Depending on the variety, lettuce can be a good source of folate and a good to excellent source of vitamin C**

VARIETIES

Four basic types of lettuce are grown: head lettuce, loose leaf, butterhead, and romaine. The most common, popular variety is iceberg, a compact head lettuce that is pale green and has a delicate flavor. Loose leaf lettuce forms rosettes of crisp, curly leaves 8 to 12 inches long and includes the red-edged varieties. The mild butterhead lettuce, including the Boston and Bibb varieties, has a softly compressed head, 8 to 12 inches across, of grass-green leaves that fade to a lighter yellowish green in the interior. Romaine lettuce forms a dark-green, tightly compressed head of leaves about 10 inches long and has a stronger, more pungent flavor than the other varieties.

ORIGIN & BOTANICAL FACTS

The cultivation of lettuce dates back more than 2,500 years. From early Rome, where many varieties were developed, its popularity spread throughout Europe and Asia. In 1885, an American agricultural report listed 87 varieties, considerably more than the 4 commonly available in today's markets. In the United States, lettuce ranks a close second to potatoes

as the most popular fresh vegetable. The four leading American producers are California, Arizona, Florida, and Colorado.

Lettuce is cultivated by direct seeding into fields or by seedling transplantation into raised beds. Loose leaf lettuce matures about 6 weeks after seeds are sown; other types take longer to mature. Romaine takes the longest (up to 12 weeks). Head lettuce is harvested when the heads reach about 2 pounds. Because lettuce is very perishable, harvesting is done by hand, and the crop is packed directly into boxes in the field. Head lettuce, which is the hardiest, can be shipped long distances without damage, but leaf lettuce is more fragile and usually is grown for local and regional markets.

USES

Lettuce should be used as soon after purchase as possible, but if it must be stored, leaves that are wilting should be removed. Unwashed lettuce can be kept 3 to 4 days at most in a perforated plastic bag in the vegetable crisper of the refrigerator. Lettuce is most often eaten raw in salads and sandwiches. Iceberg, leaf, or romaine

works well in Greek salad; romaine is often used for Caesar salad. Combinations of lettuce varieties make for tastier and more nutritious salads. A salad spinner improves the quality of salads by drying the greens quickly and completely. In addition to being used in salads, lettuce leaves can be used to hold cooked vegetables, sandwich fillings, and condiments.

NUTRIENT COMPOSITION

Romaine and loose leaf lettuce contain five to six times the vitamin C and five to ten times the vitamin A of iceberg. Romaine and butterhead lettuce are good sources of folate.

SERVING SIZE: *1 cup, shredded*

NUTRIENT CONTENT

	Romaine (56 g)	Loose leaf (56 g)
Energy (kilocalories)	8	10
Water (%)	95	94
Dietary fiber (grams)	1	1
Fat (grams)	0	0
Carbohydrate (grams)	1	2
Protein (grams)	1	1
Minerals (mg)		
Calcium	20	38
Iron	1	1
Zinc	0	0
Manganese	0	0
Potassium	162	148
Magnesium	3	6
Phosphorus	25	14
Vitamins (mg)		
Vitamin A	146 RE	106 RE
Vitamin C	13	10
Thiamin	0.1	0
Riboflavin	0.1	0
Niacin	0	0
Vitamin B6	0	0
Folate	76 µg	28 µg
Vitamin E	0	0

MUSHROOM

Mushrooms are fleshy fungi, only some of which are edible. They usually have thick stems and rounded caps with radiating gills on the underside. The caps can be smooth or bumpy, honeycombed or ruffled, ranging in size from less than 1/2 inch in diameter to 12 inches and in color from snowy white to black, with a broad spectrum of colors in between. They can be soft or crunchy, and they range in flavor from bland to nutty and earthy.

Family Fungi
Scientific name *Agaricus bisporus*
Common name mushroom

♥ **Cooked mushrooms are an excellent source of niacin and a good source of riboflavin**

VARIETIES

Mushrooms come in literally thousands of varieties. The most popular is the simple, cultivated white mushroom, the *Agaricus*, which is relatively small and has a mild, earthy flavor. Young fungi of this variety are called button mushrooms. Variations on the white mushroom are the Crimini, which is dark brown, more firm, and has a fuller flavor, and the larger Portabella, a relative of the Crimini. Shiitake mushrooms, up to 10 inches across, are a dark, umbrella-shaped variety native to Japan and Korea and have a pungent, woody flavor. Enoki mushrooms are fragile and flower-like, with tiny white caps, long, slender stems, and a mild flavor. Oyster mushrooms are mild flavored and velvety textured with large, fluted grayish caps on short stems. Chanterelles are golden to yellow-orange and have a rich, slightly almond flavor. Porcini mushrooms are thick-stemmed and nutty in flavor, with large white or reddish brown caps. Less well-known varieties, which grow mainly in the wild, include the black trumpet mushroom, which is thin, brittle, and trumpet-shaped. Because many species of poisonous mushrooms can be commonly mistaken for edible ones, no one except experienced mushroom hunters should attempt to gather wild mushrooms.

ORIGIN & BOTANICAL FACTS

Archaeological evidence indicates that humans have been eating mushrooms for thousands of years. The first cultivators of mushrooms appear to have been the Greeks and Romans. Today, mushrooms are cultivated on every continent.

Mushroom cultivation does not require darkness, as was once believed. Most important to mushrooms are a constant temperature, protection from drafts, good compost, and proper sanitation. However, in the United States, most mushrooms are grown in caves or climate-controlled, windowless buildings, because outdoor conditions are less suitable for mushroom cultivation.

USES

Mushrooms should be stored unwashed in a loosely closed paper sack or in their original packaging on a refrigerator shelf no more than 2 to 3 days and only wiped clean with a damp paper towel just before use. Mushrooms can be eaten raw or cooked. White and Enoki mushrooms can be added raw to fresh green salads. Porcini mushrooms can be cooked with pork or chicken or combined with vegetables, rice, or pasta. Shiitakes are traditionally added to stir-fries and other Asian dishes. Portabella mushrooms often are sliced, grilled, and served as an appetizer, added to sandwiches, or stuffed with any number of ingredients and baked. Some mushroom varieties are also available canned or dried.

NUTRIENT COMPOSITION

Cooked mushrooms are an excellent source of niacin and a good source of riboflavin.

NUTRIENT CONTENT

	½ cup pieces, raw (35 g)	7 medium, cooked (84 g)
Energy (kilocalories)	9	22
Water (%)	92	92
Dietary fiber (grams)	0	2
Fat (grams)	0	0
Carbohydrate (grams)	1	4
Protein (grams)	1	2
Minerals (mg)		
Calcium	2	5
Iron	0	1
Zinc	0	1
Manganese	0	0
Potassium	130	299
Magnesium	4	10
Phosphorus	36	73
Vitamins (mg)		
Vitamin A	0 RE	0 RE
Vitamin C	1	3
Thiamin	0	0.1
Riboflavin	0.1	0.3
Niacin	1	4
Vitamin B$_6$	0	0.1
Folate	4 µg	15 µg
Vitamin E	0	0

OKRA

The okra is the immature seedpod of the okra plant. The slightly curved, tapering pods range from 2 to 7 inches in length and have green, fuzzy skin. Numerous soft white seeds are clustered along the length of the pod's interior. When cooked, the pods exude a juice that thickens any liquid to which it is added and can give the vegetable a slimy texture.

Family Malvaceae
Scientific name *Hibiscus esculentus*
Common name okra, lady's finger, gumbo, bindi, bamia

♥ **High in vitamin C**

♥ **A good source of magnesium**

VARIETIES

The many varieties of okra differ in shade of green, shape (plump or slender), and surface (ribbed or smooth). The Clemson Spineless variety has medium-green, angular pods, whereas the Emerald variety is dark green and has smooth, round pods. Other varieties include Lee, Annie Oakley, and Prelude.

ORIGIN & BOTANICAL FACTS

The okra plant originated in the Near East and was brought to North Africa and the Middle East before being brought to the Americas in the early 1700s. The name "okra" is derived from the Twi (from the Gold Coast of Africa) word "nkruman." In other parts of the world, this vegetable is referred to by the African name "gumbo" and by other regional names. Whether West African slaves or French colonists of Louisiana brought the plant to the United States is unclear, but its widespread popularity in the South suggests that this region was the first in the United States to be introduced to the vegetable. Texas, Georgia, Florida, Alabama, and California are the leading producers in the United States.

The okra plant is a tropical perennial belonging to the cotton family. The plants begin to produce flowers about 60 days after germination and can grow 3 to 5 feet in height. The maroon-centered, pale-yellow flowers develop into slender seedpods that are harvested a few days after the flower petals have fallen. Okra, the vegetable, should not be confused with Chinese okra, which is also known as dishcloth gourd, sponge gourd, or loofah.

USES

Because very young or fully developed okra pods tend to be flavorless and stringy, pods that are 1 to 3 inches should be selected. Fresh okra is very perishable and should be used as soon as possible. It can be stored in a plastic bag in the refrigerator for a couple of days. Fresh okra should be washed and thoroughly dried before cooking, because moisture causes the pods to become slimy. Also, unless okra is added to soup or stew, it should never be cut before cooking, because cutting releases the fluid that acts as a thickener and gives okra the slimy texture with which it is associated. The secret to tender but crisp okra is to sauté the whole pods for no more than 5 minutes. In the United States, okra is most popularly known as an ingredient in gumbo, a stew-like dish that is a specialty of New Orleans Creole cuisine. Okra is also available in frozen and canned forms.

NUTRIENT COMPOSITION

Okra is high in vitamin C and is a good source of magnesium.

SERVING SIZE:
1/2 cup, sliced (80 g)

NUTRIENT CONTENT

Energy (kilocalories)	26
Water (%)	90
Dietary fiber (grams)	2
Fat (grams)	0
Carbohydrate (grams)	6
Protein (grams)	2
Minerals (mg)	
Calcium	50
Iron	0
Zinc	0
Manganese	1
Potassium	258
Magnesium	46
Phosphorus	45
Vitamins (mg)	
Vitamin A	46 RE
Vitamin C	13
Thiamin	0.1
Riboflavin	0
Niacin	1
Vitamin B$_6$	0.2
Folate	37 µg
Vitamin E	1

ONION

The onion is a round or oval bulb that grows in multiple layers underground and is covered by a dry, papery skin at maturity. Its flavor can range from mild and sweet to sharp and pungent.

Family Amaryllidaceae
Scientific name *Allium cepa*
Common name onion

♥ **Contains phytochemicals that promote health**

VARIETIES

The two main varieties of onion are fresh onions and dry onions. Scallions (also called green onions), the most common type of fresh onion, are pulled up before the bulb forms. Sweet onions, another type of fresh onion, grown in warmer climates, are harvested and sold during the spring and summer. They are characterized by a light-colored, thin skin and a high water and sugar content, which gives them a mild, sweet flavor. The most popular sweet onions are the mild Maui from Hawaii, the juicy Vidalia from Georgia, and the round, golden Walla-Walla from Washington State.

Dry onions, also called storage onions, are grown in cooler northern states and are available year-round. They have a darker, thicker skin, a firmer texture, and a stronger, more pungent flavor. They range from white to yellow and red. Smaller varieties of dry onions include the marble-sized pearl onions and the slightly larger boiling onions. Although red onions tend to be sweeter than yellow or white onions, the flavor of dry onions is influenced more by variety and origin than by color.

ORIGIN & BOTANICAL FACTS

Onions were grown by the ancient Egyptians, who regarded them as sacred. From Egypt, onions were brought to Rome, where they acquired their current name, derived from the Latin *unio*, meaning "large pearl." Christopher Columbus brought the onion from Europe to the Americas. Today, onions are among the world's leading vegetable crops.

Onion plants are propagated from seeds or from seedlings and are planted on raised beds, in fertile, well-balanced soil. They are hardy plants, able to withstand temperatures as low as 20° Fahrenheit, and are generally planted in early spring, 4 to 6 weeks before the last spring freeze. When first planted, young onions concentrate their growth on new roots and green leaves or tops. Bulb formation does not begin until the right combination of daylight, darkness, and temperature is achieved.

USES

Fresh onions should be refrigerated and used soon after purchase, but dry (storage) onions can be kept in a cool, dry place for weeks, or even months, without losing their nutrients. The onion's flavor and odor result primarily from sulfuric compounds. When the onion is peeled and sliced, these are released as vapors, causing the eyes to tear. Chilling the onion before use or peeling it under cold water can alleviate this problem. Onions can be served raw or cooked and can be added to numerous other foods as a flavoring. Cooking onions tends to soften them and removes any sharpness from the flavor. However, because heat makes onions bitter, they should be cooked over low to medium heat. Onions can be boiled, steamed, baked, sautéed, scalloped, or grilled.

NUTRIENT COMPOSITION

Raw onions contain phytochemicals that include antioxidants, which promote health.

NUTRIENT CONTENT

	1/2 cup raw, chopped, or 1 medium (80 g)
Energy (kilocalories)	30
Water (%)	90
Dietary fiber (grams)	1
Fat (grams)	0
Carbohydrate (grams)	7
Protein (grams)	1
Minerals (mg)	
Calcium	16
Iron	0
Zinc	0
Manganese	0
Potassium	126
Magnesium	8
Phosphorus	26
Vitamins (mg)	
Vitamin A	0 RE
Vitamin C	5
Thiamin	0
Riboflavin	0
Niacin	0
Vitamin B$_6$	0.1
Folate	15 µg
Vitamin E	0

PARSLEY

Parsley is a bright-green, multibranched biennial herb with crisp leaves and greenish yellow flowers. It is most often used as a seasoning or garnish and has a fresh, slightly peppery flavor.

Family Apiaceae
Scientific name *Petroselinum crispum*
Common name parsley

♥ **Contains some vitamin C**

VARIETIES

Although more than 30 varieties of parsley exist, the most commonly used in the United States are the curly-leaved and the Italian, or flat-leaved, varieties. Among the curled parsleys are Moss Curled, Green Velvet, and Paramount Imperial. The flat-leaved variety, most commonly used in southeastern Europe and in Asia, has a more vibrant flavor than the curled types.

ORIGIN & BOTANICAL FACTS

Parsley is believed to have originated in southern Europe, around the Mediterranean, and has been cultivated since about 320 B.C. In ancient times, parsley wreaths were believed to ward off drunkenness. According to Greek mythology, parsley sprang from the blood of Opheltes, the infant son of King Lycurgus of Nemea, who was killed by a serpent while his nurse directed some thirsty soldiers to water. Thus, Greek soldiers associated parsley with death and avoided contact with it before battle. A completely different meaning is imparted to the herb in the Jewish Seder, the ritual Passover meal. Because parsley is one of the first herbs to appear in the spring, it is used in the Seder to symbolize new beginnings. The ancient Romans ate parsley after

meals to freshen their breath, and the Roman physician Galen prescribed it for epilepsy and as a diuretic. In Europe during the Middle Ages, parsley was regarded as the devil's herb and was believed to bring disaster on anyone who grew it unless it was planted on Good Friday. Nevertheless, medieval abbess and herbalist Hildegard of Bingen and the 17th-century herbalist Nicholas Culpeper prescribed it in various forms for heart and chest pain and for arthritis. During the late 19th and early 20th centuries, parsley was prescribed for a variety of medical conditions. However, none of these uses is recognized in modern medicine. Today, parsley is used primarily as a flavoring agent and a garnish.

Parsley is a hardy plant that can be grown easily in almost any soil. It can be planted in the spring for summer use and in late summer for winter growth and spring harvesting. Soaking the seeds overnight before planting helps germination. Flat-leaved parsley is cultivated more easily than the curly-leaved types.

USES

When selecting parsley, bunches that look freshly picked, not wilted or yellow,

should be chosen. Fresh parsley can be wrapped in damp paper towels and stored in an open plastic bag in the refrigerator for up to 1 week.

The curly-leaved varieties of parsley are used as a flavoring, a salad ingredient, and a garnish. Flat-leaved parsley, with its stronger flavor, is frequently used as a seasoning in Italian cooking. Dried parsley may be substituted for the fresh herb by using one-third of the amount of the fresh herb specified by the recipe.

NUTRIENT COMPOSITION

A serving of parsley contains some vitamin C.

SERVING SIZE: *1 Tblsp. chopped, raw (4 g)*

NUTRIENT CONTENT

Energy (kilocalories)	1
Water (%)	88
Dietary fiber (grams)	0
Fat (grams)	0
Carbohydrate (grams)	0
Protein (grams)	0
Minerals (mg)	
Calcium	5
Iron	0
Zinc	–
Manganese	–
Potassium	21
Magnesium	2
Phosphorus	2
Vitamins (mg)	
Vitamin A	20 RE
Vitamin C	5
Thiamin	0
Riboflavin	0
Niacin	0
Vitamin B$_6$	0
Folate	6 µg
Vitamin E	0

Note: A line (–) indicates that the nutrient value is not available.

PARSNIP

Like the carrot and celery, the parsnip is a member of the Umbelliferae family, so-named for the umbrella-like shape of their flower clusters. It is a cold-weather, starchy root vegetable that resembles the carrot in shape but is pale yellow or ivory. It has a mild, celery-like fragrance and a sweet, but slightly peppery, flavor. The parsnip root grows up to 18 inches in length and up to 3 to 4 inches across at the top.

Family Umbelliferae
Scientific name *Pastinaca sativa* L.
Common name parsnip

 A good source of folate, vitamin C, and fiber

VARIETIES

Unlike their cousins the carrots, parsnips are not bred for variety of color or shape, and the number of varieties is small. Modern varieties are bred to grow a fat, wedge-shaped root. The most disease-resistant types are the Gladiator, a sweet, early-maturing parsnip, the Avonresister, and the Andover, a new American variety. The most popular parsnip is the All American, which has broad "shoulders," white flesh, and a tender core.

ORIGIN & BOTANICAL FACTS

Wild parsnips were eaten by the ancient Greeks and Romans. The word *pastinaca*, part of the modern scientific name for parsnips, was used by Pliny in the 1st century A.D., but it may have referred to either parsnips or carrots, or both. According to Pliny, the parsnip was so valued by the Emperor Tiberius that he had it imported to Rome from the banks of the Rhine. Sixteenth-century Germans exploited the parsnip's high sugar content to make wine, jam, and sweet flour for cakes. During the season of Lent, parsnips were eaten with salt fish. Today, parsnips are cultivated throughout Europe, Canada, and the northern United States.

Northern California and Michigan are the leading domestic growers.

Although it is a biennial, the parsnip is usually grown as an annual, harvested before the second year's leaves begin to appear. Parsnips are propagated from seeds and planted in deep, loamy, fine soil that has not been fertilized recently. In cold climates, parsnips are planted in early spring, but in areas where temperatures rarely fall below 25 degrees, seeds can be sown in early fall for a spring harvest. The roots take from 100 to 120 days to mature from seeds. The green stalks grow above ground, anywhere from 9 inches to 2 feet in height, with smooth, oblong leaflets about 4 or 5 inches long. The flowers, when permitted to develop, are deep yellow. Chilling the parsnip roots, either before or after harvesting, results in a sweeter flavor.

USES

The best parsnips are those that are small to medium in size, crisp, and plump. They should be stored in a perforated plastic bag in the refrigerator crisper. Because parsnips have a tough, fibrous core, they are usually cooked before serving. They can be peeled before or after cooking. Fresh parsnips that are not too tough can be grated or shredded and dressed like coleslaw or sliced very thin for a raw vegetable tray. They can be baked, microwaved, parboiled, or steamed until just tender and then puréed and served in place of mashed potatoes. When adding parsnips to soups, stews, and casseroles, they should be added about 15 minutes before serving time, because over-cooking can turn them soft and tasteless.

NUTRIENT COMPOSITION

Parsnips are a good source of folate, vitamin C, and dietary fiber.

NUTRIENT CONTENT

	½ cup sliced, raw (67 g)	½ cup sliced, cooked (78g)
Energy (kilocalories)	50	63
Water (%)	80	78
Dietary fiber (grams)	3	3
Fat (grams)	0	0
Carbohydrate (grams)	12	15
Protein (grams)	1	1
Minerals (mg)		
Calcium	24	29
Iron	0	0
Zinc	0	0
Manganese	0	0
Potassium	249	286
Magnesium	19	23
Phosphorus	47	54
Vitamins (mg)		
Vitamin A	0 RE	0 RE
Vitamin C	11	10
Thiamin	0.1	0.1
Riboflavin	0	0
Niacin	0	1
Vitamin B$_6$	0	0.1
Folate	44 µg	45 µg
Vitamin E	–	1

Note: A line (–) indicates that the nutrient value is not available.

PEPPERS

Peppers are thick-fleshed fruits with a smooth, waxy skin and a crunchy texture. They can be small and round, large and oblong, or almost any shape and size in between. The skin and flesh of peppers range from golden to green, bright red, orange, purple, and brown. Although all peppers are crunchy, their flavors range from sweet to extremely hot.

Family Solanaceae
Scientific name *Capsicum annuum,*
    Capsicum frutescens
Common name pepper

♥ **High in vitamin C**

♥ **Sweet and hot red peppers are a good source of vitamin A**

VARIETIES

Peppers generally are divided into two flavor categories: sweet and hot. Each type includes numerous varieties. The most common variety of sweet peppers is the round bell pepper, which is sold by color. Although all bell peppers are green when immature, most turn red when completely ripe, and some turn yellow, orange, purple, or brown. Ripe, colored bell peppers are usually sweeter than their immature, green counterparts. Other varieties of sweet pepper include banana peppers, a mild variety that is the shape and color of bananas; Cubanelle, a tapered, light-green or yellow variety that is more flavorful than bell peppers; and pimientos, thick-fleshed, heart-shaped peppers that are ideal for roasting.

Hot peppers, also called chilis, vary greatly in size, shape, and spiciness, and, like sweet peppers, they are green when immature. Cherry peppers are small, round, and red and have a mild to medium heat. Anaheims are long, tapered, and green and can be mildly pungent to hot, depending on growing conditions. The small Habañero peppers, which may be red, green, or orange and either bell-shaped or teardrop-shaped, are extremely hot. Jalapeños, the most widely available chilis, are stubby and pointed, varying from green to red and from mild to hot. Dried jalapeños are known as chipotles. Poblaños are long and pointed and have a dark-green skin and a mild to medium heat. Dried poblaños are called Anchos and are dark-purple to brown with a sweet flavor. Serraños, either green or red, are small and tapered and very hot.

ORIGIN & BOTANICAL FACTS

Peppers are native to Asia and the Western Hemisphere, where they have been a culinary staple for thousands of years, but they were unknown in Europe until Columbus brought them home from his first voyage to the Americas. In the United States, sweet peppers constitute more than 60 percent of the pepper crop. The leading domestic growers of peppers are California and Florida. Many peppers also are imported from Latin America and Asia.

Both sweet and hot peppers are perennial shrubs that thrive in tropical areas but have been adapted as annuals in colder regions. Plants grow to about 2 feet tall and equally wide. Proper flowering and fruit development require humidity. Sweet bell peppers are usually harvested when they are 3 to 4 inches in diameter and are still green and crisp. The size of hot peppers at harvest depends on the variety. Peppers are harvested by hand because they must be cut, rather than pulled, from the brittle stems.

USES

Good-quality fresh peppers should be firm and brightly colored. Sweet peppers can be stored unwashed in plastic bags in the refrigerator for up to 1 week.

PEPPERS

Most varieties of peppers can be eaten either raw or cooked. Sweet peppers frequently are julienned or chopped and added raw to salads or cooked in soups, stews, and stir-fries. They also can be roasted (which makes it easy to remove the skin and adds smoky flavor) and marinated, or they may be stuffed and baked or microwaved. Hot peppers are used in a wide variety of Latin American and Asian recipes. Raw hot peppers can be chopped and added to salsas, relishes, and salad dressings. The internal veins and seeds, which can be bitter or hot, can be removed before the pepper is used. Some milder types are used whole, stuffed with cheese or meat, and baked to make dishes such as chilis relleños. Hot peppers also can be pickled or dried, and most types of hot pepper are commercially available in their dried form.

Red peppers contain the phytochemical capsaicin, which has been approved by the Food and Drug Administration for use as a topical analgesic.

NUTRIENT COMPOSITION

Both sweet and hot peppers are high in vitamin C. Sweet and hot red peppers are a good source of vitamin A (beta-carotene).

SERVING SIZE: *hot peppers: 1 pepper (45 g)*

NUTRIENT CONTENT

	Green	Red
Energy (kilocalories)	18	18
Water (%)	88	88
Dietary fiber (grams)	1	1
Fat (grams)	0	0
Carbohydrate (grams)	4	4
Protein (grams)	1	1
Minerals (mg)		
Calcium	8	8
Iron	1	1
Zinc	0	0
Manganese	0	0
Potassium	153	153
Magnesium	11	11
Phosphorus	21	21
Vitamins (mg)		
Vitamin A	35 RE	484 RE
Vitamin C	109	109
Thiamin	0	0
Riboflavin	0	0
Niacin	0	0
Vitamin B$_6$	0.1	0.1
Folate	11 µg	11 µg
Vitamin E	0	0

SERVING SIZE: *sweet peppers: 1/2 cup, chopped (75 g)*

NUTRIENT CONTENT

	Yellow	Red	Green
Energy (kilocalories)	20	20	20
Water (%)	92	92	92
Dietary fiber (grams)	1	1	1
Fat (grams)	0	0	0
Carbohydrate (grams)	5	5	5
Protein (grams)	1	1	1
Minerals (mg)			
Calcium	8	7	7
Iron	0	0	0
Zinc	0	0	0
Manganese	0	0	0
Potassium	159	132	131
Magnesium	9	7	7
Phosphorus	18	14	14
Vitamins (mg)			
Vitamin A	18 RE	425 RE	47 RE
Vitamin C	138	142	66
Thiamin	0	0	0
Riboflavin	0	0	0
Niacin	1	0	0
Vitamin B$_6$	0.1	0.2	0.2
Folate	20 µg	16 µg	16 µg
Vitamin E	–	1	1

Note: A line (–) indicates that the nutrient value is not available.

POTATO

Potatoes are tubers, fleshy underground stems that bear minute leaves, each of which develops a bud capable of producing a new plant. Potatoes are cultivated in a variety of sizes, shapes, and colors. Their weight can range from 1 ounce to more than a pound. The skin can be smooth or rough, and tan, white, red, or any of a variety of less common colors. The flesh is usually white but can be yellow to deep orange and has a smooth to mealy texture. In flavor, they range from bland to buttery sweet.

Family Solanaceae
Scientific name *Solanum tuberosum* L.
Common name potato

♥ **A good source of vitamins C and B$_6$ and a source of potassium**

VARIETIES

Potatoes are categorized by flesh color, use, or age. The white potatoes include several varieties. The Russet has a thick, netted, brown skin and a somewhat dry, mealy texture. The Round White and the Long White have a more moist, waxy texture and smooth, tan skins. The Round Red potato has a smooth, reddish skin and creamy white, firm flesh. Yellow, or sweet, potatoes such as the Yukon Gold (not to be confused with the sweet potato) have a thicker brown skin, golden flesh, and a sweet, buttery flavor. Specialty varieties, including blue and purple potatoes, are nutty in flavor and difficult to find in most markets. White potatoes can be subdivided by use into boiling potatoes and baking potatoes. Finally, potatoes can be subdivided into new and storage types. All new potatoes are boilers by virtue of their low starch content and their smooth skins.

ORIGIN & BOTANICAL FACTS

The potato, a member of the nightshade plants, originated in the Andes Mountains of Peru, where more than 800 varieties of potatoes were once cultivated by the Incas on terraced farmland. The Spanish conquest of South America spread the cultivation of potatoes worldwide. Eventually, potatoes became a dietary staple throughout Europe. Today, potatoes are one of the most important food crops in the world. In the United States, annual potato consumption reaches 125 pounds per person. Some of the leading potato-producing states are California, Colorado, Idaho, and Maine. Potatoes can be propagated from true seeds or from pieces of tubers that contain two or more buds and some potato flesh to nourish the developing sprouts. Because exposure to sunlight can turn growing potatoes green and bitter, tuber pieces and seeds are planted deeply, and often the soil is protected from sunlight with straw mulch. Potato plants grow to about 18 inches tall and 4 feet wide. Some develop flowers and small toxic green fruits resembling green tomatoes. Potatoes are harvested about 4 months after planting by carefully prying them out of the ground to avoid puncturing or bruising them.

USES

Potatoes should be heavy for their size and free of sprouts or any greenish cast. They can be stored unwashed and unwrapped in a cool, dry, dark, well-ventilated area for weeks. The potato is a versatile vegetable that can be baked, boiled, fried, or microwaved. Potatoes can be cooked alone or in combination with meats or other vegetables. Russet potatoes are the most common variety used for baking. Russet, Round White, and Yellow potatoes are often mashed. Long White potatoes work well in potato salads, soups, and stews. Red potatoes are the type usually used in German potato salad.

NUTRIENT COMPOSITION

Potatoes are a good source of vitamins C and B$_6$ and are a source of potassium.

SERVING SIZE: *1/2 cup, baked (61 g)*

NUTRIENT CONTENT

Energy (kilocalories)	57
Water (%)	75
Dietary fiber (grams)	1
Fat (grams)	0
Carbohydrate (grams)	13
Protein (grams)	1
Minerals (mg)	
Calcium	3
Iron	0
Zinc	0
Manganese	0
Potassium	239
Magnesium	15
Phosphorus	31
Vitamins (mg)	
Vitamin A	0 RE
Vitamin C	8
Thiamin	0.1
Riboflavin	0
Niacin	1
Vitamin B$_6$	0.2
Folate	6 µg
Vitamin E	0

RADISH

The radish is a root vegetable of the mustard family, resembling beets and turnips but with a unique, peppery flavor that can range from mild to very sharp. The name comes from the Latin *radix*, meaning "root." Radishes can be round, oval, or elongated, and they range from less than 1 inch to 2 feet long. Although skin color varies from white to yellow, red, purple, and black, the interior flesh is usually white.

Family Cruciferae
Scientific name *Raphanus sativus* L.
Common name radish

♥ **High in vitamin C**

♥ **A source of phytochemicals that may help prevent cancer**

VARIETIES

Radishes are divided into spring- and winter-harvested types, with additional variations in shape and color in each category. Spring radishes, which are pulled before they reach 1 inch in diameter, include the round, red Cherry Belle; the White Icicle, which is oblong, about 6 inches long, and mild in flavor; and Rainbow Mix and Easter Egg varieties, which include purple-skinned roots. Winter radishes have a stronger, more pungent flavor and a coarser texture. They are larger than spring radishes, about the weight of a turnip, and range from white to black. White varieties include the Japanese daikon, a long, carrot-shaped, sharp-flavored radish.

ORIGIN & BOTANICAL FACTS

Radishes are an ancient vegetable, first cultivated thousands of years ago either in the eastern Mediterranean region or in the Far East and quickly spread throughout the world. The earliest radishes to be cultivated were the black varieties. Long, tapering white radishes were first mentioned in 16th-century European literature, and about 200 years later, round radishes first appeared, along with the red-skinned types. In the United States today, California and Florida are the leading radish growers.

From the top of the root, the leaves of the radish plant form a rosette that can grow to 1 foot in height. The radish is a cool-weather annual, one of the easiest vegetables to grow. Propagation is by direct seeding into a sandy soil with consistent moisture. Spring radishes can be planted as soon as the ground is soft. The fastest-growing varieties mature in about 3 weeks. Winter radishes are planted in late summer so they can mature in the cool temperatures of autumn. These larger radishes take at least 55 days to reach a reasonable size.

USES

Radishes that are firm with bright, crisp greens are the best. The leaves should be removed from the roots before storing in a perforated plastic bag in the refrigerator vegetable crisper. The roots can be stored for several weeks to a month.

In Western cuisines, radishes are eaten raw in salads or used as a colorful garnish. However, in Chinese and Japanese cuisines, radishes are a staple consumed raw, preserved, or cooked. The Japanese chop or grate daikon and use it as a condiment for sushi, sashimi, and many other dishes.

NUTRIENT COMPOSITION

Radishes are high in vitamin C and contain bioflavonoids and indoles that may help prevent cancer.

SERVING SIZE: *1/2 cup raw, sliced (13 medium) (58 g)*

NUTRIENT CONTENT

Energy (kilocalories)	12
Water (%)	95
Dietary fiber (grams)	1
Fat (grams)	0
Carbohydrate (grams)	2
Protein (grams)	0
Minerals (mg)	
Calcium	12
Iron	0
Zinc	0
Manganese	0
Potassium	135
Magnesium	5
Phosphorus	10
Vitamins (mg)	
Vitamin A	1 RE
Vitamin C	13
Thiamin	0
Riboflavin	0
Niacin	0
Vitamin B$_6$	0
Folate	16 µg
Vitamin E	0

RUTABAGA

The rutabaga, a member of the cabbage family, is a root vegetable similar to the turnip, but it is rounder, larger, denser, and sweeter and has a yellow flesh. Rutabagas have a thin, pale-yellow skin and smooth, waxy leaves. The root has a crisp texture and a sweet, peppery flavor.

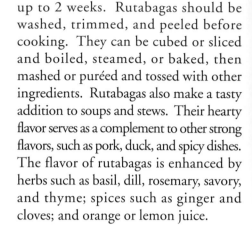

Family Cruciferae
Scientific name *Brassica napus* L.
Common name rutabaga

♥ **High in vitamin C and a source of potassium**

VARIETIES

The differences among varieties of rutabaga are primarily in the color and shape of the root. Most types have yellow flesh. The American Purple Top is purple above the ground and light yellow below, with yellow flesh. The Laurentian has a dark-purple top, a pale, smooth root, and yellow flesh. The Magre has an attractive, round root shape, and the Merrick is one of the rare white-fleshed rutabaga varieties.

ORIGIN & BOTANICAL FACTS

The rutabaga originated from a chance hybridization between cabbage and turnip plants. A relatively new vegetable compared with those that have been around for thousands of years, it probably emerged in medieval gardens where turnips and cabbage were grown side by side. The first mention of rutabagas occurred in European botanical literature of the 17th century. Rutabagas became very popular in Scandinavia (the Swedish word *rotabagge* means "round root"), from where they were brought to England in the late 18th century, acquiring the name "swede." Their cultivation in North America began early in the 19th century. Today they are grown in the northern states and Canada, because hot weather can damage the crop. Most domestic rutabagas are imported from Canada.

Rutabagas are biennials that thrive in cool temperatures and are best produced as a fall crop. Because they require a long growing season, rutabagas should be planted in the early spring in light but rich soil with good drainage. Seeds are sown directly into fields, and the plants must be irrigated regularly to produce a sweet, tender crop. The roots are ready for consumption about 3 months after sowing, when they are 4 to 5 inches in diameter. Harvesting should occur when the ground is dry so that very little soil adheres to the roots. Because they are resistant to fall frost and light winter freezes, rutabagas can be mulched, left in the ground, and harvested throughout the fall and winter. If they are pulled immediately, they can be refrigerated or stored in cool, underground cellars. Rutabagas are available year-round.

USES

Rutabagas that are smooth-skinned, firm, and heavy for their size should be chosen. They can be refrigerated in a plastic bag up to 2 weeks. Rutabagas should be washed, trimmed, and peeled before cooking. They can be cubed or sliced and boiled, steamed, or baked, then mashed or puréed and tossed with other ingredients. Rutabagas also make a tasty addition to soups and stews. Their hearty flavor serves as a complement to other strong flavors, such as pork, duck, and spicy dishes. The flavor of rutabagas is enhanced by herbs such as basil, dill, rosemary, savory, and thyme; spices such as ginger and cloves; and orange or lemon juice.

NUTRIENT COMPOSITION

Rutabagas are high in vitamin C and are a source of potassium.

NUTRIENT CONTENT

	½ cup, raw (70 g)	½ cup, cooked (85 g)
Energy (kilocalories)	25	33
Water (%)	90	89
Dietary fiber (grams)	2	2
Fat (grams)	0	0
Carbohydrate (grams)	6	7
Protein (grams)	1	1
Minerals (mg)		
Calcium	33	41
Iron	0	0
Zinc	0	0
Manganese	0	0
Potassium	236	277
Magnesium	16	20
Phosphorus	41	48
Vitamins (mg)		
Vitamin A	41 RE	48 RE
Vitamin C	18	16
Thiamin	0.1	0.1
Riboflavin	0	0
Niacin	0	0.1
Vitamin B6	0.1	0.1
Folate	15 µg	13 µg
Vitamin E	0	0

SALSIFY

Salsify is a white-fleshed vegetable root that could easily be mistaken for a yellowish gray carrot or parsnip. The salsify plant, which is 10 to 12 inches in length and has a diameter of about 2 1/2 inches, is cultivated primarily for its slender edible roots, although the young tender leaves, often called "chards," are commonly used in salads.

Family Asteraceae
Scientific name *Tragopogon porrifolius*
Common name salsify, vegetable oyster, oyster plant

VARIETIES

Salsify can be divided into three types: the most familiar white salsify, black salsify, and Spanish salsify. They are all similar in shape, flavor, flesh color, and size. However, black salsify, as its name implies, is black on the surface, and Spanish salsify is slightly wider in diameter than the other varieties. The most popular variety for the home garden is the Mammoth Sandwich Island, a subspecies of the white salsify which has French blue flowers. The Black Giant Russian is the most common of the Spanish salsify.

ORIGIN & BOTANICAL FACTS

In its wild form, salsify is believed to have been a part of the diet of the ancient Greeks, although the vegetable was not a popular culinary item until the mid-17th century. Salsify appeared in North America and in England during the 18th century. The salsify plant is also known as "vegetable oyster" or "oyster plant" because its mild, sweet flavor is reminiscent of that of the oyster, although some prefer to call it the "goatsbeard plant" because its grass-like leaves are bunched in a manner resembling a "goatee."

The plant is a biennial, alternating yearly between purple and rose-colored flowers that resemble dandelions and are closely related to lettuce and chicory. The roots grow best in rich, slightly alkaline, fine-textured, loose sand and after exposure to the cold. It is an easy vegetable to grow and is available year-round, although the peak season in the United States is from June through February. Salsify is a more popular vegetable in Europe, which may explain its greater availability in markets specializing in Greek and Italian foods. During harvest, care must be taken not to damage the brittle roots because bruised salsify loses much of its flavor. After harvest, the roots must be kept in cold storage at 90 to 98 percent humidity to retain their freshness, because dehydrated roots shrivel and also lose their flavor.

USES

A firm, well-formed, medium-size root that is heavy for its size is the best choice. Oversized roots are tough and woody and should be avoided. Salsify oxidizes very quickly when peeled and must be placed in cold lemon water to prevent darkening. If salsify is stored with the tops removed in a sealed plastic bag in a cold, moist storage area, it may keep up to 4 months.

Salsify can be baked, steamed, fried, served in soups, or cut into cubes and stewed. If roots are to be steamed, they should be scrubbed and peeled before cooking. The sliced root can be added to savory vegetable pies. In addition, young shoots and flower buds can be used as a substitute for asparagus or added raw to salads.

NUTRIENT COMPOSITION

Salsify is not a significant source of nutrients.

SERVING SIZE: *1/2 cup sliced, cooked (68 g)*

NUTRIENT CONTENT

Energy (kilocalories)	46
Water (%)	81
Dietary fiber (grams)	2
Fat (grams)	0
Carbohydrate (grams)	10
Protein (grams)	2
Minerals (mg)	
Calcium	32
Iron	0
Zinc	0
Manganese	0
Potassium	191
Magnesium	12
Phosphorus	38
Vitamins (mg)	
Vitamin A	0 RE
Vitamin C	3
Thiamin	0
Riboflavin	0.1
Niacin	0
Vitamin B$_6$	0.1
Folate	10 µg
Vitamin E	0

SCALLION

The scallion, with its long, straight, narrow green top and white base, is a true but very immature onion. Often called a green onion, in reality it is picked at an even earlier stage than true green onions, usually before it can begin to form a bulb. The scallion is crunchy and has a peppery, fresh flavor that is similar to, but milder than, that of the onion.

Family Amaryllidaceae
Scientific name *Allium cepa* var. *aggregatum*
Common name scallion, eschalot, cebollin

♥ **A good source of vitamin C**

VARIETIES

Scallion varieties may be classified by day length (the number of hours of daylight that optimizes their growth), market use, and bulb color. Sweet Spanish is an example of a long-day variety (one that develops best in areas with more hours of daylight and in midsummer). Southport White Globe onions are grown as scallions in areas with short days, where the shorter day length prevents bulb formation. Crossing onions with a variety called the Japanese bunching onion (*Allium fistulosum*), an onion that resembles a scallion but is more bitter, results in hybrids that can be grown as scallions in the summer and in areas with long days.

ORIGIN & BOTANICAL FACTS

Scallions are propagated from seeds planted in raised beds of fertile, well-prepared, well-balanced soil with good moisture retention. They are hardy plants, capable of withstanding temperatures as low as 20° Fahrenheit. However, minimal temperatures of 55° are necessary for the plants to emerge from the soil, and optimal growth occurs when the temperature ranges between 68° and 77° Fahrenheit. Growth of scallions also depends on day length. Scallions are planted in fall, spring, and summer and are ready for harvest 2 to 4 months later. Hand harvest is required, usually accomplished by undercutting the base of the plant. Scallions are available year-round, but the supply peaks from July through October. California is the leading domestic supplier of scallions. Illinois, Ohio, and New Jersey contribute a small proportion of the crop. During the winter and spring, scallions are imported from Mexico.

USES

Scallions with slender bases tend to be milder and sweeter. Those with the crispiest leaves, shiny bright-green stalks, and no yellowing or tears are the best. Scallions should be stored in a plastic bag in the refrigerator crisper.

Scallions can be consumed both raw and cooked. Like mature onions, they are used primarily to enhance the flavor of other dishes. Before use, they should be rinsed well, and the roots and any wilted leaves should be removed. The entire plant can be sliced or chopped and added raw to salads and dips. Mixed with cottage cheese or cream cheese, they make a tasty spread for bread or a dip for raw vegetables. Scallions are a flavorful addition to soups, stews, stir-fries, casseroles, cooked rice, tomato sauces, and omelets. They also can be grilled, braised, or stir-fried alone, or seasoned with ginger and garlic, to make a tasty vegetable dish to serve with rice and grilled meats. The Irish include scallions in a traditional dish called "champ," which is composed of potatoes and cooked scallions mashed together to produce a savory, green side dish.

NUTRIENT COMPOSITION

Scallions are a good source of vitamin C.

SERVING SIZE:
1/2 cup, chopped (50 g)

NUTRIENT CONTENT

Energy (kilocalories)	16
Water (%)	90
Dietary fiber (grams)	1
Fat (grams)	0
Carbohydrate (grams)	4
Protein (grams)	1
Minerals (mg)	
Calcium	36
Iron	1
Zinc	0
Manganese	0
Potassium	138
Magnesium	10
Phosphorus	19
Vitamins (mg)	
Vitamin A	20 RE
Vitamin C	9
Thiamin	0
Riboflavin	0
Niacin	0
Vitamin B$_6$	0
Folate	32 µg
Vitamin E	0

SNAP BEANS

Snap, green, or string beans are the beans most frequently consumed in the United States. Although they are members of the legume family, the long, slender green, yellow, or purple pods of snap beans are harvested while the seeds are still immature, and both pod and seeds are eaten as a vegetable. "Snap" refers to the sound the fresh pod makes when broken into pieces.

Family Leguminosae
Scientific name *Phaseolus vulgaris*
Common name green beans, string beans, snap beans, wax beans, yellow snap beans, romano beans, haricots

♥ **Yellow and green snap beans are a good source of vitamin C**

VARIETIES

The "string" in "string beans" refers to a string-like fiber that, until the late 19th century, characterized all fresh beans and had to be removed before the beans were eaten. Today, modern hybrid varieties no longer have the "string" and are referred to as snap beans. Many varieties of green snap beans are grown throughout the United States. Variants include pods that are pale yellow, called wax beans. The term "French green bean" is sometimes used to refer to small, young green beans that are cooked and eaten whole (also called haricots or haricots vertes) or to mature green beans that have been cut into diagonal strips. Purple snap beans, available in limited supply, turn green when cooked. Romano beans, also called Italian or Scarlet Runner beans, are similar to but flatter than snap beans.

ORIGIN & BOTANICAL FACTS

Snap beans, like kidney beans, white beans, pinto beans, and cranberry beans, are members of the common bean species, all of which trace their origins to the Western Hemisphere. Although remains of common beans from Central American sites have been carbon dated to 7000 B.C., the original subspecies have not been identified. Indigenous peoples of South and Central America and American Indians crossed the beans to create many subspecies and varieties. Common beans, including the snap bean, were brought to Europe by Columbus and other 15th- and 16th-century explorers. Today, many varieties of snap beans are grown throughout the world.

The growth habit of snap beans is used to divide them into two varieties: bush beans and pole beans (which must be trained to a pole or trellis). Both are warm-weather vegetables that must be planted after the danger of frost has passed. The beans are harvested when they are rapidly growing, about 8 to 10 days after flowering. At this stage, the color is bright and the pod is fleshy with small, green seeds. Leaving the pods on the plants too long decreases plant yield and results in tough, dull-colored pods.

USES

Snap beans are available year-round. The peak season in North America spans from May to October. They can be stored in the refrigerator for up to 5 days. The beans also are available frozen and canned, both whole and prechopped. Fresh snap beans can be steamed or simmered until the pods are tender. Steaming is the preferred method of cooking because it preserves nutrients. Chopped into 1- or 2-inch sections, snap beans can be tossed in salads, stir-fried, included in soups and stews, or served as a side dish.

NUTRIENT COMPOSITION

Yellow and green snap beans are a good source of vitamin C.

SERVING SIZE:
1/2 cup, cooked (67 g)

NUTRIENT CONTENT

	Green	Yellow (wax)
Energy (kilocalories)	22	22
Water (%)	89	89
Dietary fiber (grams)	2	2
Fat (grams)	0	0
Carbohydrate (grams)	5	5
Protein (grams)	1	1
Minerals (mg)		
Calcium	29	29
Iron	1	1
Zinc	0	0
Manganese	0	0
Potassium	187	187
Magnesium	16	16
Phosphorus	24	24
Vitamins (mg)		
Vitamin A	42 RE	5 RE
Vitamin C	6	6
Thiamin	0	0
Riboflavin	0.1	0.1
Niacin	0	0
Vitamin B$_6$	0	0
Folate	21 µg	21 µg
Vitamin E	0	0

SPINACH

Spinach is a leafy vegetable that grows in a dark-green rosette about 8 to 10 inches across. The leaves may be flat or curly, depending on the variety. Cooked spinach has a pungent, earthy flavor and can have a mushy texture; raw spinach is milder and crisp.

Family Chenopodiaceae
Scientific name *Spinacia oleracea* L.
Common name spinach

♥ **Raw spinach is high in vitamin A (beta-carotene) and a good source of vitamin C and folate**

♥ **Cooked spinach is high in vitamin A and folate and is a good source of vitamin C, riboflavin, vitamin B$_6$, calcium, iron, and magnesium**

VARIETIES

Spinach comes in two basic types: savoy (curly leaf) and flat (smooth leaf). Savoy has crinkly dark-green leaves. Flat-leaf spinach has unwrinkled, spade-shaped leaves and a slightly milder taste than savoy. A third type that is increasing in popularity is the semi-savoy, whose slightly curly leaves provide some of the texture of savoy but are easier to clean. All varieties have the same appearance when cooked.

ORIGIN & BOTANICAL FACTS

Spinach probably originated in southwest Asia or the western Himalayas, but wild varieties also grow in North Africa and Iran. The leafy vegetable was first cultivated by the Persians. Its cultivation reached China in the 7th century A.D. and Europe in the 9th century, when it was introduced to Spain by the Arabs, who named it. Today, spinach is grown and enjoyed in many parts of the world.

Spinach is an annual plant that requires cool, damp weather and rich, moist soil. Spinach seed can be planted in early spring or in autumn, depending on the variety. Hardier types will survive the winter in well-drained soils and can be harvested until spring. More tender varieties are planted in spring, as early as February, for summer harvest. Spinach is ready to be harvested about 6 weeks after planting, when the largest leaves are 6 to 8 inches long.

USES

Spinach leaves that are crisp and bright to dark green are best. The leaves can be refrigerated in a plastic bag for 3 days.

Spinach can be served raw or cooked. The flat-leaf variety, with its slightly milder flavor, is generally preferred as a raw salad green. Spinach should be cooked very quickly, either by steaming or by sautéing with a minimum of liquid, just until the leaves wilt. The leaves also can be added to soups, casseroles, and stews. A variety of seasonings, such as lemon juice, soy sauce, horseradish, tomato sauce, or nutmeg, add flavor to spinach dishes. Chopped, seasoned spinach also makes a flavorful stuffing for mushroom caps or a filling for savory pastries.

NUTRIENT COMPOSITION

Raw spinach is high in vitamin A (beta-carotene) and a good source of vitamin C and folate. Cooked spinach is high in vitamin A (beta-carotene) and folate and is a good source of vitamin C, riboflavin, vitamin B$_6$, calcium, iron, and magnesium. Although spinach is a good source of iron and calcium, oxalic acid (a chemical that is present in the leaves) inhibits the body's absorption of these nutrients. Absorption of iron can be increased by eating spinach with a fruit or vegetable that contains vitamin C.

NUTRIENT CONTENT

	1 cup, raw (30 g)	½ cup, cooked (90 g)
Energy (kilocalories)	7	21
Water (%)	92	91
Dietary fiber (grams)	1	2
Fat (grams)	0	0
Carbohydrate (grams)	1	3
Protein (grams)	1	3
Minerals (mg)		
Calcium	30	122
Iron	1	3
Zinc	0	1
Manganese	0	1
Potassium	167	419
Magnesium	24	78
Phosphorus	15	50
Vitamins (mg)		
Vitamin A	202 RE	737 RE
Vitamin C	8	9
Thiamin	0	0.1
Riboflavin	0.1	0.2
Niacin	0	0
Vitamin B$_6$	0.1	0.2
Folate	58 µg	131 µg
Vitamin E	1	1

SQUASH

Although thought of and eaten as a vegetable, squash is a fleshy, edible fruit related to melons and cucumbers. Numerous varieties of squash are available in a wide assortment of colors, shapes, and sizes. They can range from the patty pan (scallop) variety, which weighs only a couple of ounces, to the pumpkin, which can attain weights of up to 200 pounds. The rind can be smooth, ridged, or bumpy and can range from white or cream-colored to yellow, orange, green, and even light blue. Squash can be cylindrical, bell- or club-shaped, and simply round or oblong. Although wild squash is bitter, cultivated varieties are generally sweeter or bland and have a soft to crunchy texture when cooked.

Family Cucurbitaceae

Scientific name *Cucurbita pepo, Cucurbita maxima, Cucurbita moschata*

Common name squash

♥ **Winter squash is high in vitamin A (beta-carotene) and is a good source of potassium, fiber, and vitamin C**

VARIETIES

Squash is generally divided into two basic types, summer and winter, although seasonal distinction is no longer accurate because both types are now available year-round. Winter, or hard-shell, squash is allowed to mature on the plant and has a thick rind; large, tough seeds; and dark-yellow to orange flesh. Summer, or soft-shell, squash, which is harvested before it matures completely, has a more tender rind and lighter-colored flesh.

Winter squash types include acorn squash, shaped something like an acorn that tapers at one end, with a dark-green, ridged rind; banana squash, a large, cylinder-shaped squash with a thick, pale skin and finely textured flesh; buttercup squash, a squat, dark-green vegetable with lighter stripes and rather dry flesh; butternut squash, shaped like a long bell with a tan rind and mild flavor; and spaghetti squash, an oval, yellow variety whose mild, pale-yellow flesh forms crisp-textured spaghetti-like strands when cooked.

Among the summer squash varieties, zucchini, with its mild flavor and cucumber-like appearance, is the most popular in the United States. Although zucchini usually has a smooth, green skin, one variant, the golden zucchini, has deep-yellow skin and a sweeter flavor. Other summer squashes include the chayote, a pale-green, pear-shaped fruit with a large central seed and a fairly thick, ridged skin; the patty pan, a disk-shaped variety with a scalloped edge, white to pale-green skin, and white, succulent flesh; and yellow crookneck and yellow straightneck, which have lemon-colored skin and bulbous blossom ends, tapering to narrow stem ends that are either curved or straight. Within some of these types are further variations.

ORIGIN & BOTANICAL FACTS

The squash probably originated in Mexico or Central America. Although edible wild types are no longer known, related species with small, very bitter fruits are still found in this region. Squash was first gathered by indigenous people around 8000 B.C., but apparently only the seeds were eaten, because the fruits were unappealing. Cultivation of squash may have begun around Tehuacan, south of Mexico City, around 3400 B.C. From there, peoples throughout North and South America adopted squash cultivation. By the time squash was introduced to Europe in the 16th century, most of the modern types were already developed. Squash quickly became a staple in the

NUTRIENT CONTENT

	½ cup summer, cooked (90 g)	½ cup winter, cooked (103 g)
Energy (kilocalories)	18	40
Water (%)	94	89
Dietary fiber (grams)	1	3
Fat (grams)	0	1
Carbohydrate (grams)	4	9
Protein (grams)	1	1
Minerals (mg)		
Calcium	24	14
Iron	0	0
Zinc	0	0
Manganese	0	0
Potassium	173	448
Magnesium	22	8
Phosphorus	35	21
Vitamins (mg)		
Vitamin A	26 RE	365 RE
Vitamin C	5	10
Thiamin	0	0.1
Riboflavin	0	0
Niacin	0	1
Vitamin B₆	0.1	0.1
Folate	18 µg	29 µg
Vitamin E	0	0

diets of European colonists in America. New England settlers adapted the word "squash" from several Indian names for the vegetable, all of which meant "something eaten raw." Both George Washington and Thomas Jefferson cultivated zucchini and other types of summer squash on their Virginia estates. By the 19th century, North American merchant seamen were bringing home new varieties of squash from all over Central and South America. Today, squash, gourds, and pumpkins are grown in many parts of the world and rank 11th among the leading vegetables of the world. Although the United States does not produce a large volume of squash commercially, many people cultivate it in home gardens. California and Florida are the primary U.S. producers, and Mexico and Costa Rica contribute substantially to the U.S. supply.

Squash is a hardy, warm-weather annual that grows on vines or small bushes with trailing tendrils. Seeds can be sown in seedbeds early in spring or directly into fields later in the season, after danger of frost has passed, in hills of warm, well-fertilized soil. Vines must be trellised to provide support for the heavy fruit. Although most types of squash prefer full sun, winter varieties can tolerate light shade. For ideal growth, the plants require considerable moisture, especially after flowering. Most squash blossoms are yellow or orange, and both male and female flowers form on the same plant. Because the female blooms open for only 1 day, and only from dawn until mid-morning, pollination at the right time and place is critical. In addition, the number of seeds and the size and shape of the fruits are determined by the amount of pollen deposited. Inadequate pollination of summer squash results in small,

misshapen fruit or none at all. The first crop of summer squash is ready to harvest about 50 days after planting, and all of the fruits must be harvested at this stage in order for more to grow. If the fruits are allowed to mature, the plants stop producing. In contrast, winter squash is left on the plant until it matures, a process that takes 95 to 115 days, depending on variety. All are picked after the leaves have turned brown.

USES

When selecting winter squash, it is important to choose one that is heavy for its size and has a thick, hard shell. If stored in a cool, dry place, whole winter squash can keep well for several months. Cut pieces should be tightly wrapped and refrigerated. Winter squash is always cooked before eating, usually after the fruit has been cut open and the seeds and fibers scooped out. (The seeds of most winter squash varieties can be dried or roasted and consumed as a snack.) A heavy chef's knife or cleaver may be necessary to cut the hard shell. Halves can be baked and served plain or stuffed with cheese, meats, or other vegetables. Baking conserves the nutrients in the flesh and enhances its sweetness. Some especially tough-shelled varieties can be baked or steamed whole (after piercing the flesh) and then cut up. Squash pieces also can be boiled or steamed in broth, microwaved, or sautéed in oil. Baked or steamed winter squash is delicious mashed or puréed and seasoned with spices such as fresh ginger, curry, cinnamon, cloves, or allspice or with sweeteners such as brown sugar, maple syrup, or honey. Squash also can be mixed with

onions, garlic, and herbs or with other vegetables such as corn, tomatoes, and bell peppers. Chunks of squash can be added to soups, stews, and casseroles. Any type of mashed or puréed winter squash can be used in place of canned pumpkin in soups, pies, cookies, or quick breads. Spaghetti squash is often served as a substitute for pasta, topped with tomato sauce, pesto, or other sauces. Cooked squash also can be frozen for later use. To prevent squash from becoming watery during cooking, lightly salt the raw flesh, place it on absorbent paper to draw out the moisture, and rinse.

When purchasing summer squash, small, firm, shiny squash that are heavy for their size should be selected. Squash can be stored in perforated plastic bags in the refrigerator crisper. Summer squash can be eaten raw or cooked, and the tender skin is always left on. Raw summer squash can be sliced and added to green salads or julienned to use with dips. Grated zucchini is used to make moist breads and cakes. Quick steaming, grilling, and stir-frying are the best cooking methods for conserving nutrients. Several varieties cooked together make a colorful and tasty combination seasoned with herbs such as dill, basil, thyme, mint, tarragon, marjoram, or oregano. The mild flavor of summer squash complements soups, stews, casseroles, and mixed vegetables. Immature summer squash is used as an attractive edible garnish or side dish. In addition, the flowers are edible.

NUTRIENT COMPOSITION

Winter squash is high in vitamin A (beta-carotene) and is a good source of potassium, fiber, and vitamin C.

SWEET POTATO

The sweet potato, a smooth-skinned, oblong or elongated tropical tuber, is not related to the white potato. Instead, it is a member of the morning glory family. Nor is the sweet potato a yam, which actually is a completely different vegetable. The sweet potato's smooth skin may vary from pale yellow to vivid orange to deep purple, depending on the variety. The sweet flesh may be light yellow, pink, red, or deep orange.

Family Convolvulaceae
Scientific name *Ipomoea batatas*
Common name sweet potato

♥ **High in vitamin A (beta-carotene) and vitamin C**

♥ **A good source of vitamin B₆, potassium, and fiber**

VARIETIES

Sweet potatoes are categorized into two basic types. The orange-fleshed varieties, with tan to brownish red or purple skin, a plump shape, and sweet flavor, are the most common. The yellow-fleshed potatoes tend to be firmer, dryer, and less sweet and have a slightly mealy texture and yellowish tan to fawn-colored skins. Current varieties of sweet potato include the Beauregard, the Garnet, the Hernandez, and the Jewel.

ORIGIN & BOTANICAL FACTS

The sweet potato, a native of the tropical regions of the Americas, was important in the diet of the Aztec people of Mexico and the Incas of Peru. Remains of sweet potatoes that are 10,000 to 20,000 years old have been found in Peruvian caves. Sweet potatoes were introduced to Europe by Columbus, later brought to Asia by other explorers, and widely cultivated in the American colonies, where they became a dietary staple for early settlers and Revolutionary War soldiers. Today, sweet potatoes are cultivated in many parts of the world. Major suppliers include China, Indonesia, Vietnam, and Uganda. In the United States, sweet potatoes rank 10th among vegetables grown. North Carolina, Louisiana, California, and Georgia are the major suppliers. Although their peak season is the autumn and early winter, they are sold year-round. Sweet potato tubers can be harvested by machine, but they must be handled carefully, because their thin skin bruises easily, which can lead to rapid spoilage. After harvest, sweet potatoes can be stored for about 10 days at 85° Fahrenheit and 85 percent humidity to heal any small wounds and increase sweetness.

USES

Sweet potatoes that are firm with skin that is of uniform, bright color should be chosen. Sweet potatoes should be stored in a cool, dark, dry, well-ventilated place but not in the refrigerator, because temperatures less than 50° Fahrenheit produce a hard texture and unpleasant taste. They should be scrubbed well in cold water just before use and cooked in their skins to preserve nutrients and prevent the flesh from darkening. After cooking, the skin can be removed easily. Both types of sweet potato can be baked, boiled, or microwaved. Although cooked potatoes are naturally sweet, apple cider, lemon juice, orange peel, orange juice, pineapple, nutmeg, allspice, cinnamon, and ginger enhance their sweetness. They are a tasty ingredient in casseroles and stews, especially with apple or other fruit slices added. They also can be substituted for puréed pumpkin in baked breads, cakes, cookies, custards, pies, and muffins.

NUTRIENT COMPOSITION

Fresh sweet potatoes are high in vitamin A (beta-carotene) and vitamin C. They are a good source of vitamin B₆, potassium, and fiber. Canned and frozen potatoes are considerably less nutritious.

SERVING SIZE:
3/4 cup, baked (150 g)

NUTRIENT CONTENT

Energy (kilocalories)	155
Water (%)	73
Dietary fiber (grams)	5
Fat (grams)	0
Carbohydrate (grams)	36
Protein (grams)	3
Minerals (mg)	
Calcium	42
Iron	1
Zinc	0
Manganese	1
Potassium	522
Magnesium	30
Phosphorus	83
Vitamins (mg)	
Vitamin A	3,273 RE
Vitamin C	37
Thiamin	0.1
Riboflavin	0.2
Niacin	1
Vitamin B₆	0
Folate	34 µg
Vitamin E	0

TARO

Taro is a barrel-shaped tuber or corm with thick, brown, shaggy skin and fibrous, gray-white to lilac flesh. Its length ranges from about 5 inches to a foot or more, and it can be several inches wide. Its starchy, rather dry flesh is acrid and actually toxic when raw, but after cooking it is safe to consume and has a somewhat nutty flavor, similar to that of potatoes or water chestnuts.

Family Araceae
Scientific name *Colocasia esculenta* L. Schott
Common name taro

♥ **Cooked taro root is a good source of vitamin B₆ and fiber**

♥ **Cooked taro leaves are high in vitamin A and vitamin C**

VARIETIES

More than 300 varieties of taro are cultivated around the world, both in water and in soil, and vary considerably in color and taste. The two varieties of taro that are most important for food production are the globulifera, also called dasheen, which produces a large number of crisp, easily cut tubers (or corms), and the antiquorum, whose corms are tougher and more spongy.

ORIGIN & BOTANICAL FACTS

Because taro is an important part of many Asian diets and rituals, the tuber may have originated somewhere on that continent. Whatever its geographic origins, it is most likely one of the oldest food plants. As early as 2000 B.C., taro was brought from southeast Asia to the Pacific rim and northern Asia. Taro is believed to have been brought to Hawaii between 400 and 500 A.D. by the first Marquesan and Tahitian settlers. According to Hawaiian tradition, taro is the staff of

life. Taro also was carried westward to Arabia and was an important crop in the Nile Valley by 500 B.C. Today, taro is a staple in the diets of the people of West Africa, the Caribbean, and the Polynesian Islands. It is grown throughout tropical and subtropical Asia and the Pacific and in parts of Africa and the Americas.

Taro is a succulent perennial plant that ranges in height from about 20 inches to 6 feet. Young corms develop as off-shoots of the main corm and can produce new plants. Although taro is generally regarded as a bog plant, it can grow in a variety of environments from dry ground to wetlands and can tolerate lighting conditions ranging from deep shade to bright sunlight. Because taro seeds and seedlings do not survive well and the plants rarely flower, propagation is done primarily by planting side corms or by cutting off and planting the top of a large tuber with its shoot.

USES

Taro roots must be cooked thoroughly to neutralize the toxic calcium oxalate

crystals they contain. The most well-known use of taro is from Polynesia and Hawaii, where it is boiled, pounded into a paste, strained, and left to ferment into a potent brew called poi. Taro also can be peeled and cooked like potatoes. The young, unopened leaves of the taro plant are also edible and can be cooked and eaten like mustard or turnip greens.

NUTRIENT COMPOSITION

One serving of cooked taro root is a good source of vitamin B₆ and fiber. Cooked leaves are high in vitamin A and vitamin C.

NUTRIENT CONTENT

	½ cup root, cooked (66 g)	½ cup leaves, cooked (72 g)
Energy (kilocalories)	94	17
Water (%)	64	92
Dietary fiber (grams)	3	1
Fat (grams)	0	0
Carbohydrate (grams)	23	3
Protein (grams)	0	2
Minerals (mg)		
Calcium	12	62
Iron	1	1
Zinc	0	0
Manganese	0	0
Potassium	319	333
Magnesium	20	15
Phosphorus	50	20
Vitamins (mg)		
Vitamin A	0 RE	307 RE
Vitamin C	3	26
Thiamin	0.1	0.1
Riboflavin	0	0.3
Niacin	0	1
Vitamin B₆	0.2	0.1
Folate	13 µg	35 µg
Vitamin E	0	–

Note: A line (–) indicates that the nutrient value is not available.

TOMATILLO

The tomatillo resembles a small green, leaf-covered tomato. Indeed, the name "tomatillo" means "little tomato" in Spanish. Like the tomato, it belongs to the nightshade family and is actually a fruit. Globular in shape and between 1 1/2 and 2 inches in diameter, the slightly flattened, shiny fruits are enclosed in light-brown or green, easily removed, parchment-like coverings. The tomatillo has a tangy lemony flavor that is difficult to describe. It has a firmer texture than the tomato, and its flesh is pale green or yellow, depending on the degree of ripeness.

Family Solanaceae
Scientific name *Physalis ixocarpa*
Common name tomatillo, jamberry, strawberry tomato, Mexican green tomato, tomate verde

♥ **A good source of vitamin C**

VARIETIES

Tomatillos come in two varieties: the sweet and the sharp (or acidic), both of which are available in the United States. The more acidic variety is also known as Tomatilla de Milpa.

ORIGIN & BOTANICAL FACTS

The tomatillo is a native of Mexico, although it also grows wild in California. Plants reach heights and widths of 3 to 4 feet and have an unusual zigzag shape. The leaves are long and heart-shaped, and the flowers are bell-shaped. Tomatillos are best adapted to warm and dry climates. However, they can be grown as far north as the central midwestern United States. In North America, fruit production begins about 70 days after the plant has sprouted. Tomatillos are available year-round.

USES

The tomatillo is almost always used while it is still unripe, because the tangy lemony flavor is lost when the fruit ripens. Firm fruits that just fill the husks are best. The fruit can be stored in the refrigerator unwashed or in a paper or plastic bag for 3 weeks or longer. Before use, the fruit should be husked and washed to remove the sticky film that covers it. The tomatillo is popular in Mexican and Southwest cuisine. Although it can be used raw in salads, the tomatillo is usually cooked even when added to dishes such as salsa cruda (salsa made from raw vegetables), because cooking enhances the tomatillo's flavor and softens its skin. Salsa verde, also made from tomatillos, is a popular cooked sauce with a sharp flavor that is excellent for poultry and grilled meat or enchiladas. Canned tomatillos also are available in markets and can be used in recipes that require the cooked fruit.

NUTRIENT COMPOSITION

Tomatillos are a good source of vitamin C.

SERVING SIZE: *1/2 cup chopped, raw (66 g)*

NUTRIENT CONTENT

Energy (kilocalories)	21
Water (%)	92
Dietary fiber (grams)	1
Fat (grams)	1
Carbohydrate (grams)	4
Protein (grams)	1
Minerals (mg)	
Calcium	5
Iron	0
Zinc	0
Manganese	0
Potassium	177
Magnesium	13
Phosphorus	26
Vitamins (mg)	
Vitamin A	7 RE
Vitamin C	8
Thiamin	0
Riboflavin	0
Niacin	1
Vitamin B₆	0
Folate	5 µg
Vitamin E	0

TOMATO

Tomatoes are members of the nightshade family, related to potatoes, bell peppers, and eggplant. They can be red, pink, orange, or yellow, round to oblong, and from 1 to 6 inches in diameter. The flavor ranges from sweet to bland to tart, depending on variety.

Family Solanaceae
Scientific name *Lycopersicon esculentum*
Common name tomato

♥ **High in vitamin C**

♥ **A good source of vitamin A (carotenes)**

♥ **Contains the antioxidant lycopene**

VARIETIES

Tomatoes are available in three basic types: small, round cherry tomatoes; plump, oblong plum, or Roma, tomatoes; and round or globe-shaped slicing tomatoes, probably the sweetest and juiciest type. Within each type are numerous varieties, totaling about 4,000.

ORIGIN & BOTANICAL FACTS

Although the tomato is botanically a fruit, it is prepared and consumed as a vegetable. In fact, because of a tariff dispute, the U.S. Supreme Court officially declared it a vegetable in 1893. The word "tomato" is derived from the Mexican Nahuatl Indian word "tomatl." The wild form of the plant, which still flourishes in Mexico and Central and South America, is similar to the domestic cherry tomato. Spanish explorers to Mexico brought tomatoes back to Europe in the 10th century. The first official mention of the fruit appeared in 1544, in the work of Italian botanist Matthiolus, who described a yellow-fruited variety he called pomodoro, meaning "golden apple." Europeans initially regarded tomatoes with suspicion, because most

plants of the nightshade family were known to be poisonous. The tomato was not widely accepted as a food until the early 19th century, although even then tomatoes would be cooked for hours to neutralize the toxins they were thought to contain. Raw tomatoes were not consumed until the late 19th century. Today, tomatoes are one of the most popular vegetables in the United States.

To increase durability and shelf life, tomatoes are usually picked when they are at the "mature green" stage. In response to year-round demand, growers have developed thicker-skinned, hardy varieties of tomatoes that can withstand long-distance shipping.

USES

Unripe tomatoes can be ripened in a paper bag at room temperature. Tomatoes should be stored at room temperature. Ripe tomatoes can be kept up to 2 days. The most popular way to eat fresh tomatoes is to slice them raw and eat them in salads or sandwiches.

Tomatoes also are available in a variety of processed forms, including canned whole, diced, and puréed. Canned tomato paste is a concentrated form of the fruit's pulp. Processed tomato sauce in cans or jars is similar to purée, but with seasonings and sometimes fat added.

NUTRIENT COMPOSITION

Tomatoes are a good source of vitamin A (carotenes) and are high in vitamin C. Red tomatoes also contain substantial amounts of lycopene, an antioxidant that may help protect against cancer. The lycopene in cooked or processed tomatoes is more easily absorbed than that in fresh tomatoes.

SERVING SIZE:
1 medium, raw (123 g)

NUTRIENT CONTENT

Energy (kilocalories)	26
Water (%)	94
Dietary fiber (grams)	1
Fat (grams)	0
Carbohydrate (grams)	6
Protein (grams)	1
Minerals (mg)	
Calcium	6
Iron	1
Zinc	0
Manganese	0
Potassium	273
Magnesium	14
Phosphorus	30
Vitamins (mg)	
Vitamin A	76 RE
Vitamin C	23
Thiamin	0.1
Riboflavin	0.1
Niacin	1
Vitamin B$_6$	0.1
Folate	18 µg
Vitamin E	0

TURNIP

The turnip is a fleshy root vegetable related to broccoli, brussels sprouts, cabbage, and the mustards. Depending on age and variety, turnips can be round or shaped like a top, range in diameter from 2 inches to over a foot, and weigh up to 50 pounds. Their smooth skin can be white, yellow, green, or purple. The white or yellow flesh of the turnip has a slightly sweet, peppery flavor and a crisp texture.

Family Cruciferae
Scientific name *Brassica rapa*
Common name turnip

 A good source of vitamin C

VARIETIES

Numerous varieties of turnips are grown for harvest throughout the year. The Purple Top Milan, with flat white roots and purple markings, matures early and is good for winter production, as are the Manchester Market and the yellow-fleshed Golden Ball. The Purple Top White Globe is an old variety with round or flat roots that are reddish purple above ground and white below. The Snowball is a fast-maturing white turnip that generally is sown in spring, while the Tokyo Cross is an all-year crop. The Japanese cultivate long, carrot-shaped turnips called Hinona Kabu.

ORIGIN & BOTANICAL FACTS

Turnips are native to Europe and central Asia, where they still grow wild on open ground or next to streams. However, they were first cultivated in the Middle East about 4,000 years ago. Turnips were consumed by the ancient Romans and by Europeans during the Middle Ages. English and French settlers brought turnips to America. Today, turnips are grown in many parts of the world, including Canada and the United States, where the leading suppliers are California and New Jersey.

Turnips are economical and easy to grow. They thrive in almost any type of soil and store well after harvest. Seeds generally are sown in early spring as soon as the soil can be worked, and additional plantings are done every 2 weeks until about 5 weeks before the temperature is expected to increase to more than 80° Fahrenheit. In late summer, when temperatures begin to cool, successive plantings can be started again until 3 months before night temperatures normally decrease to less than 20° Fahrenheit. Turnips are ready for harvest about 70 days after planting or when the roots are 2 inches in diameter.

USES

Firm, unblemished turnips that are small but heavy for their size are the best and sweetest. They can be stored in perforated plastic bags in the refrigerator crisper. Although turnips can be eaten raw, the larger ones may have a strong flavor, which can be reduced by blanching in boiling water for about 5 minutes. Although the root usually is peeled before use, fresh, young turnips can be used with the skins intact. Sliced or cubed raw turnip adds a crunchy texture and a sweet,

peppery flavor to green salads. Turnips can be boiled, baked, steamed, pickled, braised in broth, microwaved, stir-fried with other vegetables, or roasted alongside meat or poultry. Cooked turnips can be mashed and served like potatoes or cut up and included in soups, casseroles, and stews. Overcooking should be avoided, because it brings out the vegetable's strong flavor. Also, turnips should not be cooked in aluminum or iron pots because the flesh can darken.

In Great Britain, turnips are carved into jack-o'-lanterns for Halloween.

NUTRIENT COMPOSITION

Turnips are a good source of vitamin C.

NUTRIENT CONTENT

	½ cup, raw (65 g)	½ cup, cooked (78 g)
Energy (kilocalories)	18	16
Water (%)	92	94
Dietary fiber (grams)	1	2
Fat (grams)	0	0
Carbohydrate (grams)	4	4
Protein (grams)	1	1
Minerals (mg)		
Calcium	20	17
Iron	0	0
Zinc	0	0
Manganese	0	0
Potassium	124	105
Magnesium	7	6
Phosphorus	18	15
Vitamins (mg)		
Vitamin A	0 RE	0 RE
Vitamin C	14	9
Thiamin	0	0
Riboflavin	0	0
Niacin	0	0
Vitamin B$_6$	0.1	0.1
Folate	9 µg	7 µg
Vitamin E	0	0

WATER CHESTNUT

The Chinese water chestnut is not a chestnut or even a nut, but the edible tuber of an aquatic plant. The walnut-sized tuber is rounded with a pointed top and consists of a tough but papery brown skin covering crisp, white meat. Whether raw or cooked, the meat has a crunchy texture, similar to that of raw potato, and a subtle, almost sweet flavor.

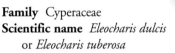

Family Cyperaceae
Scientific name *Eleocharis dulcis*
    or *Eleocharis tuberosa*
Common name water chestnut, Chinese
    water chestnut, Ma Ti,
    Ma-Tai, Chee-Chang

♥ **Raw water chestnuts are a good source of potassium and vitamin B$_6$**

VARIETIES

Numerous varieties of Chinese water chestnut exist, but only a small number are cultivated. A Chinese variety, Hon Mati, is known for its size and sweetness. A Florida variety, derived from the Chinese variety, is larger, but the flavor is more bland. Another type of edible water chestnut is the fruit of the aquatic herb *Trapa natans*, unrelated to the Chinese water chestnut. *Trapa* grows so abundantly in the waterways of the northeast that attempts are under way to eradicate it.

ORIGIN & BOTANICAL FACTS

The Chinese water chestnut is native to the Far East and grows in many parts of India, southeast Asia, New Guinea, northern Australia, and Polynesia, although it is cultivated mainly in China and Japan as a food and folk medicine. Attempts have been made to establish crops in the state of Florida, but at this time no commercial growth occurs in the United States. The plant is a sedge, similar in appearance but unrelated to grass, that grows in freshwater swamps or in shallow ponds. The rush- or reed-like leaves are bright-green hollow stems that grow to 3 feet in height. The plant, which is best grown in rich, fertile, pulverized soil covered with 6 inches of water, is planted in spring and propagates by producing spreading rhizomes throughout the summer months. One rhizome can propagate to 3 square feet. During the late autumn, the leaves yellow and chestnuts form at the ends of the rhizomes. Within 3 to 4 weeks, the leaves die back, and the corms can be harvested, although exposure to cold winter air is thought by some growers to improve the flavor. Chinese chestnuts are harvested by hand or scooped from the bottoms of ponds with forks to avoid bruising the skins.

USES

Freshwater chestnuts should be firm. They can be refrigerated in a paper bag for up to 2 weeks and should be washed and peeled just before use. To prevent discoloration from exposure to air, the peeled whole or sliced water chestnuts can be dropped into acidulated water (dilute lemon juice) if not cooked immediately. Canned water chestnuts can be drained and stored in fresh water in a sealed container up to a month if the water is changed daily. Freshwater chestnuts can be eaten raw or cooked. They can be added to stir-fries, soups, casseroles, or vegetable dishes, where they will retain their crisp texture even after heating. For use in salads and other cold dishes, they can be added raw or boiled 5 minutes, drained, and chilled.

NUTRIENT COMPOSITION

Raw water chestnuts are a good source of potassium and vitamin B$_6$.

NUTRIENT CONTENT

	½ cup, cooked (70 g)	½ cup, raw (62 g)
Energy (kilocalories)	35	60
Water (%)	87	74
Dietary fiber (grams)	2	2
Fat (grams)	0	0
Carbohydrate (grams)	9	15
Protein (grams)	1	1
Minerals (mg)		
Calcium	3	7
Iron	1	0
Zinc	0	0
Manganese	0	0
Potassium	83	362
Magnesium	4	14
Phosphorus	13	39
Vitamins (mg)		
Vitamin A	0 RE	0 RE
Vitamin C	1	2
Thiamin	0	0.1
Riboflavin	0	0.1
Niacin	0	1
Vitamin B$_6$	0.1	0.2
Folate	4 µg	10 µg
Vitamin E	0	1

YAM

The yam is a thick, starchy tuber that is similar in appearance to the sweet potato but is not related to it. Yams can range in length from a few inches to 7 1/2 feet and can weigh more than 100 pounds. The rough skin is pale tan to dark brown, whereas the flesh can range from off-white to yellow to purple or pink. Depending on the variety, the texture is moist and tender or dry and mealy. The flavor is rather bland, more similar to white potatoes than to sweet potatoes.

Family Dioscoreaceae
Scientific name *Dioscorea alata*
Common name yam

♥ **A good source of vitamin C, fiber, and potassium**

VARIETIES

Hundreds of species of yams, all of them climbing vines, are cultivated throughout the world. The most widely grown type is the *Dioscorea alata*, also called the winged yam, greater yam, or white Manila yam, whose tubers can grow to 30 to 40 pounds in weight. Also popular is the *D. batatas*, or Chinese yam, a smaller tuber weighing 5 to 10 pounds, with a flattened or fan-like shape.

ORIGIN & BOTANICAL FACTS

Yams derive their name from the Senegalese word *nyami*, which means "to eat." They are native to Africa, where they have been cultivated for 11,000 years, and to southeast Asia, where their cultivation extends back 10,000 years. Along with taro, they may have been the first plants to be cultivated. The slave trade brought yams to Central America and the Caribbean, where they became an important staple crop. Nevertheless, western Africa still produces about half of the almost 30 million metric tons of yams currently grown in the world annually. The rest are grown in Asia and Central and South America. Very small amounts are cultivated and consumed in the United States. Most of the vegetables that are called yams in the United States are actually sweet potatoes. True yams thrive in frost-free, preferably tropical or subtropical climates whose rainfall totals at least 40 inches during the 8-month growing season. The ideal soil is rich, fertile, and moist but well drained. Like potatoes, yams are easily propagated from the tubers themselves, which are cut into wedges containing two or three buds each and allowed to dry for a day before planting. Plants can be grown in pots or planted directly into hills of compost mixed with sandy soil. Because the tubers do not store well, it is best to leave them in the ground as long as possible before harvest in fall or late winter.

USES

Yams can be found in Latin American and Asian specialty markets, usually sold in chunks. Yams should be stored in a cool, dry, dark place but should not be refrigerated. Because they grow underground, they should be scrubbed well to remove any clinging soil. Yams must be cooked before eating. Like potatoes, they can be baked, boiled, fried, or microwaved. They can be substituted for sweet potatoes, cut into chunks or mashed and seasoned with apple juice, orange peel, or sweet spices. They also are a flavorful addition to soups, casseroles, and stews and can take the place of mashed pumpkin in pies, cakes, and pastries.

NUTRIENT COMPOSITION

Yams are a good source of vitamin C and dietary fiber and are high in potassium.

SERVING SIZE:
1/2 cup, baked (68 g)

NUTRIENT CONTENT

Energy (kilocalories)	79
Water (%)	70
Dietary fiber (grams)	3
Fat (grams)	0
Carbohydrate (grams)	19
Protein (grams)	1
Minerals (mg)	
Calcium	10
Iron	0
Zinc	0
Manganese	0
Potassium	456
Magnesium	13
Phosphorus	33
Vitamins (mg)	
Vitamin A	0 RE
Vitamin C	8
Thiamin	0.1
Riboflavin	0
Niacin	0
Vitamin B$_6$	0.2
Folate	11 µg
Vitamin E	0

GRAINS

Includes: Bread, Cereals, Flour, and Pasta

Cereal grains are the fruit of plants belonging to the grass family (Gramineae). Ten thousand years ago, wheat underwent spontaneous mutations causing this grass to hang onto its seed rather than scatter it to the wind. Although this change was not in the interests of the wheat from the standpoint of its own reproduction, it enabled humans to store seed for the winter. The calorie density of grains prevented starvation and so played an important role in human history worldwide. In China, rice was domesticated about 6,500 years ago, and in the New World corn was domesticated about 3,500 years ago.

Until the past century, most grains were consumed as "whole" grains. In other words, the grain kernels were intact — not stripped of their vitamins, minerals, and fiber. Whole grains provide fiber, protein, complex carbohydrates, lignans, phytates, other phytochemicals, vitamins, and minerals. Fat-soluble vitamins are found in the germ of the grain, and B vitamins and phytochemicals are found in the husk or bran. Grain and grain products are also naturally low in fat.

Whole grains can be consumed plain as hot cereals, used in pilafs, added to baked goods, and eaten in dozens of other ways. Hundreds of products are made from grain. Two of the main ones —bread and pasta — are diet staples in nearly every culture on every continent, from couscous in North Africa to soba (buckwheat noodles) in Japan. Grains and grain products literally feed the world, providing most of the calories and much of the protein consumed by the world's population.

Basics

Grains are the seeds of plants. Although the grains eaten by humans belong to a wide range of botanical families, they have the same basic structure (see the illustration on page 20, Chapter 2) and contain these components, from which plants begin to grow:

Bran — This is the outer layer of the grain seed. It's full of B vitamins, trace minerals, and, especially, fiber.

Endosperm — Sometimes referred to as the kernel, the endosperm contains the majority of the material within the seed and is meant to nourish a seedling. The endosperm is where most of the protein, carbohydrates, and small amounts of vitamins are located. It is composed mainly of starch, and often it is the only part of the grain that is eaten.

Germ — The germ is the part of the seed from which the new plant sprouts. As the embryo within the seed, it has the highest concentration of nutrients, including B vitamins, trace minerals, and some proteins. It also contains fat, which increases its perishability.

Grains are usually milled before they are used as food. Milling usually means that the bran and any husk surrounding the grain seed (along with the nutrients they contain) are removed. Then the seeds are ground in a process that converts the grain into flour or other products. The product's use and name often depend on how much of the bran is left. Wheat, for example, can be milled into whole-grain flours that contain all parts of the seed. Refined flours, which are used most often by Western nations, contain only the endosperm: the bran and germ are removed (along with much of the grain's nutrition).

Nearly any grain can be milled and made into products such as bread, cereal, or pasta—food staples worldwide—with varying degrees of success depending on the chemistry of the grain. History has taught us which grains work best and in what combinations.

Gluten, which is found in grain protein, gives bread its springy texture. It becomes stretchy and thickens when liquid is added to the flour and the combination is kneaded. The resulting gas from the fermentation of the carbohydrate in the flour is trapped by the dough, causing the gluten to stretch and, thus, the bread to rise. Because wheat and rye contain the highest quantities of gluten, flour made from these grains has proved best suited for making bread. Other grains, such as corn, have less gluten, and products made from them are more crumbly.

In most countries, the highest proportion of cultivated land is devoted to grains. Crops such as wheat, rice, corn, barley, oats, and millet remain critical components in the diets of people worldwide. Wheat is the most widely grown grain. But, because multiple crops of rice can be grown in a year in tropical areas, a nearly equal amount of rice and wheat are grown each year.

Nutrition

The mix of nutrients supplied by grains varies. But because grain seeds are composed mostly of starch, between 65 and 90 percent of the calories supplied by grains are carbohydrates. Between 8 percent and 15 percent of calories come from protein, and fat contributes the remainder of the calories. Because grains are plant-based proteins, they do not supply all of the amino acids (the building blocks of protein) that your body needs. However, grains also do not have the twin disadvantages of animal-based protein—saturated fat and cholesterol, both of which are linked to cardiovascular disease. Grains can be eaten in combination with other foods—such as legumes, small servings of meat or poultry, and dairy foods—to provide the complete balance of amino acids.

Important minerals found in whole grains are iron, phosphorus, magnesium, and zinc. Whole grains are also a source of B vitamins (niacin, thiamin, riboflavin) and antioxidants, such as vitamin E and selenium. Scientists

are just beginning to explore the health role of substances called "phytochemicals," which are also found in whole grains.

Grains are rich in dietary fiber, both insoluble fiber (which helps bowel function and may reduce the risk of some kinds of cancer) and soluble fiber (which may have a role in lowering blood cholesterol levels).

The refining and processing of grains remove many of the nutrients grains naturally provide. The bran and the nutrients it contains are often removed during the milling process. Refined wheat flour has both the bran and the germ removed. Many grain products are enriched (see Chapter 4, "Fortified" and "Enriched," page 92), meaning nutrients originally found in the grain are added back to the product during processing. But not all the nutrients are returned. Insoluble fiber is milled away when the bran is removed, and antioxidants usually are not added back into refined flours. Phytochemicals also may be missing in refined products.

The bottom line is that even though refined grains are a good source of many nutrients, whole grains are better. You can tell whether a product is made from whole grain by checking the ingredient listing of the label. Look for the words "whole grain," indicating that the product contains the endosperm, bran, and germ, and all the benefits they bring.

Selection

Grains, even those that have been processed or lightly cooked, contain some of their natural oils. Over time, they may go rancid, which is why ensuring freshness when buying them is key. Look for grains that are in sealed packages. This protects them from air, moisture, and spoilage. Some may have freshness or "best if used by" dates to help ensure quality. If you are buying in bulk, check whether the store has a rapid turnover of that particular product. There may not be a great demand for bulgur in supermarkets, for example, and so the product may have been on the shelf for some time. In contrast, specialty food markets or those specializing in natural foods may have greater demand for the product, sell more of it, and therefore have a fresher product. Grains also should smell fresh and appear clean and free of debris.

Storage

Because grains may attract insects or may become moldy if they become moist, keep them in tightly closed, moisture-proof containers. Grains can be stored at room temperature, but they will remain fresh longer if stored in the refrigerator, where they will keep for several months. Most grains can be kept much longer if stored in the freezer. They do not need to be thawed before cooking. Cooked grain may be stored in the refrigerator for several days and then reheated.

Preparation

Whole grains are hard and dry. Thus, cooking involves not only heating them but also rehydrating them. For that reason, whole grains, with few exceptions, are cooked in liquid. Here's a traditional method for doing so:

- Bring water (or other liquid, such as a stock) to a boil. Many cooks use a ratio of two to three parts water to one part grain.
- Add grains and other seasonings.
- Cover mixture and reduce to a simmer.
- Simmer until most of the liquid is absorbed.
- Remove mixture from heat and drain excess liquid if necessary. Let sit for approximately 5 minutes, then fluff with a fork.

BEYOND THE BASICS

Innovative uses of grains include the production of modified starches, caloric sweeteners, and fat substitutes. Beta-glucan, a fat-like gel made from enzyme-treated oat bran, is sold as a cholesterol-lowering fat replacement, although more testing is needed to determine its effectiveness. Several food enzymes are produced by fermentation based on grain. Wheat gluten may be used to produce flavor enhancers such as glutamate or diet supplements such as glutamine. There is also a vast range of industrial applications. None of these, however, are likely to overtake in importance the role of grains in feeding the world.

Cooking times vary depending on the kind of grain used, how it has been processed, and whether it has been pre-cooked (bulgur and kasha are often lightly cooked, then dried before they are sold in stores). Most cooks recommend cooking whole grains as you would pasta. Simply cook them until tender—a time that may range from 8 minutes for "instant" types of white rice to more than an hour for whole wheat or other unmilled grains.

Some grains that are particularly tough—such as wheat or rye—may be easier to cook if they are first soaked. Rinsing whole grains before cooking also is advised to remove debris or other residues.

The text that follows provides more detailed information about specific grains and the foods made from them. The chapter is organized into two sections:

- Grains
- Grain Products

GRAINS

Amaranth

Amaranth was one of the main food sources for the Aztecs, who also used it in religious rituals. Cultivation ended almost completely after Spanish conquistadors made growing the plant a punishable offense. Today, both farmers and anyone interested in nutrition are showing increasing interest in the plant because it has more protein (15 percent to 18 percent of calories) than most other grains (8 to 15 percent of calories). It also contains more lysine and methionine, amino acids not provided by many common grains. Combined with other grains, it can provide a complete balance of amino acids. Amaranth is also a source of calcium and magnesium and contains more iron than almost any other grain.

The amaranth plant has long clusters of red flowers and grows to a height of 1 to 3 feet. It produces tiny seeds—up to 500,000 per plant. These seeds can be cooked and eaten as a grain or popped, sprouted, or ground into flour that has a strong, nutty flavor. Amaranth flour can range from a light yellow to dark violet, although most amaranth flour sold in stores is buff-colored.

Pasta can be made from amaranth flour, and amaranth oil is obtained from the plant's seeds. The green leaves and stalk of amaranth (also called pigweed) can be cooked and eaten. The leaves have a taste similar to that of spinach.

Preparation Tips

Amaranth flour does not contain gluten, which means baked goods containing it will not rise as desired and will be crumbly. It can be used in baked goods, but it should be combined with wheat flour (which contains gluten) in recipes for muffins, bread, cookies, or pastries. Because it has a nutty, assertive flavor, you may want to experiment somewhat with how much amaranth flour to add to recipes. Amaranth's nutritional advantages, however, make adding it to baked goods worthwhile.

Amaranth seeds also can be cooked (see Preparation Tips, page 273) and eaten as a cereal. Or, they can be popped by adding them a tablespoon at a time to a hot, ungreased skillet. They take just a few minutes to pop.

Serving Suggestions

In addition to using amaranth in baked goods as described above, amaranth can be substituted for flour in pancake or waffle recipes. Cinnamon particularly complements its flavor in both of these breakfast favorites. Amaranth leaves can be substituted for spinach in salads or cooked dishes. Popped amaranth seeds can be used as a garnish or topping or in breading recipes.

Barley

Barley, an annual plant that grows to a height of 1 to 4 feet, is hardy enough to withstand various growing conditions. For this reason, throughout history it has been cultivated as a food crop and remains a staple in many nations, particularly in North Africa, the Middle East, and Asia. Today, however, much of the barley produced in Western nations is fed to animals or used in the production of beer or distilled liquor. An enzyme in malt made from barley transforms the starch in beer or liquor mashes into sugars that alcohol-producing yeast can feed on and, therefore, ferment, a process that leads to the production of alcohol.

Given the health benefits of barley, its versatility, and its pleasing, lightly nutty taste, its banishment from many nations' kitchens is something to reconsider. Barley has a tough husk surrounding the grain seed which must be removed before it is edible. Barley is a source of soluble fiber, niacin, phosphorus, magnesium, and iron.

Barley grains are usually off-white, but the color may range from black to purple. How the grain is milled determines its nutritional content. Nutrients are most concentrated near the bran. Therefore, the more milling the barley undergoes, the less nutritious it is.

Types of barley sold in stores include:

Flaked barley—As its name suggests, barley flakes are grains that have been flattened. They resemble rolled oats.

Pot barley or Scotch barley—This type of barley is coarsely ground, but it loses most of its nutrients because almost the entire husk is removed.

Pearled barley—So-named because processed barley grains are the same size as pearls and ivory-colored, pearled barley is processed multiple times to scour or polish off the outer husk and the bran. The result is a barley that cooks much more quickly than other types of barley.

Hulled barley—Hulled barley has only the outer husk removed and still contains most of its bran. This makes it one of the most nutritious types of barley available.

Barley flour—Barley flour is simply barley grains ground very fine. It is darker than refined white flour and has a delicate, nutty flavor.

Supermarkets typically do not carry a wide variety of barley products, although pearled barley and prepared barley soups are easy to find. Health food stores or specialty food markets are more likely to carry less refined barley products.

GRAIN GLOSSARY

Whole grains—The least processed grains. The outer husk is removed (hulled)

Pearled or polished grains—Grains that have had the brown bran outer coating of the kernel wholly or partially removed

Steel-cut or cracked grains—Grains that are cut into small pieces, from fine to very coarse

Flakes or rolled—Grain kernels that are sliced and then flattened between rollers

Meal—Grain that is coarsely ground to a gritty consistency

Bran or polishings—The coarsely ground or finely shredded outer husk

Germ—A coarse meal made from the sprout, or embryo, found inside the kernel

Flour—Grain ground into a powder

Preparation Tips

Less refined barley should be soaked several hours before cooking. Generally, these types of barley are cooked for about an hour over low heat in 3 or 4 cups of water for each cup of barley. Refined types of barley, such as pearled barley, do not need to be soaked and can be cooked in about a half hour. Barley flour has a low gluten content. It must be combined with higher-gluten flours (such as wheat) in baked goods or they will not rise as desired.

Serving Suggestions

Barley can be substituted for rice in many recipes or combined with beans and vegetables to provide a high-protein meal without meat. Barley readily absorbs the flavors of the liquid it is cooked in, and thus it is an excellent addition to soups and stews, where it also acts as a thickening agent. Barley also can be served on its own as a hot cereal—it is excellent topped off by plain yogurt and fruit. Barley makes an excellent base for an entrée at lunch or dinner when cooked in chicken, beef, or vegetable stock and then mixed with steamed vegetables.

Buckwheat

From a botanical standpoint, buckwheat is a fruit and is in the same plant family as rhubarb. However, it is processed, prepared, and consumed like cereal grains such as wheat, rye, and oats.

Buckwheat products have a strong, nutlike flavor and include the following:

Buckwheat flours—As the name suggests, these are flours ground from the buckwheat seed. Supreme buckwheat flour is milled from whole buckwheat. Fancy buckwheat flour is milled from hulled buckwheat seeds. Buckwheat flour is commonly used to make pancakes, but it has a variety of other uses.

Farinetta—This is simply a product made from the bran of buckwheat seeds.

Buckwheat groats—Groats are hulled buckwheat kernels that have been crushed. Groats that are roasted are known as kasha, a name given to this product in Eastern European countries, where it has been a staple for centuries.

Nutritionally, buckwheat has unique characteristics. The protein quality of buckwheat is higher than that of wheat, soy, oats, or brown rice. For this reason, buckwheat is added to other cereal flours to improve nutritional quality, and it is often an ingredient in snack foods. Recent studies have linked various phytochemicals in buckwheat with potential health benefits. Rutin, a flavonoid found in buckwheat bran, is being studied for a possible role in managing blood cholesterol levels. Fagopyritols found in buckwheat may have a favorable effect on blood glucose levels in people with type 2 diabetes. Regular consumption of buckwheat also has been shown to lower blood pressure. However, further research is necessary to confirm these benefits.

Preparation Tips

Because buckwheat flour is gluten-free, it must be mixed with flours that contain gluten, such as wheat flour, in baked goods. Otherwise, the foods will not rise as desired. To prepare kasha, add 1 cup of buckwheat groats to a heated skillet and add to it a beaten egg white. The egg separates the kernels as they cook, which prevents the groats from sticking together. This ensures that kasha will have a consistency that is similar to that of rice. Stir the kasha and egg mixture until each grain is separate and dry. Then, add 2 cups of boiling liquid—either stock or water—and a dash of salt. Simmer the mixture for 30 minutes or until the liquid is absorbed.

Serving Suggestions

Traditionally, buckwheat has been used mainly as a flour in pancake mixes. Ways to enjoy buckwheat are in the forms of soba, which is a Japanese noodle, and cooked buckwheat groats as a salad or pilaf.

Corn (Maize)

Although considered by many to be a vegetable, corn is actually one of the few

Preparation Tips

The key to serving corn on the cob is to buy the freshest possible. When the ear is plucked from the stalk, the natural sugar in it begins a gradual conversion to starch, which makes the corn less sweet and, therefore, less tasty. Look for husks around the ear of corn that are green, plump, tightly wrapped, and free of any obvious insect infestation. Before buying corn, peel back the husk slightly to check for plump, pale, and moist-looking kernels. At home, the green husk and silk are usually removed before cooking. Traditionally, corn on the cob is cooked by placing ears in a pot of boiling water for 4 to 7 minutes or in a vegetable steamer for 4 to 6 minutes. Corn on the cob also can be cooked in a microwave oven. To do so, wrap each husked ear in waxed paper and place on a paper towel. Cook on the highest power setting for 3 to 5 minutes for one ear, 5 to 7 minutes for two ears, and 9 to 12 minutes for four ears. Corn on the cob can be roasted in its husk on the grill or in the oven. (The silk must first be removed, however, and the husk replaced after this is done.) Before roasting, soak the ear in water for about 5 minutes. Then place the corn on the grill or in the oven. Cooking times vary but range from 10 to 15 minutes on a hot grill or 20 to 30 minutes in an oven set at 350° Fahrenheit.

Avoid corn that is sold in displays exposed to direct sunlight or high temperatures because heat speeds up the process of converting sugar to starch.

Serving Suggestions

Instead of flavoring corn with butter or salt, try other seasonings. Pepper, herbs, or lemon juice complement corn's flavor without adding unnecessary sodium, fat, or calories. Corn also mixes well with other

grains native to the Western Hemisphere, where it has been cultivated for centuries. In Europe, the word "corn" is the common term used to describe many cereal grains. However, after coming to the New World, Europeans began to use the word "maize" to refer to corn itself. The word "maize" is derived from the American Indian word "mahiz." To this day, Europeans call corn "maize," and Americans call it "corn."

Corn plants grow to a height of 6 to 10 feet. The tall plants, with their long, drooping leaves, are a common sight throughout the U.S. Midwest, where most of the world's supply is grown. The plant produces ears of corn that measure 6 to 12 inches, and each ear has numerous long, slender threads called silk. Corn kernels can be white, orange, red, purple, blue, black, or brown, according to the variety. Most of the corn grown today is a golden yellow.

Corn has a wide range of uses and, although less nutritious than many grains, it is extremely versatile and still a good food choice. Nutritionally speaking for humans, corn provides a good source of fiber, phosphorus, vitamin C, and thiamin. Numerous hybrid varieties of corn have become available in the past decades.

Essentially, the advances in breeding have made corn sweeter by converting its starch to sugar.

Corn can be eaten in several ways. It can be served fresh and still on the cob. Canning or freezing can preserve fresh kernels. Dried kernels can be roasted or popped.

Corn can be ground into coarse meal or flour that is made into cornbread, tortillas, pancakes, or waffles. Oil derived from corn is used widely for cooking. Starch derived from corn is often used as a thickening agent in gravies, soups, and other dishes. Tiny baby corn is popular in Asian dishes and is sold in cans or jars. Corn is also fermented and is the basis for bourbon and whiskey.

The ever-popular cornflake cereal was invented by the Kellogg brothers of cereal fame in 1894. According to legend, they discovered the process of making cornflakes by accident when they passed corn kernels that had been left too long in cooking water through rollers, resulting in flakes.

Although corn is the fundamental food plant of the United States, most of the crop is used for feeding animals or for manufacturing purposes. Corn grown for those purposes is often referred to as "field" corn.

vegetables. Popped corn (see sidebar: Putting the "Pop" in Popcorn, this page) makes an excellent snack as long as it is not drenched in butter and salt.

Flax

Flax is an ancient crop. Native to Eurasia, its first recorded use was in Babylon about 3000 B.C. There, it was cultivated for food, and its seeds were usually ground into flour or meal. The plant also was used to make fabric for clothing. (These days, linen comes from flax.) Hippocrates, the ancient Greek physician, wrote of using flaxseed for the relief of abdominal pain. The greatest of all medieval kings, Charlemagne, considered flax so healthful that he passed laws requiring its consumption.

Nutrition researchers have identified several substances in flaxseed that appear to have health benefits: lignans, fiber, and omega-3 fatty acids. Lignans are phyto-estrogens that are thought to bind to estrogen receptors in the body. Phytoestrogens may have a role in preventing hormonally related cancers of the breast, endometrium (lining of the uterus), and prostate. Populations with higher intakes of phytoestrogens appear to have a lower incidence of and mortality from these cancers. Although lignans are found in most unrefined grains, soybeans, and some vegetables (broccoli, carrots, cauliflower, and spinach), flaxseed is the richest source of lignans. Flaxseed also contains both soluble and insoluble fiber (about 3.3 grams of total fiber in 1 tablespoon of flaxseed). About one-third of the fiber is soluble. Studies have found that the soluble fiber in flaxseed, like that found in oat bran and fruit pectin, can help lower cholesterol levels. Soluble fiber also has been found to help regulate blood sugar levels. The remaining two-thirds of the fiber in flaxseed is insoluble, which aids in digestion and waste elimination.

Flaxseed is rich in alpha-linolenic acid, which is both an essential fatty acid and an omega-3 fatty acid. Researchers are interested in omega-3 fatty acids for their roles in proper infant growth and development, in reducing risk factors for heart disease and stroke (regulation of cholesterol, triglycerides, blood pressure, blood clotting), and in immune and inflammatory disorders.

Ground flaxseed is usually available in most large supermarkets and in specialty markets. The small, reddish brown whole seeds have a nutty taste. Look for flaxseed or flaxseed meal in tight packaging that does not allow light to pass through. Protecting flaxseed from light helps keep the product fresh and preserves the omega-3 fatty acids and polyunsaturated fats.

Preparation Tips

Incorporating flaxseed into a diet is simple and can add a tasty twist to routine foods and dishes. Whole (or ground) flaxseed can replace some of the flour in bread, muffin, pancake, and cookie recipes. Because of its high fat content, it also can be used to replace part or all of the fat in baked goods recipes. (One cup of flaxseed may replace 1/3 cup fat.) Time in the oven should be adjusted to allow for more rapid browning when flaxseed is used in baked goods. Flaxseed oil also is readily available and may be substituted for other oils.

Serving Suggestions

Whole flaxseeds have a nutty taste and can be sprinkled over salads, soups, yogurt, or cereals. Flaxseed meal particularly complements the flavor of bran muffins.

PUTTING THE "POP" IN POPCORN

Popcorn is a special hard variety of dried corn that pops open and puffs when it is heated. The kernel has enough internal moisture to become steam, and the kernel explodes because the steam has nowhere to go. Different types of popcorn can be different colors. Once popped, though, they're all white or yellow.

Popcorn is an excellent snack food. It is high in complex carbohydrates, a source of some fiber, and, depending on how it is served, low in fat and calories. When purchasing microwave popcorn, check labels for saturated fat content. Try to buy reduced-fat or "lite" varieties. Try popping your own and seasoning it with a pinch of salt instead of adding unnecessary fat and calories by drizzling butter over the popcorn. Alternative seasonings include onion or garlic powders or reduced-fat grated cheese.

DID YOU EVER ASK YOURSELF...

What Is Hominy?

Hominy is the starchy endosperm of maize (corn) kernels. It can be thought of as the "naked" kernels that remain after the tough hull (pericarp) and oily germ have been removed.

The hull is removed by soaking corn kernels in water mixed with lime, lye, or wood ashes. This process not only loosens the hulls but also unbinds the vitamin niacin and makes it absorbable in the digestive tract. In its dry form, hominy is sold either cracked (samp) or ground (grits). Hominy also is sold canned and ready to eat.

In Mexico, a form of corn similar to hominy is used to make tortillas. Annual consumption of corn in Mexico is about 400 pounds per person (about 1 pound per day), and it provides up to 70 percent of a person's daily calorie intake.

In the United States, the most common form of hominy is called grits. Grits are cooked with water or milk until thick and mushy. Often eaten as porridge or a side dish or in a casserole, grits are served hot or chilled and sometimes cut into squares and fried.

Hominy is a good source of complex carbohydrates and soluble fiber.

Millet

Millet is the oldest of grains. And although it is often used in the United States as bird feed, millet is one of the main food sources for many developing nations.

There are many different species of millet, which is tiny, has an oblong-spherical shape, and ranges from pale yellow to reddish orange. Unlike most grains, which form ears, most varieties of millet form panicles, or berry-shaped heads. Millet berries are small and range from white, gray, or yellow to red or reddish brown. Common millet is grown worldwide and is used mainly for human consumption and animal feed. Foxtail millet is also grown in a variety of areas and is often used for birdseed or, in Russia, to make beer. Pearl millet is grown primarily in India.

Varieties that are more familiar to North Americans include sorghum and teff. Sorghum is widely grown in the American South and included there in regional cuisine. In the United States, sorghum molasses is used as syrup at the table and in baked goods. Teff was grown almost exclusively in Ethiopia until the past decade, when it was introduced in Western markets. Often thought of as a "famine food," teff is also grown and distributed by humanitarian agencies to relieve world hunger.

Millet has a strong, nutty flavor that may take some time to appreciate. Finely ground millet is used by Ethiopians to make fermented, spongy flat bread, by Indians to make crepe-like roti, and by the Masai in Africa to make beer. Nutritionally, millet is a good source of niacin, thiamin, phosphorus, and zinc and provides a fair amount of iron. It also is easy to digest.

Preparation Tips

Millet's preparation is similar to that of rice—it is boiled in water. Ground millet is used as flour to make puddings, breads, and cakes. Because millet produces no gluten, it cannot be used on its own to make raised breads.

Serving Suggestions

Millet can be used as a substitute for many other grains. It can be served as a hot cereal and in dishes such as pilaf. Like barley, millet can be added to soups and stews.

Oats

Oats grow best in cool, moist climates and thrive in poor soils. Given these advantages, it is understandable why oats have been a food source for both humans and animals for centuries. There are several hundred varieties of oats, which are divided into two classes: winter and summer oats. Whole oat grains are usually "hairy" and can vary from white and yellow to gray, red, or black. The grains, which are small and shaped like a thin, elongated football, are cleaned, dried, and roasted, and the hull (the tough outer covering) is removed. The bran and the germ, however, are left intact, which means that they keep most of their original nutrients. Oats are especially rich in the soluble fiber beta-glucan, which may play a role in reducing blood cholesterol, blood pressure, and blood sugar levels. Oats are also a source of antioxidants.

Different types of processed oats and oat products include the following:

Oat groats—whole oats that have been hulled and roasted. Groats take about 30 to 40 minutes to cook.

Steel-cut oats—Whole oats that have been roasted and then cut into bits. This reduces cooking time to about 15 minutes. They are sometimes known as Scottish or Irish oats or pinhead oats.

Old-fashioned rolled oats—These are oat kernels that are steamed and then flattened into flakes to allow them to be cooked more quickly.

Quick-cooking oats—These are flattened oats cut more finely to reduce cooking time.

Instant oatmeal—Cooked merely by adding boiling water, these oats of convenience are pre-cooked in the manufacturing process and rolled very thin to make cooking even quicker. A trade-off may be that these oats are less flavorful. In addition, flavorings, salt, and sugar are often added to instant oatmeal.

Oat flour—Flour made from oats is also available and can be combined with wheat flour in breads and other leavened foods.

Oat bran—This is a fine meal made from the outer layers of the grain. It can be purchased and used separately, but it may be a part of other oat products.

Preparation Tips

Old-fashioned oats and quick-cooking oats usually can be interchanged in recipes. Instant oats, however, are not interchangeable because the additional processing they undergo softens the oats so much that, when combined with liquid, they can make baked goods mushy or gooey. Most types of processed oats are cooked by simmering them in water (steel-cut oats require two parts water for one part oats; for all other types of oats, use one part water for one part oats) until they are softened and cooked. Oat flour contains minimal gluten and must be mixed with other types of flour so that baked goods will rise as desired. Using oat flour results in baked goods that are often more dense than similar products made with other flours.

Serving Suggestions

All types of processed oats make a terrific hot cereal. Oats are also one of the main grains in granola (and many snack foods). Unfortunately, granola and many snack foods are high in fat, particularly saturated fats, the type linked to cardiovascular disease. This can offset the health benefits of oats. Oats can be added to muffins, cookies, and bread for a different texture. Some types of European soup and stew recipes call for oats as thickening agents. Steel-cut oats can be added to some scone recipes for additional crunch and a nutty flavor.

Quinoa

It is hailed as the super grain of the future, but the accolades are a little misleading. Quinoa (a name supposedly derived from the Spanish word for "fantastic") is not really a grain. It is the fruit of a plant that belongs to the same botanical family as beets. The quinoa plant reaches a height of 3 to 10 feet and produces flat, pointed seeds that range from buff to russet to black.

So why all the praise for quinoa? Quinoa is relatively easy to cultivate and withstands poor soil conditions and altitude. It also packs a nutritional punch in its tiny seeds. It contains more protein than most grains and offers a more evenly balanced array of amino acids, the building blocks of protein. It is higher in minerals, such as calcium, phosphorus, magnesium, potassium, copper, zinc, and iron, than many grains.

Quinoa seeds can be cooked or ground into flour. Several types of pasta are made from quinoa flour. The leaves of the plant also are edible, and the seeds can be sprouted and eaten.

Preparation Tips

Quinoa is cooked in the same way as rice, although it cooks in about half the time. Its flavor is delicate, and some describe it as hazelnut-like. Before cooking, it is important to rinse quinoa seeds until the water runs clear. They are covered with a bitter, powdery resin that can result in an unpleasant taste if it is not removed. Quinoa flour has a low gluten content. It cannot be used alone in baked goods because they will not rise properly.

Serving Suggestions

Quinoa is cooked like rice and makes an excellent substitute for it. "Toasting" the quinoa grains in a hot skillet before boiling gives it a roasted flavor. Adding cooked vegetables and fresh herbs also complements its delicate flavor. Quinoa flour can be used in many baked goods. Quinoa also makes an excellent hot cereal and can be added to soups and stews. Quinoa pasta is cooked and used like traditional types of pasta.

Rice

Most typically viewed as a side dish in Western nations, rice is the main entrée when the rest of the world sits down to eat, providing up to half the calories in a typical daily diet in many Asian countries. Rice is also a staple in Africa.

Most of the world's rice is grown in Asia. The plant, which grows from 8 to 12 feet in height, can withstand a wide range of climate conditions, but it grows best in hot, humid areas. Branching stems from the plant produce flowers, which form the rice grains when fertilized. There are more than 8,000 varieties of rice.

Rice is commercially classified by its grain size:

Short-grain (round-grain) rice—Short-grain rice is round or oval and less than 1/5 inch long. It has a higher starch content that results in this rice being sticky after cooking. This kind of rice is also called pearl or glutenous rice (although there is no gluten in it).

Medium-grain rice—Medium-grain rice is up to 1/4 inch in length. It remains firm and light when cooked and retains more moisture than long-grain rice.

Long-grain rice—This rice is more than 1/4 inch in length. The grains are much longer than they are wide. They tend to remain separate when cooked and are drier than shorter-grain rice. Most rice grown in the United States is long-grain rice.

Each type of rice comes in both brown and white forms. Brown rice has only the tough, fibrous hull removed from the rice grain during processing. Because of this, it retains most of its nutrients, which include fiber (bran), potassium, phosphorus, and trace minerals. It also is chewier, has a stronger flavor, and takes longer to cook. Quick-cooking forms of brown rice, which

have been partially cooked and then dried, are available.

White rice is the most popular form of rice. One reason is that the milling process, which removes the husk, bran, and germ, makes the rice cook more rapidly and extends its shelf life. Unfortunately, it's also the least nutritious form of rice. In Western nations, rice is enriched, meaning nutrients such as iron, niacin, riboflavin, and thiamin are returned to the rice before it is sold to consumers. Parboiled rice has been processed to preserve some of the nutrients during milling.

Instant rice is white rice that has been milled, cooked, and then dehydrated. It takes about 5 minutes to cook, and its nutritional content is generally equivalent to that of white rice because most instant rice is enriched.

Specialty types of rice include arborio rice, which is round white rice used in

The delicate flavor of quinoa mixes well with cooked vegetables and fresh herbs.

Extra flavor can easily be added to rice as it cooks. One way is to take the desired spice or herb, mix it with a small amount of cooking oil (about a teaspoon or less), and add it to the water just before adding the rice. Spices and herbs also can be added to rice after it is cooked. Spices and herbs that pair well with rice include cumin, caraway, basil, cilantro, mint, and parsley.

Italian dishes, and aromatic rices. Generally, aromatic rices—sometimes referred to as perfumed rices—are long-grain types of rice and have a distinct flavor typically compared to that of popcorn or nuts. Basmati rice has a nut-like aroma and a rich, buttery flavor. It is widely used in Indian and Pakistani cuisine. Jasmine rice is cultivated primarily in Southeast Asia and has a soft texture. Glutenous, or sweet, rice has a high starch content and is used in Asian cooking to thicken sauces and make dumplings and for some types of desserts. Rice also can be milled into fine, powdery flour, which can be used in baked goods.

Store rice in an airtight container. Brown rice, which naturally contains oil, keeps for about 6 months, but it is subject to rancidity. Its shelf life can be extended, however, by storing it in the refrigerator in an airtight container. White rice can keep for up to a year. Cooked rice can be stored in the refrigerator for several days, but it will last for 6 to 8 months when kept in the freezer.

Preparation Tips

Preparation varies according to the type of rice used. In general, however, rice is cooked by adding it to liquid (two parts water to one part rice) and then simmering it until the liquid is absorbed. Both brown rice and parboiled rice may require longer cooking times; brown rice may need to simmer for up to 40 minutes, for example.

Cooks differ about whether rice should be rinsed before cooking. Some believe that rinsing rice prevents stickiness. However, rinsing domestic rice can wash off nutrients added during processing. A quick tip: to prevent stickiness, don't stir the rice while you are cooking it. Cooking rice in a vegetable, beef, or chicken stock is a low-fat, low-calorie way to add flavor to rice before serving. Rice flour is gluten-free. It must be combined with higher-gluten flours (such as wheat) in baked goods or they will not rise as desired.

Serving Suggestions

Few grains are as versatile or well loved as rice. It can be served as a pilaf, which can be jazzed up by adding cut-up vegetables and seasonings. Rice can be added to soups, stuffings, and salads. In particular, it makes an excellent replacement for potatoes. Rice also serves as the base for vegetarian dishes. Used as the foundation for stir-fry dishes, it offers the perfect opportunity to make a grain the centerpiece of a meal, instead of having the usual—and less healthful—focus on meat, poultry, or seafood. Rice flour can be used to thicken sauces.

A note about prepared rice mixes: although popular, they're often more expensive than plain rice and typically contain a significant amount of sodium. Rice is easy to cook on its own. Try adding your favorite herbs and seasonings; you can choose how much (if any) salt you use. It will likely taste just as good as or better than the boxed rice dishes, and it will almost always be healthier.

Varieties of rice (clockwise from the top): jasmine, basmati, arborio, white, and brown (center).

Rye

Rye belongs to the same botanical family as wheat and barley. The plant is indigenous to Europe and Asia. The plant itself is often bushy at the base and stands 5 to 8 feet high. Rye grows well in areas where the soil is too poor and the climate too cool for wheat.

Rye grains have the same elongated shape as wheat grains. The color ranges from buff to gray. The husks are removed during milling, but usually much of the germ and bran remain, which enhances the nutritional value of rye products. The grains may be used whole, cracked, or rolled like oats, but they are generally ground into flour.

Rye flour comes in several different varieties. Light rye flour has most of the bran removed. Dark rye flour retains most of the bran and germ and is a source of magnesium, trace minerals, folic acid, thiamin, and niacin.

Dark rye flour is traditionally used to make the dark, strongly flavored German bread called pumpernickel. Rye also is used in many alcoholic beverages, including whiskey and some types of vodka.

Like most grain products, rye should be stored in airtight containers to retard spoilage.

Preparation Tips

Whole rye grains are often cooked in the same way as rice, that is, simmered in water until they are tender. However, rye grains should be pre-soaked in water several hours to speed up cooking. Rye flour has little gluten, and therefore breads in which it is the main flour are denser because they do not rise as well. However, rye contains several long-chained 5-carbon sugars (pentosans), which have a high water-binding capacity. This trait helps rye bread retain moisture better than wheat bread.

Serving Suggestions

Use dark rye flour to make your own pumpernickel bread. Also, substitute dark rye flour in place of some of the other flour in baked goods to add a nutty flavor.

Wheat

Wheat is among the oldest of grain crops. Major wheat-producing areas include the United States, whose Great Plains are considered the "breadbasket of the world," Canada, China, western Europe, Ukraine, Kazakhstan, Russia, India, Pakistan, and Australia.

Wheat is an annual plant that grows between 2 and 4 feet high, depending on the variety of wheat grown and growing conditions. The grains are contained in a bearded spike developed by the plant. Unlike many other grains, it can be difficult to cultivate. Nevertheless, it is easily one of the most common grains grown around the world. Only rice rivals it in production.

Most of the wheat grown in the world is eaten by humans—often in the form of bread. Wheat's high gluten content makes it particularly well adapted for this use. Because of this, wheat flour is the frame-

THE RYE AND ERGOT CONNECTION

The poor soil and moist climate in which rye grows are also favorable conditions for the growth of a fungus called ergot, which can grow on moisture-laden rye kernels. Ergot has long been known to have medicinal properties, and today it is used in medications given during childbirth to promote contractions of the uterus and control bleeding. It also is used in some migraine headache medications. Ingesting too much of it, however, may lead to hallucinogenic effects. A derivative of ergot is lysergic acid diethylamide (LSD). Modern milling processes clean rye grains and remove any that may be contaminated with ergot.

Traditional methods for growing rice include the manual tasks of sowing the seeds, transplanting, and harvesting by hand.

work for almost all baked goods and pasta. Wheat also is used in the manufacture of beer and whiskey.

The three major types of wheat are:

Hard—Hard wheat is high in protein (10 to 14 percent). It is also high in gluten content, which gives the flour elasticity and makes it particularly suitable for yeast breads.

Soft—The low-protein (6 to 10 percent) flour is lower in gluten. It's often used to make "softer" baked goods, such as cakes.

Durum—This is the hardest wheat grown and is highest in gluten. Despite that, it is not used in baked goods. Instead, it is used to make semolina (see sidebar: What Is Semolina? page 287), the main ingredient of pasta.

Wheat also is classified according to the time of year it is planted. Spring wheat, as its name suggests, is sown in the spring. Winter wheat is grown in areas with more moderate winters and is sown in the late fall.

Literally thousands of products are made from wheat—flour, bread, and pasta are just the fundamental items. Whole-wheat products, however, are the most nutritious because they have not been milled as extensively as more refined products. Whole-wheat products include the following:

Cracked wheat—As the name implies, this is wheat seed that has been broken into small pieces. It can be finely or coarsely cracked.

Bulgur—A popular ingredient in many Middle East dishes, bulgur is a type of cracked wheat that has been steam-cooked and dried. Because of this, it does not require as much cooking time as other whole-wheat products. Bulgur is used for making tabbouleh (a well-known Middle East cuisine favorite), cereal, and pilaf. It is available in a variety of grinds, from fine to coarse.

WHAT IS TRITICALE?

Triticale is a 20th-century hybrid that is a cross between wheat and rye. It combines the nutritional benefits of both: the high protein content of wheat and the high lysine content of rye. Several varieties are under cultivation. Researchers are studying ways to improve the yield and adaptability of this unique crop.

Farina—A breakfast favorite, farina is perhaps better known as Cream of Wheat. It is made from the endosperm of the wheat seed, which is then milled very finely. Farina can be used in dumplings, main dishes, or desserts, particularly in Indian and Greek cuisine.

Wheat flakes—Also known as rolled wheat, these are wheat seeds that have been flattened. They look like rolled oats, although they are slightly larger and thicker. Although it might be easy to picture these as the main component of many cold breakfast cereals, wheat flakes are not the product used in these cereals.

Groats or wheat berries—These are other names for wheat seeds sold whole. Because they have undergone very little milling, these are among the most nutritious of wheat products.

Preparation Tips

If buying whole-wheat kernels in bulk, wash the product before use to remove debris. Those that are packaged typically do not need to be rinsed.

The various forms of whole wheat that are used in hot cereals or served as side dishes are generally cooked by adding water to them and simmering until the water is gone. Cooking time varies, although a general rule is that cracked wheat requires less cooking time than whole wheat. Most whole-wheat products have cooking instructions on their labels. Because the wheat germ contains fat, whole-wheat grain and products made from it should be refrigerated to prevent spoiling.

Serving Suggestions

Whole-wheat products can be served on their own as a hot cereal or pilaf. They also can be added to other dishes for extra nutrition and taste. Cracked wheat, for example,

is added to bread for extra crunch and nutrition. Wheat flakes make excellent additions to hot cereals or baked goods. Wheat groats have a strong, nut-like flavor. They are used in bread doughs or soups or are served on their own as a pilaf, for example. Wheat groats also can be eaten sprouted.

ANATOMY OF A WHEAT SEED

The wheat seed, sometimes referred to as a wheat berry, is comprised of three parts:

Wheat germ—This is the sprout, or embryo, found inside the wheat seed. It is oily and is a highly concentrated source of nutrients. Products made from wheat germ (an excellent source of vitamin E) include wheat germ oil and toasted wheat germ.

Wheat bran—The wheat bran is the tough outer covering of the wheat seed. The bran is typically removed during milling. This processing is unfortunate because the bran is an excellent source of insoluble fiber. Cooks often use it to add taste and nutrition to meat loaf, casseroles, and baked goods.

Endosperm—The endosperm constitutes the majority of the material within the wheat seed and is meant to nourish a seedling. Most of the protein and carbohydrates and small amounts of vitamins are in the endosperm. It is composed mainly of starch and is the main ingredient in wheat flour and other wheat products.

GRAIN PRODUCTS

Bread

Bread is such a fundamental food that the word "bread" itself is often equivalent to "food" or "money" in many parts of the world. Although it is a simple food, bread requires the conversion of grain into flour, leavening ingredients, and a means of baking. Bread also plays a role in many customary rituals, such as the breaking and blessing of bread in religious rites.

Although there are hundreds of different types of bread, the main types are leavened (meaning raised) and unleavened breads. There are also quick breads, in which baking powder or baking soda is used as a leavening agent.

The main ingredients in most breads are the following:

Flour—The powdery material from ground grain, flour is the main ingredient in bread. Because of its high gluten content, wheat flour lends itself best to bread making. The gluten, when mixed with water, gives the bread dough elasticity. This allows the dough to expand when the yeast ferments, yet it is strong enough to contain it. The result is light and airy bread. Any grain can be used to make bread. In countries where wheat is less readily available, grains that are used include millet, barley, rye, and oats.

Liquid—Water is the most common liquid in bread making, but beer, milk, and fruit juice also can be used. Liquid is needed in raised bread to allow the gluten in flour to do its work. The type of liquid used can result in the bread having different properties. Water, for example, will result in a thick crust.

Yeast—Yeast is a one-celled organism that is used to leaven bread. Unleavened breads and quick breads do not contain

Wild Rice

Not really a rice, although closely related to it genetically, wild rice is actually an aquatic grain. It is the seed of a marsh grass that grows in the northern Great Lakes area of the United States. It is known for its hazelnut-like flavor, dark-brown color, and chewy texture. American Indians, who once waged wars over areas where wild rice grew abundantly, have harvested it for centuries. These indigenous peoples called wild rice "mahnomen," meaning "precious gift from the gods."

First domesticated successfully in the 1950s after many attempts to mimic the moist, murky conditions where the wild rice-producing grass grows best, wild rice is now grown commercially in the Upper Midwest and in California. The crop is now worth more than $20 million annually. Most wild rice is eaten whole, but it also can be ground into flour.

Preparation Tips

Wash wild rice before cooking. The basic preparation recipe for wild rice is to place 1 cup of wild rice in a saucepan with 4 cups of water (chicken or beef stock can be added for flavor). Bring the mixture to a boil and then reduce the heat. Simmer the rice covered for 40 to 50 minutes until it is tender and most of the grains have split evenly.

Serving Suggestions

Wild rice is mixed into a multitude of dishes, from everyday cooking to gourmet creations. Use cooked wild rice in place of pasta products in casseroles or salads. It can be used in place of bulgur in tabbouleh salads or as the base for stir-fry dishes instead of white rice. Mixed with vegetables or small amounts of meat, poultry, or fish, wild rice provides the base for an excellent entrée.

yeast. When yeast ferments the substances naturally present in flour, it produces a gas called carbon dioxide. Bread rises as the gluten in the dough traps this gas. Yeast is also responsible for bread's delicious aroma and gives it its flavor.

Salt—Bread can be made without this staple, but salt does several things when it is added to dough. It adds flavor, helps strengthen the gluten, and helps regulate yeast production.

Optional ingredients—Two ingredients that do not have to be added to bread but often are include sugar and fat. Sugar provides a ready food source for the multiplying yeast, adds flavor to bread, and helps it stay moist. Fat is often used in commercial bread making. It adds flavor and tenderness. In addition, it gives the dough more elastic qualities, allowing it to expand more.

The most common type of bread eaten in the United States is made from refined white flour. Although enriched during processing and baking, not all of the nutrients lost when the flour is refined are returned to it. A more nutritious choice is whole-wheat bread. Whole-wheat bread is made from flour ground from whole-wheat grains—meaning the bran and the germ also are used. Make sure the label indicates that only whole-wheat flour was used. Otherwise, whole wheat or cracked wheat may have been added to white flour.

Common types of breads are as follows:

Bagels—Once only an ethnic delicacy, bagels have gone mainstream and are now enjoyed by just about everyone as a breakfast main course, the foundation for a sandwich, or a nutritious snack. These donut-shaped rolls are made from flour, yeast, and salt. Tradition calls for them to be boiled before they're baked—a process that gives them a characteristic shiny appearance and chewy texture. Bagels can be flavored in many ways, although favorites include onions or raisins. A word of caution: generally a nutritious food choice, bagels can be high in calories depending on their size, ingredients, and choice of topping. Check the label or ask the deli for nutrition information so you know what you are getting.

Flat breads—Named for their shape, flat breads are rolled out and allowed to rise only minimally. They are baked only until they are soft. Pita bread (sometimes referred to as pocket bread) is a common type of flat bread. Other types include crackers and tortillas, which are commonly made from corn.

French bread—Sometimes referred to as baguettes because of the traditional elongated shape of the loaf, French bread has a thick, shiny crust and a chewy texture. Coating the dough with egg whites before baking gives the crust its characteristic properties. The bread is traditionally made without preservatives.

Pumpernickel—Rye flour gives this bread its hearty flavor and its dense, chewy texture. Rye flour does not contain as much gluten as wheat flour and, therefore, the bread does not rise as much as bread made with wheat flour. Caramel or molasses gives it its dark color.

Rye bread—Rye bread also is made with rye flour, but the flour is usually mixed with wheat flour during preparation. Look at labels to determine how much rye flour a bread contains. As little as 3 percent is all that is needed to call it rye bread.

Sourdough—Bread aficionados know this bread for its characteristic tangy taste and smell. The leavening for sourdough breads is a type of bacterium, such as *Lactobacilli*, that produces carbon dioxide (to leaven the bread) and lactic acid (for a sour taste). Sometimes both bacteria and yeast are in the bread starter, in which case two types of fermentation occur. This is the type of starter used for San Francisco sourdough breads.

Always buy the freshest bread possible. Look at expiration dates on packages, and check to ensure there is no mold growth on the bread. Check labels to find out how much fat or salt has been added. If purchasing whole-grain or so-called multi-grain bread, check labels to find out how much

whole-grain flour has been used. Often, it is mixed with more refined types of flour.

Preparation Tips

Experimenting with different flours is an excellent idea, allowing you to add both taste and nutrition to regular recipes. Keep in mind the gluten content of the flour you are working with, however. Otherwise, the result may not be as desired. It may take a few tries to find out how much rye flour, or wheat bran, you should add. Loaves of bread should be stored wrapped to keep them from drying out. Sliced bread will keep for up to a week at room temperature and for about 2 months in the freezer.

A quick note about nutrition: toasting bread can reduce the amount of some key nutrients by as much as 20 percent.

Serving Suggestions

Bread can be used as slices, cubes, or crumbs. Hollowed-out loaves of bread make simple, edible containers for soups or dips. Bread cut into small pieces was the most common thickening agent in early European cooking, particularly poultry stuffing. Bread crumbs are often used to add body to foods, such as steamed pudding, sausages, and meat loaf. Bread croutons, or "little crusts," are added to soups and salads for texture and taste. Try using different breads—such as pita breads—to give sandwiches a different twist.

Cereals

Although cereal has traditionally been a synonym for grain, most people think of it today as the food that comes packaged in a bright box and is poured into a bowl of milk in the morning for breakfast.

Hundreds of cereals are available in just about any American supermarket today. Dr. John Harvey Kellogg, inventor of a

flattened, toasted wheat flake in the late 1800s, would probably be surprised at the enthusiasm for his invention and the derivatives of it.

For several decades after Kellogg's wheat flake was introduced (followed closely by corn flakes and shredded wheat), marketers touted the cereals as the key to both health and vitality. Today, corn, wheat, oats, and even some lesser-known grains such as amaranth and quinoa are puffed, popped, baked, shredded, or processed into "Os," letters of the alphabet, and even in the shape of popular cartoon characters.

Unfortunately, cereal's reputation as a healthful food has suffered. Many cereals are a wise breakfast choice, but too many of them (particularly children's cereals) also have sugar and artificial flavorings and colors added to them. Granola-based cereals and many that purport to be rich in oat bran or fiber also may be high in fat. Checking the labels to see what you are getting is always a good idea. One ingredient to look for in particular is hydrogenated fat, which contains a type of fat called trans fatty acids. Trans fatty acids have been linked to an increased risk of cardiovascular disease.

Preparation Tips

Consider the nutrition of the other main ingredient in a bowl of cereal: milk. The type you choose can have a major effect on the nutrition of a meal based on a bowl of cereal. Choose skim milk. Skim milk reduces calories and saturated fat significantly yet provides the same nutrients as whole milk.

Whole grains for hot cereal can be purchased at specialty stores and in the natural food aisle of many supermarkets. Follow label instructions for cooking.

PALMITATE IN CEREALS

Many people, particularly those who are watching fat intake closely, wonder whether palmitate, a common ingredient in cereals, is the same as palm oil. Palm oil is high in saturated fat—the type of fat linked most closely with cardiovascular disease. Palmitate is a form of palmitic acid, one of the saturated fatty acids found in palm oil. Vegetable fats, such as soybean oil, also contain palmitic acid. However, cereal manufacturers add a very small amount of palmitate to stabilize vitamin A and maintain the nutrient's potency. You will often see the ingredient listed as "vitamin A palmitate." There is no need to worry about the tiny amounts of palmitate added to cereals.

Serving Suggestions

Try adding cold cereal to reduced-fat or nonfat yogurt for a crunchy, creamy treat. Or, top off cold cereal with sliced fruit of your own choosing. Fruit added to cereals by manufacturers is often high in sodium and sugar, and usually there is not a lot of it in the box. Add nutrition to a bowl of hot or cold cereal by sprinkling wheat bran or germ over it.

Flour

Flour is the powdery substance made from grinding grains. Flour has been used by nearly every culture in the world for making foods—usually breads—that are staples in the diet. Wheat is usually used, but flour can be ground from almost any grain and sometimes is ground from potatoes, peanuts, chickpeas, lentils, and edible roots of plants.

Flour was traditionally ground by hand or by stone, but today's flour undergoes an extensive process in which grain seeds are pulverized by steel rollers. More refined flour has the bran and the germ—and thus, the nutrition—removed from the seed. Flour is mostly composed of the seed's starchy endosperm. However, the germ and bran are returned to the flour at the end of the milling process in whole-grain flours. For other types of flour, nutrients are returned at the end of the process, although not all of them are returned. Check the label to see whether the flour has been enriched.

The characteristics of flour depend on the type of material used to make it. Because most flour is used to make bread and other baked goods, most flour is ground from wheat. The high gluten content of wheat works well in leavened bread, leading to a light and airy finished product. Most wheat flour contains a combination of flour from hard and soft wheat. Hard wheats contain more protein and gluten, and soft wheat flours make for a more delicate texture.

Types of flour include the following:

All-purpose flour—This is what is typically on the shelf at the supermarket. A blend of hard and soft wheat flour, all-purpose flour has a wide range of uses and works well in breads or pastries. Look at

CELIAC DISEASE

Imagine what it would be like if eating pizza, pasta, most breads, cookies, cakes, candy bars, canned soup, or luncheon meats or drinking a beer left you with cramps, diarrhea, anemia, and even osteoporosis. For many people with celiac disease, that is a reality.

Celiac disease, also called celiac sprue, is a hereditary disease that occurs when a protein called gluten found in wheat, barley, rye, and possibly oats generates an immune reaction in the small intestine of genetically susceptible people. As a result, tiny hair-like projections in the small intestine, called villi, shrink and sometimes disappear. The villi then are not able to absorb nutrients from food, and the result is abnormally colored, foul-smelling stools and weight loss. This malabsorption also can deprive the brain, nervous system, bones, liver, and other organs of nourishment and cause vitamin and mineral deficiencies that may lead to other medical problems.

About 1 in 500 people in the United States has celiac disease—about 500,000 Americans. Some speculate that celiac disease has affected humans since they first switched from a foraging diet of meat and nuts to a cultivated diet that included high-protein grasses such as wheat. Physicians have gained an understanding of the disease and how to treat it in only the past 50 years. Today, people with celiac disease are able to lead nearly normal, healthy lives.

A gluten-free diet—a lifelong and complete avoidance of wheat, rye, barley, and oats and any foods that contain them—is the only way to treat this disease. Following such a diet is not as easy as it seems because many processed foods and medications contain gluten.

Once gluten is removed from the diet, the digestive tract begins healing within several days. Significant healing and regrowth of the villi may take several months in young people and as long as 2 to 3 years in older persons.

Foods allowed in a gluten-free diet include fresh meats, fish, poultry, milk and unprocessed cheeses, dried beans, plain fresh or frozen fruits and vegetables, and gluten-free grains such as corn and rice.

Identifying gluten-free foods can be difficult. People with celiac disease should discuss their food selections with their physician and a registered dietitian. A dietitian also can advise how best to improve the nutritional quality of a diet. Food manufacturers can be contacted to find out whether a product contains gluten. Celiac disease support groups and Internet sites also may have information on the ingredients found in food products.

the label to see whether the flour has been bleached. Manufacturers often bleach it to whiten it. One result is that the flour may have more gluten. Unbleached flours, however, may have more flavor.

Bread flour—A specialty flour used for bread making, this flour has a higher gluten content.

Cake flour—Made exclusively of soft wheat, this very refined flour gives cakes a light, soft texture. Because it is so refined, it has a low gluten content and cannot be used to make raised breads. Pastry flour is a less refined version of cake flour.

Durum flour—Made from hard wheat, durum flour is often used in pasta because it is high in gluten.

Gluten flour—This flour undergoes a manufacturing process so that its gluten has about twice the strength of regular flour. It is useful for adding to recipes to balance flours that are low in gluten.

Self-rising flour—This flour contains salt and a leavening agent, such as baking soda. It should not be used in yeast breads. In addition, leavening agents in this flour can lose strength with age.

Whole-wheat or whole-grain flour—This is flour that has the wheat germ and bran (or the bran and germ from the grain being used) that were removed during milling added back before it is packaged for consumers. Sometimes this is called graham flour. This type of flour is higher in nutrients.

Preparation Tips

Proper storage of flour is essential because flour can spoil—sometimes quickly under the right conditions. The result is an objectionable odor and inferior baked products.

How to store flour depends on the type being used. It is best to store whole-grain flours in airtight containers in the freezer. Whole-grain flours will stay fresh for up to a year this way. Whole-grain flours include the germ of the grain, which contains polyunsaturated fat. This fat is susceptible to oxidation and rancidity the longer it is exposed to air.

Refined flours have only the starchy endosperm of the grain. Such flours can be stored at room temperature up to a year or in the freezer for up to 2 years. Airtight containers will keep refined flours tasting fresh.

Because flour can spoil, it may be wise to purchase flours in small quantities so you use them up more quickly.

Serving Suggestions

Flour is the basis for most baked goods. Although wheat flour is typically used, other types of flour can be used to boost flavor and nutrition. Flour also typically is used as a thickening agent in soups, stews, and creams.

Pasta

The origins of this popular and versatile food are lost in the mists of history; several countries (China, Japan, and Italy, just to name a few) claim credit for pasta. Nearly every country, however, has some pasta variation to claim as its own.

The term "pasta" is used broadly and generically to describe a wide variety of noodles made from dough. The word "pasta" itself is thought to be derived from the Italian word for paste.

The main ingredients in pasta dough are flour—which is usually made from durum wheat and is called semolina—and a liquid. The dough is rolled out, cut or pressed into the desired shape, and readied for sale. Pasta is sold fresh, frozen, or dried. Imported dried pasta is considered superior to American-made products, mainly because the imported pasta is made with only semolina, which does not absorb as much water and is pleasantly firm when cooked al dente (slightly firm).

Pasta also may include other ingredients. Some doughs have a little egg added. Other ingredients may include soybean and mung bean flour, vegetables (spinach, tomatoes, beets, carrots), gluten, whey, herbs, spices, and flavorings. Color can be provided by vegetable purées or food coloring.

Pasta comes in literally hundreds of shapes, sizes, thicknesses, and colors. The U.S. Food and Drug Administration (FDA), however, groups it all into two main categories:

Macaroni—This includes just about every pasta shape and size. The FDA requires that macaroni be made from durum wheat flour or semolina. Ingredients such as salt, eggs, and flavorings also may be added.

Noodles—Noodles are generally made with softer durum wheat flours than semolina and contain egg. In addition, the amount of egg they can contain is limited to 5 1/2 percent of weight or less.

The shape and choice of the pasta you choose depend on what you like and how it will be served. A general rule is that thinner pastas are best in soups and stews.

What Is Semolina?

Semolina is a yellow, granular flour that is ground from durum wheat. The word is derived from the Latin "simila," which means fine white flour. Semolina is made from the endosperm of the durum wheat seed. It has a high protein content. Although it can be used in a variety of baked goods, semolina mainly is used to make pasta.

PASTA COOKING TIMES

Pasta	Type	
	Dried (in minutes)	Fresh (in minutes)
Cannelloni	8	1/2-2
Shells	10	3
Farfalle (bowties)	7-10	2-3
Fettuccine	8-9	1 1/2-3
Lasagna	10	2-4
Linguine	5-7	1 1/2-2
Macaroni	9-10	Not available
Ravioli	Not available	7-9
Spaghetti	7-9	2-3
Tortellini	9-11	7-9

Pasta that is curved or tubular is thought to soak up creams and sauces better.

The color and crispness of dried pasta determine quality. White pasta should be slightly golden and translucent, not grayish or cloudy. Spaghetti should have the springiness of fresh twigs. A good-quality flat noodle will fracture in a jagged line when broken and not look starchy. Check fresh pasta for expiration dates.

Dried pasta should be stored airtight in a cool, dry place and can be kept almost indefinitely. Fresh pasta should have a pleasant aroma. It is highly perishable and will keep for several days in the refrigerator and for up to a month in the freezer. Cooked pasta will keep in the refrigerator for up to 5 days.

Preparation Tips

Pasta is cooked by adding it to boiling water. If desired, add a pinch of salt for flavor and a small amount of oil to the water. The oil will help prevent the pasta from becoming sticky. Then, cook the pasta until it is done. Doneness is mostly a matter of taste—how firm or soft do you like it? Many cooks use the term "al dente" in reference to pasta doneness. Al dente simply means cooking the pasta until it is firm to the bite.

Cooking time varies, however, depending on whether the pasta is fresh or dried. It also depends on whether the pasta is made from soft or hard flour. Generally, pasta made from hard wheat flour is cooked longer than pasta made from soft wheat flour. Fresh pasta cooks much faster than dried pasta.

If desired, rinse pasta with cold water after removing it from heat. Some pastas used in baked dishes—such as lasagna, manicotti, and cannelloni—do not require precooking, but they usually require a greater amount of sauce, which is absorbed by the pasta as it cooks. Pasta that is cooked for a long time loses slightly more of its water-soluble B vitamins than pasta cooked al dente.

Serving Suggestions

Pasta itself is low in calories and fat, but sauces that are heavy and fatty, as well as other additions, can negate pasta's nutritional advantages. Fortunately, healthy options abound. Supermarkets offer a wide variety of reduced-fat pasta sauces or those that are vegetable- and herb-based. Tomato-based sauces are also easy—and quick—to make from scratch. Simply use several cans of whole, peeled tomatoes, crush them, and then simmer them in a skillet until they turn "saucy." Add desired seasonings (garlic, pepper, and salt work well) and a small amount of olive oil to the cooking mixture. Top with reduced-fat cheeses. Pasta is also excellent served cold when tossed with a little oil, vinegar, garlic, and fresh herbs.

Asian Noodles

Noodles have been part of Asian cuisine for centuries and continue to play a central role in many well-loved dishes. They are served in ways most Westerners are familiar with: either chilled or hot, covered with a sauce or dressing, or added to soups or stews.

Asian noodles come in varying lengths and widths. Chinese wheat noodles are usually made from wheat, water, and salt, and eggs are sometimes added. In Japan, wheat noodles are classified by size: thin noodles are called somen, and thick noodles are called udon. Other types of noodles are the following:

Buckwheat noodles—Called soba by the Japanese, these noodles can be purchased fresh or dried.

Mung bean noodles—These noodles are transparent and sometimes referred to as cellophane noodles. They are added to various dishes, including many soups.

Rice noodles—Made from rice flour and water, rice noodles vary widely in size and texture.

Preparation Tips

Noodles should be soft and look fresh. Your best bet for fresh noodles is to buy them from an Asian specialty food store. Cooking methods vary, as do cooking times. Most noodles, however, are cooked initially by adding them to boiling water. Some noodles may need to be soaked before cooking. Follow package instructions.

Serving Suggestions

Although fried Asian noodles are added to various dishes, healthier alternatives include serving chilled buckwheat noodles with low-sodium soy sauce or adding various Asian noodles to soups and stews. Asian noodles can also be added to stir-fry dishes or mixed in salads with cooked vegetables and strips of cooked chicken or beef.

HIGH-PROTEIN FOODS

Includes: Poultry, Eggs, Meat, Fish, Shellfish, Legumes, Nuts & Seeds

Compared with other populations, Americans have always found plentiful sources of proteins from both animal and plant sources. Ever since the 1800s, fueled by a growing American cattle industry and refrigerated railroad cars, meat has been viewed as the key component to a meal. If you doubt that statement, think of the last time you were asked, "What's for dinner?" Even if the menu included a balanced meal of green salad, glazed carrots, brown rice, and a small lean steak, many people would respond, "We're having steak."

So important is the idea of meat — or protein of any kind — that it's considered a meal in and of itself.

Protein is an important nutrient in the diet. That's undisputed. The trouble is that most Americans consume too much of it, particularly animal-based proteins, which are high in saturated fat and cholesterol, both of which are linked to cardiovascular disease.

Most people need just 5 to 6 ounces a day of high-protein foods. That may sound like a lot, but it isn't. A 2- to 3-ounce serving of meat is about the size of a deck of cards. Now think of the giant portions of meat and other high-protein foods often served at restaurants. Eating two to three times as much protein as you need in a day can easily be accomplished in just one restaurant meal.

Nutritionists have long recognized that Americans generally eat too many animal-based proteins. That's one of the reasons the "meat group" makes up just a small segment of the U.S. Department of Agriculture's (USDA) Food Guide Pyramid. It's also why the Food Guide Pyramid no longer refers to proteins as "meat" alone.

Although most people associate protein with only meat, there are many other ways to add this critical nutrient to the diet. This chapter covers the traditional animal sources of protein in the diet. It also introduces you to alternative plant-based sources and provides tips on incorporating a variety of these foods into a healthful diet.

Because of the wide variety of high-protein foods, this chapter has been organized into the following sections to make it easier to learn about them:

- Poultry
- Eggs
- Meat
- Fish
- Shellfish
- Legumes
- Nuts and Seeds

POULTRY

When nomadic hunters and gatherers first became farmers, they realized the importance of raising birds. The egg could be eaten, the feathers were used for bedding and clothing, and the flesh made a fine roast. It also was economical to keep poultry; cattle needed miles of grazing land, but a chicken could peck around a yard and keep itself fed. Until mass production techniques, poultry meat was fairly expensive; new technology has made poultry more affordable through breeding and production techniques.

Today, poultry is defined as any domestic bird used as food. There are many domesticated varieties of poultry, including chicken, turkey, duck, goose, Rock Cornish hen, guinea fowl, and pheasant. Generally, all types of birds are sold fresh, frozen, or cooked. They can be purchased whole, halved, or in pieces, such as boneless breasts, strips, or medallions. Buying a whole bird is typically least expensive because additional processing adds cost.

Poultry is a versatile addition to any meal and can be prepared with just about any cooking method. A recommended serving, no matter how it is prepared, is still about 2 to 3 ounces ready to eat, without bone or skin.

Nutrition

All poultry, which is defined as any domestic bird used as food, is a nutritional star. Classified as a complete protein, poultry meat is a good source of phosphorus and zinc and an excellent source of niacin.

The fat content of poultry depends on the bird. Goose and duck are the fattiest types of poultry. (See the Appendix: Dietary Reference Intakes, page 460.) However, poultry generally contains less fat than meat from other animals. To eliminate significant fat, do not eat the skin. Cooking the meat with or without the skin makes little difference in fat content. The meat will be more moist if the skin is left on during cooking.

Selection

When buying fresh poultry, look for meat that is supple and moist and has no dry or discolored patches. Another important criterion is odor. Avoid birds that have an objectionable smell. Do not buy frozen poultry that is dried out or is covered with frost. A sure sign that a bird has been frozen, thawed, and refrozen is the presence of pinkish ice on the carcass.

Storage

Always keep poultry refrigerated at 40° Fahrenheit. Poultry can be stored in its original wrapping. If freezing, over-wrap with airtight foil or freezer bags to prevent freezer burn. Frozen poultry can be stored for up to 12 months if it is purchased whole and 6 to 9 months if it is purchased as poultry parts. Frozen poultry should always be defrosted in the refrigerator and cooked in less than 24 hours after thawing. Refrigerated poultry should be cooked within 2 to 3 days.

Safety Issues

Poultry is a particular concern when it comes to food-borne disease. The gastrointestinal systems of poultry frequently harbor harmful microorganisms. Mass slaughtering processes offer ample opportunity for these organisms to contaminate the meat. Outbreaks of *Salmonella*, a bacterium that causes vomiting and diarrhea, are often associated with eating contaminated poultry. Another microorganism found on chicken is *Campylobacter*, a bacterium that can cause severe diarrhea and stomach cramping.

At the store, avoid cross-contamination by putting poultry in plastic bags to prevent leakage onto other foods. At home, prevent raw poultry from coming into contact with other foods by using separate cutting boards and utensils during meal preparation. Always quickly disinfect any surface or utensil that has come in contact with any raw poultry. In addition, wash hands with hot water and soap, lathering for 20 seconds

or more, after handling raw poultry. In addition, disinfect any surface or utensil that came in contact with the raw bird.

Thorough cooking destroys any disease-causing organisms (pathogens) in meat. Use a meat thermometer and cook poultry until the internal temperature reaches 180° Fahrenheit. The center of the chicken should be white with no sign of pink.

The following pages provide an overview of the main types of poultry: chicken, duck, game birds, goose, and turkey.

Chicken

Thanks in large part to chicken's low price, low fat, and high versatility, the amount of chicken consumed in the United States has increased exponentially.

That wasn't always the case. Even though chicken is one of the oldest living species of animal, it was a rarity on the dinner table. King Henri IV of France stated in his coronation speech that he hoped each peasant in his realm would have "a chicken in his pot every Sunday." At one time, only the rich (and chicken farmers) could manage the proverbial Sunday chicken. Today, thanks to modern production methods, almost anyone can afford chicken. In fact, adjusted for inflation, chicken is only a third the price it was 40 years ago.

Chicken consumption also has increased because of increasing awareness of the need to reduce fat in the diet. Chicken, as long as the skin is not eaten, is generally lower in fat than most other types of meat. At least half of the fat in a chicken is in the skin.

The government grades chicken quality with USDA classifications A, B, and C, which are based on meatiness, appearance, and how intact the skin and bones are.

Grade A chickens, the highest grade, are usually found in markets. Grade B chickens are less meaty, and grade C birds are scrawnier yet. B- and C-graded chickens often are used for processed and packaged foods. The grade stamp can be found within a shield on the package wrapping, or sometimes on a tag attached to the bird's wing. Many ungraded chickens find their way to stores because grading is not mandatory.

Chickens called "broilers" are butchered at about 7 weeks of age, when they weigh between 2 and 4 pounds. The term "fryer" is often given to larger birds from this age range. "Roasting chickens" generally weigh more than 4 pounds and are slaughtered when they reach 10 weeks. "Stewing chickens" — also known as hens or boiling fowl — range in age from 10 to 18 months. They can weigh between 3 and 6 pounds. Generally, they are used for stews and soups because their meat is tougher.

Shoppers may also encounter other terms to describe chicken. A Rock Cornish hen (or game hen) is a chicken hybrid that weighs about 2 pounds when butchered. Because there is relatively little meat on the carcass, each hen is typically considered 1 serving.

Another type of chicken in stores is called free-range chicken. According to the USDA, this term means that the chicken was allowed to roam outdoors. Depending on the manufacturer, the chicken may or may not have been fed a vegetarian diet free of hormones, growth enhancers, and antibiotics. Some believe that this special treatment results in a fuller-flavored chicken. One thing certain is that it adds to the expense. Most free-range chickens are far more expensive per pound than regular chicken.

A tip for shoppers: larger chickens are a better buy because there is more flesh on the bones. With smaller chickens, you do not get as much meat and you pay for bones.

Roasted chicken

Preparation Tips

Keep chicken refrigerated until you are ready to use it, or freeze it and then thaw it in the refrigerator. Cut away any excess fat, but keep the skin on while cooking to provide flavor, then remove the skin for a healthier entrée. Chicken lends itself to a variety of cooking preparations, including baking, broiling, boiling, roasting, frying, braising, barbecuing, stir-frying, and stewing. Boneless chicken requires less cooking time. However, this type of chicken will taste more bland because the bones and the skin add that real "chickeny" flavor. Yet, boneless chicken picks up the flavors of other foods, herbs, and spices it is cooked with, such as tarragon, ginger, garlic, and vegetables.

Serving Suggestions

Chicken is extremely versatile. Because of its popularity, entire cookbooks have been written focusing on only this bird. It seems almost every ethnicity has its own way to use chicken — Indian curry chicken, Chinese stir-fry, Mexican chicken enchiladas, Spanish paella, and Italian chicken parmesan. Chicken's flavor is enhanced by almost any herb, spice, or condiment.

Frying is also a popular way to serve chicken. However, this cooking method adds extra fat and calories, detracting from the health benefits of eating chicken. If eating at a fast-food restaurant, choose grilled chicken instead of chicken that has been breaded and deep-fried.

Duck

Duck, or duckling, includes any of the 80 different species of wild or domestic birds with webbed feet. Today, several major breeds of duck are raised for their meat. These domesticated ducks may be three times larger than their wilderness-reared relatives.

Duck is very popular in Europe and China, where more duck is eaten each year than chicken. In the United States, however, duck is usually perceived as a special holiday dish, if eaten at all, because Americans annually eat less than a pound of duck per person. In contrast, Americans eat close to 50 pounds of chicken per year.

The consumption level may be lower because Americans consider ducks to be fatty and scrawny. The perception of ducks as fatty rings true, because ducks are one of the highest-fat types of poultry. The fat (mainly found within and beneath the skin) helps keep the duck buoyant. Duck is also rich in protein. It is an excellent source of riboflavin, niacin, and phosphorus and is a good source of iron, zinc, and thiamin. Duck's reputation as scrawny may be undeserved, however. They do have a large skeleton and thus a relatively high proportion of bone. But specialty breeds, such as Muscovy ducks, have increased breast size for more meat. The USDA grades duck quality with classifications A, B, and C, similar to other poultry.

The majority of ducks are really ducklings, 6 to 8 weeks old. Broilers and fryers are less than 8 weeks old. Roasters are birds that are slaughtered when they are no more than 16 weeks old. Domestic ducks can weigh between 3 and 7 1/2 pounds. Older ducks are generally larger. Almost all ducks sold in stores are frozen so that they are available year-round.

Preparation Tips

Fresh duck should have a broad, fairly plump breast; the skin should be elastic, not saggy. For frozen birds, the packaging should be tight and unbroken. Frozen duck should be thawed in the refrigerator, a process that takes from 24 to 36 hours, depending on the size of the bird. Duck should not be refrozen once it has been thawed.

Before cooking, all visible fat should be removed. Also, prick the skin with a fork to help the fat melt and drip away from the bird during roasting. Some cooks also remove the skin and underlying fat. Cook whole birds to 180° Fahrenheit, breasts to 170° Fahrenheit, and legs, thighs, and wings to 180° Fahrenheit.

Serving Suggestions

Duck can be prepared in a variety of ways, including roasting, braising, and broiling. Ducklings are best roasted in the oven on a rack so that as much fat as possible can drip away from the bird. A citrus sauce nicely complements a duck's flavor; accordingly, the dish called duck à l'orange is one of the most popular ways to eat this bird.

Allow about 1 to 1 1/2 pounds of raw bone-in duck per person. This will yield 3 to 4 ounces of fully cooked duck. (One pound of boneless raw meat, when cooked, will serve 3 people.)

Game Birds

Game birds is a broad category that includes any wild bird hunted and eaten as food. These birds include the following:

Peking duck — a Chinese delicacy in which the duck is roasted until the skin is very crisp. Small slices are served with mandarin pancakes, scallions, and hoisin sauce.

- Large birds such as wild turkey and goose
- Medium-sized birds, including pheasant, guinea fowl, and wild duck
- Small birds, such as grouse, hazel hen, lark, mud hen, partridge, pigeon, and quail

Game birds that have domestic counterparts, such as turkey and duck, can provide an overwhelming taste if you have grown used to the supermarket variety. However, these birds are usually leaner than mass-market birds and even leaner than farm-raised "wild" game. If you are fortunate to have a hunter in the family, you may have good access to game birds. For most people, however, finding game birds will be difficult, although specialty stores may be of help. Sometimes, game birds are sold frozen, and smaller birds may be sold canned.

Preparation Tips

When buying game birds, avoid those with an "off" odor. If you are preparing birds you have hunted yourself, check with a local hunting organization or gaming officer, because preparation tips differ from region to region according to weather conditions. Because wild birds can be the leanest of any poultry, they may benefit from marinating. They usually are also basted during roasting. Older birds are best cooked with slow, moist heat such as braising or used in soups or stews.

Do not overcook game birds. Test for doneness by plunging a fork into the fleshiest part of the thigh and by using a meat thermometer. For light-fleshed birds, the juices should run clear. For dark-fleshed birds, the juices should be light pink.

Serving Suggestions

Game birds can be used in place of traditional poultry in many dishes. In meals with game birds as the main course, a traditional side dish is potatoes or sauerkraut. Small birds may be stuffed with a few green olives and garlic.

Goose

Geese, those graceful, large birds with their long necks, large beaks, and signature cry, have long been a favorite target for hunters. One reason is that the birds, which often weigh about 25 pounds, provide a lot of meat. In addition, their flesh is tender and flavorful. Goose has long been a traditional holiday dish and remains a popular dinner in England and northern and central Europe.

Geese have not achieved the same popularity in the United States. This may actually be an advantage because geese are one of the fattiest types of poultry. Up to half the calories in a serving of goose are derived from fat. There are about 13 grams of fat in a skinless 3 1/2-ounce serving and up to 22 grams for a serving with the skin on.

Geese are often advertised at holidays in the United States, but they are typically available frozen all year. Popular products made from goose include smoked goose breast, goose liver pâté, and goose liver sausage.

Preparation Tips

A goose should be plump and have a good fatty layer, skin that is clean and unblemished, and pinkish or light-red flesh. A frozen bird's packaging should be tight and unbroken. The goose should be thawed in the refrigerator, a process that can take up to 2 days, depending on the size of the bird. The goose should not be refrozen once it has been thawed, a general rule with all frozen products.

Domestic geese

Goose is federally inspected for wholesomeness. Grading is similar to that of other poultry. One pound of boneless goose will serve 3 people. Each serving is about 3 to 4 ounces.

During the holidays, it may be possible to find fresh goose in stores. At other times of the year, it is usually sold after it has been frozen.

Because geese have so much fat, they are best roasted. Piercing the skin while roasting will allow fat to escape, reducing the fat content of the bird that is eaten. Monitor the bird closely, however. The large amount of fat that cooks from a goose can smoke and catch fire. Larger, older birds are tougher and should be cooked (after the skin and fat are removed) with a moist-heat method, such as braising.

Serving Suggestions

Like duck, cooked goose benefits from being served with a tart fruit sauce, which helps offset any fatty taste. Roast goose is often served surrounded by sweet potatoes. A salad featuring sliced oranges and onions complements the flavor of the goose and the sweet potatoes nicely.

Turkey

Once, turkey was for Thanksgiving and Christmas only. In fact, 90 percent of all turkeys were sold during November and December. It all started when wild turkey was served as part of a feast of thanksgiving in 1621, and the tradition of turkey on the Thanksgiving table endures.

Americans now embrace the turkey for more than just special-occasion dining. Today, turkey is sold in supermarkets in parts, just like chicken, for ease of use. Breeders also have created smaller versions of turkey. Served whole, these new birds can weigh in at 5 to 8 pounds instead of the standard sizes, which can be 20 or so pounds at holiday time.

Turkey is very similar to chicken in many regards, both nutritionally and with respect to USDA grading and storage requirements. The meat is high in protein, niacin, and vitamin B_6. It also provides a good source of phosphorus and zinc. Like chicken, turkey is a low-fat poultry choice, containing about 5 grams of fat per serving after roasting and removing the skin.

Turkeys are available in supermarkets year-round. The skin on fresh turkey should be off-white to cream-colored and the meat should be pink. Self-basting turkeys have butter or vegetable oil injected under their skin to increase their flavor and moistness, cutting down on the possibility of an overly dry dinner. Turkeys also are available smoked — as whole or breast only — or canned.

Preparation Tips

If you are buying a frozen turkey, make sure it is rock-hard and free of any cuts or tears in the protective wrapping. Then, defrost it using the same methods and precautions as recommended for chicken.

If you are buying a fresh turkey or fresh turkey parts, rinse the flesh with cold water and pat it with paper towels before cooking. For thawing, which is always done in the refrigerator, allow about 1 day for every 5 pounds of turkey.

Check to make sure you have removed the giblets, heart, or other organs, which may be packaged in the breast cavity. Keep and cook the giblets separate from the turkey. Regardless of when you cook the turkey, the giblets should be cooked or frozen within 24 hours.

Serving Suggestions

Although turkey is traditionally stuffed and roasted, it can be cooked in several ways. Parts such as breasts, legs, and cutlets can be prepared in the same way as a favorite chicken recipe. Cooked and smoked turkey also goes well in cold salads and sandwiches.

If you are preparing the traditional turkey meal, estimate 3 hours for an unstuffed turkey of 8 to 12 pounds and 4 1/2 hours for an unstuffed turkey of 18 to 24 pounds. Use a food thermometer. Turkey breasts should be cooked to 170° Fahrenheit, and drumsticks, thighs, and wings to 180° Fahrenheit. To ensure uniform cooking and safety, cook stuffing outside the bird. If it is cooked inside the turkey, the center of the stuffing must reach 165° Fahrenheit.

EGGS

In addition to being a popular breakfast food, eggs are a symbol of beginnings. Primitive humans recognized the egg as the

Roasted turkey

beginning of life, and it became a symbol of spring and fertility.

Eggs have four main parts:

Shell — As the name suggests, this is the fragile and porous outer covering. The shell is made mostly of minerals — calcium carbonate, magnesium carbonate, and calcium phosphate.

Shell membranes — These are layers of protein fibers that stick to the shell. They provide additional protection for the egg's insides, preventing mold and bacteria from getting in, for example.

Albumen — This is the white of the egg. It is almost all protein and water.

Yolk — The yellow bull's eye of the egg, the yolk is made of a substance called "vitellus." It can be a pale yellow or dark yellow. About 30 percent of the yolk is fat, and about 16 percent is protein. The remainder is made up of solids.

EGG PRODUCTS

Table-ready pasteurized liquid eggs are found in the refrigerated section of the supermarket. The white and yolk of the eggs are mixed, then pasteurized at a temperature high enough to kill any bacteria without cooking the eggs. They can be refrigerated unopened for up to 12 weeks from the pack date. They can be used like eggs already scrambled. Pasteurized eggs in their shells are also now available.

Egg substitutes are a blend of egg whites and other ingredients such as food starch, corn oil, skim-milk powder, tofu, artificial coloring, and various additives. Because they are almost all protein, egg substitutes can become rubbery if overcooked.

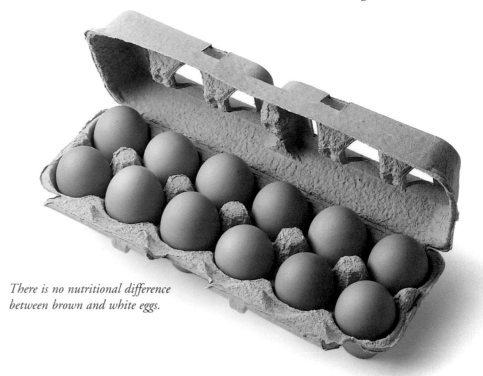

There is no nutritional difference between brown and white eggs.

Eggs provide an inexpensive source of high-quality protein, vitamins, and minerals, including vitamins A and B_{12}, folic acid, and phosphorus. They are an excellent source of riboflavin. The egg is also said to be a complete protein, because it contains a complete count of essential amino acids. The white supplies more than half the protein in an egg. The yolk supplies the fat, along with the remaining protein, and most of the calories.

Eggs are also graded and are classified by both size and quality. The best are grade AA or A, both of which are related to the level of freshness and the quality of the shell at marking. Most recipes are based on large eggs. Nutritionally, there's no difference between brown eggs and white eggs. Different colored eggs come from different varieties of hens.

The most common egg used for food today is the chicken's egg, although eggs from other fowl can be bought in specialty stores.

Preparation Tips

When selecting eggs, check the container for any cracked or broken eggs and elimi-nate them from the carton. Place the carton in the refrigerator for up to 5 weeks. Do not place eggs in the designated egg holders in the door of older refrigerators. It is too warm for the eggs there because they get a blast of hot air each time the door is opened. The egg carton helps keep eggs from absorbing odors from other foods and helps keep the eggs fresh.

When adding eggs to a mixing bowl, break the egg in a separate dish to make sure the egg is not rotten.

Serving Suggestions

Eggs serve many purposes in cooking and preparing food, including a leavening agent in baked goods, a base for mayonnaise, and a thickener in sauces and custards.

Served alone, eggs can be poached, boiled (soft or hard), fried, scrambled, or made into an omelet. From a safety standpoint, it is recommended that both the yolk and the white be cooked until firm. Because they are made mostly of water and protein, eggs are best cooked over low heat.

MEAT

Meat is the general term for any muscle from any animal. However, most people think of meat as meaning "red meat." This section describes red meats, including the common American staples of beef, pork, lamb, and veal. Also included are ostrich, rabbit, buffalo, and game meats. Rabbit, buffalo, and game meats were once common to the American menu but are now eaten much less often. The same is true of so-called variety meats — brain, heart, kidney, liver, tripe, and tongue. Whereas Americans once needed to make use of every part of a slaughtered animal, the abundance of food, along with changing tastes and attitudes, has made variety meats only rare additions to meals.

Meat is composed of three basic materials: water, protein, and fat. On average, lean muscle tissue is about 75 percent water, 18 percent protein, and 3 percent fat. The eventual texture and taste of the cooked meat depend on the amount of fat and water in the tissue and on the kinds of proteins. In general, the most tender cuts of meat have more fat and less fibrous muscle. Tougher tissue from older animals often has more flavor. Both limitations can be overcome by using the proper cooking methods.

Roughly 50 percent of the protein in meat comes from fibers that contract the muscle and 30 percent from oxygen-storing pigments called myoglobin and various enzymes. The remaining 20 percent comes from connective tissues that hold the muscles together.

Nutrition

From a nutritional point of view, because the composition of most animals is similar to that of humans, animal tissues supply us with complete proteins. More simply, the protein from animals is similar to that produced in our own bodies. Thus, it contains the full count of essential amino acids. (See Chapter 2, Dietary Protein and Body Protein, page 24.)

Red meat is an important source of minerals, including iron, phosphorus, zinc, and copper, and vitamins, including B vitamins and vitamin A (see the Appendix: Dietary Reference Intakes, page 454). However, there are some nutritional trade-offs to be made. Red meat can be the main source of fat — particularly saturated fat — and cholesterol in the American diet. Both have been linked by numerous studies to cardiovascular disease. Too much fat and cholesterol can contribute to the buildup of plaque in arteries, which in turn increases the risk of heart attack and stroke. In addition, diets high in fat have been linked to some types of cancer, particularly colon cancer.

Americans have traditionally enjoyed an abundance of meat in their diets compared with other populations, and tastes in meat continue to evolve. As more people have become aware of the risks of eating too much red meat, consumption of beef has declined. Recently, it has started to increase again and remains top on the consumption list. An important reason is that as health concerns have grown, both the meat industry and grocers have responded by introducing leaner cuts of meat and lower-fat options, such as lean hamburger. Poultry consumption has grown significantly since the 1970s.

Selection

Many factors influence how tender and tasty meat will be after cooking. These factors include:

- The type of muscle (or cut)
- The amount of fat and connective tissue in the meat
- The age of the cut
- The manner of preparation

Cuts from more exercised muscles (leg, hip, and shoulder) are tougher than those from the center of the animal — the rib, loin, or breast.

Fresher meats are generally tougher than aged meats. Aging causes meats to go

BEEF CUTS: WHERE THEY COME FROM

Section	Names of cuts
Rib:	*Rib roast, rib steak, rib eye, back ribs*
Short Loin*:	*Tip loin, T-bone, porterhouse, tenderloin*
Sirloin*:	*Top sirloin, sirloin, Boston sirloin*
Round*:	*Round steak, top round, bottom round, eye of round, tip steak, rump roast*
Flank*:	*Flank steak*
Plate:	*Skirt steak*
Chuck:	*Chuck roast, arm pot roast, shoulder pot roast, short ribs*

*Cuts that are lower in fat.

through "rigor." With time, the muscles relax and increase in acidity — a natural way to tenderize meat — and muscle proteins increase their water-holding ability, which enhances juiciness.

Aging is a process that relaxes the tissues, making the beef easier to chew and enjoy. The original process of aging, now called dry-aging, demanded that a carcass hang in a cooler for 3 weeks or more. The surface would dry out and be thrown away. The remaining beef made excellent but expensive steaks. This process is now reserved for steaks bound for only the best steakhouses.

Less fatty cuts are preferred from a nutritional standpoint. However, they are less tender than those with fat marbled through the muscle fibers. "Choice" and "select" grades have less fat and marbling and can be tougher than the "prime" grade unless they are prepared to maximize tenderness.

Storage

Meat is perishable, but there are ways to maximize its shelf life. To do so, minimize its exposure to light, oxygen, enzymes, and microorganisms. All of these can hasten the breakdown of muscle tissue and the decay process.

Keep meat securely wrapped — preferably with opaque butcher's paper, available at grocery stores — and refrigerated at less than 40° Fahrenheit in a dark place. Bacteria and molds thrive in higher temperatures and humidity. Refrigerating meat inhibits the growth of these microorganisms, and freezing meat at 0° Fahrenheit or below actually halts their growth. Meat wrapped in butcher's paper can be stored in the freezer for up to 12 months.

Safety Issues

Any kind of meat can harbor harmful microorganisms that can cause food-borne

QUICK TIP

Freezer burn (the discoloration and change in flavor that meat can acquire after it has been frozen) can be minimized by wrapping meat as tightly as possible with waterproof packaging. According to the USDA's Food Safety and Inspection Service, overwrap the supermarket wrapping with heavy-duty foil, plastic wrap, a plastic bag, or freezer paper to ensure that air is kept out.

illness. In recent years, *Escherichia coli* (*E. coli*), a potentially deadly bacterium that can cause severe, bloody diarrhea, has become a concern with red meat, particularly ground beef.

Fortunately, proper cooking kills all microorganisms in meat. Preventing meat from contaminating other foods before cooking is important. Good meat-handling practices include thawing meat in the refrigerator, making sure juices from raw meat do not get on other food, and disinfecting cooking and preparation surfaces after handling raw meat. For more information about cooking meat and other protein foods properly, see Chapter 5, Cooking It Safe, page 149, and Food Safety, page 148.

The following pages contain more specific information on different types of meat and how to include them in a healthful diet.

IRRADIATED HAMBURGER

Consumers in a growing number of states can now purchase irradiated hamburger at the supermarket. So what does "irradiated" mean, and should you buy the meat?

Irradiation is simply the name given to a process that uses radiation or electricity to kill disease-causing organisms (pathogens) in food. Food is passed briefly through either type of energy. It does not become radioactive. Instead, the energy kills most of the pathogens in food. The food still looks and tastes the same. By law, food that is irradiated must have a label that clearly states it has undergone this process. The symbol is called the radura. (See Chapter 4, What About Irradiated Foods? page 91.)

Irradiation has been used for decades to kill pathogens in spices and other foods and also to sterilize medical materials. In February 1999, the U.S. government cleared the way for irradiated red meat to be made available in stores. Because red meat, and beef in particular, can harbor several pathogens, including the potentially deadly *E. coli* bacterium, scientists and experts in food-borne disease have long advocated irradiation as an important step in making this food group safer.

Irradiation has its critics, who say that the method needs to be researched more. However, it is endorsed by nearly every major medical organization in the world, including the World Health Organization, the U.S. Centers for Disease Control and Prevention, and the American Medical Association.

Beef

Beef comes from the adult bovine, including cows (females that have had a calf), steers (males castrated when very young), heifers (females that have not been bred), and bulls younger than 2 years. Beef from an animal slaughtered after age 2 years is generally classified as "well-matured beef." Meat from these animals begins to toughen and becomes more of a purplish red. Baby beef, in contrast, is from a 7- to 10-month-old calf.

Of all the animals domesticated for food, cattle reign supreme. This animal, once only a beast of burden, became a source of meat and milk only after feed from improved agricultural practices became plentiful.

Even though cattle were first introduced to the New World in the 1500s, beef did not become popular in the United States until the Civil War, when other meats and poultry were in short supply.

After the Civil War, the abundance of grazing land and emerging transportation systems in the United States made it easy to move large amounts of beef throughout the country. For years, Americans have been one of the world's top consumers of beef. But in response to concerns about red meat's link to cardiovascular disease and cancer, consumption has declined significantly. Since 1978, beef consumption has dropped 28 percent, whereas poultry and pork consumption has grown rapidly. Beef producers have launched several programs to provide leaner beef products. The industry has developed leaner beef breeds, let the animals forage from grasslands instead of grain lots, and developed economic incentives for producing leaner cattle.

Retailers also have reduced the average thickness of fat around the edge of steaks and roasts from 3/4 inch to 1/10 inch, and sometimes no external fat is present at all. In addition, low-fat ground beef and other meats have hit the shelves. The result is that beef is 27 percent leaner than it was 20 years ago.

When eaten in moderation and prepared with low-fat cooking methods, beef remains a nutritious addition to meals.

Preparation Tips
When shopping, pick a cut of beef whose tenderness and level of flavor appropriately fit the recipe you are using. In general, the more expensive the cut of meat (rib, loin, breast, filet, sirloin), the greater the flavor and tenderness. Cheaper cuts (leg, hip, shoulder) may be less tender but are just as flavorful. They are best prepared with slow, moist-heat cooking, such as boiling, poaching, stewing, steaming, or braising. Any of these moist-cooking methods are also best for any leaner, tougher cut of beef (leg, hip, shoulder), and slow cooking provides the best results.

Using a tenderizer also makes tough cuts more palatable. Acid ingredients such as vinegar, yogurt, cider wine, citrus juice, and tomatoes often are used in marinades because they tenderize the meat. Natural enzymes such as papaya, figs, and pineapple also can be used for the same purpose. Cover meat with the marinade and place it in a non-metallic container in the refrigerator for 6 to 24 hours.

Aged beef that you might find at your local supermarket probably has undergone wet-aging. The carcass is vacuum-packed in plastic bags and then placed in coolers for a week to a month. This process tenderizes the beef, but it does not have the dramatic improvement in flavor that occurs with dry-aging.

Ground beef contains enough fat (about 15 to 20 percent) to give it flavor and make it juicy without excess shrinkage. The most flavorful hamburgers are made with ground chuck, but that is not the only kind of

MAKING THE GRADE

All meat sold in the United States must be inspected for wholesomeness by the government, usually by the USDA nationwide team of meat inspectors. Many people are familiar with this agency's stamp of inspection, which is placed on the outside of the carcass.

The USDA also grades meat, although this is a voluntary process. Grading meat essentially means using categories to give consumers an estimation of quality. Beef cuts are judged by the palatability characteristics of the meat. Inspectors look at the marbling (flecks of fat in the meat that give it flavor) and the firmness. The USDA grades for beef, from best quality to lowest quality, are as follows:

- Prime
- Choice
- Select
- Standard
- Commercial
- Utility
- Cutter
- Canner

The cuts given the highest grade also have the most marbling. To help make leaner, more healthful cuts appealing to consumers, the USDA came up with the term "select" for good quality cuts of meat that were lower in fat. Previously, the agency had used the term "good."

Most consumers will not find the five lowest-quality grades of beef in stores. These generally are used only for sausages and in cured and canned meats.

hamburger available. The leanest (around 11 percent fat) and most expensive of the ground meats are ground round and ground sirloin. New processing of ground meat is now making available even leaner choices (around 5 percent fat). The leaner choices are ideal for calorie and fat watchers, but they have to be cooked carefully to avoid their becoming too dry.

For more flavorful cuts of meat, dry-heat cooking — roasting, baking, broiling, or grilling — is more common. Try these methods with steaks, tenderloins, and filets. To reduce the fat content of beef sauces or soups, refrigerate them and gently spoon off the fat layer that forms on the surface.

Serving Suggestions

Although beef can be part of a healthful diet, the key is to think of it as a side dish and not the main entrée. Dishes such as sirloin kabobs fulfill the craving for beef without the need to eat a lot of it. Stir-frying, in which thin strips or chunks of beef are combined with large quantities of vegetables, is also an excellent way to have beef play a supporting role at mealtime. Stews and casseroles with vegetable-based sauces also offer this advantage.

Another simple option is to cut down on the amount of ground beef called for in a recipe. Reducing the amount of ground beef by half cuts the calories, fat, and cholesterol by half.

Buffalo

Bison, the shaggy, humped member of the cattle family, is the scientific name for what we generally refer to as American buffalo. Buffalo once roamed the West in great herds. Westward expansion and the introduction of cattle contributed to the decline of the roaming herds, but a new popula-

GROUND BEEF SUBSTITUTES, AND HOW THEY STACK UP

Three ounces ready-to-eat	Calories	Fat (grams)	Saturated fat (grams)	Cholesterol (milligrams)
Hamburger	245	20	7	75
Extra-lean ground beef	215	15	5	70
Ground turkey (meat and skin)	200	11	3	87
Ground turkey (breast only)	120	1	Trace	70
Textured soy protein	95	Trace	Trace	0
Tofu, firm	52	2	Trace	0
Tempeh	170	9	3	0

tion of ranch-raised buffalo is bringing bison meat back to the American table.

Buffalo is lower in fat than most cuts of beef and chicken — as well as some fish. It also tastes like lean beef and has no gamey flavor. Because of these characteristics, it is viewed by some food experts as the "gourmet food of the future." Although many ranchers and farmers have begun raising buffalo, so far the demand for it is outstripping supply.

Preparation Tips

Trim any excess fat before cooking. Because buffalo meat is so lean, it should be cooked slowly at a low heat. Otherwise, the meat will be chewy and flavorless.

Serving Suggestions

Grill buffalo steaks at least 6 inches from the heat source, basting often. Cuts of buffalo are similar to cuts of beef and can be substituted in most beef recipes, as long as the meat is not cooked past medium-rare. Remember, however, that buffalo cooks more rapidly than beef because it has less marbled fat.

Lamb

Spanish padres and English pioneers brought the first sheep to the United States. Using sheep for meat and clothing, the padres expanded their missions for the next 3 centuries. Today, lamb is produced in every state in the United States, even though the average citizen eats only 1 to 2 pounds of lamb a year.

In other countries, lamb is prized for its hearty flavor. The American aversion

WHAT IS BEEFALO?

Beefalo is what its name suggests: a cross between buffalo and cattle. The result is a dark red meat that is very lean and has a flavor that is slightly stronger than that of beef. Most grocery stores do not carry beefalo. It is available in some specialty markets. Although introduced some years ago, beefalo has been slow to gain popularity. At the same time, demand for ranch-raised buffalo is increasing.

A serving of meat is about 3 ounces — about the size of one lamb chop.

to the meat may be because sheep were originally bred for wool first and then consumed for meat. The older lamb resulted in a stringy, tough piece of meat. Today, animals are bred for only one purpose, and the meat has a finer consistency and a better flavor.

To qualify as a lamb, a sheep must be younger than a year. However, most lambs are sent to market at 6 months or younger. Baby lamb and spring lamb are both milk-fed and slaughtered before they are weaned. Regular lamb is slaughtered before it reaches 12 months of age. Lambs between 12 and 24 months are yearlings.

Mutton, the meat of sheep more than 2 years old, has a much stronger flavor and less tender flesh. Mutton has gained a reputation among Americans as inedible, but with proper slow-cooking methods, that reputation is unwarranted. Even so, mutton is

difficult to buy in the United States. Most lamb available at your local supermarket will be the more tender spring lamb. Lamb is leaner, too, because of selective breeding practices and because the meat is trimmed of more excess fat before packaging.

There are five USDA grades for lamb, based on the proportion of lean meat to fat: prime, choice, select, utility, and cull. Most of the lamb sold today is choice.

Use color as a guide to purchase lamb. A general rule is that the color gets darker as the animal gets older. For example, baby lamb is pale pink, and regular lamb is pinkish red.

You also can use the weight of a leg of lamb to judge its age. The more a leg weighs, the older the animal was at slaughter. A large leg, nearing 10 pounds, then, will have strong flavor and tougher flesh. Mutton legs are around 12 pounds.

Preparation Tips

Many cuts of lamb are sold with the fat already trimmed, but on some cuts, such as the leg or shoulder, thick external fat remains. Have your butcher remove this layer, or do it yourself before you are ready to cook the lamb. Leaving the fat in place will cause the amount of saturated fat to skyrocket, and the lamb will take on a stronger taste that many Americans, unaccustomed to lamb, might find displeasing. Lamb fat burns at a much lower temperature than other animal fat, and the lamb is left with a smoky, fatty aftertaste.

You also need to remove the fell, a membrane that encases the surface fat. The fell is left on some larger cuts to help trap the natural juices. The fell is inedible, and neither heat nor seasonings can penetrate it.

The cuts you most often find at your grocery store — leg of lamb, rack of lamb, chops, and loin — can be cooked by roasting, broiling, sautéing, or grilling. But if you are planning to serve the lamb rare or medium-rare, which is recommended, rib chops surpass shoulder chops. A quick-read thermometer inserted in the thickest portion of the meat should read 145° Fahrenheit (medium-rare). Chops cut from the shoulder are best slowly braised.

The size of a rack of lamb depends on the kind of sheep. An American-raised sheep will have a rack with seven to eight ribs, weighing approximately 2 pounds and feeding three or four people. A rack from a New Zealand sheep may weigh only 1 pound and feed two people.

Serving Suggestions

Leg of lamb is a traditional dish at Easter, and lamb and mutton are common ingredients in Middle Eastern couscous dishes. Lamb also makes a wonderful meat for kabobs, the best cut being leg meat cut into

1-inch cubes. Complementary seasonings include garlic, mustard, basil, mint, rosemary, and sage.

Lamb fat solidifies once the meat cools, so the meat should be served on warm plates. Lamb stew meat usually is cut from the shoulder, neck, breast, or leg (this might also be called a lamb shank). Use it in place of beef, pork, or veal in any slow-cooked stew or braised meat recipe. Use ground lamb as you would use ground beef.

The less tender cuts of lamb (shoulder, breast, shank) are best marinated.

Ostrich

With origins in Africa and parts of southwest Asia, the ostrich is a huge, flightless bird that can weigh up to 250 pounds and reach up to 7 feet in height. For centuries its meat and eggs were sought, but now ostrich can be raised like other forms of livestock.

Although still more exotic than a rib eye steak, ostrich meat is showing up on more and more restaurant menus, a direct correlation to the hundreds of ostrich ranches now in the United States. Some specialty meat markets may carry, or will special order, ostrich meat. The meat is deep red and looks like very lean beef. Ostrich is low in saturated fat and has very little cholesterol. Because ostrich is a bird, it technically is classified as poultry, but it tastes more like venison and is similar to lean beef in its color and texture.

Preparation Tips

The best ways to cook ostrich are by sautéing (pan frying or cooking) and quick grilling. Because it is such a lean meat, it easily becomes dry if overcooked. The best cuts of ostrich are the fan fillet, inside strip, tenderloin, and oyster.

Serving Suggestions

Rub olive oil on ostrich before grilling, then season with herbs or a touch of salt and pepper. Substitute ostrich in any recipe calling for lean beef or venison.

Pork

Pork was popular early in American history because pigs offered large litters of offspring and meat that could be preserved by smoking and pickling for long winter months. Pigs also would eat anything available. A typical farmer owned four or five hogs. An early governor of Virginia was one of the first to introduce swine to the New World. The state is famous for the quality of hams and other pork products produced there.

Although pork generally refers to swine younger than 1 year, most pork today is slaughtered at a younger age (6 to 9 months) to produce meat that is more tender and mild-flavored. The diet of a hog before slaughter has changed. What the colonists once thought was a positive — that a pig would eat anything — caused trichinosis, a food-borne disease that was once acquired almost exclusively from undercooked pork. Today's hogs are fed a diet of grain, proteins, vitamins, and minerals, and trichinosis is thus rare.

Until recently, pigs were bred to be hefty and fat, but pork has been gradually transformed by concerted breeding efforts to produce leaner meat. In general, today's hogs provide meat that is lower in calories and higher in protein than just 10 years ago. On average, pork is 31 percent lower in fat and 14 percent lower in calories than it was in 1983. However, not all pork cuts are lean, depending on the part of the pig used. For example, bacon still has 14 grams of fat per ounce (about 4 slices), but extra-lean cured ham may have less than 2 grams of fat per ounce.

Pork is an extraordinarily versatile meat. Pork comes to market in two basic forms: fresh and smoked. Only about a third of all pork is sold as fresh pork. The majority is cured, smoked, or processed into items such as bologna and hot dogs. The rump and hind legs of the pig are usually cured and smoked as hams. The same is true of the belly, or what becomes bacon after curing and smoking.

Most fresh pork comes from the pork loin and the shoulder, an area of the animal that is also known as "Boston butt." This is cut into chops, steaks, roasts, cubes, and strips. The loin section has been so popular, hogs have been bred with one more rib (compared with lamb, beef, or veal) to

PORK CUTS: WHERE THEY COME FROM

Section	Names of cuts
Shoulder butt:	*Cubed steak, blade steak, boneless and bone-in blade (Boston) roast, ground pork*
Loin:	*Sirloin, rib chop, loin chop, country-style ribs, back ribs, tenderloin, Canadian bacon, center rib roast*
Leg:	*Center ham slice, boneless ham, ham shank, leg cutlets*
Side (belly):	*Spareribs, slab bacon, sliced bacon*
Picnic shoulder:	*Arm picnic, arm roast, arm steak, ground pork, pork hocks*

increase the loin's length (see sidebar: Pork Cuts: Where They Come From, page 303).

Pork is a good source of thiamin, a B vitamin humans need to convert carbohydrates into energy. It is also a good source of zinc. The following pages provide an overview of the different types of pork.

Fresh Pork

Some of the more popular fresh pork cuts are pork chops, pork loin, and pork ribs. There are three types of pork ribs. Spareribs are from the breast and rib sections and provide little meat. Back ribs, or baby-back ribs, are cut from the loin, so they have more meat. Country-style ribs, from the shoulder end of the loin, have the most meat, but not necessarily the most flavor.

When purchasing fresh pork, look for meat that is pale-pink with a small amount of marbling and white (not yellow) fat. The darker the pink flesh, the older the animal.

Preparation Tips

Fresh pork was once cooked to an internal temperature of 170° to 185° Fahrenheit to avoid trichinosis. But with the new leaner pork, such a temperature is no longer necessary, nor is it advised. Cooking meat to this temperature will dry out the pork, making it chewy and hard to cut. Some older cookbooks on your shelf still may advise this higher temperature as a guide, but a better internal temperature is 160° to 170° Fahrenheit, which will produce juicy, tender meat. At this temperature, the inside of a fresh pork cut may still be pinkish. This tinge of color is nothing to worry about as long as the internal temperature has reached at least 160°. That temperature destroys any organisms that could cause trichinosis.

Fresh cuts of pork can be prepared with dry-heat cooking methods of grilling, broiling, and roasting, but marinating or basting may be necessary to keep the meat tender.

Serving Suggestions

Marinades made from citrus fruits add a nice accompanying flavor to pork. The acid from the fruit also helps tenderize the meat. Pork holds up under some strong sauces, such as barbecue sauce. Pork can be sliced into medallions and added to stir-fry dishes or served with steamed vegetables for an elegant entrée. Sweeter foods, such as applesauce or sweet-and-sour sauces, also complement the flavor of fresh pork.

Ham

Ham comes from the rump and hind leg sections of the hog and is available in either fresh or cured forms. Fresh ham commonly is roasted, but cured ham, which is often ready-to-eat, can be quickly baked, pan-fried, or microwaved.

The meat is usually cured in one of three ways: dry salt curing, brine curing, or brine injection curing. For dry curing, the surface of the ham is heavily salted, and then the ham is stored to allow the salt to saturate the meat. In brine curing, the ham is immersed in a sweet, seasoned brine. If sugar is added to the curing mix, the ham may be labeled sugar-cured. Most mass producers of ham use the injection-curing method, in which the brine is injected directly into the ham, shortening the curing process.

After curing, a ham may be smoked to add flavor and aging capability. Gourmet hams are heavily smoked for a month or more. A wide selection of specially cured hams are also imported from many European countries, including German Westphalian ham, which is smoked with juniper berry and beechwood. Other specialty hams include English York ham and French Bayonne.

The smoked flavor will vary depending on the type of wood used (usually hickory or maple) and the addition of unusual ingredients such as juniper berries and sage.

Fresh pork cooked to an internal temperature of 160° Fahrenheit will be juicy, tender — and safe.

Once curing and smoking are completed, gourmet hams are usually aged to develop flavors further, sometimes for up to 2 years.

Hams are sold in several forms, including boneless (with the hip, thigh, and shank bones removed), partially boned (with the hip or shank bones removed), and bone-in. Most producers of gourmet ham leave some bone in to enhance the flavor during cooking. Canned hams may be a whole piece of boneless meat or they may be created from bits and pieces of meat and held together with a gelatin mixture.

Preparation Tips

A fresh or cured ham should look firm and have a white layer of fat and pink to rose-colored flesh. Country-style hams (dry-cured hams) such as Smithfield hams are coated with salt, so they should be scrubbed with a stiff brush, soaked in cold water for 48 hours, and then scrubbed again.

Labels on hams should be checked for cooking and serving instructions, because hams are available fully cooked, partially cooked, or uncooked. Fully cooked hams, sometimes labeled "heat-and-serve" or "ready-to-eat," do not require additional cooking.

Serving Suggestions

Keep in mind that the curing process for ham makes it high in sodium. Persons following a low-sodium diet may want to save ham for special occasions and minimize how much is eaten. Those whose diets can tolerate a high-sodium food can add lean or extra-lean ham to a wide variety of dishes. Ham goes particularly well with pasta and rice dishes and in combination with vegetables. Ham can be baked, grilled, sautéed, broiled, or simmered. Try adding lean or extra-lean ham to make a hearty salad that serves as the centerpiece of a meal.

Bacon

Bacon is meat from the side of a hog — the pork belly — that is cured and smoked. Fat imparts the crispness and flavor to bacon and is usually half to two-thirds of the total weight, making it more of a fat selection than a meat selection. Bacon is also high in sodium and contains nitrates and nitrites, which are chemical preservatives that have been shown to cause cancer in animals. The amount in bacon is not great, and therefore it is not clear that they are harmful in the amounts normally consumed. For all of these reasons, the regular inclusion of bacon at breakfast should be reconsidered.

Compared with American bacon, Canadian bacon is more like ham, because it comes from the tender eye of the pork loin. It is more expensive than regular bacon, but it is leaner and precooked (less shrinkage), providing more servings per pound.

Sliced bacon comes in thin slices (about 35 strips per pound), regular slices (16 to 20 strips per pound), or thick slices (12 to 16 strips per pound). Canned bacon is popular with campers because it is precooked and needs no refrigeration.

Bacon bits are crisp pieces of bacon that are preserved and dried. They should be stored in the refrigerator. (A popular imitation made from vegetable protein may be kept at room temperature.) Bacon grease, the fat rendered from cooked bacon, is sometimes used as a cooking fat in regional dishes, as are salt pork (salt-cured cuts from the sides and belly of the pig) and fatback (fat from the pig's back, that can be salted and made into "cracklings" or unsalted and made into lard).

Preparation Tips

You can reduce the fat in bacon by broiling it on a rack, allowing excess fat to drip away. Microwaving is another option for reducing the fat in bacon. Put paper towels under and over the bacon to absorb the fat as the bacon is cooked.

Serving Suggestions

Bacon is too fatty and high in salt for everyday use. On occasions when it is eaten, reduce the amount of bacon strips used in the food. Or, add flavor without using a lot of bacon by using bacon bits. Even better, try imitation bacon bits, which are not made from bacon at all but from a vegetable protein. The result is bacon flavor without bacon's nutritional drawbacks.

Rabbit and Hare

Rabbit meat is often compared to chicken, but it has a sweeter, milder taste. Rabbit is also a little leaner than chicken. Three ounces of roasted rabbit has 7 grams of fat, whereas the dark meat of a roasted chicken has 8 grams and the white meat 3 grams, skin removed.

Rabbit meat for the table can be from either wild or domesticated animals. Farm-bred rabbits are somewhat fatter and blander in flavor than their free-roaming counterparts. They also have a fine-textured flesh that is almost all white meat, unlike their wilder counterparts. A mature rabbit weighs between 3 and 5 pounds, an amount that will serve four people.

More than double the size of its rabbit relative, the hare can weigh as much as 12 to 14 pounds. Hares have longer ears than rabbits, a notched lip, and powerful hind legs. Whether wild or domesticated, hares have a darker flesh and heartier flavor than rabbits.

Preparation Tips

Fresh and frozen rabbit is available dressed either whole or cut into pieces. Young rabbits weigh around 2 pounds and should have a

JUGGED HARE

You might have heard of this classic English dish. The main ingredient is a hare that has been soaked in a marinade of red wine and juniper berries for a day or more. The marinated meat is browned and then made into a casserole that includes vegetables, seasonings, and stock for baking. Juices from this mixture are poured off after cooking and combined with cream, blood from the hare that was set aside at butchering, and the hare's liver, which has been pulverized. The strained sauce is served over the meat and vegetables. Because the dish was historically served in a crock or jug, the dish has been referred to as "jugged hare."

light-colored flesh. They are considered the most tender. When cooking aged or wild rabbit, use moist heat to cook it (such as stewing, braising, or marinating) to tenderize and whiten the meat. Wild hare, also called jackrabbit and snowshoe rabbit, generally needs marinating to tenderize it before cooking. This process also whitens the meat. Young animals (1 year or less) can usually be roasted, but older animals are best cooked with moist-heat methods, such as slow cooking in a casserole or stew.

Serving Suggestions

In general, recipes for rabbit, especially young rabbit, are similar to those for chicken. French and Italian recipes use rabbit with excellent results. Rabbit is traditionally served in sauces to minimize a flavor that can be intense.

Veal

Credit the American love of Italian and French food to the increase in veal appreciation. Dishes such as veal scallopini and grilled veal chops introduced the meat to palates that were much more aware of the heavier, heartier taste of beef.

The name veal is derived from the Latin *vitellus*, which means calf. The meat is garnered from a young calf, usually 1 to 3 months old, that has been fed only milk.

Milk-fed veal comes from calves up to 12 weeks old that have not been weaned from their mother's milk, but veal of this quality is rare in today's supermarket. Shoppers are more likely to find veal fed a nutritionally balanced milk or soy-based diet that is fortified with essential nutrients. Color is the most important criterion when choosing a good cut of veal. The flesh should be a creamy white to ivory tone — barely tinged with grayish pink — and the fat should be white and creamy. The pinker the meat, the older the animal was at slaughter, and the meat is tougher and stronger-flavored.

If the meat is a reddish tone but still marked as veal, it may be a calf between 6 and 12 months and should more appropriately be called baby beef. Or, the calf may have been allowed to eat grains or grasses, which also darken the meat.

Animals were once confined to limit their movement; hence, the meat would be more tender and pale. For that reason, the consumption of veal was a source of controversy. In recent years, veal producers have attempted to make their modes of production more humane.

The USDA can grade veal in five different categories, but it usually does not. If the veal has been categorized, it will carry marks of (from highest quality to lowest) prime, choice, good, standard, and utility. The last three grades are rarely sold in retail outlets.

Preparation Tips

Although veal is supposed to be leaner and more tender than beef, not all veal is made equally, and not all cuts carry the same level of quality.

If you are interested in preparing veal, your first step should be to locate a supermarket or butcher who carries veal on a regular basis. Because Americans tend to eat veal in restaurants, the retail market is much smaller than beef's beefy share. You may need to order the cut of veal you want and the amount you need in advance.

The best cuts are from the rib and the loin of the calf. They are the most tender and the most expensive and can be bought as ribs, chops, or scallops.

Veal scallops have many names — scallopini, cutlets, schnitzel. However, all these words describe thinly cut veal slices. The best scallops come from the top round, cut against the muscle fibers. Cutting against the grain ensures the thin scallops will not buckle when they hit the heated pan.

If you are buying scallops, look for a smooth surface, which indicates they have been cut properly. Storage of veal is similar to that of beef.

Serving Suggestions

Veal scallops can be used to make veal scallopini. Scallops are dredged in flour and lightly cooked in oil or butter for 1 to 2 minutes. Veal roast can be cooked with either low or high heat, but high heat works better with more expensive cuts of meat. For cuts from the shoulder, breast, or leg, roasting at a lower heat with rubbing of added fat creates a more tender entrée.

Tougher cuts of veal from the lower leg and shoulder can be braised for the Italian dish osso buco. The name means "bone with a hole." The bone is a marrow-filled round in the center of the cut that adds flavor and is considered a velvety treat.

Veal chops and medallions are best prepared by pan frying or grilling. Chops should be at least an inch thick, and medallions should be about 3/4-inch thick. Otherwise, the veal will dry out.

Because veal is such a lean meat, many recipes may call for adding fat of some sort: from a health standpoint, a small amount of olive oil is best. Watch closely, because veal cooks quickly and it is very easy to overcook it.

Variety Meats

Variety meats is a category that includes brain, heart, kidney, liver, and various other meats. Because they are the most perishable parts of the animal, they were traditionally eaten first. And often, these meats were the centerpiece of some great feast.

Many variety meats are too high in fat and cholesterol for more than an occasional indulgence, although they are a good source of iron and, often, vitamin A and folic acid. However, some variety meats, such as calf's tongue and heart, may be unfairly neglected. They are a good low-fat source of protein, vitamins, and minerals. They also can be very economical, yet they are scarce in your local market. You may need to order variety meats from a butcher or ethnic food store. The following is an overview of the main kinds of variety meats.

Brain

Brain is very soft and porous and is considered a delicacy in many parts of the world. Although eaten far less often in Western nations, beef, calf, pork, and lamb brains are available in many supermarkets. Calf brain

usually is recommended most often for use in dishes because of its flavor and texture.

Brain should be a bright pinkish white, plump, firm, and absolutely fresh. Brain that is shriveled and dry should be avoided. It is perishable and should be used the day of purchase.

Preparation Tips

Brain should be washed well, then blanched. Although they differ in flavor and texture, brains and sweetbreads (see below) are used interchangeably in most recipes.

Serving Suggestions

Brain can be poached, fried, baked, or broiled, and is sometimes served as "beurre noir." This is a sauce that has a clarified butter base that is heated until dark brown. Brain also can be combined with scrambled eggs for a Southern delicacy. One brain

usually feeds two people. Brain also may be added to salads or stuffing or served in soups.

Heart

When the Indians of America hunted bison, the heart went to the warrior who brought home the beast because this organ was thought to contain the essence and therefore the power of the slain animal.

Because heart consists almost entirely of a hard-working muscle, it tends to be tough. In general, hearts from young animals are more tender. Your supermarket may stock veal, lamb, and pork hearts, but beef heart is more commonly available. The heart is an excellent source of protein, iron, zinc, riboflavin, folic acid, and vitamin B_{12} and a good source of niacin. It has more cholesterol than regular types of meat. However, it has less cholesterol than kidney, liver, or brain.

Veal picatta

When purchasing this organ, choose a product that appears fresh, and avoid those that have begun to turn gray. Lamb and beef hearts should be reddish brown. Pig and chicken hearts should be bright red, and calf heart should be light red.

Preparation Tips

Trim fat from the heart and remove veins and membranes. Rinse and clean the heart thoroughly. The heart can be cooked and served whole or sliced. Cooking times vary depending on the method used. Sliced heart is fried for 5 to 7 minutes. When braising this meat, hearts of young animals should be cooked 2 to 3 hours. Larger hearts may need to be braised for 4 to 6 hours.

Serving Suggestions

Hearts make excellent additions to stews and casseroles. Small hearts, such as those from young lambs and pigs, are often stuffed and sautéed. They also can be roasted. Typically, small hearts are served one per person.

Kidney

Kidneys typically used for cooking are beef, veal, lamb, and pork. The shape of the kidney depends on its source. Beef and veal kidneys are multilobed and elongated. In contrast, lamb and pork kidneys have just one lobe that resembles a giant bean. Kidneys from young animals tend to have a more tender texture and more delicate flavor. In addition, young animals' kidneys are usually pale. Those from older animals are a deep reddish brown.

In selecting kidneys, look for those that are firm and have a glossy, even color without dry spots. Kidneys should be used the day they are purchased, or stored loosely wrapped in the refrigerator for up to 1 day.

Preparation Tips

Remove the white membrane around the kidney by using a pair of scissors to snip the membrane from the core. Then, peel the membrane back with your fingers and remove any excess fat. Kidneys may be soaked in vinegar or lemon water to reduce the strong odor. Pork and large beef kidneys should be soaked in milk or cold salted water for 2 hours to minimize their strong taste.

Serving Suggestions

Kidneys are eaten braised, broiled, simmered, or cooked in casseroles, stews, and dishes such as the well-known British dish beefsteak and kidney pie. This dish is also commonly served in areas along the Canadian border in the United States. Nearly every comprehensive cookbook has a recipe for this meat pastry. One calf kidney or two lamb kidneys is considered a serving.

Liver

After beef and calf liver, the most common animal livers eaten are lamb, pork, poultry, and goose. Goose and duck livers are used mainly to produce the famous pâté de foie gras. Fresh American foie gras must come from ducks, but imported foie gras can come from either goose or duck. These birds are specifically bred with enlarged livers.

An important note is that the liver processes most substances that enter the body. This includes any chemicals that an animal might be fed or given as a medication. The older the animal, the more likely it is that there may be accumulations of unwanted residues in this organ. For this reason, liver from younger animals is generally preferred.

Fresh liver should have a bright color, a moist but not slick surface, and a clean smell. Loosely wrap it and refrigerate it immediately. Liver does not keep for more than a day.

Preparation Tips

Liver is encased in a thin membrane, which toughens in the cooking process and should be peeled off before cooking. Be careful not to overcook liver because it toughens quickly when overcooked. One popular solution is to sauté liver. Leave the liver a little pink in the center when you use this method and let the residual heat finish the job.

Serving Suggestions

Although liver typically is not considered a popular food, this organ meat can be tasty. It is commonly served with gravy made from it and onions. For a more elegant meal, broil or pan cook cubed liver and serve with mustard and boiled new potatoes. Liver also can be roasted (at 325° Fahrenheit for about 15 minutes per pound) and also served with boiled new potatoes. Soaking the liver in cognac for several hours before cooking adds a rich flavor to roasted liver.

Sausage

Born of economic necessity, sausage making has now become an art. When farmers slaughtered their own animals, they wanted to make sure that they used every part of the animal. So, the lesser quality cuts and scraps were ground up and made into sausages. Sausage essentially is any kind of chopped or ground meat that is stuffed into a casing. Most often, pork is the main ingredient in sausage, but poultry or fish can be used.

After countless decades of experimentation in ingredients — which include meat, spices, and fillers — sausages are often now considered a delicacy. They can be different in taste from one to the next. Sausages are fresh (made of raw ground meat and spices), precooked (such as hot dogs or bologna), or partially dried and fully cured (such as salami or dried pepperoni).

Preparation Tips

How you use the sausage depends on the type you buy. Fresh sausages need to be cooked, often by pan frying. Precooked sausages may need no preparation at all, or they can be grilled, broiled, or poached in hot water.

Fresh sausage can be kept for only about 2 days in the refrigerator. Dried or semi-dried will last 2 to 3 weeks. Cooked sausage can be kept for about a week.

Serving Suggestions

Partially dried and fully cured sausages may be used in sandwiches or cut up for pizza or paella. One thing to keep in mind is that sausages are laden with sodium, calories, and fat. Reduce the amount of fat in the sausage you eat by draining the excess fat during cooking or by selecting the reduced-fat and reduced-sodium varieties of sausages now available in stores.

Minimize the amount of sausage you eat by using it as a flavoring in meals instead of a main course.

Game Meats

Game meat was once classified as wild animals hunted for human consumption. However, because many "wild" animals are now being raised as livestock on farms or hunted in protected environments, the definition has changed slightly to include animals once hunted in the wild for meat.

The graying of the term also has made nutritional generalizations difficult. Because the diets and activity levels of domesticated game animals are different from those of their roaming counterparts, their meat has a different flavor and is often described as having a milder, less gamey taste. Meat from wild animals also may be lower in fat than meat from farm-raised animals.

However, farm-raised animals may have more fat and be more tender than their wild counterparts.

Game animals are categorized as large game and small game. Large game animals include deer, elk, moose, caribou, and antelope. Other popular large game animals include wild boar and bear. Other varieties eaten around the world include camel, elephant, kangaroo, zebra, and wild sheep. The most common small game animal is wild rabbit. Squirrel is also popular. Beaver, muskrat, opossum, raccoon, armadillo, and even porcupine are also eaten.

Preparation Tips

Any game meat found in commercial markets is federally inspected. The skin of a hunted animal should be removed and specific steps for butchering should be followed. The animal also can be taken to a butcher who specializes in game preparation.

For maximum tenderness, most game meat should be cooked slowly and not overdone. Cooking can be done with moist heat by braising or with dry heat by roasting.

Meat from older animals can be tenderized in a marinade that contains an acid such as vinegar to break down tough fibers, oil to add succulence, and wine, herbs, and spices to permeate the meat with flavor to reduce the "gamey" taste. Meat should always be marinated in the refrigerator because it takes quite long to mellow the flavor and tenderize (3 to 5 days).

Serving Suggestions

Roasts and steaks may be served for special occasions. When ground, they may substitute for more traditional patties, meat loafs, or casseroles. Small pieces may be stewed.

Liver pâté

FISH

Health-conscious Americans are figuring out what much of the world already knows. When prepared well, seafood tastes good, and it is good for you. Fish consumption increased 3.5 percent in 1998 and 3.7 percent in 1999.

The variety and convenience of fish may be causing the increase, because markets across the country — even those in the landlocked states — are stocking live shellfish and fresh fillets from both coasts. What once seemed intimidating to meat-eating Americans can now be cooked with the same ease as a pork chop or pot roast.

Nutrition

Seafood is a good source of high-quality protein, usually with a low number of calories. One 3-ounce serving of most fish or shellfish provides an excellent source of protein, at often fewer than 100 to 150 calories for many lean fish and most shellfish. Even oil-rich fish, such as salmon, tuna, and mackerel, generally have fewer than 225 calories in a 3-ounce portion, comparable to the calories of lean meats.

The fat in fish is mainly polyunsaturated and monounsaturated rather than saturated, as in meat. Furthermore, fish have a unique polyunsaturated fatty acid called omega-3 that is believed to have a beneficial effect in reducing blood clots, lowering blood cholesterol levels, and minimizing heart disease (see Chapter 3, Coronary Artery Disease, page 61). The fish that are particularly good sources of omega-3 fatty acids are sardines, herring, mackerel, Atlantic bluefish, tuna, salmon, pilchard, butterfish, and pompano.

Most shellfish were once considered high in cholesterol. However, new research has shown that although shrimp and squid have high levels of cholesterol, other shellfish, including clams, mussels, oysters, scallops, crab, and lobster, actually have amounts comparable to those of most fish. Shellfish are also low in fat. Most fish have levels of cholesterol comparable to that of the white meat of poultry and of lean, well-trimmed red meat.

Fish also contain important vitamins and minerals, especially some of the B vitamins, iron, potassium, magnesium, and phosphorus. Saltwater fish supply iodine. Canned salmon, sardines, and herring, with their soft bones, which are mostly edible, are also a good source of calcium.

In general, fish are low in sodium, similar to the amount in red meat and poultry. Persons following low-sodium diets should limit their intake of processed salted or dried fish, pickled herring, smoked fish and shellfish, sardines, surimi products, and anchovies.

Selection

What to look for depends on the type of fish being purchased:

Whole, fresh fish — The criteria for buying this type of fish boil down to this: look for the fish that appears to have been pulled from the water most recently. Scales should be shiny and stuck firmly to the skin. The flesh should feel firm and should not pull away easily from the bone. Gills should be moist and red, and the eyes should be shiny and not sunken. As always, the odor is a telltale sign of fish freshness. The fish should have a mild, fresh smell. The more fishy a fish smells, the less likely it is fresh.

Fresh fillets and steaks — These, too, must pass the smell test. In addition, check the flesh to make sure it does not pull away from the bones. Don't buy fish that appears to be discolored or dried out.

Frozen fish — Avoid dried-out fish. Packaging should be intact and free of frost.

Commercial fishing allows consumers who are land-locked to enjoy fresh fish.

Salted and smoked fish — Avoid fish with an "off" odor.

Storage

Whether the fish you buy is fresh or frozen, make it the last thing you purchase before heading home. If you will be delayed, have the market pack fresh fish on ice. Immediately refrigerate fresh fish in the coldest part of your refrigerator (usually the lowest shelf at the back or in the meat keeper), and use it within a day or two. Freeze your fish quickly to keep cell walls intact, but thaw the fish gradually so that fewer juices leak out of cells. The best way to defrost fish is overnight in the refrigerator. If you must thaw fish quickly, seal it in a plastic bag and immerse it in cold water, allowing 1 hour to thaw a 1-pound package. You can also microwave frozen fish on the "defrost" setting, stopping when the fish is still icy but pliable. Most fish will keep in the freezer for about 6 months. Never refreeze fish.

Canned fish, such as tuna, salmon, and sardines, will keep for about a year or less. However, because you cannot be sure about the conditions in which canned goods have been stored in the warehouse, it is best to buy only what will be used within a few months.

Preparation

Moist-heat cooking methods (poaching, steaming, or stewing) are best-suited for lean fish such as cod, flounder, or sole. Dry-heat cooking methods such as baking, broiling, and grilling are best suited for moderate-to high-fat fish, such as bluefish, butterfish, catfish, or salmon. En papillote, the French technique of cooking fish enclosed in parchment paper or foil, is an elegant way to keep fish moist. To determine whether fish is done, evaluate the color and

Fish steak, fillet, and butterfly fillet

flakiness of the flesh. Slip a knife into the fish and pull the flesh aside. The edges of the flesh should be opaque and the center somewhat translucent. The flesh should just barely flake. For large fish, a meat thermometer can be used to determine whether the fish is cooked thoroughly. The flesh should reach 145° Fahrenheit to be considered done. Fish continue to cook after they are removed from heat. For that reason, it is often recommended that you stop cooking fish just before it appears to be done. Overcooked fish looks dry, falls apart easily, and does not have many of its natural juices left over.

Marinades are an excellent way to add flavor with little fat to fish. Make sure that you marinate the fish in the refrigerator to prevent harmful microorganisms, if present, from multiplying.

Safety Issues

Although most seafood that reaches the consumer is safe, fish spoilage and contamination do occur. Bacteria, viruses, and other microorganisms can contaminate fish. Without careful handling, these can spread to humans through undercooked fish or cross-contamination, which occurs when surfaces used to prepare fish are not disinfected.

Like all living organisms, fish can occasionally carry various parasites. These parasites are easily destroyed by normal cooking procedures. In pickled products, such as pickled herring, the acidity of the vinegar used in pickling, often in combination with salt, preserves products and destroys parasites and harmful bacteria.

To prevent food-borne illness, the USDA advises avoiding raw seafood or lightly marinated raw seafood of any kind.

Are There Harmful Chemicals in Fish?

Because of pollution in oceans, lakes, and rivers, fish may contain harmful chemicals such as mercury or polychlorinated biphenyls (PCBs). This may be a particular problem in areas where there are manufacturing plants that burn fossils fuels or, in the past, have dumped mercury-containing waste into nearby streams and lakes. Fish at the top of the food chain (generally, the bigger types such as shark or swordfish) may contain higher amounts of mercury or other harmful chemicals because they feed on lesser fish and cumulate these substances in their fat and flesh. Some species such as large tuna (typically sold as fresh steaks or sushi) can also contain higher amounts. Various health agencies have issued advisories on how much fish is safe to eat, particularly for pregnant or nursing women and children. The Food and Drug Administration recommends that if you are pregnant or may become pregnant, you should avoid shark, swordfish, king mackerel, and tilefish. If you are considering having a child or are pregnant or nursing, check with your physician if your diet includes a lot of fish.

This advice includes sushi. It is particularly important for people in high-risk groups, which include older people, pregnant women, infants, and persons with liver disease, diabetes, immune disorders, or gastrointestinal conditions.

Fish are generally a low-fat source of protein. Not all fish have equal amounts of fat, however. Instead, fish are grouped into three main categories according to their fat content: lean, moderate-fat, and high-fat. The next section introduces you to the types of fish in each group.

Lean Fish

Lean fish have less than 2 1/2 percent fat. The flesh of these fish is lightly colored and has a mild taste. Commonly available varieties of lean fish include cod, flounder, halibut, and perch.

Cod

The cod is a popular saltwater fish that comes from the cold, deep waters of the North Atlantic and the North Pacific oceans. The cod has been one of the most intensely fished species because it is easy to preserve for transport and storage. It is part of the larger Gadidae family that also includes the haddock, the silver hake, the whiting, the black pollock, and the tomcod. "Scrod" is a term used to describe small cod (1 to 2 pounds), haddock, or pollock. Cod is a medium-sized fish. It is usually caught at weights between 4 and 9 pounds and measures between 15 and 52 inches. It has a firm, white flesh.

Preparation Tips

Cod is a versatile fish and is available fresh or canned. It can be cooked in a variety of ways. If you poach it, make sure that it does not boil. Instead, simmer it for 8 minutes in a bouillon or add it to an already boiling liquid and then remove the pan from the heat immediately. Set it aside until the fish is cooked (about 15 minutes). The roe, or eggs, of the cod are eaten fresh, smoked, or salted. The oil extracted from cod livers is an important source of vitamins A and D.

Serving Suggestions

Garlic and dill are two seasonings that complement this light, flaky fish. Another popular dish is cod cakes, which are simply a variation on the widely available recipe for crab cakes. Cod also can be cured in lye to make lutefisk, a Scandinavian delicacy popular in the Midwest for which one acquires a taste.

Round and Flat Fish

The term "fish" includes thousands of different species. For preparation purposes, it is necessary to know only that they are generally divided into two main groups: round fish and flat fish.

Round fish. As the name suggests, round fish have a plump shape. Their eyes are on either side of their head. Here is the critical part: their backbone runs down the center. Because thick fillets lie on either side, round fish are usually used as fillets or steaks. Examples include salmon, red snapper, and striped bass.

Flat fish. This group's name also gives away its anatomy. This fish swims horizontally and is shaped like a flat, thin disk. Its eyes are on the top of the head. The backbone is located in the center of the fish. Fillets do not come from either side, as they do in round fish, but from the top and bottom. Fillets are typically cut from flat fish, but larger fish may be cut into steaks. Examples of flatfish include flounder, halibut, and sole.

Flounder

This very common flat fish is found in the waters off nearly every part of the American coastline. Flounder is actually a large family of fish. Sometimes flounder is referred to as sole or turbot, which are members of the flounder family.

Preparation Tips

Flounder can be purchased whole or filleted and fresh or frozen. Nearly any type of cooking method can be used with success.

Serving Suggestions

Sautéeing flounder in a skillet with a little sesame oil, ginger, and garlic is a quick and tasty way to cook this versatile fish. Or, braise the fish with shallots, fish broth, and white wine.

An assortment of seafood

Halibut

One of the largest saltwater flat fish, the halibut is found in the cold waters of the North Atlantic and Pacific oceans. The fish usually weighs between 10 and 155 pounds. However, some halibut hauled in by fishermen have weighed more than 600 pounds. Halibut is usually available in fillets or steaks. When cooked, it has a firm, white flesh and a mild flavor.

Preparation Tips

Halibut can be grilled, poached, baked, broiled, or sautéed. The Atlantic halibut, which is usually sold as steaks, is a bit more flavorful than the Pacific halibut, which is more often available as fillets.

Serving Suggestions

Light marinades, particularly those made from red or white wine, enhance halibut's flavor without adding fat.

Perch

Perch is a freshwater fish, although sometimes the rockfish, a saltwater fish, is referred to by this name. Common types of perch include the yellow perch and the walleye, which is usually found in lakes, streams, and rivers of the Great Lakes states.

Preparation Tips

Another versatile fish, perch responds well to most traditional cooking methods. Sautéed, steamed, baked, broiled, or fried, perch's light, flaky, mild-tasting flesh responds well. One caveat: perch in general are very bony fish.

Serving Suggestions

In some regions of the United States, a walleye sandwich is considered a delicacy. Marinades also enhance perch's flavor. And perch is excellent served baked or steamed.

Moderate-Fat Fish

Moderate-fat fish contain about 6 percent fat. Their flesh ranges in color from white to beige and their taste from mild to muddy. Common kinds of moderate-fat fish include striped bass, catfish, swordfish, and tuna.

Striped Bass

The striped bass was once abundant on both the East and West coasts. But it has become less common because of overfishing and pollution. Fish farms have become one of the primary suppliers of this type of fish.

Preparation Tips

Striped bass has a pleasantly sweet flavor. Its sturdy fillets also lend themselves to a variety of cooking methods. It can be baked, broiled, or steamed.

Serving Suggestions

An increasingly popular and healthful method to cook this fish is to drizzle a fillet with a little olive oil, rub it with garlic, and then grill until done.

Catfish

Long a Southern delicacy, this fish has become popular in all regions of the country. Because of that, it is now often raised on farms instead of being gathered out of only rivers and lakes.

Preparation Tips

The fish has a tough skin that must be removed. To do this, use a sharp knife and cut the skin just behind the gills. Pull off the skin with the fingers or a tweezers. Catfish can often be bought at stores cleaned and dressed.

Serving Suggestions

Catfish is often served breaded or rolled in cornmeal or flour before it is fried. It can be deep fried in fat — not necessarily the healthiest way to cook a fish already considered higher in fat — until it is brown on both sides.

A healthier alternative is to poach catfish fillets in a chicken stock, white wine, and ginger. This cooking method results in a pleasing dish that is lower in fat than the traditional frying.

Swordfish

There is only one species of swordfish, named for its unusually long and slender upper jaw. It is found in temperate waters throughout the world and is common in the Atlantic Ocean, North Sea, Baltic Sea, and the Mediterranean Sea. Capable of moving through the water at up to 60 miles an hour, it is known for its power and muscular, scaleless body. It usually weighs between 200 and 500 pounds and measures between 6 and 10 feet.

Preparation Tips

Swordfish is a rich-flavored fish with firm, meaty flesh. It is commonly purchased as fresh steaks and fillets, but it also can be found smoked, frozen, or canned. Because the flesh of swordfish is so firm, it can be prepared in almost any manner, including sautéeing, grilling, broiling, baking, and poaching.

Serving Suggestions

The rich flavor of swordfish stands up to sauces with pronounced flavor, such as tomato-caper or basil-garlic. Swordfish is also a good fish for grilled kabobs, another low-fat, healthful way to serve fish.

Tuna salad sandwiches can be an elegant luncheon.

Tuna

The tuna is a migratory fish found in the temperate marine waters of the Mediterranean Sea, the Atlantic Ocean, and the Indian Ocean. The most common species include the bluefin, the albacore, the bonito, and the yellowfin tuna. The tuna is a powerful fish that is very agile and a fast swimmer. The larger bluefin tuna usually weighs between 200 and 400 pounds and measures between 3 and 7 feet. The smaller bonito tuna rarely is more than 20 inches long and weighs less than 5 pounds.

Tuna is also the most popular canned fish, in large part because of its low cost and versatility. Several varieties are canned and packed in oil, but the fish packed in water provides the same protein without the extra fat and calories. Canned tuna is sold as white tuna (albacore) and light tuna (bluefin and yellowfin). The bonito is labeled skipjack. Tuna comes in three grades, the best being solid or fancy (large pieces), followed by chunk (smaller pieces) and flaked or grated (pieces).

Tuna is sold fresh and frozen as steaks, fillets, or pieces. It is also available canned in oil or in water and either salted or unsalted.

Preparation Tips

All tuna has a distinctively rich-flavored flesh that is firmly textured yet flaky and tender. Tuna flesh tends to be soft before cooking, but it has a firm texture when cooked. Tuna lends itself to several cooking methods, including poaching, braising, grilling, or baking.

Serving Suggestions

Fresh tuna is especially good with spicy and aromatic seasonings. Marinate it in teriyaki sauce or in lemon juice with cracked black pepper, then grill it. Mixed with reduced-fat mayonnaise and seasonings, canned, water-packed tuna makes an excellent filling for sandwiches.

Higher-Fat Fish

Up to 60 percent of calories of higher-fat fish may be derived from fat. On average, though, fish in this group have only about 43 percent of calories from fat. The good news is that much of the fat that these fish contain is omega-3 fatty acids, which many studies suggest play a role in protecting against cardiovascular disease and enhancing brain function. Flesh from these fish is darker and firmer and often has a stronger flavor. Common types of higher-fat fish include mackerel, salmon, smelt, and trout.

Mackerel

This oily fish is related to the tuna and is an excellent source of omega-3 fatty acid.

Tuna — Make sure the edges are opaque and the center slightly translucent with flakes just beginning to separate. Let stand 3 to 4 minutes to finish cooking.

Most mackerel caught is canned. It is also salted or smoked.

Persons who live near the coasts, however, may be able to purchase it fresh from fish markets. If purchasing it fresh, look for fish that are firm and stiff. Limpness is a key indicator that the fish is not fresh. The fish should be cooked and eaten as soon as possible because it spoils quickly. A bitter taste is a key sign that the fish has begun to spoil. Freezing the fish is not an option because this seems to drain the flavor from this fish.

Preparation Tips

Remove the bones if filleting the fish. Because of mackerel's high fat content, try to use methods that do not add fat. Mackerel can be baked, broiled, or poached.

Serving Suggestions

Traditionally, mackerel is served with gooseberries. Other methods include using a citrus marinade. Mackerel can be used as a substitute for canned tuna in many recipes. It is also available smoked.

Salmon

Salmon can range from moderate- to high-fat, depending on the species. There are several species of salmon in the Pacific (chinook, sockeye, coho, pink, chum) and one in the Atlantic. It has even been acclimated to freshwater. A once threatened species, it has survived thanks to salmon farms and species management. Salmon are "anadromous," meaning they migrate from their saltwater habitat to spawn in fresh water. The females of some species can lay up to 13,000 eggs. The larger chinook salmon (also called king salmon) weighs between 30 and 40 pounds and measures between 34 and 36 inches. The smaller freshwater, or landlocked, salmon is shorter, between 8 and 24 inches, and it rarely weighs more than 13 pounds.

Salmon is sold in many forms, including fresh, frozen, smoked, salted, dried, and canned. Fresh and frozen salmon can be purchased whole or as steaks, pieces, or fillets. Whole salmon is usually sold cleaned, with the head on or removed.

Salmon is an excellent source of heart-healthy omega-3 fatty acids.

Smoked salmon is usually sealed in plastic or frozen. Salmon's moist flesh is flaky and tender, and the flavor varies by species from delicate and mild to rich and distinctive. Salmon spoils quickly because the flesh is fatty. The chinook is the fattiest salmon, and the pink and chum are leaner.

Preparation Tips

The bones often included in canned salmon are edible and serve as a good source of calcium. Make sure to crush the bones well. To remove the small bones (called pin bones) from fresh fillets or steaks, use a tweezers or pry them out of the flesh from the side.

Serving Suggestions

Smoked salmon (also called lox) is often served on bagels with cream cheese (choose reduced-fat or fat-free cheeses), capers, and onions, or it is added as a final touch to sandwiches, omelets, salads, and even dips and spreads. Salmon roe is becoming more popular as "red caviar," but real caviar comes from sturgeon roe. There are good salmon cuts for almost every cooking method. Fillets are delicious when grilled and served with a wedge of lemon. Salmon is good served hot or cold with a variety of sauces.

Smelt

A relative of the salmon, this small, thin fish can be both a freshwater and a saltwater fish. Some species are found in the Pacific and Atlantic oceans, and others inhabit freshwater sources. The main types of smelt are American smelt, European smelt, and capelin. All three types grow to about 7 to 8 inches long.

Preparation Tips

Smelt is sold in a variety of ways, including fresh, frozen, smoked, or dried. Smelt is also commonly sold without the head or innards. Some people eat the bones, but it is easy to remove bones from this fish — either before eating it or while you are doing so, as you would a sardine. The whole fish can be consumed, however.

Serving Suggestions

A common method of cooking is to coat smelt with breading and then fry it in a skillet until cooked (about 2 to 3 minutes). Smelt also can be grilled.

Trout

The trout is also part of the salmon family. It is found mainly in the cold, fresh waters of lakes and rivers, but it also can make its home in saltwater. Like the salmon, trout that live in the sea (the steelhead trout) return to freshwater to spawn. A favorite of sports anglers, the trout was the first fish to be raised in captivity to forestall its extinction. The most common species include the brown trout, the rainbow trout, the lake trout, the brook trout, the arctic char, and the common grayling. The trout is a smaller fish and ranges in weight from 1 1/2 to 13 pounds in the market and measures up to 20 inches.

Preparation Tips

Trout has thin skin and minute scales, so scaling or skinning is not necessary. It is available whole or in fillets, fresh or frozen. Trout is also smoked, and a very small quantity is canned. In general, trout has tender, flaky flesh with a mild flavor. Its delicate flavor varies slightly from one species to another, as does its color, which can be white, pink, or reddish.

Serving Suggestions

Although it is often served fried, trout also can be poached, baked, steamed, grilled, and broiled. Whole trout is sometimes stuffed before being cooked. Any seasonings should be kept mild so as not to mask trout's delicate flavor.

SHELLFISH

The shellfish family is divided into two basic categories: crustaceans and mollusks. Crustaceans have elongated bodies and jointed shells. These include crab, lobster, and shrimp. Mollusks are divided into three groups:

Gastropods (or univalves) — These have a single shell and single muscle. An example is the abalone.

Bivalves — Like the clam and oyster, bivalves have two shells hinged together by a strong muscle.

Cephalopods — Examples include the octopus and the squid, which have tentacles and ink sacs.

Nutrition

Like fish, shellfish are rich in protein yet low in fat and calories. Compared with other types of meat, however, the amount of cholesterol in most shellfish is about equivalent to that of a lean piece of beef or a chicken breast with the skin removed. Shrimp, squid, and crayfish, however, are very high in cholesterol. They have about twice as much as lean beef, making them something to savor on occasion.

Selection

When selecting shellfish, remember that fresh is best. The best way to guarantee freshness is to buy live. If your geographic location limits your opportunities for live shellfish, check out the freezer case for quick-frozen items, which can be almost as good.

Just as in picking out fish, use your nose to select shellfish. Shucked oysters and clams should be plump and free of any sour aroma. Fresh scallops should have a slightly sweet aroma and, when packaged, be practically free of liquid. Fresh shrimp are firm and have a mild, faintly sweet smell. Crab or lobster should move and not have any noticeable odor.

Next, view the shellfish's appearance. The liquid packed with oysters should be clear, not pink or opaque. Scallops, lobster, and shrimp flesh should be free of any black spots. When you buy frozen shrimp, crab, or lobster tail, make sure any exposed meat is white, not yellow, and not dried out.

Storage

Once you leave the market, it is important to keep shellfish cool and refrigerate it as soon as possible. At home, shellfish should be rinsed under cool, running water, placed in a container, and covered with wet paper towels. Use shellfish the day of purchase, if possible, or the next day.

Do not seal live oysters, clams, or mussels in a plastic bag. They need to breathe, so store them covered with wet paper towels. Use them as soon as possible after purchase. Shucked oysters that are refrigerated in the original container should stay fresh for a week. Plan to use commercially packaged frozen seafood stored in your freezer within 2 months for best flavor.

FISH AND SHELLFISH ALLERGIES

Seafood is a common source of food allergies. About 250,000 Americans experience allergic reactions to fish and shellfish each year.

People with seafood allergies can have symptoms that range from mild to life-threatening. Even tiny amounts of fish substances can trigger a reaction in some people. What's more, these allergies are rarely outgrown.

Examples of shellfish that are common causes of allergic reactions include shrimp, crab, lobster, oyster, clam, scallop, mussel, and squid. Fish that can trigger allergic reactions include cod, salmon, trout, herring, sardine, bass, tuna, and orange roughy.

Symptoms of an allergic reaction include nasal congestion, hives, itching, swelling, wheezing or shortness of breath, nausea, upset stomach, cramps, heartburn, gas or diarrhea, light-headedness, or fainting.

If you suspect that you have any food allergies, see an allergist for a careful evaluation. This generally includes a medical history, physical examination, and skin or blood testing. If you are found to have a fish or shellfish allergy, the best advice is to avoid fish or shellfish altogether. That may be harder than it sounds. You may not be aware that seafood is an ingredient in a dish that you are eating. Be sure to check the labels of any product you buy. In addition, make sure that persons close to you are aware of the potential for this kind of allergic reaction. Many people with a food allergy wear medical alert bracelets.

Safety Issues

Persons who eat shellfish raw should beware. Raw shellfish may contain the organisms that cause hepatitis and other diseases. Cooking will kill any microorganisms in shellfish. In addition, because shellfish filter large amounts of water each day, they may contain residual amounts of any pollutants in the water near them. Concerns about potentially harmful chemicals are also a concern about shellfish. Women who are pregnant or thinking about having a child should check with their physician about how much shellfish they should eat.

Another important safety note is that shellfish are a common trigger of allergic reactions in some people (see sidebar: Fish and Shellfish Allergies, page 317).

Abalone

Abalone is actually a large snail that lives in the sea. It has a single shell and a tough, muscular "foot" with which it clings tenaciously to rocks as it grazes on seaweed. The edible portion is this foot.

Abalone is found primarily along the coastlines of California, Mexico, and Japan. Four of California's seven species constitute most of the commercial catch. From largest to smallest are red, pink, green, and black abalone, with red reaching a legal market size at 7 3/4 inches and black at 5 3/4 inches. Black abalone is the most affordable wild species today. Its meat is tougher and requires more pounding than the other species, but it has an excellent sweet flavor. It is an excellent source of protein and a good source of iron, magnesium, and phosphorus.

Abalone is known by many names: "ormer" in the English Channel, "awabi" in Japan, "muttonfish" in Australia, and "paua" in New Zealand. Its iridescent shell is a source of mother-of-pearl.

Preparation

Like all fresh shellfish, abalone should be alive when purchased — it will move when it is touched — and smell sweet, not fishy. Choose those that are relatively small for best flavor. Refrigerate abalone immediately and cook within a day. Abalone is also sold canned (once opened, refrigerate, covered with water in a sealed container, for up to 5 days), dried (store tightly wrapped in a cool, dry place indefinitely), and frozen (store for up to 3 months).

Widely used in Chinese and Japanese cooking, abalone must be washed first to remove sand that may be caught in the flesh. Abalone must be pounded before cooking to make it tender, because the edible portion is a muscle. Use a mallet to flatten the meat to 1/8- to 1/4-inch thickness. Overcooking abalone can toughen it. Sauté abalone briefly, for not more than 20 to 30 seconds per side. Prevent abalone from curling during cooking by scoring the meat at 1/2-inch intervals with a sharp knife.

Serving Suggestions

Abalone is an excellent addition to appetizers and salads. Tough or overcooked abalone can be added to chowders or soups.

Clam

Clams are double-hinged mollusks that fall into two broad categories — hard shell and soft shell. The parts eaten are the muscles clams use to close their shells, although the siphon (the valve used to intake water) and the foot (which nudges this creature along on the ocean's floor) are also edible. Clams generally are chewy and have a mild, even sweet, flavor. Taste and characteristics vary by the type of clam and its size. All are a lean seafood choice and an excellent source of vitamin B_{12} and iron.

The quahog (pronounced CO-hog) is a hard-shell species from the East Coast and is the largest of Eastern clams (1 1/2 to 6 inches across). Quahog is also a family name for hard-shell clams that can include cherrystones (less than 3 inches) and littlenecks (2 to 2 1/2 inches). Also popular in the East are soft-shell clams called steamers. As the name indicates, they are great for steaming. On the West Coast, soft-shell Pacific geoducks (pronounced GOO-ey-ducks) weigh between 2 and 4 pounds, although they can be 3 feet long and up to 5 pounds. They have an enormous siphon that extends from the shell; this is often sliced for sushi because it is also quite sweet.

As with all filter feeders, clams sometimes ingest toxic levels of planktonic microorganisms during a condition called

Cherrystone clams. American Indians called them "quahog."

red tide and also can become contaminated from pollution.

When buying hard-shell clams such as the littleneck or cherrystone (other varieties include chowder, pismo, or butter clams), make sure the shells are tightly closed. If a shell is slightly open, tap it lightly. If it does not snap shut, the clam is dead and should be discarded. The shells should be whole, not broken or cracked. To test a soft-shell clam, such as geoducks or steamers (another popular variety is razor clams), lightly touch the neck; if it moves, it is alive.

Store live clams in an open container covered with a moist cloth for up to 2 days in the refrigerator. Shucked clams should be plump and the liquor (liquid) surrounding them clear. Store shucked clams in their liquor in the refrigerator for up to 3 days and in the freezer for up to 3 months. If there is not enough liquor to cover the shucked clams, make your own by dissolving 1/2 teaspoon salt in 1 cup water.

Preparation Tips

Get rid of sand by soaking live clams in cold, salted water (use 1/3 cup salt per gallon water) for an hour. Sand also can be removed by covering the clams with water, then sprinkling liberally with cornmeal and letting stand for about 3 hours. Any clams that are floating after these two methods should be discarded.

Live clams should be well scrubbed under cold, running water. Clams are much easier to open if they are put into the freezer in a single layer for 30 to 45 minutes. A quicker method for relaxing clams so that they are easier to open is to drop a few at a time into boiling water. Retrieve with a slotted spoon after 15 seconds and open. Clams also can be microwaved until they open.

Crab, lobster, and the tiny crayfish are all excellent steamed.

Serving Suggestions

Clams should be cooked at low heat to prevent toughening. Clams cooked in their shells are done just when their shells open. Soft-shell clams are best for this method. Use only fresh or frozen clams for soups and stews, because the texture of canned clams is too soft for long-cooking dishes. Clams are often added to soups and stews. A famous dish in which they are featured is clam chowder. When adding clams to this type of dish, do so at the last minute so they do not lose their texture. Clams are also excellent additions to dips, salads, and quiches.

Crab

There are dozens of different crabs on the market today, varying widely in size and shape. Therefore, it is important to know what kind of crab you are buying before bringing it home.

Crabs have 10 legs, the front 2 of which have pincers. There are freshwater and saltwater crabs. Saltwater crabs are the most plentiful. Soft-shell crabs are not a distinct variety of crabs. Rather, they are crabs that have shed their hard outer shell. Before growing a new shell, these crabs (usually a variety called blue crabs) have a 6-hour window during which they have a soft shell. This window can be extended if the crustaceans are removed from water. The benefit of a soft-shell crab is that it can be eaten whole without bothering with a pesky, rigid exterior. Alaskan king crab is another term crab buyers may encounter. Meat from this large crab species is frozen soon after the catch. Then, the sweet, meaty flesh is frozen so it can be shipped to far-flung locations.

Crab is a lean source of protein and an excellent source of vitamin B$_{12}$.

Preparation Tips

Crab is sold whole, cooked or alive. Choose live crabs that are active and heavy for their size. Hard-shell varieties should not have softening shells. Refrigerate live crabs covered with a damp towel until just before cooking. Live crabs should be used the day they are purchased.

Cooked whole crabs and crabmeat should have a fresh, sweet smell. Cooked crab is available in whole pieces (lump white meat from the body) or in smaller pieces (flaked light and dark meat from the body and legs).

For ease of use, consider using canned crabmeat. Or, if you prefer, choose pasteurized crabmeat, which has been heated in cans but has not been subjected to the higher temperatures of the canning process. For this reason, pasteurized crabmeat should be stored unopened in the refrigerator no longer than 6 months. Use it quickly after opening. Always use your fingers to pick over crabmeat, fresh or canned, to make sure there are not tiny pieces of hidden shell. Refrigerate leftover cooked crabmeat, tightly covered, for not more than 2 days.

Serving Suggestions

Crabmeat has a tender, flaky texture and a delicate, sweet flavor. For the best possible flavor, choose live crabs. If only cooked crab is available, add it to hot dishes during the last minutes of cooking and cook just until hot.

Crabs can be cooked in a variety of low-fat ways, including steaming and broiling or in broth-based soups and gumbos. Crab cakes, traditionally held together with mayonnaise and bread crumbs, are leaving the confines of New England for the rest of the states. A lower-fat version uses egg whites in place of mayonnaise. Instead of floating crabs in a pan filled with oil, try dusting them with flour and reducing the amount of oil you use.

Lobster

One of the largest crustaceans, lobster has a jointed body and limbs covered with a hard shell. The coveted edible portion of the lobster is generally the sweet flesh inside its tail, and sometimes in its claws. Its liver (tomalley) and eggs (coral) also can be eaten. Lobsters are prized by diners for their firm flesh and flavorful, yet not overpowering, taste.

Lobsters can be divided into two groups: large, clawed (American) lobsters and spiny (rock) lobsters. American lobsters hail from around Maine, whereas rock lobsters can be caught in the waters off the southeastern United States and in the Pacific. American lobsters boast claws with yet more meat, but only the tails of spiny lobsters are eaten. These tails are often sold frozen throughout the United States. The meat is more dense and fibrous than that of the American lobster. It lacks some of that characteristic sweetness, but the tails are far more inexpensive than buying a whole American lobster.

Lobster is an excellent source of zinc.

Preparation Tips

When selecting a live lobster, look for an active one whose tail curls under its body when lifted. If lobsters have been stored on ice, they may be sluggish. Bacteria form quickly in a dead lobster, so it is important that you purchase it live. Live lobsters will die quickly, so they must be kept in seawater, wrapped in a wet cloth, or wrapped in several sheets of wet newspapers. Use these methods for only a few hours (24 hours at most), then cook the lobster. It is best to cook lobster the same day you buy it.

Whole, cooked lobsters should have their tails curled tightly under their bodies, a sign they were alive when cooked. Cooked lobster meat should be sweet-smelling and snow-white. Frozen tails should be in an untorn package with no sign of frost; the visible meat should be free of dry-looking spots.

Lobsters can be prepared in a variety of ways, including steaming, broiling, and grilling. Another common preparation method is boiling. A general rule for cooking lobster with any method is to allow 12 minutes per pound. The lobster is done as soon as its shell becomes red or the meat turns opaque. A sign that a lobster has been overcooked is that its meat is stringy and tough.

Serving Suggestions

The firm flesh and pleasant flavor of lobster make it a versatile seafood. It can be eaten plain, seasoned with lemon juice, garlic, and a drizzle of butter, or added to pâtés, salads, or sauces. Lobster bisque, essentially a cream-based lobster stew, is a classic lobster dish enjoyed by diners the world over. Because it is high in calories and fat, it should be savored on special occasions. Lobster meat also can be mixed with curry sauce. It is also an elegant, tasteful addition when added to salad greens and sprinkled with lemon juice or a spicy lower-fat vinaigrette.

Mussels

Perhaps it is the dark shell that accounts for its foreboding reputation. Perhaps it is the name, which makes it sound too much like a stringy, tough muscle. Whatever the case, mussels do not have the audience in the United States that other forms of seafood do, even though people throughout the world have been eating them for more than 20,000 years.

Although there are dozens of mussel species, only two reach American markets with any regularity. The blue mussel (which is actually dark blue to black) and the blue-green mussel can be used in recipes interchangeably, but the blue-green type from New Zealand is larger and more expensive. The blue mussel is most abundant, found along the Mediterranean, Atlantic, and Pacific coasts. It is usually 2 to 3 inches in length.

Many mussels are being cultivated instead of harvested because of the dangers posed by microscopic organisms (of red tide notoriety) that make mussels unsafe to eat during the spring and summer months.

When choosing mussels, make sure they are still alive by tapping on their shells. If they slam shut, they are still alive. Those with tightly closed shells also are a good choice. In general, smaller mussels are more tender. Cultivated mussels are usually smaller, and they also lack some of the sand and grit of harvested mussels. Choose shucked mussels that are plump and have a clear liquor (liquid). Or, opt for plain and smoked mussels packed in oil.

Store live mussels in a single layer on a tray in the refrigerator covered only with a damp towel or wet newspapers for up to 2 days. Store shucked mussels in the refrigerator up to 3 days. They should be covered completely with their liquor (liquid). If you need more liquor, combine 1 cup water with 1/2 teaspoon salt and pour this brine over the mussels to cover.

Preparation Tips

Before cooking, use a stiff brush to scrub mussels under cold, running water. Pull out the dark threads (beard) that protrude from the shell. Mussels die when debearded, the term given for removal of these threads, so do not remove the dark threads until just before cooking. Get rid of sand by soaking mussels in cold, salted water (use 1/3 cup salt per gallon of water) for 1 hour. Like all shellfish, mussels should be cooked gently to prevent toughening. Mussels cooked in their shells are done when their shells pop open. Discard any with unopened shells.

Mussels steamed with white wine and onion

Serving Suggestions

An elegant way to serve them is simple: boil them. Add garlic, onion, a splash of olive oil, and white wine to the water while cooking — generally, about 8 to 10 minutes. Saffron also provides an excellent flavor. Serve with French bread. This can be broken into pieces to be used to soak up the flavorful broth you created while cooking the mussels. A general rule: about 1 pound of mussels per person is needed.

Octopus

Although popular in Japan and Mediterranean countries, the octopus in North America is more likely to be a villain of "scary" movies than the star of a seafood meal. But octopus is relatively inexpensive and versatile, and it provides highly flavorful, meaty (albeit sometimes chewy) chunks. Most octopuses weigh only about 3 pounds and reach a size of only 1 to 2 feet before being caught for food.

Fresh octopus already dressed and ready for dishes can be found in many supermarkets and specialty fish markets. Frozen, smoked, and canned octopus is also available. Shop for octopus the same way you shop for fish, using your nose as a guide. Octopus should smell fresh, not fishy, and if it has eyes, they should be bright, not cloudy. As with most aquatic species, octopus that is smaller is younger and more tender. Both the body and all of the tentacles are edible. The ink sac contains an edible black liquid that can be used to color and flavor foods such as pasta, soups, and stews.

Preparation Tips

Always rinse octopus well before cooking. Although many cooks believe that an octopus must be tenderized before cooking (such as by beating it with a mallet or even hurling it against a hard surface), octopus can simply be cooked until it is tender. Cooking time varies by size. Generally, an octopus that weighs more than 4 pounds should simmer for about 45 minutes. This seafood can be cooked in a variety of ways, including grilling, frying, steaming, and poaching.

Serving Suggestions

Octopus is an excellent addition to seafood stews and soups, adding an unusual texture to the dish. Its flavor is enhanced by garlic, tomatoes, red wine, and soy sauce.

Oysters

Oysters have many different names, which are clues to where they come from and their varied taste and appearance. The marine plants they feed on account for these differences. The hard, rough, gray shell contains meat that can vary in color from creamy beige to pale gray, in flavor from salty to bland, and in texture from tender to firm.

Oysters are rich in zinc, iron, copper, and vitamin B_{12}. Interestingly, it is oysters' relatively high zinc content that gave them their reputation as an aphrodisiac when eaten raw.

Purchase the freshest oysters you can find. Gauge freshness by tapping on the oyster shells with your fingernail. If they snap shut quickly or are already shut tightly, they are fresh. In general, smaller oysters will be more tender than larger ones. Buy shucked oysters of uniform size and color encased in a clear liquor (liquid).

You may store oysters for up to 3 days in the refrigerator using the same method as for mussels, but the sooner you use them, the better. If any shells open during storage, tap them. If they do not close, throw them out. Store shucked oysters, covered by their liquor, for up to 2 days in the refrigerator and up to 3 months in the freezer. If there is not enough liquor to cover the shucked oysters, make your own by dissolving 1/2 teaspoon of salt in 1 cup of water.

Preparation Tips

Scrub live oysters under cold, running water before opening. Oyster knives specially designed for removing the flesh from the shells — which can be a tricky operation — are available at cooking supply stores. Oysters and other bivalves will open more easily if they are heated briefly — such as by steaming them for several seconds or microwaving them for a minute.

Serving Suggestions

When fresh and safely available, raw oysters are a unique treat. "Raw bars" serve the traditional "shooter," a New Orleans-inspired hot sauce covering a raw oyster. If you serve them raw, keep them chilled (and prevent the growth of microorganisms) by putting them on a bed of ice.

Oysters in the shell also can be cooked in numerous ways: baked, steamed, or grilled. Oysters Rockefeller, a hot hors d'oeuvre that is baked and broiled with a spinach topping, is a traditional, elegant favorite. Use shucked oysters in dressings or poultry stuffings. Oyster stew is often made with only heavy cream, oysters, and butter, but a healthier alternative can be created with evaporated milk and added vegetables, such as potatoes, pearl onions, carrots, or mushrooms.

Scallops

Like clams and oysters, scallops are bivalves (two-shelled mollusks). Scallops propel themselves along the ocean floor by clapping their shells together using a well-developed muscle called the eye, the part of the scallop that is eaten.

If you avoid fish and other types of shellfish, you may still like scallops because its eye is firmer than most shellfish meat and its flavor is mild yet sweet. Scallops are rich in vitamin B_{12} and potassium.

Scallops are classified in two broad groups: bay scallops and sea scallops. Bay scallops, with meat about 1/2 inch in diameter, are sweeter, more succulent, and more expensive. Sea scallops, with meat about 1 1/2 inches in diameter, are more widely available. A third type is the calico scallop, a small sea scallop from Florida, but it is often sold as bay scallops. These scallops have to be steamed to be opened, so they are partially cooked when sold. They are the cheapest of the scallops.

Because scallops cannot close their shells tightly, they spoil quickly out of water, and they are usually shucked at sea. The muscle is iced and the rest discarded. Sometimes the coral-colored roe is left attached to be eaten with the muscle as a delicacy.

Look for scallops with a sweet smell and a fresh, moist sheen. Avoid any with a strong sulfur odor. Scallops can range in color from pale beige to creamy to orange. Avoid those that are stark white, a sign that they have been soaked in water as a marketing ploy to increase the weight. Refrigerate shucked scallops immediately after purchase and use within 1 to 2 days.

Preparation Tips

Scallops must be opened like oysters if purchased unshucked. Some sea scallops still have a small piece of tough connective tissue attached to one side, which should be pulled off and discarded. Scallops need only brief cooking. Overcooking or reheating cooked scallops will toughen them. The roe, if attached, can be cooked right along with the muscle. Large scallops should be sliced into smaller pieces.

Serving Suggestions

Scallops are suitable for a variety of preparation methods, including sautéeing, grilling, broiling, and poaching. They also are used in soups, stews, and salads. Scallops cook very quickly — about 1 to 3 minutes — and no scallop needs to cook longer than 6 to 8 minutes, no matter how large.

Shrimp

Shrimp is the most widely consumed seafood worldwide, and it is the second most popular in the United States, after tuna. The harvest of both wild and farmed shrimp is increasing rapidly to keep pace with the soaring demand for this crustacean. It has become popular internationally because its sweet, mild, and yet distinct flavor is similar to that of lobster, but it is easier to cook and costs half as much.

Although hundreds of species are caught, shrimp generally can be classified by origin — either warm water or cold water. Cold-water varieties — such as the well-known small, pink, peeled shrimp — tend to be sweeter than the larger shrimp, which typically develop in warmer waters. These larger varieties are often sold uncooked and unpeeled under the name "prawns," but the name can be used for shrimp of any size.

The names of shrimp can be confusing. Many varieties of shrimp are named after a color (white, pink, brown), but white shrimp are not necessarily white. They may look pink, brown, or gray. Confounding the problem, most shrimp shells change color when cooked.

Shrimp are marketed according to size (number per pound), but market terms vary greatly. Because shrimp freeze well, most, even those classified as "fresh" in the supermarket, have been frozen at some point. Neither taste nor texture suffers much.

Shrimp is sold shelled or unshelled, raw or cooked, and fresh or frozen. Shrimp also can be bought breaded or stuffed or as a spread for crackers or bread. Dried shrimp and shrimp paste can be found in Asian grocery stores.

Preparation Tips

Choose raw, shelled shrimp that are firm, moist, and translucent. Avoid any shrimp that smells like ammonia. Unshelled shrimp should have shiny, firm shells. Avoid those with black spots and those with yellow or gritty shells, which result from a bleaching process to remove the spots.

Before storing fresh, uncooked shrimp, rinse them well under cold, running water and drain thoroughly. Refrigerate, tightly covered, for up to 2 days.

Shrimp are usually shelled before they are eaten, which is easily accomplished by peeling back the head first, then the rest of the shell. Shelling is easier when shrimp are cold or even frozen. Shrimp are also usually deveined, although this is not necessary before eating small to medium-sized shrimp. To devein a shrimp, cut into the shrimp lengthwise with a sharp knife. Then use the knife to remove the dark vein — which is actually the intestine that runs down its back.

Although there are slight differences in texture and flavor, shrimp of various sizes (except the miniatures) can usually be substituted for each other.

As with all shellfish, shrimp should be cooked briefly or it becomes tough and rubbery. Cook only until the flesh turns opaque. To test, cut a shrimp in half, and check whether the flesh has turned from translucent to opaque. Whole shrimp should just begin to curl. If the shells are on, they should turn pink.

For a tasty and healthful change, grill scallops and serve with a variety of vegetables (grilled zucchini, white beans, and scallions).

Serving Suggestions

One reason for the popularity of shrimp is its versatility. Name a cooking technique, and you probably will be able to find a shrimp recipe for it. Shrimp is served cold in shrimp cocktail. It can be grilled, boiled, broiled, roasted, curried, or stir-fried. It goes well with pasta and rice. It is also an excellent ingredient in soups and can be used interchangeably with other shellfish in recipes. Garlic is a seasoning that particularly complements shrimp's flavor. The two are used together in well-known dishes such as shrimp scampi.

LEGUMES

Legumes have historically been part of meals throughout the world. There is abundant evidence that the peanut and lima bean have been used for centuries in South America. Soybeans and mung beans, among others, have been a key part of Asian dishes throughout history. The Middle East is the origin of broad beans, chickpeas, and lentils. Because of this, cooking with legumes can provide an education and an introduction to international cuisine.

Most legumes are annual plants that can grow as vines or bushes. The shape, size, and color of the stems, leaves, and flowers differ according to species. After fertilization of the flowers, pods develop. These contain seeds of varying sizes, shapes, and colors.

The plants of the legume family share two main features. First, they produce single-chambered, flattened seedpods with seeds inside. Either the pod or the seeds inside (or both) are eaten.

A second common feature is that legumes are capable of an important biological process called nitrogen fixation, which actually enriches or naturally fertilizes the soil in which the plants grow. Accordingly, legumes (such as soybeans and alfalfa) are grown in rotation with other crops (such as corn) that only take nutrients from the soil.

Legumes are sold in many forms. They are available as whole fresh pods, such as green beans, and in dried pods, such as the tamarind. Lima beans are examples of legumes available as fresh seeds, and dried seeds include black-eyed peas. There are also seed sprouts, such as alfalfa sprouts and soybean sprouts. Although technically considered legumes, tamarind and jicama are treated as a fruit and vegetable, respectively.

Legumes are generally easy to prepare and can be either the main entrée or the side dish. Generally, dried legumes are rehydrated before cooking, which is done by soaking them in water for about 6 to 8 hours. An alternative and shorter method is to put them in a pressure cooker or to simmer them in a pan

where they can soften in less than an hour. Soaking shortens cooking time by 30 minutes to up to an hour. It also has the advantage of reducing flatulence (intestinal gas) by making them more digestible.

Legumes that are sold as "quick-cooking" have been presoaked and redried before packaging and thus do not need to be soaked.

When several types of legumes are required for a dish, it is best to cook each type separately because it is difficult to cook them uniformly together.

Nutrition

Legumes are great sources of nutrition because they carry the embryonic necessities for starting a new plant. They are high in protein, folic acid, potassium, iron, magnesium, and phytochemicals (see the Appendix: Phytochemical Contents of Selected Foods, page 484). Legumes are not complete proteins like meats (soybeans are an exception), but they can be paired with complementary foods, such as grains, to ensure a meal provides a complete source of amino acids for building proteins. And unlike meat, they are low in fat, high in fiber, and inexpensive.

Because of their low cost, legumes were once considered "poor man's meat." But with the increased popularity of ethnic cuisines (such as Mexican, Chinese, and Mediterranean), the growing popularity of vegetarianism, and the recognized health benefits of legumes, they have shed this outdated perception.

Selection

There are expensive "designer beans," but most legumes are widely available at grocery stores and are an inexpensive addition to meals. Look for a uniform appearance to the product you are buying. Legumes should also have a deep, almost glossy color. Avoid buying products that are cracked, broken, dry-looking, or faded. These are most likely to have been on the shelf for a while.

Storage

Dried legumes are easily stored. Putting them in a covered container or closed plastic bag can help maintain their freshness and extend their shelf life.

Legumes that are commonly used are described on the following pages.

Alfalfa

Also known as lucerne, alfalfa is the common name of a legume that once was thought of only as animal feed. The plant is believed to have originated in southwestern Asia around the area of Asia Minor and the Caucasus Mountains. Spanish explorers brought the plant to the Americas, where, in the United States, it was first established as a crop in California in the 1850s.

Alfalfa grows up to 5 feet in height and bears spiral-shaped pods containing six to eight small brown or yellow seeds. Because the roots are capable of extending as much as 30 feet into the soil, the alfalfa plant can reach stores of water and nutrients that allow it to survive periods of extreme drought.

Alfalfa became part of the human diet on a wide scale in the 1970s, when many people began to enjoy alfalfa sprouts. Today, alfalfa sprouts are sold in most grocery stores.

Preparation Tips

Alfalfa sprouts should be washed thoroughly before they are eaten. Buy sprouts that look healthy and green. Avoid those that are off-color, smell moldy, or look soggy. They can be kept for about a week in the refrigerator (see sidebar: Is It Safe to Eat Raw Sprouts? this page).

Serving Suggestions

Usually used raw, alfalfa sprouts can be added to salads, sandwiches, omelets, tacos, and hors d'oeuvres. They also can be used to garnish soups and stews. In parts of China and Russia, the plant's tender, young leaves are eaten as a vegetable.

Broad Beans (Fava Beans)

Commonly called fava beans in the United States, broad beans are also known as haba, English, Windsor, tick, cold, horse, or field beans. Broad beans get their name from the seeds, which are large and flat and look like very large lima beans. The seeds range in length from about a half inch to 2 inches. They can also vary in color from white, green, buff, brown, and purple to black.

Broad beans are an excellent source of nutrition. They are rich in fiber, folic acid, potassium, magnesium, and thiamin, to name just a few nutrients.

Preparation Tips

Broad beans are commonly available dried or precooked in cans, although they can sometimes be found fresh in the pod. The beans have a very tough skin that can be

IS IT SAFE TO EAT RAW SPROUTS?

All plant sprouts that are eaten raw may pose a health risk. The reason, according to the U.S. Food and Drug Administration, is that they could be contaminated with food-borne pathogens such as the *Salmonella* bacterium or another harmful bacterium, *Escherichia* (*E.*) *coli*. Pregnant women, children, the elderly, and persons with compromised immune systems (such as people with cancer) may be harmed by ingesting sprouts.

If you are one of the people in these groups, avoiding sprouts when eating out is advised. However, even the seeds that are used for sprouts that are grown at home can be contaminated.

removed by blanching (plunging the beans into boiling water briefly, then running them under cold water). This process loosens the skin, which can then be easily removed. Buying beans that are split and dried will help the beans cook faster and eliminate the need to get rid of the tough outer skin.

Sprouting at home

Serving Suggestions

Broad beans are starchy and strong-flavored and are great puréed or mashed and in salads. In Italy, broad beans are combined with other strong-flavored ingredients, and in France they are considered a good cocktail food.

A simple but elegant serving suggestion is to add a little butter to the cooked beans and a sprinkle of salt and pepper to taste and then serve topped with chopped parsley.

Chickpeas (Garbanzo Beans)

Botanical and archeological evidence reveals that chickpea plants were first domesticated in the Middle East in ancient times. Today, however, India supplies 80 to 90 percent of the world's supply of chickpeas.

The many names that chickpeas go by are a nod to the many regions of the world where they are now grown and eaten. In India, they are referred to as Bengal gram. In Spanish-speaking countries, they are garbanzo. The Arab world refers to them as hamaz (or hummus), and in Ethiopia they are called shimbra.

The plants grow in tropical to temperate regions and reach about 2 feet in height. Plants bear inflated inch-long pods enclosing one or two irregularly shaped seeds.

The seeds are about one-quarter to one-half an inch in diameter and can be buff-colored, yellow, brown, black, or green. The plant's young, green pods and sprouts can also be eaten.

Like many legumes, chickpeas are an excellent source of fiber. In addition, they are a good source of magnesium.

Preparation Tips

Chickpeas are available at most grocery stores both canned and dried. They can be eaten fresh, fried, roasted, or boiled. Generally, chickpeas should be soaked overnight before cooking, which is usually done by boiling them. Dried chickpeas may take as long as 2 hours before they are soft and ready to eat. A pressure cooker is also an option and can reduce cooking time by half.

Serving Suggestions

Chickpeas have a mild, slightly nutty flavor and a firm texture. They can be used in appetizers, salads, soups, or main dishes. Flour made from ground chickpeas can be made into breads or used as batter for deep-fat frying. They also can be combined with pasta or simply served by themselves. Sometimes they are served roasted and salted like peanuts. They are part of cuisine worldwide. In the

Middle East, they are mashed and used as the main ingredient in hummus, a thick sauce made with lemon juice, olive oil, and sesame seed paste. Hummus is becoming a popular dish in the United States. Falafel, a Middle Eastern croquette, is another dish that draws on the chickpea as its main ingredient. In the Mediterranean region, chickpeas are added to Spanish stews and Italian minestrone soups.

Common Beans

Beans provide a good source of dietary fiber. Beans or peas, eaten together or at separate meals, form a high-quality protein essentially equivalent to that from animal sources. (For more information on green beans, also known as snap or string beans, see page 257.)

Common beans originated in Central America, where archaeological remains of these beans, found in association with the remains of maize and squash, have been carbon dated to more than 7,000 years of existence. Columbus and subsequent Spanish and Portuguese explorers carried beans back to Europe and eventually introduced them throughout Asia and Africa.

A few of the most familiar beans are described here.

Black Beans

Many people have become acquainted with black beans after eating at Mexican restaurants, where they are served boiled or refried. Also known as turtle beans, black beans are a common part of the cuisine throughout Central and South America, the Caribbean, and the southern United States. As their name indicates, they have a completely black skin. They have a mild, somewhat sweet taste.

Black beans are an excellent source of folate and a good source of iron, magnesium, phosphorus, potassium, and thiamin.

Broad beans (fava beans)   *Chickpeas (garbanzo beans)*

Preparation Tips
Presoak beans. Black beans are relatively thin-skinned and cook quickly (about 30 minutes) if you want to keep them somewhat firm. For soups and stews, they may need to be boiled for 1 1/2 hours or longer.

Serving Suggestions
Black beans are a delicious dish all by themselves, served with a dollop of low-fat sour cream and bits of diced avocado or guacamole. They are also often stewed, accompanied by rice. They are the key ingredient in frijoles, refritos, or refried beans (but go easy on the fat), and they are the star of black bean soup.

Cranberry Beans
Cranberry beans are about a half-inch long and are brownish with pink splotches that disappear when cooked. Cranberry beans have a nutty flavor. Their creamy, red-streaked flesh has a smooth texture.

Nutritionally speaking, cranberry beans are an excellent source of folate and a good source of iron, magnesium, phosphorus, potassium, and copper.

Preparation Tips
Presoak beans. Cranberry beans can be substituted in any recipe that calls for red beans or white beans, and they can be prepared like pinto beans (simmering in water, covered, for 50 to 60 minutes).

Serving Suggestions
Cranberry beans can be added to pasta dishes and salads. In Europe, they often are used in stews.

Kidney Beans
As their name suggests, kidney beans are kidney-shaped. They are available in an assortment of colors. Chili lovers will readily recognize the most common kidney

Why Do Beans Cause Gassiness?

Beans do cause flatulence in many persons who eat them. The gassiness is the result of fermentation of the seeds' complex sugars, or oligosaccharides, by bacteria in the large intestine. Persons who eat beans frequently find that they do not develop gas as much.

To reduce the flatulence effect, try these strategies:
• The flatulence-producing effect of beans can be further reduced by changing the water several times during soaking and during cooking and by simmering the beans slowly until they are tender.
• If a recipe calls for salt, lemon juice, vinegar, or tomatoes, these ingredients should be added near the end of cooking because acidic ingredients stop the process by which legumes absorb liquid and soften.
• Use commercially available products that can be added to dishes before serving. These products contain an enzyme that breaks down the complex sugars before they start causing problems.

bean: the red kidney bean, which has a deep-red color and a full flavor.

Another type of kidney bean is the flageolet; its seeds are small, thin, and pale-green. Mainly available dried, canned, or frozen in the United States, flageolets are more popular in Europe. The cannellini bean is another type of kidney bean. It is large and white and has a more delicate flavor than the red kidney bean.

Kidney beans are an excellent source of folic acid and a good source of iron, magnesium, phosphorus, potassium, copper, and thiamin.

Preparation Tips
Kidney beans commonly are canned. If you choose to use fresh beans instead of canned, prepare red kidney beans by presoaking and then simmering them in water, covered, for 1 1/2 hours. Flageolets and cannellini beans can be cooked with the same method for 25 to 30 minutes.

Serving Suggestions
Consider red kidney beans an all purpose bean. They make good additions to chili

and can be baked, puréed, or refried. Cannellini beans can be added to salads or soups. Flageolets are often served in a white sauce seasoned with shallots, thyme, bay leaf, and clove. In this way, they are said to be served "French style." In France, they also may be served with a roast leg of lamb.

Pinto Beans
Pinto means "painted" in Spanish. It is an apt word to describe this bean because they are pink or beige with reddish brown spots and streaks. Pinto beans are popular in the American Southwest and in Mexico, where they are an essential part of everyday cuisine.

From a nutrition perspective, pinto beans are an excellent source of folate and a good source of iron, magnesium, phosphorus, potassium, copper, and thiamin.

Preparation Tips
Pinto beans usually are sold dried but are also available cooked and canned. Presoak dried beans and simmer them, covered, for 50 to 60 minutes.

Serving Suggestions

Pinto beans are great served with rice, as refried beans, puréed, or in chili.

White Beans

White bean is a term given to varieties of beans that have light-colored seeds. There are several varieties of white beans:

Marrow beans — The largest and roundest of white beans, marrow beans are often grown in the eastern United States. They are creamy but firm after cooking and usually are available as a dried bean.

Great Northern beans — Great Northern beans are smaller than marrow beans and have a more delicate flavor. They are typically grown in the Midwest. They stand up well to baking.

Navy beans — Most people know these as the beans found in canned pork and bean products. Navy beans require lengthy, slow cooking, which makes them excellent additions to soups and baked dishes.

White beans are an excellent source of iron and folate and a good source of magnesium, phosphorus, potassium, and copper.

Preparation Tips

Presoak beans. Cook them in water, covered for marrow beans, 35 to 45 minutes; for Great Northern, 1 to 1 1/2 hours; and for navy beans, 1 1/2 to 2 hours.

Serving Suggestions

Mild-flavored white beans can be puréed, baked, added to soups and stews, or combined with other vegetables and served with pasta. Navy bean soup is a hearty dish made from the navy bean. Baked beans, usually made of navy beans, are enjoyed cold or hot.

Slow cooking with molasses gives "white" navy beans their color.

Dolichos Beans

Derived from the Greek word *dolikhos*, meaning long or elongated, the general term "dolichos beans" is used to refer to legumes of the genus *Vigna* and the lablab bean. Common beans considered dolichos beans include the adzuki bean, the black-eyed pea, the lablab bean, and the mung bean.

Adzuki Beans (Azuki Beans)

Adzuki beans have been cultivated and enjoyed for many centuries in Asia. The Chinese have attributed mystical power to these beans and believe that they bring good luck. This is one reason they are a part of many foods used at celebrations.

The bean has a rich, somewhat sweet flavor. It has a reddish color. The plant that bears adzuki beans has 5-inch-long cylindrical pods that contain 4 to 12 oblong seeds with flat ends. Adzuki beans are usually sold dried, but they also may be sold as young pods that are eaten like green beans.

Adzuki beans are an excellent source of folate and a good source of iron, magnesium, phosphorus, potassium, zinc, and copper.

Preparation Tips

Presoak dried beans. Simmer in water, covered, for 30 to 40 minutes.

Serving Suggestions

Adzuki beans have a mild, delicate flavor and grainy texture. Sometimes they are eaten with rice. However, they are often made into a flour or paste, which is used in desserts or candies. Combined with sugar, water, starch, plant gums, and other ingredients, adzuki beans are used as a filling for bread, steamed cakes, and dumplings. In addition, they can be puffed like corn or sprouted. Adzuki beans may be roasted to make a substitute for coffee.

Black-Eyed Peas

The black-eyed pea has many names, including the cowpea, callivance, cherry bean, frijol, China pea, and Indian pea. It gets its name from the circular black hilum, or "eye," on the seed's inner curve, where it is attached to the pod. The hilum may also be brown, red, or yellow. The seeds can be wrinkled or smooth and range in shape from round to kidney-shaped. The plant that bears black-eyed peas is grown in warm regions of the world and can grow to 3 feet in height.

Black-eyed peas are an excellent source of folate and a good source of magnesium and potassium.

Preparation Tips

Because black-eyed peas have thin skins, presoaking is optional. Soaked or unsoaked versions of fresh and dried black-eyed peas cook in about 30 to 60 minutes, covered, over low heat. Black-eyed peas are also available frozen and canned.

Serving Suggestions

Black-eyed peas can be used to make soups, salads, fritters, and casseroles and are often served with meat. They are the key ingredient in the Southern dish called "Hoppin' John," which consists of black-eyed peas cooked with salt pork and seasonings and served with rice. According to southern U.S. tradition, eating black-eyed peas on New Year's Day brings good luck for the year.

Lablab (Hyacinth Bean)

Although grown in the United States mainly as an ornamental plant, the lablab is a popular food in Africa, Asia, and Central and South America. Seeds are less than a half-inch long. Their color ranges from white to brown, red, and black. The plant itself grows as a vine, with large broad leaves.

Lablab is an excellent source of iron and magnesium and a good source of phosphorus, zinc, copper, and thiamin.

Preparation Tips

Lablab pods can be eaten fresh. Dried beans can be prepared in the same way as other legumes.

Serving Suggestions

Lablab sprouts can be eaten. Often, however, lablab seeds are ground into flour and used to make bread or an oatmeal-like dish. In India, lablab seeds are dried, split, and then cooked.

Mung Bean

Although its name suggests Chinese or Asian cuisine, the mung bean has been grown in India for centuries. India is still one of the leading producers of this legume.

Mung beans are also grown in the United States, where they are sometimes referred to as a "chickasaw pea." Sometimes this bean is also known as green gram, golden gram, and chop suey bean (mung bean sprouts are an important ingredient in this dish).

Mung bean seeds can be green (the most common), yellow, brown, or mottled black. The seeds themselves are tiny — about one-eighth inch in diameter.

Nutritionally speaking, the mung bean seed is an excellent source of folic acid and a good source of magnesium, phosphorus, and thiamin.

Preparation Tips

Mung beans are available as dried beans or as sprouts. Wash sprouts thoroughly before use.

Adzuki

Lablab

Mung

Black-eyed peas

Dolichos bean varieties

Beans do not need to be soaked before cooking. Whole beans cook in about an hour.

Serving Suggestions

Mung bean sprouts can be used fresh in salads or stir-fried with vegetables, noodles, and meat, poultry, seafood, or tofu.

Beans can be ground into flour to make noodles (called bean threads or cellophane noodles because of their thinness and transparent appearance) or candy. Puréed mung beans may be used to fill breads and pastries. They are also used to make moog dal, an Indian spread eaten with rice or bread.

Lentil

The lentil was probably one of the earliest legumes to be domesticated. Now cultivated in many parts of the world, the lentil is known as ads in Arabic, merimek in Turkey, messer in Ethiopia, heramame in Japan, and masoor, dal, or gram in India, which is now the leading producer.

This bushy plant reaches a maximum of 2 feet in height. The pods are short, flat, and oblong. Seeds can be red, orange, yellow, brown, or green. The seeds are classified as large (macrospermae) or small (microspermae), with each type containing dozens of varieties.

Lentils, like many legumes, are an excellent source of folate and a good source of potassium, iron, and phosphorus.

Preparation Tips

Lentils are available in a variety of forms. They can be purchased whole, husked, and split like peas. They cook quickly and, thus, dried lentils do not need to be soaked before preparation. Different varieties call for different cooking times — from 5 minutes for yellow lentils to 30 minutes for brown or green lentils. Lentils should always be

Lentils

washed before cooking to remove dirt, dust, and, possibly, tiny stones.

Serving Suggestions

Because lentils do not hold their shape well, they are popularly used to make soups and stews. They also can be added to salads or mixed with grains to make breads and cakes. Lentils are particularly popular in India, where they are frequently made into a spicy dish called dal. Dal is made with lentils, tomatoes, onions, and other seasonings. Dal also can be made from many other legumes.

Lima Beans

Lima beans were originally cultivated in South America, from where they were brought to Europe, Asia, and Africa by European explorers. In the southern United States, lima beans are called butter beans, and the mottled purple varieties are called calico or speckled butter beans. Lima beans also may be referred to as Madagascar beans.

Lima bean plants bear flat, oblong pods about 2 to 4 inches in length that contain two to four smooth, kidney-shaped seeds. There are numerous varieties of lima beans, and their seeds vary in size and color. The commonly sold seeds are pale green, but purple, red, brown, black, and mottled ones are also available. The two most common varieties are the Fordhook and the baby lima, which is smaller and milder.

Lima beans are a good source of iron, magnesium, phosphorus, and potassium.

Preparation Tips

Fordhook and baby lima beans are sometimes available fresh in their pods. Immature lima beans can be eaten fresh with or without the pods. Although mature pods are too tough to be edible, the seeds are available year-round in frozen, canned, and dried forms that are usually labeled according to size rather than botanical variety. Presoak dried beans, then simmer in water, covered, for 60 to 90 minutes.

Unlike many other types of beans, lima beans can be easily overcooked; they quickly become mushy if cooked longer than necessary to make them soft.

Serving Suggestions

The taste of cooked lima beans is starchy but delicate. They can be boiled and served whole or mashed, or they can be added to soups and salads. In succotash, a traditional Southern dish that includes peppers, tomatoes, and corn, lima beans are the main ingredient.

Peanuts

Despite their name, peanuts are not nuts at all, but the seeds of a legume. They are commonly thought of as nuts because of

Lima beans

how they are used and because of their nut-like shells. The "shells," however, are actually the fibrous seed pods of a legume, encasing one to three seeds wrapped in an edible, papery thin seed coat. These seed pods are easy to crack and range from less than an inch to about 2 inches long and have the same contours as the round seeds underneath.

Peanut plants are separated into either bunch or runner types. The bunch type bears seed pods close to the base of the plant, whereas the runner type has seed pods scattered along the branches. Runner types were introduced in the 1970s and are now more popular than bunch types, probably because runner peanuts are primarily used to make peanut butter, for which half of all peanuts are produced.

Spanish peanuts — a bunch-type peanut with small, round seeds covered by a reddish brown skin — are usually roasted, salted, and vacuum-packed. Virginia peanuts, which can be a runner or a bunch plant, are larger and more oval and are usually sold roasted in the shell.

Peanuts contain quite a bit of fat, but the fat in them is primarily monounsaturated fat. Peanuts are an excellent source of magnesium, phosphorus, zinc, copper, niacin, and folate and a good source of iron.

George Washington Carver, an African-American botanist who worked in the late 19th century, is well known as the "Father of the Peanut Industry" for having ingeniously developed more than 300 uses for the peanut, including as an ingredient in shoe polish, soap, bleach, medicine, ink, paint, and ice cream. In 1890, an American physician invented what we now know as peanut butter to provide an easily digestible, nutritious food for his elderly patients. However, long before this, other cultures made similar edible paste from peanuts.

Preparation Tips

Peanuts are available in a variety of forms, including raw, dry-roasted or honey-roasted, salted or unsalted, shelled or unshelled, peeled or unpeeled, whole or chopped, and as peanut butter. The young pods, leaves, and plant tips can be cooked and eaten in the same manner as a green vegetable. Unshelled peanuts can be refrigerated in an airtight container for up to 6 months, and shelled peanuts for up to 3 months. Peanuts also can be cooked, a process that generally takes about 30 minutes.

Serving Suggestions

Although peanuts are usually consumed as a snack, turned into peanut butter, or used to make candy or baked goods in the United States, they are frequently used as

Peanuts

a vegetable in African, Indian, South American, and Asian cooking. Peanuts can be cooked with fish, meat, and poultry and used to flavor sauces, soups, salads, and desserts. Peanut soup, a southern U.S. favorite, is a creamy, spicy-hot dish.

Peanut butter — not just for sandwiches

Peas

Like peanuts, peas may not be immediately recognizable as a legume, because they are marketed and consumed as a vegetable. But appearances are deceiving because, like other legumes, peas are an excellent source of protein. A serving of peas (about 3/4 of a cup) contains as much protein as a tablespoon of peanut butter, but with far less fat.

The pea is an annual plant that grows from 1 to 5 feet high. It requires a cool, relatively humid climate. There are more than 1,000 different types of peas. Different plants produce smooth-seeded peas, wrinkled-seeded peas, field peas, snow peas, and sugar snap peas.

Smooth-seeded peas are commonly sold frozen, whereas wrinkled-seeded peas are used for canning because they are sweeter. Field peas are grown mainly for drying. Snow peas and sugar snap peas are grown for their edible crisp, sweet pods.

Peas are a good source of iron.

Preparation Tips

Peas are available fresh, canned, frozen, or dried. Dried peas, which can be yellow or green, are sold whole. Or, they may be split. Whole dried peas need to be soaked before cooking and may take up to 1 to 2 hours to become soft. Split dried peas do not need to be soaked; however, they do not hold their shape during cooking and so are generally used for sauces and soups. A familiar dish made of split peas is, of course, split pea soup. Snow peas and sugar snap peas are usually sold fresh. Before eating, rinse them off and then cut the top from a snow pea pod. Remove the string from both sides of a sugar snap pea's seams by pulling the attached fibrous string upward from the bottom. Although they are delicious raw, pea pods also can be cooked in the same way as green beans. Many Asian dishes call for them as a key ingredient.

To shell fresh peas (usually green peas or English peas), use the same stringing technique on only one side of the pod, then use your thumb to push out the peas. Rinse thoroughly. When buying fresh peas, look for bright-green, smooth, uniform pods that are free of spots, dryness, or other blight.

Serving Suggestions

Peas can be added to meals in a variety of ways. They can be braised, boiled, steamed, or stir-fried and added to pasta dishes and casseroles. Pea pods can be substituted in any recipe that calls for green beans.

Raw, fresh peas, although difficult to find, are tasty additions to salads. So are snow peas and sugar snap peas. Fresh and frozen peas should be cooked only briefly to preserve their color and flavor.

Soybeans

If the only soy in your diet comes from the soy sauce you sprinkle on chow mein, you may be missing out on more than just a tasty and versatile food. Soy is an inexpensive way to add protein to your diet and may also help reduce fat when substituted for meat in traditional dishes.

Soy products come from the soybean, a legume native to northern China. The United States now produces much of the world's soybeans. There are more than 1,000 varieties of soybeans. They range in size from a pea to a cherry. Colors include red, yellow, green, brown, and black. The protein in soy is a "complete" protein — the most complete you can get from vegetable sources — and just as good nutritionally as animal protein. In fact, there is more protein in 1 cup of soybeans than in 3 ounces of cooked meat. In addition, soybeans are an excellent source of a variety of nutrients, including iron, vitamin B_6, and phosphorus, and a good source of potassium and calcium. Also, they are rich in the phytochemicals called isoflavones.

Soybeans are usually processed into other products. For example, soy oil is used to make ink for newspapers. Soybean products are added to a variety of foods during processing. Many foods, though, are made almost entirely from soy (see sidebar: "Soy" Many Products to Choose From, page 333).

For cultures in which soy is the main source of protein, rates of cardiovascular disease and some kinds of cancers are relatively low. Researchers are also looking into whether soy plays a role in preventing osteoporosis and easing hot flashes associated with menopause. Some studies suggest there may be a link (see Chapter 2 sidebar: Soy What? page 34).

Although it may be too early to make specific health claims for soy, there is evidence that adding soy to your diet makes good nutritional sense. And, you may just discover a whole new range of healthful food products to enjoy.

Preparation Tips

Generally, the soybeans now being incorporated into food are already processed, such as the soybean oil in margarines and salad dressings and the soy protein in baby formula and meat substitutes. Or, they may already be incorporated into food products, such as tofu. Dried soybeans are often available at health food stores. They need to be soaked before cooking to soften them. They are usually simmered, sometimes up to 9 hours, before they are softened enough to eat. Fresh and frozen soybeans are now appearing in markets. Many cooks recommend cooking soybeans with full-flavored items because they are rather flavorless by themselves.

Serving Suggestions

Although Americans generally eat soy as part of other products, soybeans can be eaten fresh, roasted, ground into flour, or pressed into oil. The Chinese first invented soy sauce, then the Japanese borrowed the process to make their own sweeter version. Dark and light varieties are available. Serve dark soy sauce with red meat and light soy sauce with chicken or seafood. Because of its high sodium level, it should be used sparingly if you are watching your salt intake.

Tofu, made from coagulating soy milk until it forms curds, is available in soft, firm, and extra-firm consistencies, depending on how much liquid was left in the pressed curds. Tofu can be used in salads, soups, and stir-fried dishes.

NUTS AND SEEDS

Nuts are high in fat and calories, but taken in moderation they can be part of a healthful diet. Nuts are not only flavorful, but, ounce for ounce, also full of nutrients.

Most nuts are seeds or the dried fruit from trees. Peanuts, which are commonly thought of as nuts, are actually legumes. They belong to the same family as peas and beans.

The word "nut" can be confusing. The term originally referred to an edible kernel surrounded by a hard shell. In its most scientific definition, the term now refers to a single-seed fruit with the seed surrounded by a dry, tough fruit. This definition works for hazelnuts, beechnuts, and chestnuts, but it does not "fit" almonds and walnuts (because their surrounding fruits are theoretically edible) or peanuts (which are legumes). Adding to the confusion, Brazil nuts and pine nuts are not nuts, either. They are actually seeds, and sunflower seeds are actually fruits.

"SOY" MANY PRODUCTS TO CHOOSE FROM

Soybean-based products take many forms. Besides oil that is pressed from soybeans, soy food products include the following:

Miso — Miso is a salty, strong-flavored paste made by fermenting soybean meal and a grain, such as rice or wheat. It is used as a flavor enhancer and thickening agent in many Asian dishes. Colors range from light yellow to dark orange, depending on the type of soybean used to make the miso.

Soy flour — Many prepared foods include soy flour, which can be made from whole soybeans or soybean meal.

Soy nuts — Roasted soybeans, commonly eaten as a snack.

Soy sauce — Another product made from fermented soybeans, soy sauce is an essential condiment and cooking ingredient in Asian cuisine. To make soy sauce, soybeans or soy meal is mixed with ground wheat. Fermentation may take 1 to 3 months.

Soybean sprouts — As the name suggests, these are germinated soybeans. They are used in salads and as a garnish.

Soybean milk — Soybean milk is made from soybeans that are soaked, ground, heated, and then filtered. Soy milk is a part of many baby formulas for infants who are lactose-intolerant. Soy milk can be drunk as a beverage. Dried soy milk is sometimes added to products such as ice cream.

Tempeh — An essential component in Indonesian cooking, tempeh is a meat substitute that is made from soybeans that are soaked, dehulled, cooked, and then fermented. Tempeh can be formed or made into patties. Tempeh is available in the United States in health food stores.

Tofu — Made by coagulating the protein from soy beans, tofu is a high-protein curd that is used in many Asian dishes. Tofu is sold in blocks and is often used in place of meat. It generally has a spongy texture. It can be cooked in a variety of ways to make dressings, dips, and shakes. It picks up the flavor of marinades and thus will taste like meat in many dishes.

Yuba — A product made of the protein-rich skin that forms on the top of soybean milk when heated just to boiling, yuba has a stringy, chewy texture. It is made of protein and oil and is often used to produce imitation meat products.

Common soybean products, such as soy milk, tofu, and fresh soybeans

Nuts are a versatile food. They can be eaten fresh, cooked, and, sometimes, with their shell. They are available whole or chopped, salted and unsalted. In addition, products made from them include butters, oils, and spreads.

Nutrition

The protein in nuts and seeds lacks an essential amino acid called lysine, which can be gained from legumes and animal products. Although nuts are high in calories for their size, they are also considered a "nutrient-dense" food. They contain a lot of nutrients in relation to their calories.

Nuts are also rich in different plant compounds. Flavonoids, for instance, are found in all nuts. These antioxidants help reduce the formation of free radicals in the body that may contribute to cancer and cardiovascular disease. Relative to their size, nuts are also among the best plant sources for protein.

Nuts are generally high in fat. In most cases, more than 75 percent of their calories comes from fat (the exception being chestnuts — only 8 percent of the calories are from fat). But, on the plus side, it is the "right kind" of fat. Most of the fat in nuts is monounsaturated and polyunsaturated, with the exception of the coconut and palm kernel. Unlike saturated fats (typically found in red meats and dairy products), these fats do not appear to increase blood cholesterol levels. In small amounts, monounsaturated and polyunsaturated fats may actually lower cholesterol levels.

Watch the salt, however. Nuts do not come by that naturally — it is an added feature. If you need to limit your salt intake, look for products with no added salt.

Selection

Purchase nuts with a clean, uniform appearance. When buying nuts in the shell, look for whole, unbroken shells. To ensure maximal freshness, look for nuts that are vacuum-sealed in bags, jars, or other containers.

Storage

Because of their high fat content, nuts and seeds should be stored in dark, cool, dry conditions in closed glass or plastic containers to prevent rancidity. Unshelled nuts keep better than shelled nuts, which can become rancid in a few weeks unless frozen. Most unshelled nuts will keep 2 months to a year in the refrigerator or a year or more in the freezer.

Safety Issues

Allergies to nuts are one of the most common kinds of food allergies. Symptoms of an allergic reaction include nasal congestion, hives, itching, swelling, wheezing or shortness of breath, nausea, upset stomach, cramps, heartburn, gas or diarrhea, lightheadedness, or fainting. If you suspect that you have any food allergies, see an allergist for a careful evaluation.

The following pages are an overview of nuts and seeds available, along with tips to include them in your meals.

Almond

Although almonds are native to the warmer regions of western Asia and northern Africa, they grow well in California, where 99 percent of domestic almonds are grown.

The almond fruit is the edible seed of sweet almond trees. Almonds are teardrop-shaped and surrounded by a shell and an outer fleshy hull. They can be either sweet or bitter. The sweet types have a delicate yet distinctive flavor, but the bitter almond is inedible in its raw form because it contains traces of the poison prussic acid, a cyanide compound. During the late 19th century, cross-breeding among sweet almonds created new varieties that include the California, Mission, Price, Carmel, and the Nonpareil (the most popular variety today). Since then, approximately 30 additional varieties have been developed and are grown commercially.

Nutritionally, almonds have more calcium than any other nut and are an excellent source of iron, riboflavin, and vitamin E. More than 60% of the fat comes from monounsaturated fat.

Preparation Tips

Almonds are available whole (both shelled and unshelled), sliced, slivered, diced, and chopped. If you buy whole almonds that are not already blanched, you need to blanch them yourself by pouring boiling water on them, draining them after a minute, and then pouring cold water on them, draining again after a minute. After blanching, rub off the skins. Toasting almonds before adding them to dishes enhances their flavor and texture.

Serving Suggestions

The mild flavor of almonds makes them a popular addition to a wide variety of dishes. They are popular as fillings and as ingredients in pastries and other baked goods. But almonds also can be used in sauces, stuffing, and pasta dishes. A well-known vegetable

Almonds

dish is green beans amandine, which essentially is green beans topped with slivered almonds. Marzipan, a German candy, is made from sweetened almond paste. Almond oil may be added to salads, and almond extract or essence may be used as a flavoring in baked goods and drinks.

Beechnut
Beech trees grow in the temperate forests of Europe, North America, Asia, and North Africa and can be up to 120 feet tall. Beechnuts are simply the seeds (or nuts) enclosed in prickly burrs that fall to the ground in the autumn. Beechnuts look like small chestnuts and taste like hazelnuts. More than 10 species of beech trees produce nuts.

Beechnuts are an excellent source of thiamin and riboflavin and a good source of iron. Monounsaturated and polyunsaturated fats provide more than 80% of the fat content.

Preparation Tips
The thin coat surrounding each cream-colored nut must be removed before it is eaten.

Although beechnuts were once used as feed for farm animals, they were also commonly eaten during famines.

Serving Suggestions
Beechnuts taste best roasted, a process that mellows the flavor of these nuts. Ground beechnuts also may be used as a coffee substitute. In Europe, oil is made from beechnuts and is used for cooking.

Brazil Nut
The Brazil nut grows on an evergreen tree, mainly along the Amazon, Orinoco, and Rio Negro rivers of South America. About 6 inches in diameter and resembling a coconut, the fruit of the Brazil nut tree has 18 to 20 nuts within its hard shell. Inside, the triangular nuts are arranged like the wedges of an orange. It is the white kernel inside this shell that is eaten. This kernel is

rich, creamy, and sweet and is a good source of phosphorus and thiamin and contains some calcium. This nut also has a high fat content, although most of it is unsaturated.

Only after the fruits have fallen to the ground are they harvested and chopped open to obtain the nuts. Because the trees grow wild in the heart of the Amazon jungle, the nuts can be shipped only during the rainy season, when streams and rivers are navigable. Although the nuts are obtained only in Brazil, most are exported, and native Brazilians rarely eat Brazil nuts.

The Brazil nut is also known as the para nut, butternut, cream nut, and castanea. Oil from the nut is used to make soap or as an industrial lubricant.

Preparation Tips
Brazil nuts are available as raw, unshelled nuts or shelled and roasted or dry-roasted. Because of their high fat content, both

A variety of nuts and seeds

Brazil nuts

shelled and unshelled nuts should be refrigerated or frozen to maintain freshness. Buy only vacuum-packed nuts to maintain their freshness.

Serving Suggestions
Brazil nuts add a crunchy texture to cakes and breads. Because of their sweetness, they are also used in ice cream, cookies, and candies. Brazil nuts can be difficult to shell. Keep a nutcracker handy if planning to serve them whole and in the shell, often a tradition in many families at holidays.

Cashew Nut
India is the world's leading producer of cashew nuts, although other important producers include Mozambique, Tanzania, Kenya, and Brazil.

Cashew trees produce a fruit-like stalk called the cashew apple (even though it is the shape of a pear). Attached to the end of the cashew apple is the cashew nut, which has a smooth, ash-colored outer shell. The edible kernel of the nut varies between three-quarters of an inch and an inch in length. Its generic name, *Anacardium*, refers to a heart shape, but cashews have more of a kidney shape.

It is impossible to buy cashews in the shell because the edible kernel is covered with an inner shell, and the space between the inner and outer shells is filled with a thick, caustic, toxic oil (related to poison ivy). Unshelled, unroasted cashew nuts will burn the mouth and lips and raise blisters on the skin.

Nutritionally, cashews are an excellent source of iron, phosphorus, and zinc and a good source of riboflavin, thiamin, and potassium. The primary source of fat in cashews is monounsaturated fat.

Preparation Tips
Cashew nuts have a sweet, buttery flavor that is further enhanced by roasting. Cashews are bought ready to use. Look for cashews sold in vacuum-packed cans to guarantee maximal freshness.

Serving Suggestions
Although Americans view the cashew as a nut for baking or dipped in chocolate for a treat, it also can be added to stir-fry dishes and salads. Cashews also can be made into a delicious nut butter.

Chestnut
Chestnut trees are found in Asia, Europe, and North America, and their fruit has been made popular in the lyrics of a Christmas carol. The trees that bear chestnuts were first cultivated in China about 5,000 years ago. Mount Olympus, home to the gods of ancient Greece, was said to have had an abundance of chestnut trees.

Cashew nuts

Chestnut trees are also long-lived. Some trees that were grafted more than 500 years ago are still alive. The trees are also adaptable to poor land. The nuts are harvested by hand once they have fallen from the burr, the name for the spiny outer covering. Trees typically start bearing nuts after about 25 years.

Chestnuts can range in size from a half inch to more than an inch, depending on the variety. They have one rounded side and one flat side and one rounded end and one pointed end. The roasted kernel is soft, meaty, and sweet.

The common American chestnut, native to the eastern United States, once dominated the forests of Maine to Georgia. But this tree is now nearly extinct because of a tree blight caused by a fungus. Today, only a small number of trees survive.

Preparation Tips
The most traditional use of chestnuts is to roast them whole, with shell and skin, and to peel and eat them while they are still warm. Before roasting, cut an "x" into the flat side of the nuts to prevent them from exploding. Chestnuts also can be boiled or steamed.

Serving Suggestions
Use chestnuts as an ingredient in stuffing, casseroles, and baked goods. A glazed type of candy (marrons glacés) is made from sugared chestnuts. Ground chestnuts also can be used as a coffee and chocolate substitute.

Ginkgo Nut
The ginkgo nut grows on what is considered the oldest known living species of tree. It has remained virtually unchanged for 250 million years and is considered a living fossil.

The ginkgo nut is the seed of an inedible, apricot-like fruit that is well known

Roasted chestnuts

for its strong, some might say offensive, odor. Nuts borne from female trees have hard, buff-colored shells that are pointed on either end. The meat inside has the same shape and color, is soft, and has a delicate, sweet taste. Although the ginkgo tree is commonly grown in U.S. cities because it is resistant to air pollution, many people are unaware that it bears nuts and that they can be eaten.

In Asia, ginkgo nuts and the fan-shaped leaves from the tree (*Ginkgo biloba* is the scientific name) have been used in traditional medicine for thousands of years. Extracts and powders derived from ginkgo leaves are among the top-selling herbal supplements sold in the United States, where they are promoted as being a memory aid.

Scientists in the United States are studying the efficacy of ginkgo extracts, but results so far have been inconclusive. Additionally, because scientists do not know how the extract works or its potential side effects, they are not recommending ginkgo extracts be used for medicinal purposes at this time.

From a nutrition standpoint, ginkgo nuts are an excellent source of phosphorus, potassium, copper, thiamin, and niacin.

Preparation Tips
Before ginkgo nuts can be consumed or cooked, the hard shells must be removed. The kernels must then be soaked in hot water to loosen the thin skins. Fresh, dried, or canned ginkgo nuts are widely available in Asian grocery stores.

Serving Suggestions
Ginkgo nuts can be eaten as a snack after they are roasted. They are used in many Asian main dishes and desserts. Ginkgo nuts are a key ingredient in an egg-custard-like dish called chawanmushi.

Hazelnut
The United States produces about 9,000 tons of unshelled hazelnuts, and Oregon and Washington account for 5 percent of the entire world's production. The grape-sized nuts have a thick, woody, brownish red shell that readily separates from the ker-

nel. Nuts grow from leafy husks that open as the nut ripens. The nuts begin to drop from the trees in August. In September or October, they are swept up, washed, and then dried. They are sorted for size and then sold for processing. Hazelnuts are referred to by a variety of names, including filbert or cobnut.

There are 10 major species of hazelnut trees. The most commonly grown are the European, the American, and the Turkish hazelnut. Various hybrids are based on these types. These hybrids include the Winkler, Duchilly, and the Barcelona, which is thought by many hazelnut fanciers to produce the finest nuts.

Nutritionally speaking, hazelnuts are an excellent source of iron, magnesium, phosphorus, potassium, and thiamin and a good source of niacin. Monounsaturated fat is the primary source of fat in hazelnuts.

Preparation Tips
The hazelnut kernel is sweet. It is wrapped in a thin, slightly bitter-tasting, brown skin. This skin must be removed before the nuts are used. Heat the shelled nuts in a 350° Fahrenheit oven for 10 to 15 minutes until the skin begins to flake. Then, wrap the

Hazelnuts

warm nuts in a tea towel and rub them with the towel after about 5 minutes to remove the skin.

Serving Suggestions

Hazelnuts can be eaten raw or they can be roasted. Chopped hazelnuts add crunch to salads, soups, sauces, breads, cakes, and cookies. Finely ground nuts can be used as a replacement for flour. Hazelnut paste, made by grinding the nuts to a fine butter and mixing with sugar, is used for bakery fillings and toppings.

A fragrant oil made from pressed hazelnuts adds a nutty flavor to dressings, sauces, pastries, and coffee.

Lotus Seed

The lotus plant and especially its flowers have been used as a motif in the arts of the ancient cultures of India, China, and Egypt for centuries. Although many people know it is grown for its ornamental flowers, the lotus plant also has edible roots and seeds. The creamy seeds add a mild flavor to cuisine in many Asian countries.

In addition to the seeds, the roots of the lotus plant can be eaten. When sliced and cooked, the light brown, starchy roots have a crisp texture. They are commonly used as a vegetable in Asian dishes. Even the leaves of the plant may be eaten. Young leaves can be eaten raw. Mature leaves may be used to wrap rice, meat, and fruit dishes before they are steamed.

Lotus seeds are referred to by a variety of names. Common ones include Indian lotus, hasu, and nelumbium.

Preparation Tips

You can most readily find lotus seeds in Asian supermarkets, where they are sold fresh, dried, and canned. Fresh seeds are sold with the seed coat attached or removed, but both types spoil quickly. The seed coats must be removed before the nut can be eaten.

Serving Suggestions

Lotus seeds can be eaten raw. However, most commonly, they are used in cooking. Or, they are candied and eaten as a snack or added to pastries as a filling. In Asian cooking, they may be stewed with poultry, added to stir-fry meals, or cooked in water to make a dessert.

Macadamia Nut

Native to Australia, the macadamia tree was named for the naturalist Dr. John MacAdam. According to legend, he was the first to find the nuts edible.

The macadamia nut is enclosed in a very hard, brownish shell, and the kernel is off-white in color. Each nut is about half an inch to an inch in diameter and is enclosed in a thin, fleshy husk. This husk opens as the nut matures. The slightly

Macadamia nuts

sweet, creamy, rich flavor of the nutmeat has acquired a "gourmet" reputation and a hefty asking price.

Macadamia nuts have the highest fat and calorie content of any nut, but they are an excellent source of magnesium, copper, and thiamin and a good source of iron and niacin. The fat in the macadamia nut is primarily monounsaturated fat.

The tree that bears macadamia nuts was introduced to some regions of the United States in the 1880s. It was not until the 1930s, however, that a real consumer audience made the macadamia industry possible. Historically, Hawaii has supplied 90 percent of the world's macadamia nuts. Growers in California and Florida also have begun contributing to domestic production. Macadamia nuts are also grown in Brazil, Costa Rica, Guatemala, Venezuela, Jamaica, South Africa, and Samoa.

Preparation Tips

Because of their extremely hard shell (which takes up to 300 pounds of pressure to crack) and high oil content, macadamia nuts are usually sold shelled in vacuum-packed containers. The nuts can become rancid quickly because of their high oil content. Refrigerating them, however, can extend their shelf life up to 2 months.

Serving Suggestions

Macadamia nuts may be bought raw or roasted, plain, chocolate-covered, or salted, and whole or chopped. They are great by themselves as a snack. Chopped macadamia nuts add a nice texture and flavor to salads, rice dishes, curries, cookies, cakes, candies, and ice cream. The nuts also can be ground into a creamy butter and used as a spread. Oil made from the nuts can be added to salads or used for cooking.

Pecan Nut

Belonging to the same family as walnuts, pecan trees can be found growing wild from Illinois to Texas and Maryland to Florida. In fact, more than 100 varieties of pecans are cultivated in the United States. Commercial cultivation of pecans, however, is limited to warm states. Georgia produces more than a third of the total U.S. production, which totals about 200,000 tons of unshelled pecans annually.

Pecans are one of the most widely cultivated nuts in the world. The nut itself is elongated and wrinkled, resembling a walnut, and has a buttery flavor. The shell surrounding pecans is shiny, brown, and easily cracked. Most of the fat in pecans is unsaturated. In addition, pecans are an excellent source of many nutrients, including phosphorus, thiamin, copper, and zinc, and a good source of iron and potassium. More than half of the fat in pecans is monounsaturated fat.

Preparation Tips

Unshelled pecans are most widely available during the autumn months, although packaged, shelled nuts can be found in grocery stores year-round. Tightly wrapped, unshelled nuts can be stored in a cool, dry place for up to 6 months; shelled nuts can be kept in an airtight container in a freezer for up to 2 years.

Serving Suggestions

Most people know and love pecan pie, a sugary treat popular in the South. Pecans can be eaten raw or roasted as a snack. They add a rich flavor to breads, muffins, cakes, and other baked goods or can be sprinkled on cereal or many other types of food. A candy with Southern roots is pecan praline, made from sugared pecans. Oil pressed from pecans may be used for cooking or for salads, but it is expensive.

Pecan nuts

Pistachio Nut

Pistachio nuts grow in clusters on the pistachio tree, which grows wild throughout central and western Asia and is cultivated in warmer areas of the world.

The nuts are enclosed in fleshy husks. The nut itself has a hard, thin, tan shell that partially splits open when the nut is ripe. The shells may be dyed red or blanched white by distributors to hide imperfections. Inside the shell is a smooth, pale-green kernel wrapped in a fine brownish skin. This kernel has a delicate and sweet flavor, which lends itself to desserts.

Shelled and unshelled nuts should be kept in an airtight container in the refrigerator, where they will keep for 3 months. Pistachios are available year-round.

From a nutrition perspective, pistachios are an excellent source of iron, magnesium, phosphorus, potassium, and thiamin. The main source of fat in pistachios is monounsaturated fat.

Preparation Tips

Pistachios with fully closed shells are immature and should be avoided. The nuts can be eaten as a snack either raw or roasted. In addition, they can be added to puddings, cakes, candies, and luncheon meats. Or, chopped nuts can be added to stuffing, pâtés, and sauces.

Pecan pie

Pistachio nuts

Serving Suggestions
In Asian and Mediterranean cuisine, it is common to find pistachios in meat and poultry dishes as well as in pastries. Baklava, the popular Middle Eastern dessert made with the thin pastry called filo, contains pistachios as a main ingredient, along with honey and walnuts.

Safflower Seed
The safflower plant has been cultivated in India, China, Persia, and Egypt for centuries and it was only introduced in the United States in 1925. India still remains the largest producer of safflower seed.

Common names for safflower seeds include safflower, false saffron, and saffron thistle. Safflower seeds are about a quarter of an inch long and elongated.

They are an excellent source of iron, magnesium, thiamin, and riboflavin and a good source of potassium and niacin.

Preparation Tips
Safflower seeds are enclosed in a fibrous seed coat that must be removed before they can be eaten or pressed for oil. Up to 40 percent of the weight of the seeds is oil.

Serving Suggestions
The seeds of the safflower are eaten fried or roasted. They are one of the primary ingredients in an Indian dish called chutney. The plant's shoots are also edible and often eaten as salad greens.

Sesame Seed
Sesame seeds are small, oval, and flat and have a paper-thin, edible hull. They may be white, yellow, brown, red, or black. Lighter-colored seeds are considered preferable to dark seeds.

Sesame seeds are 40 to 60 percent oil by weight. They are an excellent source of iron, thiamin, riboflavin, and phosphorus and a good source of potassium. Polyunsaturated fat is the primary source of fat.

Preparation Tips
Available hulled, unhulled, or ground, sesame seeds have a rich, nutlike flavor when roasted. Hulled seeds should be stored in the refrigerator because they quickly turn rancid.

Serving Suggestions
As anyone who has eaten a hamburger knows, sesame seeds are commonly sprin-

Sesame seeds

kled over the tops of buns. They are also used on bread, rolls, and crackers. Sesame seeds are mainly a condiment in the United States, but they can be added to stews for a delicate, nutlike flavor. In cuisine of Middle Eastern countries, sesame seeds are the main ingredient, along with honey and almonds, in a confection called halvah. Ground sesame seeds are also made into butter. In the Middle East, this is called tahini and is a popular addition to sauces and main dishes. Chickpeas and tahini can be combined to make hummus, a popular part of Middle Eastern cuisine.

Squash Seed
Cultivated worldwide, squash probably was first gathered by indigenous people around 8000 B.C., primarily for its seeds. European settlers arriving in North America found it was a common crop grown by American Indians.

Although most people are familiar with eating the flesh of squash, the seeds of squash by themselves are both tasty and nutritious. The most popular squash seed eaten in the United States is the pumpkin seed, which is flat with one rounded end and one end that tapers to a point. The seeds are off-white and approximately three-quarters of an inch long. They enclose a green kernel that is delicately nutty in flavor. In Mexico, pumpkin seeds are known as pepitas and are popular in many dishes.

The seeds of winter squash are also commonly eaten. The plants are harvested when the fruits and, hence, the seeds are fully ripe. Summer squash, in contrast, is eaten before it and its seeds are fully ripe. Seeds from acorn squash, buttercup squash, and butternut squash are also edible. The seeds are a healthful addition to your diet, being an excellent source of iron, phosphorus, and potassium and a good source of thiamin and riboflavin. A little less than half the fat comes from polyunsaturated fat.

Preparation Tips
Squash and pumpkin seeds are sold plain or salted, dried or roasted, unhulled or hulled. You also can make your own snack when hollowing out your Halloween jack-o'-lantern or cooking your winter squash. Scoop out the seeds, rinse them, dry them, and then toast them in the oven until they are golden brown. Soaking them in salt water before baking adds flavor.

Serving Suggestions
Seeds can be added to salads, sauces, and pasta dishes to add a crunchy texture or to soups for additional texture.

Sunflower Seed
With more than a hundred species, the sunflower genus is thought to be native to either western North America or South America, although it now grows worldwide. The United States and the former countries of the Soviet Union are among the largest producers of sunflower seeds.

Sunflower seeds are about a quarter of an inch long, angular, and grayish green, tan, or black. They are enclosed in thin shells. Often, these shells are striped in black and white. Two main varieties of sunflower are cultivated commercially. The Russian variety is grown mainly for oil because the seeds contain more than 40 percent oil by weight. The North American variety has larger seeds and is grown for human consumption and for bird food.

Although high in calories and fat, sunflower seeds contain mostly unsaturated fat. In addition, they are a rich source of

Sunflower seeds

nutrients, including folic acid, niacin, potassium, and zinc. They are also high in fiber. Sunflower seeds are best kept in a cool, dry place. If shelled, they should be stored in the refrigerator to prevent rancidity.

Preparation Tips
Sunflower seeds usually are shelled before they are added to dishes. Shelling by hand can be difficult, although soaking helps. Most people simply buy commercially shelled sunflower seeds.

Serving Suggestions
Sunflower seeds can be eaten raw, roasted, and salted or unsalted and are terrific snacks by themselves. They also are excellent additions to salads, stuffing, or yogurt. Ground seeds can be combined with flour to make breads, as is done in Portugal and Russia. The seeds also can be roasted to make a coffee-like drink. A note to cooks: the seeds may turn green when cooked.

Walnut
Walnut is the common name for about 20 species of deciduous trees that include the hickory and pecan tree. The two most common species are the black walnut and the English, also known as the Persian, walnut.

The fleshy green fruit of the black walnut encapsulates the nut and cushions it when it falls from the tree. Effort is required to pry the edible kernel from the nut's thick, woody shell. The inside of the fruit may stain your hands.

Roasted seeds from squash are also edible.

The kernel of the English walnut however, is more easily removed from it shell, which is generally thinner and easie to crack. In addition, the husks of Englisl walnuts separate when the fruit is mature allowing the nut to drop out. Partly fo these reasons, the English walnut is th more commercially cultivated species.

Nutritionally, walnuts are an excel lent source of phosphorus, zinc, coppe and thiamin and a good source of iro and potassium. More than 70 percent of the fat comes from polyunsaturated fat.

Preparation Tips

Walnuts purchased in the shell should be free of cracks or holes. Shelled nuts should

Walnuts

be plump, meaty, and crisp. Unshelled walnuts can be stored in a cool, dry place for up to 3 months. Shelled kernels can be refrigerated in a tightly covered container for up to 6 months. Shelling walnuts can

be difficult, but commercially shelled walnuts are readily available. Toasting walnuts enhances their flavor, helps maintain crispness, and makes them easier to chop.

Serving Suggestions

Walnuts are a versatile cooking ingredient that can be added to salads, pilafs, cookies, muffins, breads, cakes, and ice cream. Whole young nuts can be pickled in vinegar. Walnuts also can be pressed to make an oil that is typically used in salads.

Walnuts also have a variety of nonculinary uses. The shells are used as an antiskid agent for tires and as blasting grit. Ground nutshells are sometimes added to commercial spices as a filler.

The traditional Waldorf salad is made with walnuts.

DAIRY FOODS

Includes: Milk, Yogurt, and Cheese

Milk is a bundle of nutrients, all contained in a nondescript white liquid. Although milk's presence as a beverage at meals may not be as popular as it used to be, milk is used in many products that are consumed throughout the day.

On the Food Guide Pyramid, milk and dairy products are placed near the top because, although they are part of a healthful diet, they should be consumed in moderation. Adults should consume 2 servings of low-fat or nonfat dairy products daily; 1 serving equals 1 cup of milk or yogurt or 1 1/2 ounces of cheese. Children and pregnant or lactating women should add an extra serving each day. Milk and other dairy foods are rich in calcium, a mineral important for developing strong bones and teeth and for nerve transmission. They are also an important source of many vitamins and minerals. Large quantities of these foods, however, are not needed to ensure that you are getting adequate amounts of these nutrients. Just three 8-ounce glasses of skim milk, for example, provide nearly all of the calcium you need each day.

Some people do not include enough dairy foods in their diets. One reason is the mistaken belief that all dairy products are high in fat. Some are, but there is an abundance of low-fat and nonfat dairy products, from milk to yogurt to cheese.

Other people do not consume dairy foods because of intolerance to milk sugar or allergy to milk proteins. However, those with intolerance to milk often do not need to follow a diet that is completely milk-free (see sidebar: Lactose Intolerance, page 347). People with allergy to milk must avoid dairy foods and may want to get help with adjusting their diets to ensure nutritional adequacy (see sidebar: Milk Allergy: Hidden Ingredients, page 349).

Another reason people do not consume dairy products is the growing consumption of soda pop. The average American drinks about a half gallon of milk a week but, in comparison, about 11 cans — or a gallon — of soda pop a week. Taking calcium supplements or eating calcium-enriched food can help you obtain needed calcium, but dairy foods are an easy way to get the calcium and other essential nutrients you need.

Basics

Milk can be consumed in its fluid form, in a more solid form (such as yogurt), as cheese, or as a major ingredient that is added to other foods. Dairy cases now abound with milk-based products and their reduced-fat and nonfat versions. The cornucopia of dairy products includes the following:

Fluid milk — Although cow's milk is generally consumed in the United States, other cultures use milk from goats, camels, llamas, reindeer, sheep, and water buffalo. Milk is a staple in diets worldwide.

Dried and concentrated milk — These products include powdered milk, evaporated milk, and condensed milk.

Cheese — Cheese is made by coagulating and draining milk or cream or a combination of both.

Yogurt — Yogurt is made by adding bacteria to milk to ferment it.

Ice cream and other dairy desserts — Ice cream and other frozen desserts are simply milk or cream to which sugar, flavorings, and, often, eggs have been added.

Cream and sour cream — Cream is the fat that rises to the top of the surface in unprocessed milk. Sour cream is simply cream that has been fermented or thickened. The cream is usually "soured" by adding bacteria to it, much in the way that yogurt is created.

Butter — This yellowish substance is essentially fat that has been separated from cream. For that reason, it is discussed in Fats, Oils, & Sweeteners, page 389.

Processing of Milk and Milk Products

Virtually all milk and milk used in dairy products is pasteurized. Pasteurization is a process invented by French chemist Louis Pasteur. It uses heat to destroy harmful bacteria in milk, but it retains the nutritional value of milk.

Pasteurization kills bacteria that have been responsible for major plagues such as tuberculosis, polio, scarlet fever, and typhoid fever. It is also advantageous because it destroys many of the bacteria that cause spoilage and many of the enzymes that promote rancidity. Pasteurization, therefore, increases both shelf life and safety of milk.

A common term that consumers see when purchasing milk is "homogenized." Homogenization is a process introduced in the 1950s in which fat globules in the milk are broken down so they are evenly dispersed throughout the milk. Most milk at the supermarket is homogenized.

During homogenization, milk is forced through a small opening at high pressure. The product has a smoother, richer texture and a whiter color than nonhomogenized milk.

Nutrition

Milk and dairy products provide many of the key nutrients needed daily, particularly calcium (for more specific information, refer to milk in the nutrient table on page 472).

Milk and dairy products also supply high-quality protein. Because of its animal source, milk protein is complete — meaning it provides a sufficient amount of the nine essential amino acids (see Chapter 2, Protein, page 23).

Dairy products are also naturally rich in B vitamins and most of the minerals considered to be essential in the diet, including calcium, magnesium, phosphorus, zinc, iodine, and selenium. In addition, milk also contains several vitamins and minerals that have been added to meet the requirements of the Food and Drug Administration. Low-fat and nonfat milk may be fortified with vitamin A because this fat-soluble vitamin is lost when the milk fat is removed. Vitamin D is added to all milk to help the body better use calcium.

Milk also is a good source of carbohydrates. With the exception of cheeses and butter, milk products are higher in carbohydrates than protein or fat. Milk's carbohydrate is lactose, a sugar unique to milk that is actually two sugars (glucose and galactose) linked together. Food scientists call this type of sugar a "disaccharide."

Lactose is not as sweet as other sugars. It helps the body absorb calcium and phosphorus and may even help in the growth of friendly bacteria needed in the intestines. In addition, galactose, one of the sugars in lactose, is a vital part of brain and nerve tissue. It is released when the body digests lactose. Lactose is a bit of a paradox, however. Although it has these beneficial properties, many people have difficulty digesting milk (see sidebar: Lactose Intolerance, page 347).

The processing of milk begins at the farm.

LACTOSE INTOLERANCE

As many as 50 million Americans are estimated to have lactose intolerance — an inability to adequately digest ordinary amounts of dairy products such as milk and ice cream.

Worldwide, nearly 70 percent of the adult population is thought to be lactose intolerant, and the condition is very common among American Indians and those of Asian, African, Hispanic, and Mediterranean descent.

Lactose is the sugar that is naturally present in milk and milk products. It must be broken down by lactase (an enzyme found in the intestine) before the body can use it. If there is not enough lactase, undigested milk sugar remains in the intestine. Bacteria in the colon then ferment this sugar. Gas, cramping, and diarrhea can follow.

Most of us begin to lose intestinal lactase as we age. However, this occurs to varying degrees. Thus, people with lactase deficiency vary in their ability to comfortably digest milk and milk products.

As obvious as the symptoms of lactose intolerance may be, it is not easily diagnosed from the symptoms alone. Many other conditions, including stomach flu and irritable bowel syndrome, can cause similar symptoms.

See your physician to determine whether you are lactose intolerant. Measurement of the hydrogen in your breath after you have taken in lactose is a useful test because large amounts of hydrogen indicate that lactose is not being fully digested and that you are probably intolerant.

Persons with milk allergies should avoid milk, but those with lactose intolerance often do not need to follow a diet that is completely lactose-free. The following suggestions may help:

- Avoid eating or drinking large servings of dairy products at one time. (Several smaller servings over the course of a few hours are much easier to digest.)
- Drink milk or eat dairy products with a meal.
- Choose hard or aged cheeses, such as Swiss or cheddar, over fresh varieties. Hard cheeses have smaller amounts of lactose and are more likely to be tolerated.
- Take lactase tablets or drops, such as Lactaid or Dairy Ease. These types of products contain the enzyme that breaks down lactose, reducing the amount that your body must digest on its own.

For help with meal planning, you may want to see a registered dietitian.

Despite all the nutrients in milk, the nutritional advantages of dairy products must be weighed against the potential health drawbacks of two key components in milk: sodium and fat. Whole milk, cream, and cheeses contain substantial amounts of fat, especially saturated fat. These fats add calories and have been tied to higher cholesterol levels and cardiovascular disease (see Chapter 3, Coronary Artery Disease, page 61). However, it is important to note that low-fat and nonfat milk varieties are available and are significantly lower in fat than whole milk. In addition, depending on how much is consumed, milk or products made from milk may be a major source of sodium — a special concern for anyone following a low-sodium diet.

Selection

Unless dried or canned, milk and dairy products are perishable. For that reason, most have an expiration date printed on the packaging. The date often states, "Sell by . . ." and is a good indicator of freshness. Look for the date before buying and before consuming a product. Usually, dairy products will keep about a week beyond that date.

Storage

Keep milk in the coldest part of your refrigerator. Avoid storing milk in the refrigerator door unless it has a special compartment designed to keep the milk colder than in the rest of the refrigerator.

Keep yogurt and fresh cheeses in airtight containers in the refrigerator. Loosely wrapped, these foods will pick up smells in the refrigerator, possibly leaving them with an undesirable taste.

Cheeses such as cottage cheese, ricotta, and cream cheese will keep for 1 week after the sell-by date. Soft cheeses — such as Brie, Camembert, Muenster, and mozzarella — and blue-vein cheeses can keep from 1 to 3 weeks. Semi-firm and hard cheeses, such as cheddar and Monterey Jack, will keep as long or longer. Generally, the harder the cheese, the longer it will remain fresh when carefully stored.

Shredded cheese will not keep as long because it has more surface exposed to the air. Soft cheese that has mold on it should be discarded. Firm cheese that has mold can sometimes be used as long as 1/2 inch to 1 inch of cheese near the molded spot has been cut away and discarded. If any milk or milk product has a strange odor, throw it out.

Safety Issues

Some small markets or independent farmers still sell raw milk. Because it has not been pasteurized, this milk may contain germs that make you ill. For that reason, the sale of raw milk is often prohibited by law, depending on location.

For some people, proteins in cow's milk may trigger allergic reactions. Whey proteins (beta-lactoglobulin and beta-lactalbumin) and casein are the primary proteins that trigger allergic reactions. Symptoms of a milk allergy may include nasal congestion, hives, itching, swelling, wheezing, shortness of breath, nausea, upset

stomach, cramps, heartburn, gas or diarrhea, light-headedness, and fainting.

It is easy to confuse a milk allergy with another common health concern related to dairy foods — lactose intolerance. Lactose intolerance (see page 347) also can lead to nausea, vomiting, cramping, and diarrhea. However, if you have lactose intolerance, you usually can eat small amounts of dairy food without problems. In contrast, a tiny amount of a food to which you are allergic can trigger a reaction.

If you suspect that you have any food allergies, see your physician. You may then be referred to an allergist for a careful evaluation. This generally includes a medical history, physical examination, and skin or blood testing. If the diagnosis is a milk allergy, it is essential to eliminate milk and foods made with milk from your diet (see sidebar: Milk Allergy: Hidden Ingredients, page 349).

MILK

Milk has a wide variety of uses and thus is one of the most basic items found in kitchens worldwide. It is consumed as a beverage, poured on cereals, and used in many different ways in cooking. In response to consumers with different nutritional demands, modern food science has made many different types of fluid milk available.

Although milk can be less inviting to people who are concerned with their weight because of its high fat content, some types of milk contain no fat at all.

Whole milk — Containing 3 1/2 percent milk fat, this type of milk is often simply labeled "milk" or "vitamin D milk" if that particular vitamin has been added. Of all types of milk, whole milk is among the highest in fat and calories. One cup has 150 calories and approximately 8 grams of fat.

Reduced-fat milk — Often referred to as two percent, this type of milk has had some milk fat removed from it. Two percent reflects the amount of fat in the milk by weight. It does not refer to the percentage of calories from fat. One cup of 2 percent milk has 130 calories and 5 grams of fat.

Low-fat milk — Also known as 1 percent milk, this type of milk contains about 100 calories and 2.6 grams of fat in 1 cup.

Nonfat or skim milk — Skim milk, which contains less than 0.5 percent milk fat, is now more often labeled nonfat milk. It contains the same amount of nutrients, such as calcium, as its higher fat counterparts, but it has no fat and just 90 calories.

Buttermilk — Buttermilk was once the residue left from churning butter, but today's version is made from adding a lactic acid culture to milk. The result is far less rich than the original "natural" buttermilk, but it still retains the thick texture and acidic tang of old. Some manufacturers add flecks of butter for an authentic look or stabilizers to prevent separation. Because of its name, buttermilk may sound high in fat. Yet, in most instances it is not. Buttermilk derives its fat content from the milk used to make it, and in the United States low-fat or nonfat milk is used most often. Calories and fat in buttermilk depend on what type of milk was used to make it. Check the label for fat content.

Acidophilus milk — Normally killed during pasteurization, the healthy bacteria culture *Lactobacillus acidophilus* is reintroduced into whole, low-fat, or nonfat milk to create sweet acidophilus milk. In a/B milk, both acidophilus and bifidobacteria cultures are added. Acidophilus occurs naturally in the body and is found mainly in the small intestine. Many factors can alter the level of this intestinal bacterium, including diet, alcohol consumption, illness,

MILK ALLERGY: HIDDEN INGREDIENTS

If you are allergic to milk, it is important to check the labels of the foods you eat. Milk or components of milk are common ingredients in many different foods, some of which you might not be aware of.

Foods that likely contain milk include the following:
- Butter, including artificial butter flavor, butter fat, ghee, buttermilk
- Chocolate, caramel, nougat
- Cheese, including cream cheese, cottage cheese, and cheese curds
- Simplesse (a fat substitute)
- Luncheon meats, hot dogs, and sausages
- Cream, including half-and-half, nondairy creamers (containing casein)
- Margarine
- Yogurt
- So-called nondairy items, including coffee lighteners, whipped toppings, imitation cheeses, and frozen or soft-serve dessert items. These may contain casein, a milk protein.

Ingredients on the label that indicate milk or a dairy product is in the product include the following:
- Lactalbumin
- Lactoglobulin
- Rennet casein
- Lactose
- Casein
- Hydrolysates (casein hydrolysates, milk protein hydrolysates, protein hydrolysates, whey or whey protein hydrolysates)
- Evaporated, dry, or milk solids
- Whey
- So-called natural flavorings (check with manufacturer for more information)

and medications. Alterations in levels can sometimes lead to poor digestion, diarrhea, and bloating. Drinking acidophilus milk products may reduce intestinal infection and diarrhea and improve milk digestion and tolerance. Studies are under way to evaluate whether these bacteria can help regulate blood cholesterol levels and prevent cancer. Calories and fat in acidophilus milk depend on what type of milk was cultured with the acidophilus bacterium. If whole milk was used, for example, acidophilus-treated milk contains the same amount of fat and calories as whole milk.

Lactose-reduced and lactose-free milk — These products are tailored to people who have trouble digesting lactose, a sugar found in milk. An enzyme called lactase is added during the processing of this milk. The result is that lactose in the milk is reduced by at least 70 percent (lactose-reduced) or up to 99.9 percent (lactose-free). Calories and fat in lactose-reduced and lactose-free milk depend on what type of milk was cultured.

Ultrapasteurized milk (UHT) — This milk has been popular in Europe for many years, but it has only recently appeared in U.S. supermarkets. The "ultra-heat treatment" (UHT) sterilizes milk by quickly heating it, sometimes as high as 300 degrees Fahrenheit, and then quickly cooling it before packaging it in vacuum-packed, aseptic containers. UHT milk can be stored for 2 to 3 months without refrigeration, until opened. Once opened, UHT milk should be refrigerated and quickly consumed. It can spoil, but unlike other milk, it does not curdle as a warning sign of spoilage. The ultra-heat process makes the milk taste slightly scalded, but it is thought that the treatment does not substantially affect the nutrient value. The amount of fat and calories in UHT milk depends on the type of milk from which it was made.

Flavored milk — Flavorings, sugar, or other sweeteners are added to fluid milk. Although some milk is flavored with strawberry, vanilla, and even peanut butter flavorings, chocolate is used most commonly. Flavored milks are higher in calories than their unflavored counterparts. Calories and fat in flavored milk range from 150 calories and no fat for 1 cup of chocolate milk made from skim milk to more than 210 calories and 8 grams of fat for 1 cup of whole chocolate milk. When purchasing flavored milks, look for products that are made from skim or reduced-fat milk. Avoid those labeled "premium." They often are made from whole milk, which is high in fat.

Preparation Tips

It is easy to overcook milk when heating it. When milk is heated to a temperature that is too high, its proteins clump together and curds appear in the milk. When heating milk, always use low heat and stir frequently. Using a double boiler when heating milk also helps prevent overheating.

Anyone who has ever had homemade hot cocoa knows that heated milk can develop a "skin" (a thickened surface). An easy way to prevent this is to mix a little cornstarch into the milk before heating it.

Serving Suggestions

Nonfat milk has the least amount of fat and calories but still provides all of milk's nutrients. Many people prefer its lighter texture and taste to the heaviness of whole milk. Even the staunchest fan of whole milk can easily be converted to using this healthier alternative. Make the change gradually. Start by mixing equal parts of whole milk with 2 percent milk. Then, in stepwise progression, use just 2 percent, next a combination of 2 percent and 1 percent, then just 1 percent, then a combination of 1 percent

SOY AND RICE "MILK"

Nondairy beverages made from soybeans and rice are available for persons who have milk allergies, are lactose-intolerant, or prefer not to eat animal products. Both soy milk and rice milk are similar in look and taste to milk and can be used for many of the same purposes, such as pouring on cereal or making smoothies and other dairy-based beverages.

Soymilk is made from water, soybeans, malted corn and barley extract, seaweed (used for thickening purposes), salt, and, depending on the brand, a fat or oil. Rice milk is made from water, brown rice, salt, and, depending on the brand, safflower oil and various vitamins and minerals. Soy and rice milks have fewer calories and less fat than whole milk, but they have more of both than skim milk. One cup of soymilk has about 135 calories and 5 grams of fat. One cup of rice milk has about 120 calories and 2 grams of fat.

and skim milk, and eventually only skim milk. If you dislike skim milk, 1 percent or 2 percent milk is a reasonable option, especially if it keeps you drinking milk.

Buttermilk can be substituted for cream in many recipes: a half cup of buttermilk has 1 gram of fat, but the same serving of light cream has 31 grams.

Whenever possible, lower the amount of fat in a recipe by substituting a lower-fat milk. A cream soup made with low-fat milk is just as rich tasting, especially if you thicken the soup with a bit of flour. A cup of cocoa made with skim milk provides more nutrients and fewer calories than the average chocolate dessert, and it is just as effective for satisfying a sweet tooth.

DRIED & CONCENTRATED MILKS

Dried and concentrated milks offer convenience and increased shelf life. Such products also can be used in numerous recipes to give the taste of milk without all the water volume of fluid milk. Dried and concentrated milk products include the following:

Powdered milk — As its name suggests, powdered milk is milk that has had nearly all the water removed from it. Mixing it with water (follow package directions) results in fluid milk. Powdered milk is usually made from skim milk because having less fat helps the product resist rancidity. Recipes sometimes call for powdered milk to thicken sauces or to add calcium and protein to foods. Powdered milk does not taste like fresh milk. However, its stability and portability make it a frequent ingredient in convenience foods or prepared mixes. These qualities also make it an excellent choice for travelers. Opened packages of low-fat, nonfat, and butter-

SELECTING MILK FOR CHILDREN

Milk recommendations for children older than 1 year are a practical concern for parents. Can everyone in the family drink the same milk, or should you buy certain milk for only the youngest members?

Breast milk, of course, is recommended during the first year of life. If the infant is weaned during the first year, the best alternative is to use iron-fortified formula. Formula-fed infants should remain on iron-fortified formula until 1 year of age. After age 1, the American Academy of Pediatrics recommends using whole milk if the use of breast milk or formula is discontinued. Until age 2, fat should not be limited in an infant's diet.

After age 2 years, children can begin to consume fat in moderation, just as the rest of the family does. However, whether to use low-fat milk will vary for each child. If a child has had poor growth or a chronic medical condition, discuss the appropriate fat content of milk — and diet in general — with the child's physician. For otherwise healthy children older than 2 years, low-fat milk (skim, 1 percent, or 2 percent) is fine — and it means you have to buy only one type of milk for the entire family.

milk forms of powdered milk are good for up to 6 months. Reseal opened packages and store in a cool, dark, and dry place to prevent the product from absorbing moisture in the air. All opened packages of dried milk keep better in the refrigerator because temperature and humidity are better controlled.

Evaporated milk — Packaged in sealed cans, evaporated milk is milk from which about 60 percent of water has been evaporated. The milk is homogenized and packed in heat-sterilized cans that help extend shelf life. Unopened cans can be stored at room temperature for up to 6 months but should be refrigerated and used within 5 days after being opened. Evaporated milk has a slightly darker color than regular milk because it was heated during the evaporation process.

Sweetened condensed milk — Thick and sold in sealed cans, condensed milk is often used to make desserts because it is usually sweetened. Like evaporated milk, more than half of the water in sweetened condensed milk is removed by heating it. Up to 40 to 45 percent of this product by weight is sugar. In its undiluted form, sweetened condensed milk contains nearly 13 tablespoons of sugar and 1,000 calories per cup. For that reason, it is best to eat small servings of desserts and dishes made with this ingredient.

Preparation Tips

Dry milk can be reconstituted according to package directions, or it can be mixed with cool water in a blender for a more pleasing consistency. When using sweetened condensed milk, cut down on the amount of other sugar you use in the recipe to save on calories.

Serving Suggestions

Nonfat dry milk can add a boost of calcium to recipes for meatloaf, hot cereal, gravy, or canned cream soups. A tablespoon contains

94 milligrams of calcium but adds only 27 calories. Many brands are fortified with vitamins A and D for a nutrient bonus.

Evaporated milk often is used in soups and sauces for a smooth, creamy texture. Well chilled, it also can be whipped and used as a dessert topping with only a tenth of the calories of heavy whipped cream.

CREAM AND SOUR CREAM

Cream is made from milk fat. It is extremely high in fat and calories, but it creates the creamy, rich, indulgent taste in desserts and sauces.

A century ago, cream was skimmed from the top of milk that was set in a cool place. Today, machines separate commer-

cially made cream. Many types of cream are sold:

Heavy cream and light whipping cream — Heavy cream has between 36 and 40 percent milk fat by weight. The thickest of the "sweet creams," heavy cream is used mainly for whipping cream and for desserts. A lower-fat version, with 30 to 36 percent milk fat, is called light whipping cream. For this cream to whip properly, emulsifiers and stabilizers are added to the cream. Both products double in volume when whipped. For this reason, whipping cream is also called double cream.

Light cream — Light cream contains 18 to 30 percent milk fat by weight and cannot be whipped. Instead of doubling in size, it remains the same volume. Hence, it is called single cream. Other names for light cream include table cream or coffee cream, because it is often the cream used to fill coffee creamers.

Half-and-half — This is a mixture of equal parts of whole milk and light cream, homogenized to prevent separation. It contains from 10 to 12 percent milk fat by weight and can be substituted in many recipes calling for cream. The product cuts calories and fat, but it lacks some of the velvety qualities of heavy or light cream. Half-and-half commonly is added to coffee, although a far healthier alternative is skim milk.

Sour cream — Real sour cream contains 18 to 20 percent milk fat by weight. It is created commercially by introducing a bacterial culture to cream that converts the milk's sugar, lactose, into lactic acid. The acid gives sour cream its distinctive, tangy flavor. Stabilizers such as sodium alginate, carrageenan, locust bean gum, or

WHAT GOES INTO YOUR COFFEE COUNTS

What goes into the brew you purchase at your local espresso bar matters. For example, here's how your choice of milk affects two popular espresso-based beverages:

Latte (12 ounces)	Calories	Fat (grams)	Carbohydrates (grams)
Made with whole milk	190	11	14
Made with skim milk	95	0	14
Cappuccino (12 ounces)			
Made with whole milk	155	9	11
Made with skim milk	55	0	9

gelatin are sometimes added to make sour cream thick and smooth, and rennet and nonfat milk solids are added to give it more body. Low-fat and light sour creams are both made with half-and-half according to the same process to create a similar product with 60 percent less fat than regular sour cream. Fat-free sour cream substitute is made with the same process, and skim milk is used as the base.

Nondairy creamers and toppings — These imitation dairy products sometimes contain coconut oil, palm kernel oil, or other highly saturated and hydrogenated vegetable oils mixed with casein (a milk protein) and lactose (a milk sugar). These ingredients create a high level of saturated fat without providing any of the vitamins and minerals found in milk or cream. Fat-free and sugar-free versions of regular coffee creamers, lighteners, and nondairy whipped toppings have various amounts of sugars and fats. Pressurized whipped cream, packed in cans under pressure, uses gas to expand the cream. Aerosol dessert toppings do not contain any milk or cream.

Preparation Tips

Because cream is highly perishable, it should be stored in the coolest part of the refrigerator and used quickly. To whip cream, chill the cream thoroughly and place the beaters and bowl in the freezer for 10 minutes before whipping. Whip at medium speed until the cream thickens.

Serving Suggestions

Although air is added to whipped cream, a dollop atop a special treat adds extra fat (mostly saturated): about 3 grams for 1/4 cup. Cream sauces served over pasta are also high in fat. Instead, try a vegetable-rich marinara sauce. Substitute a lower-fat version of cream in recipes when possible, or use milk or yogurt. For recipes calling for sour cream, try buttermilk or yogurt.

Cream Glossary

Chantilly cream — Named for the place in France where it is believed to have originated, chantilly cream is made simply by adding sugar and vanilla to whipped cream.

Clotted cream (also called Devonshire cream) — A sour cream that originated in Devonshire, England, clotted cream is thicker than regular sour cream. After being heated and cooled, the cream is skimmed and then eaten on scones with jam.

Crème fraîche — A product that falls between fresh cream and sour cream, crème fraîche is used often in French cooking, in which it is served lightly whipped and sweetened. It is made by adding a small amount of buttermilk to cream and heating the mixture. It is then stored in a warm place until it thickens. This usually takes between 12 and 36 hours. The mixture is then refrigerated and can be kept up to a week.

Smetana — A dense Russian sour cream, smetana is traditionally served on borscht and salads. It is also known by the names smitane, smatana, or sliuki.

If substitutions leave you yearning for the real thing, or if you eat more of the lower-fat item than you would have eaten of the higher-fat ingredient, you may want to stick to the recipe and eat a smaller serving as an occasional indulgence.

CHEESE

Despite the high fat content of most forms of cheese, cheese remains an American favorite. Dairy cases are filled with different varieties of cheeses, and classic foods such as pizza, cheeseburgers, and tacos, all of which use some form of cheese, guarantee generations of cheese lovers.

The first cheese was said to have developed by accident, when milk was allowed to ferment. Whether the first cheese was formed from Mongolian yak's milk, the African camel's milk, or the Middle Eastern ewe's milk is unknown and still debated. But the results, after thousands of years, remain the same: the earliest coagulating curds of milk carried in a shepherd's pouch have become a tempting treat, with many different types from which to choose.

Cheese can be made from various milks. Milk from cows is typically used in the United States, but milk from sheep, goats, camels, and other animals is used worldwide. In fact, some of the world's finest gourmet cheeses are made from sheep's milk.

No matter what type of milk is used, the process is essentially the same. The first step is to curdle the milk, essentially causing proteins in the milk to clump. Bacterial cultures or certain enzymes are used to curdle the milk. Next, the liquid surrounding the curds, which is called the whey, is drained.

Then the curds are pressed into shapes. Salt may be added at this point. The freshly made cheese is then allowed to age, a process that develops its flavor. Other ingredients also may be added at this point. In general, 11 pounds of milk are needed to make 1 pound of cheese. Knowing that, it is easy to see why cheese is dense in both calories and fat.

Like any dairy product, cheese is perishable. A general rule is that the harder the cheese, the longer it keeps. Categories of cheese are determined by the method used to make it, the type of milk used, the texture, or even the appearance of the rind. This classification system groups cheeses with common characteristics.

Fresh Cheeses

These cheeses were once made on the farm from surplus cream and quickly served. Today these cheeses are made with pasteurized milk, but they still have a short shelf life and must be consumed quickly. Fresh cheeses are not allowed to ripen or ferment very long, so they have a high moisture content, a mild flavor, and a smooth, creamy texture. They generally keep for 1 week after purchase or the "use by" date.

Common types of fresh cheeses include the following:

Cottage cheese — Usually thought of as a "diet" food, cottage cheese is a healthful food choice when it is made from skim or low-fat milk. Cottage cheese is only a few steps from milk. It is essentially the separation of milk into curds and whey. The curds are partially drained before cottage cheese is packaged and sold.

Cream cheese — The mild white spread often used for bagels, cream cheese is a better choice than butter, but it still has a lot of fat. Up to 90 percent of the calories in cream cheese are from fat. One tablespoon has about 50 calories and about 5 grams of fat. Even reduced-fat cream cheese is high in fat, with up to 75 percent of calories from fat. From a calorie and fat standpoint, fat-free cream cheeses are the best choice.

Farmer's cheese — Often used in baking, farmer's cheese is essentially cottage cheese that has had most of the liquid

Fresh mozzarella cheese, tomato, and basil on crackers.

pressed out of it. It is usually sold formed as a loaf and is relatively low in fat.

Mozzarella — The pizza topping of choice, mozzarella is a soft, bland cheese. Unlike other fresh cheeses, mozzarella has undergone a heating and kneading process. Whole-fat, skim, low-moisture, and fat-free versions of mozzarella cheese are available. Fresh mozzarella, sold in specialty and ethnic stores, is usually made from whole milk and, therefore, is higher in fat than other types.

Ricotta — A common ingredient in Italian dishes, ricotta is similar to cottage cheese but has a finer texture. Ricotta was once made from whey left over from making other cheeses. Today, it is made from whey and milk.

Semisoft Cheeses

Semisoft cheeses are firm on the outside yet soft and moist on the inside. Because they are aged for just a few weeks, they have a soft, moist texture and mild flavor.

Semisoft cheeses are used widely in cooking because they melt smoothly and easily. They are also easy to slice and so are excellent for hors d'oeuvres or for more ordinary uses, such as sandwich toppings.

Because these cheeses are soft, they are often coated with wax or another material to keep them intact. Some types of semisoft cheeses are aged. Others are "washed" in brine, which causes them to develop a rind on the outside. These processes also intensify the cheese's flavor and, in some instances, sodium content.

Some common types of semisoft cheeses include:

Brick — The shape of this cheese is the origin of its name. When aged, it has an assertive flavor, like cheddar cheese. When it is young, it is mild.

GOAT CHEESES

Goat's milk has a little more calcium than cow's milk but it is deficient in vitamin B_{12} and folate. Goat's milk has smaller fat globules than cow's milk, so it does not need to be homogenized. However, it does need to be pasteurized for the same reasons cow's milk should — to kill any germs that might be harmful to humans.

Goat's milk is not an acceptable alternative for persons who are lactose-intolerant, because it contains lactose in the same percentage as cow's milk. Nor is goat's milk or goat cheese a lower-fat alternative. Some people prefer the taste and texture of goat cheese, but 1 1/2 ounces of hard goat cheese has 13 grams of fat — about the same as cheese made from whole cow's milk, such as cheddar.

Soft goat cheeses such as Montrachet and some types of feta — a salty, white cheese originating in Greece which can be made with goat or sheep's milk — contain more moisture and are comparable to cheeses made from part-skim cow's milk. One and one-half ounces of feta cheese has 9 grams of fat, about the same amount as in an equal amount of mozzarella or Neufchâtel cheese.

Edam — A Dutch specialty, this cheese has a mild, buttery taste. It is often sold in balls or blocks coated with red wax. It is also available smoked.

Gouda — Another Dutch cheese, Gouda is sold in wedges and wheels usually covered in red wax. Like other semisoft cheeses, it has a mild flavor that becomes sharper as it ages. Gouda can be purchased as a smoked cheese.

Jarlsberg — A Norwegian specialty, this cheese is often compared to Swiss cheese. It is softer, however, and milder. Jarlsberg is also typically sold in wedges.

Limburger — Famed for its characteristic aroma, Limburger is one of the strongest-flavored semisoft cheeses. Limburger is eas-ily sliced and can add a different twist to ordinary foods, such as sandwiches.

Provolone — The taste of this cheese depends on its age and how it is processed. Young provolone has a mild taste and ivory color. With age, its flavor becomes stronger, its texture drier, and its color darker. The cheese is sometimes smoked or has had a smoke flavoring added to it. Provolone is often sold in loaves.

Semisoft cheeses are generally higher in calories and fat than soft cheeses. For example, 1 1/2 ounces of Edam or provolone cheese has about 150 calories and about 12 grams of fat. In contrast, the same amount of cottage cheese (made with 2 percent milk) contains about 40 calories and 1 gram

REDUCED-FAT AND IMITATION CHEESES

Reduced-fat cheese usually is made from nonfat milk, but additives are needed to create the creamy texture of full-fat versions. Imitation cheese does not necessarily mean the product contains no milk: it may use casein (a milk protein) and emulsifiers, enzymes, and artificial flavorings and colors. Other imitation cheeses are made from soybean derivatives. Both reduced-fat and imitation cheeses can be used as you would use regular cheese if you do not plan to cook the cheese. Both reduced-fat and imitation cheeses tend to have a denser, more rubbery texture when heated and may not melt in the way traditional cheese does.

of fat. However, semisoft cheeses generally contain less fat and calories than hard cheeses because less milk is used to make semisoft cheeses than hard cheeses.

Soft, White-Rind Cheeses

Soft, white-rind cheeses are descendents of natural-rind cheeses, in which gray, green, and even red molds are allowed to grow on the surface of the cheese as it ripens. Most North American cheese consumers are put off by the colored mold growth, so the colorful natural-rind cheeses are nearly impossible to obtain outside of France.

Soft, white-rind cheeses are readily available, however. Instead of allowing natural mold growth, these cheeses are sprayed with white mold spores that seal the outside while allowing the interior of the cheese to main-

tain a soft, butter-like consistency at maturity. These cheeses garner their characteristic flavor from bacteria that grow on the outside and move inward. The result is a rich, creamy texture and full flavor. These cheeses often have fewer calories than hard cheeses.

Soft, white-rind cheeses include:

Brie — A cheese originating in northern France, Brie is often sold in wedges and has a tangy, buttery flavor.

Camembert — Also originating in northern France, Camembert has a velvety texture and a soft, light-yellow interior. Camembert is often wrapped in foil and sold in wooden boxes.

Blue or Blue-Veined Cheeses

Blue-veined cheeses are created by the introduction of a blue mold into the milk before it thickens. The blue color, however, would not appear as the characteristic blue-green veins in cheeses without exposure to air. Therefore, the cheese is pierced with steel rods to let air circulate.

Most blue cheeses are made in the style of classic European blue cheeses. They can be firm or creamy and any color from chalk-white to golden-yellow. The flavor of these cheeses grows stronger with age. Although these cheeses are high in fat, only a small amount is typically used because of their strong flavor. Blue cheeses keep for 1 to 4 weeks after purchase.

Classic or blue-veined cheeses include:

Gorgonzola — Sold in wheels, Gorgonzola is an Italian specialty. The interior of the cheese is white with veins that are usually more green than blue.

Roquefort — Named for the area in France where the cheese is said to have originated, Roquefort has a crumbly texture and a sharp flavor. It is made from sheep's milk.

Hard and Firm Cheeses

This category is what most people think of when it comes to cheese. So-named because they become hardened with age, hard and firm cheeses include the well-known cheddar and Parmesan varieties. They have a strong flavor and are widely used in cooking. They are also richer in calcium than softer cheeses because more milk is used in their production. However, this also means they are higher in fat and calories and so should be used in moderation in a healthful diet.

Hard and firm cheeses are divided into these categories:

Hard grating cheeses — Hard grating cheeses include Parmesan and Romano. As the name suggests, they are often grated before use, but they can be served as chunks. Both cheeses originated in Italy. Parmesan cheese takes its name from the Parma region, where this cheese may be aged up to 4 years. Romano cheese probably originated in Rome. Italian versions are made from sheep's milk; American versions, in contrast, are made from cow's milk. Both types are common toppings for Italian favorites such as spaghetti. In general, these cheeses have a tangy flavor and pleasing aroma. During production, they are heated to set the curd and reduce moisture. Aging enhances their flavor and results in their texture becoming more crumbly.

Cheddar-type cheeses — Cheddar cheese originated in the English village of Cheddar and has since been adopted by cheese lovers all over the world. Cheddar's distinct bite can range from mild to sharp, and the cheese is often seasoned with wine or spices. Cheddar cheese's characteristic color is orange — the result of adding a natural vegetable coloring called annatto during production. Other cheddar-type cheeses include Cantal, Cheshire, Gloucester, Wensleydale, and Leicester.

Colby — A blander, more moist cheese than cheddar, Colby was developed in Wisconsin a century ago.

Gruyère-type cheeses — Carbon dioxide gases trapped inside the cheese while it is ripening create the characteristic "eyes" of this type of cheese. The cheese usually is a straw-yellow color and has a mild to rich, full flavor.

Monterey Jack — A mild, light-colored cheese, Monterey Jack also may be spiced up with bits of jalapeno peppers, pepperoni, or herbs and spices.

Swiss cheese — Known for the holes in it, Swiss cheese is a golden-yellow cheese and has a tangy flavor. The holes in it are caused by pockets of gas that develop when the cheese is made.

Processed Cheeses

Processed cheese is the most common type of cheese eaten in the United States, where it was originally developed. In processed cheese, one or more types of cheese are heated (which stops the aging process) and melted. An emulsifier is then added as a binding agent. Additional dairy ingredients may be added, such as cream, whole or skim milk, buttermilk, or dried milk. Depending on the process, other thickeners or emulsifiers may be added for firmness and smoothness. A common type of processed cheese is American cheese, which is usually derived from cheddar cheese.

Processed cheeses have a mild flavor and melt easily and smoothly. They have a number of uses, from spreads to pasta toppings to dips. However, there are nutritional trade-offs. Processed cheeses are often higher in sodium than traditional cheeses and are somewhat lower in protein and other nutrients.

MIGRAINES AND CHEESE

Migraines, also called vascular headaches, are thought to involve blood vessels in the brain, although the exact cause is unknown. Some cheeses contain a naturally occurring compound called tyramine, which, in susceptible people, can cause an increase in blood pressure, an increase in the size of blood vessels in the brain, and headache pain. For people who take drugs called monoamine oxidase inhibitors (MAOIs), avoidance of all foods containing tyramine — including aged cheeses — is essential.

Tyramine is found naturally in food. It is formed from the breakdown of protein as foods age. Few studies have measured tyramine content in cheese. However, the research that has been done indicates that the longer a cheese has aged, the greater its tyramine content. Compared with other foods, aged cheeses have the highest tyramine content. The amount of tyramine in cheeses differs greatly because of the variations in processing, fermenting, aging, degradation, or even bacterial contamination. The following types of cheeses are aged or have been reported to be high in tyramine and should be avoided if you are susceptible to migraines or if you take MAOIs:

- Blue cheeses
- Brie
- Cheddar
- English Stilton
- Feta
- Gorgonzola
- Mozzarella
- Muenster
- Parmesan
- Swiss

The U.S. Food and Drug Administration (FDA) closely regulates the composition of processed cheese. Label terms that indicate you are using a processed cheese include pasteurized process cheese, cheese food, and cheese spread.

Preparation Tips

To get the most flavor from your cheese, it should be allowed to warm to room temperature. Therefore, take cheese out of the refrigerator an hour in advance of when you plan to serve it. Keep the wrapper intact so the cheese does not dry out.

The opposite is true if you plan to grate cheese. It grates better when it is cold, and 10 minutes in the freezer speeds the process.

When melting cheese, use a gradual, medium heat, because it can turn rubbery when heated at a high temperature.

Serving Suggestions

Because most cheese is high in fat — about 40 percent of which is saturated — it should accompany other foods rather than be the centerpiece of a meal. Also, most cheeses are high in sodium because of the salt used for curing and flavoring.

That said, the flavor and texture of cheese mean that large quantities are not needed to enjoy it. Cheese is excellent as a garnish for soups and salads. Or, crumble bits of real blue cheese on your salad instead of pouring on fat-filled blue cheese dressing. The

result is a more authentic blue cheese taste
with less fat and fewer calories. Top pasta
with a small serving of grated cheese, such
as Romano or Parmesan, but lean heavily
on a vegetable-based sauce. Use a single slice
of cheese atop a veggie-filled sandwich, or
trade the cheese for a lean slice of turkey.

When making nachos, sprinkle baked
tortilla pieces lightly with sharp cheddar
cheese and then top the chips with plenty
of healthier options: vegetables, beans, and
salsa. If you have a craving for pizza, make
your own. Take-out pizzas, especially if
ordered with extra cheese, can supply the
entire day's fat supply with one piece.

YOGURT

Yogurt is the result of milk that has been fermented and coagulated. Its inception, probably by accident, was thought to be around
4,000 years ago when nomadic Balkan tribes
stumbled on the process as a way of pre-
serving milk. In this age-old process, milk
is left at 110 degrees Fahrenheit for several
hours to be invaded by friendly bacteria.

Today's yogurt, however, is created in
a much more sophisticated manner. The
process starts with the milk. The type of
milk used defines the fat, calories, and, consequently, the richness of the yogurt. For
example, nonfat yogurt comes from nonfat
milk, and low-fat yogurt is derived from
low-fat milk.

The milk is then pasteurized. One of
the two milk proteins — the whey — is
coagulated to create yogurt's characteristic
glutinous consistency, and the substance is
then homogenized and cooled.

The true yogurt-making process then
begins with the introduction of the starter
bacteria cultures. In North America, the
two most common bacteria strains used are
*Streptococcus thermophilus and Lactobacillus
bulgaricus.* These two types of friendly bacteria change the milk's sugar (called lactose)
into lactic acid. The lactic acid is responsible for the tangy, acidic taste of yogurt.
The more bacterial strains used, the stronger
the acidic flavor.

Some of the bacterial cultures survive
the yogurt-making process. This type of
yogurt will list "active yogurt cultures" or

"living yogurt cultures" on the label.
Other types of yogurt are pasteurized again
after the cultures have sufficiently
fermented the yogurt. This type is labeled
"heat-treated."

Active yogurt cultures help to digest
casein, a protein found in milk. There is
also some evidence that active yogurt
cultures replenish the "friendly" bacteria in
our intestines after the supply dwindles.
This decrease in bacteria happens because
of normal aging, illness, or use of some
medications.

Yogurt did not attract Americans' attention until the health-food movement of the
1960s. Even now, Europeans still consume
5 times as much yogurt as North Americans.
Consumption in America is growing with
the increased marketing of yogurt that has
additional flavors added. One caution:
added flavors can add calories and fat.

Preparation Tips

Plain yogurt can be used in place of cream
or mayonnaise in recipes, but the result will
be less creamy and more tart. Cooking with

yogurt can be a challenge because it curdles so easily. Make sure to allow yogurt to warm to room temperature before slowly heating it.

Low-fat and nonfat versions of plain yogurt can be used in many recipes that call for sour cream. The acidity from the lactic acid creates a taste similar to that of sour cream, but the texture is a bit compromised. However, a half cup of sour cream has 214 calories and 21 grams of fat, and the same amount of low-fat yogurt has 63 calories and 2 grams of fat. From a health standpoint, therefore, the texture becomes secondary. Another option would to be to use half sour cream and half yogurt.

Because yogurt contains an acid that can work as a tenderizer, it makes a wonderful marinade for meats. Tandoori chicken, an Indian dish, is probably the most well-known yogurt-marinaded meal. The basic marinade consists of lemon juice and plain yogurt.

Yogurt can be used for fruit and vegetable dips, atop baked potatoes or cold cereals, or in stroganoff recipes. It can even replace the sour cream used with nachos or be added to guacamole to reduce the fat.

Serving Suggestions

Yogurt is the ultimate convenience food. Most supermarkets stock single-serving containers tailor-made for lunch boxes or quick snacks. Some producers have taken this idea a step further, presenting yogurt in push-from-the-bottom tubes that can be eaten without a spoon.

Honey, fruit, and granola are popular additions to yogurt, but yogurt can stand up to bolder additions such as chili.

Yogurt also offers two great dessert opportunities: frozen yogurt and smoothies. Frozen yogurt can be found in the super-market freezer section or you can make your own by adding ingredients to yogurt and then freezing it. Smoothies can be made with low-fat yogurt and fruit or fruit juice and then whipped in a blender as a healthier alternative to malts and shakes.

ICE CREAM AND DAIRY DESSERTS

Early versions of the frozen confection we know as ice cream probably used snow and sweeteners to please the palates in ancient China and the Roman Empire. The dessert made its pilgrimage to North America in 1774, when caterer Phillip Lenzi told a New York newspaper that he would be selling a dessert he had discovered in London called "ice cream."

Many items available in the "ice cream" section of your local super-market are not derived from dairy products. Referred to as nondairy frozen desserts, these ice cream substitutes are derived from either soybeans or rice.

Soymilk and tofu are the base of soybean-based products. Water, fructose or other sweeteners, vegetable oil, and flavorings are added. Rice-based desserts are treated with a special process that enhances the rice's sweetness while breaking down the proteins and starches. This base also needs sweeteners and additives.

These ice cream substitutes contain no milk or lactose, so they provide an alternative for persons who have milk allergies or who are lactose-intolerant. They also contain no cholesterol. But some types, especially the items made from tofu, can have just as many calories as ice cream.

However, it was not until 1926 that refrigeration allowed the mass production of ice cream. With subsequent decades came better freezers and an increase in ice cream consumption. Today, the average American eats more than 15 quarts of ice cream in a year.

Part of ice cream's popularity can be attributed to its smooth, creamy texture. The process of homogenization helps create its unique taste by breaking down the size of the fat globules in the milk, making a smoother product.

Adding air also makes ice cream smoother. After flavors and colors have been added, but before any mix-ins such as fruit or candy pieces are added, the mixture is whipped to increase its volume by 150 percent. Without this added air, the density of ice cream would resemble that of an ice cube. Too much air, though, creates an ice cream that is too mushy and unsatisfying in texture to serve. Air does not determine the difference between soft-serve and hard ice creams, however. Soft-serve ice cream is not allowed to freeze fully, so it maintains its "soft" consistency and can easily be manipulated by machine into cones or containers. Hard ice cream is allowed to freeze so it can be scooped or spooned out of containers.

Ice cream can be made at home with an ice-cream maker, milk, cream, sugar, and flavors. Many recipes also call for eggs.

Homemade ice cream does not contain the stabilizers used in commercial ice cream to increase body and stave off melting. Nor does it contain artificial flavorings, as many commercial products do. The result is a texture and taste that are very different from those of commercial ice cream.

Most commercial ice cream has around 10 percent milk fat and added sweeteners and so a high calorie count. The count increases with the number of high-calorie mix-ins: pieces of fruits, nuts, candy, and cookies with flavored syrups in ribbons, swirls, and ripples. Many of these ingredients have added fat.

Types of frozen dairy desserts include:

Ice milk — Ice milk has fewer calories from fat because it is prepared in the same manner as ice cream but with only 3 to 5

percent milk fat. (Regular ice cream has around 10 percent milk fat, and premium varieties can be as high as 16 percent.) Most varieties of ice milk also have fewer calories overall, but some brands add more sweetener and flavoring to compensate for the less creamy texture.

Sherbet or sorbet — Sherbet's main ingredients are frozen, sweetened fruit juice and water, but it also can contain milk and egg whites. Therefore, sherbet is not a safe alternative to ice cream for persons with milk or egg allergies. Sorbets and ices may be better choices, because they are supposed to be prepared without these ingredients. Their names are not regulated, so always check the label before purchasing. Both contain a liquid base (usually fruit juice), sweetener, and water, but sorbets are less creamy

FAVORITE FLAVOR

What is the most popular ice cream flavor in North America? Vanilla, of course. It is the flavor of almost 30 percent of all ice cream produced. Next on the list are chocolate and Neapolitan.

than sherbets, and ices are even less creamy than sorbets, usually with a granular texture.

Frozen yogurt — Frozen yogurt is made from fermented milk treated with a lactic acid culture (see Yogurt, page 358). It is sold in either soft-serve or hard forms. It usually contains less fat than ice cream. It also provides the vitamins and minerals commonly found in dairy products, unlike nondairy items. However, frozen yogurt does not contain the same friendly bacteria

that some forms of unfrozen yogurt do. The freezing process kills the bacteria that aid in digestion.

Preparation Tips

It is fun to make ice cream at home, and several new makers on the market need little, if any, manual cranking. Just gather the ingredients, mix, and follow the manufacturer's directions. Because homemade ice cream does not contain the stabilizers and preservatives of its commercial counterparts, it lasts only a few days in the freezer before it starts to form ice crystals and pick up the ambient smells. Therefore, it is best to make ice cream when the amount made will be quickly consumed.

For the occasional snacker, commercially available desserts provide no real preparation time: allow the container to soften slightly on the counter (perhaps only a minute or two), and then scoop out a 1.5-cup serving.

Serving Suggestions

Ice cream is delicious, but it should be an occasional indulgence, given the amount of calories and fat in it. When it is time to splurge, limit the amount of ice cream or other dessert treat to a small serving instead of filling a bowl. In addition, choose a frozen dessert that has less than 3 grams of fat per half cup, and use it as a foundation to enhance other, healthier foods. For example, add fruit — such as strawberries or bananas — to ice cream in equal ratios instead of pouring on the chocolate syrup.

To lower your fat intake, choose sorbets and fruit ices or low-fat ice cream and ice cream substitutes. Watch your serving size, however, so calories do not become a concern.

HERBS & SPICES

Herbs and spices have been added to foods throughout history for preservation and flavor. Although they are plentiful and inexpensive today, herbs, spices, and other flavorings were considered as valuable as gold or jewels for many centuries. Quests for them helped shape human history, influencing explorers to set out for the New World in the 15th century and also leading to the establishment of trade routes between Europe, Asia, and Africa.

If you think about it, it is easy to understand why people long ago placed such an emphasis on flavorings for their food. Easy refrigeration of food has been only a recent development. Before its development, food perished rapidly and thus had a bad taste when eaten. Some food was preserved with large quantities of salt. In addition, the wide variety of foods—particularly fresh fruits and vegetables—that we enjoy today was not available. As a result, the daily diet of people throughout much of history was bland and unexciting. It is no wonder that herbs and spices were valued.

Today, modern technology, agriculture, and transportation systems make our diets full of abundance and variety. Herbs and spices, although no longer worth their weight in gold, nevertheless still play a crucial role in shaping cuisine and adding interest to foods. They may also play a role in health, offering an opportunity to add flavor without adding the health drawbacks of excess fat or salt.

Basics

Although many people think herbs and spices are one and the same, they are not.

The definition of herbs has varied throughout the ages. Generally, herbs are now considered to be the aromatic leaves of plants that grow in a variety of climates. The leaves are used fresh, dried, chopped, or crushed to add a subtle taste to foods or oils. Sometimes they are steeped in water for teas and other beverages. Herbs are usually added at the end of the cooking process because long cooking times can erode their flavor.

Spices have many of the same uses as herbs. They are usually grown in tropical areas. Often, they have a more intense flavor and are derived from a wider range of plant parts: the fruit, seed, roots, flower bud, or bark. Spices are usually added at the beginning of the cooking process.

Both herbs and spices can be crushed for nonculinary uses: for medicinal purposes (see Chapter 2, sidebar: Herbal Products, page 37) or for use as fragrances in perfumes or lotions.

Nutrition

Herbs generally contain less fat and carbohydrates than spices. Both do contain some nutrients. For example, basil and cloves contain calcium and potassium. The small amounts of herbs and spices used in cooking, however, minimize the nutritional contributions they might make in this way.

Herbs and spices add only a negligible number of calories to the foods to which they are added. As a result, they are an excellent replacement for both fat and salt when it comes to flavoring food. Creative use of herbs and spices can make it far easier to enjoy your meals while maintaining a healthful diet.

Selection

Fresh herbs and spices deliver the most pleasing flavor. For that reason, many serious cooks buy spices whole (such as the whole seed or stem) and grow their own. In addition, many cooks have their own herb gardens to have fresh herbs on hand. Herbs are both easy to grow and attractive. They are almost always perennial plants, so they come back year after year.

Fortunately, for those without a green thumb, fresh herbs and spices are increasingly available at the supermarket. When buying them, look for products that appear the freshest — those whose appearance and aroma indicate that the time since they were harvested has been minimal. Avoid products that have mold on them or are discolored.

Dried herbs and spices are also widely available. A general rule is that 1/4 teaspoon ground leaves or 1 teaspoon dried leaves should be used for each tablespoon of fresh product.

Storage

Both herbs and spices can lose their potency over time, which is why proper storage is critical. How this is done depends on the type of product purchased.

Whole spice seeds keep longer than ground spices. Both should be kept in tightly closed containers in a cool, dry place. Dried herbs should be stored in the same way. Exposure to light and heat can cause leaves to deteriorate. A rule for determining whether the product is still good is to test its aroma. If there is not much aroma,

the product probably has lost most of its flavor and should be replaced. Make sure to date each container when you put it in your pantry. A rule of thumb is to keep dried herbs no longer than 6 months. It is ideal to refrigerate dried herbs after 3 months.

Fresh herbs are highly perishable and may last only a few days in the refrigerator. Maximize their shelf life by managing their moisture. They need some to avoid wilting. However, too much can cause them to rot. A solution is to pack fresh herbs in a perforated plastic bag in the refrigerator crisper, which will help keep the air around the product humid. Pat excess moisture off the leaves with a paper towel before putting the bag in the crisper.

This storage method works for most herbs, but there is a better one for basil, cilantro, sage, mint, parsley, and other leafy herbs. They are best stored like a flower bouquet. Tie the leaves together, stems down, and put them in a container filled with cold water. Store in the refrigerator

and change the water every few days. Fresh herbs stored in this manner will keep for about a week.

COMMON HERBS

Basil

Basil is a member of the mint family. Most types have shiny, light-green leaves and a pungent, slightly sweet aroma. There are many types of basil, each slightly different in aroma and taste. One of the most widely used herbs for seasoning tomatoes and tomato sauces, basil plays a key role in Mediterranean, Asian, and Middle Eastern cuisine. Basil particularly complements the flavor of garlic and olives. It is also pleasing combined with lemon.

Bay Leaf

The green, pointed leaves, usually sold dried, are grown on a small tree belonging to the

MEDICINAL USES OF HERBS AND SPICES

Both herbs and spices have been used throughout history as medicines. In fact, some of the earliest medical manuscripts highlight the theoretical healing properties of these plants. The best known of these ancient compilations are *Inquiry Into Plants* and *Growth of Plants*, written around 320 B.C. by Theophrastus, a Greek philosopher who studied with Aristotle. Another herbal medicine tome, *De Materia Medica*, written around 60 A.D., includes descriptions of more than 600 herbs. It is said to have influenced medicine for more than 1,500 years.

With the growing interest in alternative medicine, many physicians and scientists have begun studying the therapeutic potential of these plants. In some cases, they have found that some plants or extracts made from them have promise in fighting disease or maintaining health. However, some of the most promising plants — such as saw palmetto for prostate enlargement — typically are not found in kitchens. Nor is it

likely that the small amounts of herbs and spices used in cooking deliver any medical benefits.

More study is needed to establish the role of these plants in health. Until research is complete, it is best to avoid relying on them for medical purposes or, at the very least, you should see a qualified medical practitioner before you use them. Using them in place of traditional medicines may have harmful effects. Side effects can result from improperly using just about any so-called natural product. Or, these products may interact dangerously with a medication you are already taking.

One other caution is that regulation of the manufacturing and marketing of herbal supplements in the United States is far less strict than that of prescription medications. There are no guarantees that the herbal supplement contains the amount of active ingredient the label claims. In fact, testing of these products has found wide variation among supplement brands.

laurel family. They add a pungent, almost evergreen, flavor to foods, and the whole leaf must be removed before the food is served. Bay leaves are used to season a variety of foods. Their use is especially called for in simmered dishes, soups, stews, sauces, and tomato dishes. Bay leaves can easily overwhelm a food, so use them with caution. The longer they cook, the more flavor they add.

Chervil

A member of the parsley family, chervil has dark-green curling leaves. Its delicate flavor is similar to that of parsley, with a hint of anise and lemon. Considered essential in French cooking, it is excellent in salads, soups, and vinaigrettes and with seafood. Chervil can be used to replace parsley. It should be used fresh, when its flavor is best. A caution: chervil can lose flavor when it is overcooked.

Chives

A member of the onion family, chives have long, slender, hollow green stems and are usually sold in bundles. The mellow, delicate onion flavor of chives is useful for sauces, soups, baked potatoes, salads, omelets, pasta, seafood, and meat. They are also commonly used as a garnish. In addition, their light-purple flowers are edible. Use fresh chives when possible because dried chives have little flavor. Chives can lose flavor when cooked too long.

Cilantro

Cilantro is also related to the parsley family. Its lacy green leaves have a pungent, juniper-like spicy flavor. A popular seasoning used worldwide, cilantro particularly complements spicy foods. It is commonly used in Mediterranean, Latin American, and Middle Eastern cuisines. Cilantro is often found in salsa. It can overpower a dish.

Use it judiciously, adding a little at first. Cilantro may be referred to as "fresh coriander" because its seeds are ground into the spice coriander.

Dill

Another member of the parsley family, dill has feathery leaves and flat, oval-shaped

brown seeds. Both are used for seasoning. The leaves have a pungent, tangy taste, and the seeds have a bitter flavor with caraway overtones. Dried leaves are sold as dill weed. Dill complements the flavor of fish, chicken, eggs, salads, and a variety of vegetables. It is also used as a pickle flavoring. Use fresh dill leaves whenever possible; drying

flat parsley

bay leaf

thyme

basil

lemon grass

curly parsley

tarragon

lemon balm

fennel

lavender

causes them to lose their flavor. The leaves also make a lacy garnish. Dill weed should be used at the end of cooking so it will keep its flavor. Alternatively, heat intensifies the flavor of dill seed.

Fennel

Fennel has long, green, feathery leaves on celery-like stems. Both stem and leaves have a delicate, anise-like flavor and may be eaten as a vegetable. Flowers produce seeds that may also be used as a flavoring. Raw fennel stems

and leaves may be added to salads. Fennel also can be used to flavor cheese, sauces, mayonnaise, and bread. Fennel is a traditional seasoning for fish. Use fresh fennel whenever possible to maximize flavor. Overcooking can cause fennel to lose its flavor.

Lavender

Spikes of pungently aromatic purple flowers and gray-green leaves make identification of this herb easy. It is a traditional flavoring for teas, candies, and desserts. Lavender

also can add a subtle and different flavor to custard and ice cream.

Lemon Balm

A member of the mint family, lemon balm imparts a strong lemony flavor to foods. Use it in fruit or vegetable salads, to garnish fish, or to freshen drinks. Lemon balm can stand in for lemon peel in most recipes. In ancient times it was used to "balm" (comfort) wounds and to flavor alcoholic beverages such as claret and mead. It even

cilantro

rosemary

dill

mint

chives

oregano

sage

marjoram

sorrel

chervil

served as one of the first "air fresheners" — during the Middle Ages it was strewn onto floors and when walked on would release its lemon scent.

Lemon Grass

Also known as citronella grass, lemon grass has long, green stalks and serrated leaves. The stalks have a lemony aroma and flavor tinged with ginger. Only the lower 4 to 6 inches of the stalk is used. Lemon grass is common in Thai and southeast Asian cook-

ing. It enhances the flavor of curries, stews, soups, chicken, and seafood. Make sure to remove the lemon grass before serving. Fresh lemon peel and grated ginger can be used in place of lemon grass.

Marjoram

Marjoram is made from the short, pale-green leaves of a shrub cultivated through-out Europe for centuries. Closely related to oregano, marjoram has a flavor resembling that of mint and basil. Marjoram is

used in many tomato-based dishes, but it has a flavor that complements just about any food. It is used widely in Mediterranean cooking and makes an excellent flavoring for oil and vinegar. This herb can be used interchangeably with oregano. Buy it fresh when possible.

Mint

Mint's cool, aromatic menthol taste and smell are instantly recognizable. There are hundreds of varieties of mints: peppermint

Herb focaccia bread

and spearmint are the most popular. Often used as a dessert or candy flavoring, mint also adds an interesting flavor to sauces, meat dishes, salads, and iced tea. Lamb is traditionally served with mint jelly. Herbs that mix well with mint include cilantro, basil, and marjoram.

Oregano

Closely related to marjoram, oregano has a woody stalk with small green leaves. Considered less sweet than marjoram, it is thought to have a stronger, more peppery flavor. A ubiquitous pizza sauce flavoring, oregano is also widely used in Mediterranean cuisine (mainly Italian and Greek) and in meat and poultry dishes. Oregano retains its flavor when dried.

Parsley

Parsley has long, slender stalks and feathery leaves. It has a tangy, fresh, sometimes lemony flavor. There are two main types: curly leaf and flat leaf. The flat-leaf type

has a more intense, peppery flavor. A popular garnish, parsley also can be added to soups, marinades, and salads. Both types are often used to bring out the flavor of other herbs. The flat-leaf variety is typically used for cooking because of its more intense flavor. The flavor of both types is stronger in the stalks.

Rosemary

Rosemary has needle-shaped evergreen leaves and a piney, lemon flavor. Rosemary is used to season a variety of dishes, particularly in Mediterranean cuisine. It may be sprinkled on another of this region's specialties: focaccia bread. Rosemary branches can be burned under grilled meat or fish for a more subtle flavor. Use rosemary judiciously. It can be overpowering.

Sage

The soft, somewhat furry gray-green leaves of this herb have a pungent and camphor-like taste and aroma. A strongly flavored

herb, sage enhances poultry stuffing, sausage, veal, and tomato sauces. Excellent for flavoring oils and vinegar, sage goes well with thyme and oregano.

Sorrel

The slender, arrow-shaped leaves of sorrel impart a sharp or acidic flavor to creamed soups, meats, omelets, vegetables, or breads. Its flavor is due to oxalic acid, which should be avoided by people with a history of oxalate kidney stones. Young tender leaves are mildest in flavor and may be cooked and served as a vegetable.

Tarragon

Native to Siberia, this herb has narrow and pointed, highly aromatic dark-green leaves. It has a delicate anise flavor with undertones of sage. A staple in French sauces, tarragon is also widely used in chicken, fish, and vegetable dishes. It may be used as a flavoring for wine vinegar. There are two types of tarragon: French and Russian. The French variety has a delicate flavor, whereas the Russian tarragon has a stronger, slightly bitter taste.

Thyme

There are several varieties of this herb, which has small, gray-green leaves and tiny purple flowers. Garden thyme is the most widely used for cooking. It has a strong, somewhat bitter flavor. Thyme is often used in herb butters, stuffing, soups, and dishes with potatoes or beans as the main ingredient. It is excellent in pasta sauces and is considered an important culinary herb in Europe, particularly in France. Lemon thyme is best suited for fish and egg dishes. Lemon thyme also can be used to make herbal tea.

COMMON SPICES

Allspice

Also known as Jamaican pepper, allspice is ground from the hard brown berries of the allspice tree, which grows in Mexico and throughout the Caribbean. Allspice is so named because it imparts the flavor of nutmeg, cloves, and cinnamon. An excellent addition to marinades, allspice is also used to flavor cured and jerked meat, desserts, and sauces. It is also an ingredient in gingerbread. The finest allspice trees are thought to be grown in Jamaica. Use the spice sparingly to avoid overpowering other ingredients.

Anise Seeds

The tiny gray-green anise seeds come from a plant belonging to the parsley family. They have a licorice flavor (anise is used to make licorice), but one that also imparts a feeling of warmth when the seeds are eaten. Used to flavor breads, candies, and alcoholic drinks, anise seeds also can be added to cabbage or braised beef. Anise seed is a common ingredient in Indian vegetable and fish curries. Anise complements the flavor of cinnamon and nutmeg in baked goods. Anise leaves can be used to make herbal tea or added to salads.

Caraway Seeds

These small, crescent-shaped brown seeds have a nutty, peppery flavor. Caraway is often used whole in rye bread or sprinkled over the top of baked goods, particularly in Germany and many northern European countries. It can also be added to potato salad or meat loaf or sprinkled over pasta. Ground caraway seed can have a very strong flavor. Use sparingly. Caraway also is used to flavor aquavit, a Scandinavian liquor.

Cardamom

Cardamom is made from the seedpods of a perennial plant that is part of the ginger family. The seeds have a warm, sweet, slightly peppery flavor and an aroma that combines ginger, coriander, and nutmeg. A popular ingredient in Asian cuisine, cardamom also is used in Scandinavian cooking as a flavoring for fruit compotes, gingerbread, and meatballs. It goes well with sweet potatoes and squash. Green cardamom pods are the most flavorful and need to be ground before use. Pods may be bleached or lightened, however, and this processing may affect the flavor. The lightened pods are preferred over the ground seeds, which lose their flavor quickly. Ground cardamom can be mixed with other spices and therefore have a blunted flavor. Cardamom can be expensive. However, a little goes a long way. Less costly versions may have less flavor.

Cayenne (Crushed Red Pepper)

Made from ground dried hot chili peppers, cayenne adds warmth to whatever foods it is added. Capsaicin found in the chili's seeds and membranes gives this pepper its fire. It is popular in Mexican, Caribbean, Chinese, and Indian cuisines. It is also widely used in barbecue sauces. Cayenne is the main ingredient in chili powder. Go easy on the use of cayenne if you are not accustomed to hot foods.

Celery Seed

Celery seeds are the small brown seeds of the celery plant. They give a strong celery flavoring to foods. The seeds may be somewhat bitter in taste. Added to casseroles, fish, poultry, and sauces, celery seed is also good in potato dishes and stuffing. Celery salt is called for in many recipes. Instead, you can use a small amount of plain celery seed along with some lemon zest.

Tandoori chicken

Chili Powder

A blend of dried chili peppers, chili powder varies in intensity, flavor, and color. It is typically rusty red. The powder also may contain cumin, garlic, oregano, or salt. This spice is used to flavor Southwestern cuisine. It is a common ingredient in chili with beans and in chili con carne. It adds heat to dishes with a dash of flavor. Because chili powder ranges in flavor, many people like to grind their own powder from the chili pepper they prefer. Add the powder sparingly while cooking until the flavor and heat you desire are achieved. The best chili powders are ground from only chili peppers.

Cinnamon

Cinnamon is ground from the curled bark of the evergreen cinnamon and cassia trees throughout Asia, India, and Sri Lanka. Cinnamon is sold ground or in sticks. A popular flavoring for cookies, pies, desserts, candies, and coffees, cinnamon can be used to season meats, pasta, and marinades. It is excellent with sweet vegetables. Ground cinnamon has more flavor than cinnamon sticks. Its flavor deteriorates more quickly, however. Depending on the type, cinnamon may range from strong and spicy to sweet and mellow.

Cloves

Cloves are the oily unopened buds of the clove tree. They have a pungent flavor and aroma. Cloves add flavoring to roasted meat and can be used in pies and baked

licorice root   *cinnamon*   *saffron*   *poppy seed*

salt   *celery*   *caraway*

Szechuan peppercorn

gingerroot   *sesame*   *cloves*   *anise*

fruit dishes, cakes, cookies, and gingerbread. Cloves complement the flavor of nutmeg and cinnamon. Ground cloves lose their flavor quickly.

Cumin

The dried seeds of a plant belonging to the parsley family, cumin has an earthy, nutty flavor and smell. Used in many cultures, cumin is a seasoning for chickpeas, the background flavor for chili, or added to couscous, vegetable dishes, or yogurt. Cumin is often mistaken for caraway. If you grind your own cumin, toast cumin seeds in a dry skillet first to intensify its flavor.

Curry

Curry powder is a mixture of spices that may contain coriander, cumin, pepper, chili peppers, ginger, fenugreek, onion, cinnamon, paprika, saffron, cilantro, or turmeric. Ingredients may depend on the area of the world in which the curry was mixed.

Curry is a staple of Indian and southeast Asian cuisine. It is used to flavor many meat-based dishes and soups. Curry powder adds a sweet, distinctive, and sometimes hot flavor to foods. A common use is in stews. Curry also complements the flavor of lamb. Because the flavor of curry may vary, many cooks prefer to grind their own. Beware of store-bought curry powders, which may contain mostly turmeric. This gives dishes a yellow color, and the flavor may be bitter.

fenugreek

cardamom

mace

turmeric root

cumin

nutmeg

vanilla

star anise

allspice

juniper

Fenugreek Seeds

Ground from the seeds of a plant belonging to the pea family, fenugreek has a bitter-sweet flavor but leaves a caramel or maple-like aftertaste. A component in many Indian dishes, fenugreek also can be added to curry powders. Or, it may be used to flavor artificial maple syrups. Use sparingly; the flavor of fenugreek can be overpowering.

Ginger (Gingerroot)

In fresh form, the knobby gingerroot's peel is discarded and its flavorful flesh is sliced, chopped, or minced and added to dishes for its peppery, sweet, and pungent flavor. The dried form is ground from the ginger root. This brownish gold spice has a warm, slightly sweet, slightly citrus flavor. Fresh

ginger is popular in Asian and Indian cuisine. Ground ginger also is used in many baked goods and desserts. Ginger is the basis for ginger beer and ginger ale. Pickled ginger root is an Asian delicacy and is often served with sushi.

Gingerroot is sold at many supermarkets, and it is easy to add fresh, grated ginger to dishes. Although fresh ginger can be substituted for the dried form, do not substitute dried in place of fresh because the flavors are quite different.

Juniper Berry

The hard purple berries of an evergreen bush, juniper berries have a turpentine-like flavor. Juniper berries add a spicy, pungent flavor to game, red cabbage, or meat stews.

Juniper berries give gin its flavor. Berries should be crushed before they are used.

Licorice Root

From this woody plant licorice flavor is extracted and used in candy and medications. The extract also may be used to color and thicken stout or porter beers. The sweet taste of natural licorice extract comes from glycyrrhizia, a naturally occurring chemical that, if ingested in large quantities, has adverse effects on blood pressure. For this reason, in the United States, most licorice candy is flavored with anise or is artificially flavored (check labels).

Mace and Nutmeg

Both come from the same tree. Nutmeg is ground from the seed. Mace is from the seed's covering. Both have a sweet, warm flavor. Mace is somewhat more pungent. Favorites in baked goods and fruit dishes, mace and nutmeg can also enhance the flavor of stewed beef or poultry or can be added to baked vegetables. Nutmeg is a key spice in a holiday classic, eggnog. Both are excellent toppings for the foamed milk on espresso coffee drinks.

Mustard

Mustard seeds can be used in pickling foods. Ground seeds can be added to sauces or to add zip to salad dressings. The condiment can be eaten on meat and fish and added to salad dressings or mayonnaise. Keep prepared mustard in the refrigerator to preserve its flavor. For a different flavor, look for mustard varieties made with wine or vinegar.

Paprika

Bright russet-orange in color, paprika is made from ground sweet red peppers. Depending on the variety, paprika may add either a mild, sweet flavor or hot

Vinegar

Vinegar is used in almost every culture as a condiment. It can be made from a wide range of foodstuffs — from grains, fruits, wine, or even ethyl alcohol. Essentially, the process to turn any of these into vinegar is the same. Bacteria is added to an alcohol solution to convert the alcohol in acetic acid. The liquid is then processed and pasteurized to kill any organisms in it that might be harmful to humans. It also may be distilled before it is bottled for consumer use.

Vinegar's tart, acidic flavor makes it a versatile ingredient. It is often used to make vinaigrette dressings, mustards, or marinades, as a condiment for seafood, or to flavor dishes in which beans are the primary ingredient. In addition, it can be used to pickle and preserve foods.

How vinegar is used depends on what type it is:

Balsamic vinegar — Considered the finest of all vinegars, balsamic vinegar is added to salads, pasta, and cooked vegetables. It has a dark color and rich flavor with herbal and wine undertones. Balsamic vinegar is traditionally made in northern Italy in the provinces of Modena and Reggio. Look for the word "tradizionale" on the label. This indicates that the vinegar has been aged at least 12 years. Balsamic vinegar from Modena that uses the term "vecchio" on the label means the vinegar has been aged 12 years or more. "Extra vecchio" is vinegar that is 25 years old or more. The vinegars from Reggio are color-coded according to quality: red label (highest), silver, and gold. These "real" balsamic vinegars are expensive. Commercial balsamic vinegars are not regulated and are quite different. Some chefs enhance the flavor of the vinegar by adding brown sugar or by boiling it to intensify its flavor.

Cider vinegar — As the name suggests, this vinegar is derived from fruit juices — usually apple. It retains an apple flavor and is often used for pickling.

Malt vinegar — This type of vinegar is made from malted barley. Malt vinegar is typically colorless. However, brown coloring is often added. It has a strong, sour flavor and is frequently used as a condiment for fish and chips.

Rice vinegar — Made from sake, a Japanese rice wine, rice vinegar is used in many sweet-and-sour recipes for Asian cuisine.

Wine vinegar — This type of vinegar can be made from various red or white wines. It is thought to have the most "bite" of any vinegar. Wine vinegar is most often used as an ingredient in cooking, particularly soups and stews.

Flavored vinegars are popular in gourmet shops. But it is easy to make your own. Choose your herbs, preferably fresh ones, blanch them, and pat dry. Transfer the herbs to a food processor and then add 1/2 to 2/3 cup of vinegar, one with a less intense flavor. Process until you have achieved the desired consistency. Transfer the mixture back into the bottle of vinegar. Let the mixture sit overnight, and then strain the vinegar before putting it back into the original container. A sprig of your chosen herb can be added to the bottle for decoration. Although vinegar is acidic and is pasteurized, introducing herbs does contaminate it. The risk for illness is not high. Make flavored vinegar in small amounts, refrigerate it, and use it within a week to maintain freshness. Or, better yet, purchase one of many commercial varieties, which have been heat-treated.

---

warmth to food. Widely used in Hungarian and Spanish dishes, paprika adds flavor and color to potatoes, soups, baked fish, and salad dressings. Hungarian paprika is considered the finest and can be purchased in specialty shops. A note of caution: it may be hotter in flavor than other types of paprika.

Pepper

Pepper is one of the world's most common spices and one of its most versatile. Used in nearly every culture, pepper is a condiment found on tables worldwide. Ground or whole, it can add a kick to nearly any dish, adding warmth and texture with just the right amount of subtlety. It is a popular addition to soups, stews, cheeses, marinades, and luncheon meats.

Pepper is actually the fruit of the pepper plant, which is a vine indigenous to India. Small white flowers on the plant produce peppercorns. They turn various colors — green, red, then brown — as they mature and can be harvested and used at any time during their ripening stage.

Green pepper — Typically sold canned, green peppercorns are green pepper berries that are harvested before they mature. They are mild in flavor.

Black pepper — These peppercorns are pepper berries harvested just as they are about ready to turn red. Black pepper is the most flavorful of all pepper varieties. It is often sold ground, although many connoisseurs prefer to grind their own with a pepper mill.

White pepper—This milder pepper is made from peppercorns that have turned red and therefore are ripe. The dark outer shell of the berries is removed before the pepper is processed. White pepper is also typically sold ground.

Other types of pepper you may find are gray pepper and pink pepper. Gray pepper can be a mix of white and black pepper, although it can be a black pepper that has been minimally processed. It typically has a mild flavor. Pink pepper is not made from the berries of the pepper plant. Instead, it is derived from a South American shrub that is a member of the ragweed family. Its flavor is more subtle than that of pepper.

For maximal flavor, buy peppercorns whole and grind them yourself. Ground pepper may have undesirable additives. An advantage to buying whole peppercorns is that they can be kept for months at room temperature. In contrast, ground pepper keeps its kick for 3 months or less.

Saffron

One of the most expensive spices, this golden red spice is made from the powdery stigma of a purple-flowered crocus. Saffron is used for soups, seafood, poultry, and rice dishes. It is a popular seasoning in Indian, Italian, and Spanish cuisine. It also can be added to baked goods. Saffron needs to be dissolved in a teaspoon or so of warm water before use. Powdered saffron may have other ingredients added to it, and these reduce its flavor.

Salt

Salt is made of crystallized sodium and chloride and has been used for centuries to add flavor to food and to preserve it. Today, this condiment can be found on virtually every table in Western nations.

Various salts are available today:

Table salt — Often supplemented with iodine, table salt consists of fine-grained salt crystals that may be treated to help it flow freely out of salt shakers.

Sea salt — Available in both fine and coarse grains, sea salt is made by evaporating sea water. It can be used at the table or for cooking. It has a bit more of a tang

than table salt and may contain other chemicals — such as magnesium and calcium — found naturally in sea water.

Kosher salt — This coarse-grained salt is often sprinkled over baked goods or salads. It contains no additives, and many say it tastes less salty than table salt (although it has the same amount of sodium as table salt). It is made and processed in compliance with guidelines set forth by the Jewish religion.

Rock salt — This salt is mined from natural deposits in the earth's surface. Table salt is typically refined from rock salt.

Seasoned salt — Spices, herbs, or other agents may be mixed with salt to make a seasoned, salty product. Most of these products are made up primarily of salt.

Salt plays a key role in many different functions of the human body. However, too much of it can be harmful to your health.

Eating salt in moderation can be difficult. Many foods contain some salt naturally. Prepared foods often contain high amounts of sodium, in some cases a thousand

milligrams of sodium or more. Foods high in salt include condiments, pickled foods, canned vegetables, convenience foods, and cured meats. Always look at the label to ensure you know how much salt you are getting.

One simple but important step to cut back on salt is to taste your food first before adding salt to it. Too often, salting is a reflex, not a necessity. A better strategy for reducing salt in your diet is to cut back gradually and reduce or eliminate prepared foods altogether.

Sesame Seeds

The tiny flat seeds of a plant native to India, sesame seeds have a nutty, slightly sweet taste. Sesame seeds are used as a topping for bread and crackers. A paste made from the seeds (tahini) is combined with chick-peas to make hummus. Toast seeds before using them to enhance their flavor.

Szechuan Peppercorns

Despite the name, these dried berries are not related to traditional black peppercorns. They come from a type of ash tree and have a peppery, somewhat citrus taste. Popular in Chinese cuisine, Szechuan peppercorns often are mixed with salt or used as a flavoring for cooking oils. This flavored oil also can be used as a salad dressing. Although growing in popularity, Szechuan peppercorns may be available only at stores specializing in Asian foods.

Turmeric

Deep yellow in color, this spice is made from a root related to ginger. It has a sharp, woodsy taste. Widely used in Indian cuisine, turmeric is added to potatoes and light-colored vegetables for both taste and its yellowish orange color. Turmeric can be substituted for saffron.

Vanilla

A spice usually sold in liquid form, vanilla is extracted from the dried seed pods (beans) of a tropical plant belonging to the orchid family. It has a sweet, rich scent and flavor. Added to cookies, cakes, and other baked goods, vanilla is a widely used flavoring for ice cream, desserts, and coffees. Check the label to make sure you are getting real vanilla extract. Real vanilla extract, which has far more flavor than synthetic vanilla flavorings, is also much more expensive. "Vanilla" brought back from tropical areas may contain coumarin, a harmful substance that can cause kidney and liver damage.

BEVERAGES

Beverages are used to quench thirst in everyday settings. At meals, they make a flavorful complement for foods. Special occasions are marked by toasts of a favorite beverage. Drinking coffee, tea, soda pop, juice, or water is often used to designate breaks during an ordinary workday. Beverages are as diverse as the cultural landscape worldwide.

Like the food choices we make, the choosing of beverages plays a crucial role in our health. Drinking enough water, for example, is critical for many different body processes and thus is vital for good health. In contrast, beverages high in calories, fat, or sugar can contribute to obesity, tooth decay, and other health problems. Those high in caffeine, a chemical stimulant, can have other health drawbacks if consumed in excess. Beverages high in alcohol can affect our judgment and coordination and may even be addictive.

Perhaps at no other point in time has there ever been such a wide variety of beverages available. To help you make the most healthful choice, the following sections contain information about many of the most widely used beverages (information about milk is contained in Dairy Foods, page 345).

Basics

In the world of science, water is referred to as the universal solvent. This label is clear in practical terms in any kitchen or restaurant. Served by itself or as part of something else, water is a basic ingredient in all beverages.

Throughout history, water has been heated, chilled, brewed, carbonated, had herbs and other flavorings steeped in it, or been added to other ingredients and fermented. The result is a rich array of beverages to choose from:

Coffee—Brewed from the roasted beans of the semitropical coffee plant, coffee is enjoyed around the world, and each culture offers a different variation on this basic beverage.

Tea—Another beverage that crosses cultural lines, tea is water that is heated with dried leaves from the evergreen shrub *Camellia sinensis*. A variation is herbal tea, in which herbs such as chamomile are used.

Juice—The liquid squeezed from fruits and vegetables has been enjoyed as a beverage throughout history.

Soda pop—A beverage of the modern age, this highly sweetened carbonated drink is popular despite its nutritional drawbacks. It's high in sugar and calories and has no appreciable nutrients in it.

Wine—At its most basic, wine is simply fermented grape juice, which of course is made up mainly of water. There are thousands of variations of this ancient alcoholic drink.

Beer—Another international beverage, beer is water brewed with yeast, hops, and other ingredients to yield a carbonated, alcoholic drink with a distinctive taste and yeasty smell that is enjoyed around the globe.

Of course, water itself should not be overlooked as a beverage. Served throughout history to quench thirst and complement meals, water is now available in a myriad of forms, from bottled to carbonated to straight from the tap. The numerous varieties of water sold commercially make it easy to enjoy this clear, colorless, calorie-free, and most ancient of all beverages.

Nutrition

The calories and nutritional content of beverages depend on what has been added to the water that is their basic ingredient or the process that the water has been through. For example, drinks high in sugar are also high in calories. Alcoholic drinks are also high in calories.

The most important nutritional information to remember about beverages, however, is how vital water is to your health and that most people do not drink enough water. The recommended daily amount is eight 8-ounce glasses of water.

To understand why water is so important, consider that your body is one-half to four-fifths water. Every system in your body depends on water. Water regulates your body temperature, removes wastes, carries nutrients and oxygen to your cells, cushions your joints, helps prevent constipation, aids kidney function, and helps dissolve vitamins, minerals, and other nutrients to make them accessible to your body.

Lack of water can lead to dehydration. Even slight dehydration can sap your energy, contribute to headaches, and make you feel lethargic. Dehydration poses a particular health risk for the very young and very old.

You lose about 10 cups of fluid a day through sweating, exhaling, urinating, and bowel movements. Exercising or engaging in any activity that causes you to perspire increases your daily water requirement, as does hot, humid, or cold weather and high altitudes. Some beverages, such as those with caffeine and alcohol, are dehydrating, so if you drink them, you need even more water to compensate.

Storage and Selection

Storage and selection of beverages depend on what it is you plan to drink. The following sections provide this information and more for some of the most common and widely used beverages.

WATER

Concerns about the safety of tap water have resulted in a burgeoning market for bottled water, making it one of the most widely sold commercial beverages. Within the category of bottled water, there is a vast array of water types from which to choose. The U.S. Food and Drug Administration (FDA), which regulates bottled water, classifies them in this way:

Artesian water—This type of water is drawn from a confined aquifer (a rock formation containing water that stands above the natural water table).

Distilled water—This is water that has been evaporated and then condensed, leaving it free of dissolved minerals.

Purified water—Purified water has been demineralized. It is produced by deionization (passing it through resins) or by reverse osmosis (passing it through filters to remove dissolved minerals). Distilled water is also considered purified water.

Mineral water—Water that contains no less than 250 parts per million (ppm) of totally dissolved, naturally occurring solids or minerals. Mineral water can be labeled "low mineral content" (less than 500 ppm) or "high mineral content" (more than 1,500 ppm).

Spring water—Spring water is obtained from an underground formation from which water flows naturally to the surface. It also can be collected through a bore drilled into the spring.

Sparkling water—Another name for carbonated water, sparkling water contains carbon dioxide gas that is in it naturally or has been added to it.

Seltzer water—A type of sparkling water. The name comes from the town of Nieder Selters in Germany. Seltzer was introduced in the late 1700s and is considered the forerunner of soda pop.

Soda water—A carbonated water that contains sodium bicarbonate.

Club soda—The same as soda water except that mineral salts have been added.

Tonic water—Tonic water has been carbonated and flavored with fruit extracts, sugar, or quinine.

Most water sold commercially comes in handy storage containers—usually cans or tightly sealed bottles. Water in these containers can be kept indefinitely, chilled or in the pantry. Be sure not to let containers freeze, particularly carbonated types of water. Unlike other substances, water's volume expands when it freezes. The resulting ice will break the container.

SAFETY OF TAP WATER

Many Americans are concerned about the safety of their tap water, which has led to a dramatic increase in sales of water filtration systems and bottled water.

Most tap water, however, is fine. It is regulated by the Environmental Protection Agency (EPA) for safety and purity and chlorinated to destroy most organisms that can spread disease. However, small amounts of microbiological and chemical contaminants are allowed within EPA limits of safety or when water treatment equipment breaks down.

The EPA requires public suppliers to notify consumers if water from public supplies does not meet safety standards.

Preparation Tips

Most tap water in the United States is safe to drink. Try serving tap water with ice and a slice of lemon or lime, which gives the water a pleasing taste. Prices for bottled water range a great deal. Generally, however, less expensive types of bottled water compare favorably with more expensive versions.

Serving Suggestions

Water is an excellent beverage for any occasion. Try serving sparkling water for special occasions, putting it in a champagne flute or wine goblet for added visual appeal. For everyday occasions, make water the default beverage you choose. Soda pop and other beverages should be chosen far less frequently. Drink a glass of water when you get up, one with each meal, and another when you go to bed. Keep a bottle with you during the day or take regular water breaks.

COFFEE

Coffee beans are actually the seeds of a cherry-like fruit of the semitropical coffee plant, which is grown in Brazil, Colombia, Indonesia, and parts of Africa and Central America. The seeds are separated from the coffee fruit and then roasted. During roasting they acquire the rich, dark-brown color (a result of caramelization, melting, and subsequent browning of sugars in the beans) most people associate with coffee beans. Roasting time affects both the flavor and the color of coffee beans. Generally, the darker the beans, the longer they have been roasted. Longer roasted beans also typically have the most intense flavor.

There are dozens of varieties of coffees, many of them named after their country or port of origin. However, all coffee beans used commercially are one of two main species: *Coffea arabica* and *Coffea robusta*. Most of the beans used around the world are arabica beans, which are grown at a high altitude, require plenty of rainfall, and are considered somewhat difficult to cultivate. Robusta beans flourish at lower altitudes with less intense care.

Arabica beans, however, are thought to produce the finest flavor. They are also the most expensive coffee beans. For this reason, supermarket coffees contain mostly robusta beans, although some arabica beans may be mixed in for a richer flavor. Despite the less intense flavor of robusta beans, they have about twice the caffeine content of arabica beans (see sidebar: Decaffeinating Coffee Beans, page 380).

Coffee beans can be roasted, ground, mixed, brewed, and flavored in various ways.

Common coffee drinks: café latte, perked coffee, espresso

WHAT IS INSTANT COFFEE?

Instant coffee (almost always made from robusta beans) is simply freshly brewed liquid coffee that has been dehydrated into a powder. Adding water to this powder rehydrates the coffee, thus resulting in a quick, if perhaps less flavorful, cup of coffee. Powder particles are processed to look more like traditional coffee grounds.

Some instant coffee brands are freeze-dried. This means that the coffee solution was frozen to extract the water from it. The resulting product looks more like traditional coffee grounds and has a richer flavor.

The result? Numerous specialty coffee drinks for coffee connoisseurs to choose from:

Espresso—This dark, strong, and concentrated coffee is made by forcing hot water through coffee that is very finely ground, darkly roasted, and specially blended. It is so rich that when served unadorned, only a small quantity is needed. Thus, espresso usually is served in a tiny cup, often referred to as a demitasse. Espresso also may be the foundation of other specialty coffee drinks.

French press coffee maker

French press coffee: mix fresh ground coffee and hot water; steep for 3 to 4 minutes. The lid has a screen that pushes the grounds to the bottom.

Café au lait—From the French for "coffee with milk," café au lait is made with equal parts of regular coffee and scalded milk. In France and Quebec, café au lait is considered a morning tradition.

Café latte—Very similar to café au lait, this coffee beverage is equal parts of foamy steamed milk and espresso.

Café mocha—More dessert than beverage, café mocha is made with espresso, chocolate syrup, and foamy, steamed milk. The chocolate syrup adds considerable calories to this coffee drink.

Cappuccino—This is espresso topped with the creamy foam from steamed milk. Some of this milk may be added to the espresso. Sometimes sweetened cinnamon, cocoa, or vanilla powder is sprinkled over the foamed milk.

Turkish coffee—An intensely flavored coffee beverage, Turkish coffee is made by bringing finely ground coffee to a boil several times. Sugar, water, and spices such as cinnamon or cardamom can be added. The flavor of Turkish coffee is so rich that only a small quantity is served.

The growing interest in specialty coffees has spurred an interest in the types of coffee beans used at home for the regular brew. Once, only gourmands bought whole beans and ground them at home. Now, coffee grinders are considered routine kitchen accessories. Specialty coffee makers are also becoming more commonplace.

A welcome consequence is that excellent coffee can be made at home. The outcome of home brewing, although somewhat dependent on the equipment you own, is affected mainly by the type of coffee you buy. Perhaps the best thing you can do to ensure a good cup of coffee is to buy fresh, whole beans, grinding them just before brewing. Follow grinding instructions on your grinder carefully, paying particular attention to recommended grind times.

Avoid beans or preground coffee packed in the large, traditional round canisters. This packing method almost always allows some oxygen into the container, which can cause the coffee to become stale. Instead, look for coffee sold in vacuum-packed bags, which have a mechanism that allows gas in the container to escape but does not let any in.

DECAFFEINATING COFFEE BEANS

Virtually all coffee beans naturally contain caffeine, a chemical stimulant that can increase heart rate and blood pressure and act as a diuretic and may have some addictive properties.

There are several ways to remove the caffeine from coffee beans. One method uses a solvent to remove caffeine chemically. The beans are washed afterward to remove the solvent. Roasting also removes any traces. A second method is the Swiss water process, in which steam is used to heat the beans. Then, the outer layer of the beans, which contains most of the caffeine, is scraped off. The Sanka brand of coffee, introduced in the early part of the 20th century, was the first caffeine-free coffee developed in the United States. Its name comes from the French term "sans caféine," which means without caffeine.

At home, store coffee at room temperature and try to use it as soon as possible. If you use coffee from a can and it will be open for more than 2 weeks, place it in an airtight container in the refrigerator. If the can will be open for more than 1 month, place the coffee in an airtight container in the freezer. Return the coffee to the refrigerator or freezer immediately after measuring the amount you will use.

Preparation Tips

In addition to using fresh beans and grinding them just before brewing, other ways to improve the flavor of home-brewed coffee are the following:

Start off with clean equipment—Oil residues from previous batches of coffee can cling to equipment and affect flavor—particularly if flavored coffee was used.

Use fresh, cold water—Using hot water will result in flatter-tasting coffee. Also, make sure the water is free of minerals or other chemicals. If you don't like the taste of your tap water, consider using filtered or bottled water.

Use the right amount of coffee—For traditional coffee-brewing machines, 1 tablespoon of coffee grounds is recommended for each 6 ounces of water. Adjust the amount of coffee grounds to your preference.

Don't steep coffee grounds in the coffee too long—This extracts bitter substances from the grounds that affect the taste of the coffee.

Consider the filter—Paper filters may affect the texture of the coffee. Consider using a metal filter in your coffee maker instead. Metal filters are available for most types of machines.

Avoid flavored coffees—Many of these coffees are made with lower-quality beans.

Instead of buying them, grind your own beans and add coffee flavorings of your own. Most coffee shops sell essences and flavorings just for this purpose.

Serving Suggestions

Coffee complements various foods and can be served by itself or mixed with other flavorings throughout the day. Coffees with cream or milk are typically considered morning coffees, and espresso is considered a beverage for later in the day. If you are a frequent coffee drinker, keep in mind that coffee is high in caffeine (up to 105 milligrams per 6-ounce cup). In addition, specialty coffees can be high in calories or fat if made with cream or whole milk or flavorings such as chocolate. Use skim milk instead of whole milk in coffee drinks whenever possible.

TEA

Tea is said to be the universal beverage. Made from the leaves of an evergreen plant (related to the camellia flower) steeped in hot water, this beverage is a part of daily life from Great Britain to Morocco to China, where tea is thought to have originated. Most of the leaves grown for the world's tea are still grown in Asia, although some is produced in the United States.

All types of teas are made from the leaves of the same plant, *Camellia sinensis*. Typically, finer teas use only the top leaf and bud; stronger, coarser brews use the leaves farther down on a branch. Tea leaves are typically dried after they are picked and then are broken into fragments, which bring forth the oils that give tea its flavor. Climate, soil, and processing give tea different tastes and characteristics.

Tea field

The main types of tea include:

Green tea—A favorite in Asia, green tea is so named because the leaves are dried and fragmented soon after picking. Tea made from these leaves is mild and fresher in taste than other types of tea. Because of this, green tea usually is not served with milk or sugar. Types of green tea include gunpowder, Tencha, and Gyokuro, a Japanese tea also known as pearl dew tea.

Black tea—Actually a dark reddish brown tea when it is brewed, the strongly flavored black tea is popular in Western nations. It is the most processed and strongest flavored tea. After the leaves are picked, they are allowed to ferment in the open sun before being dried. The size of the tea leaves determines the grading of black tea. Common black tea varieties include Ceylon, Assam, and Darjeeling, considered by many to be the finest black tea.

COMPARING CAFFEINE CONTENT

Brewed beverage (6-ounce cup)	Caffeine content (in milligrams)
Green tea	25
Black tea	35
Coffee	105

Oolong teas—Oolong tea has characteristics of both black and green teas. Its leaves are fermented for about half the time of black tea. Oolong tea originated in the Fukien province of China, where much of the world's production of oolong tea takes place. Formosan tea, named for the former name of Taiwan, is considered by many to be the finest oolong tea.

Blended teas—Often referred to as English teas, these are black teas that have been blended with spices and flavorings to enhance tea flavor and aroma. Thousands of blended teas are available worldwide. Popular blended teas include English breakfast and Earl Grey.

Herbal teas—Not made from tea leaves, herbal tea is a tea-like drink made by steeping herbs, flowers, and spices in heated water. Herbal tea has been made throughout history, often for medicinal purposes or simply to make water taste better. Popular herbal teas are made from chamomile, rose hips, and mint, to name just a few. In France, herbal teas are referred to as "tisanes." Herbal teas typically contain no caffeine.

Instant tea—Popular in the United States since the 1950s, instant tea is tea that has been dehydrated and granulated so that it dissolves rapidly in water. Often, it also contains sugar and other flavorings.

Store tea away from heat in a sealed container. Tea keeps for about 6 months. After that, it loses its flavor and should be discarded.

Preparation Tips

The best tea is made using whole or large fragments of tea leaves, available in many specialty tea shops. Many of these shops also sell implements, such as mesh containers, which allow tea leaves to infuse their flavor into water without leaving the leaves behind.

To make a good cup or pot of tea, start by using cold, fresh, and filtered water (if

FAST FACT

Although the English are known for their love of tea, ingenious Americans invented the tea bag and began the practice of drinking iced tea in the early 1900s.

TEAS AND POSSIBLE HEALTH BENEFITS

Tea has been consumed throughout history for its supposed curative powers, and medical research now suggests that there are health benefits from drinking green and black teas.

Several studies show an association between consumption of green tea and reduction in the risk for cancer and heart disease. Green tea naturally contains chemical compounds called polyphenols. Within this family of compounds are chemicals that appear to play a role in cell growth and programmed cell death, which could be important in preventing and controlling cancer. Polyphenols also are antioxidants that can help prevent cell damage and may help prevent formation of plaque in the arteries.

you don't like the taste of your tap water). Heat the water to a simmer—do not boil—remove from heat, and add the tea. Steeping guidelines are generally 1 teaspoon of tea per cup of water, but the amount may vary according to the type of tea used. Green teas usually need to be steeped in water for 1 to 2 minutes, and black teas may require 3 to 4 minutes. Avoid oversteeping. More than 5 minutes can make all types of tea bitter.

Serving Suggestions

Tea is an excellent beverage at any time of the day. Black teas are typically served at breakfast, often with milk and sweeteners. Herbal teas typically do not contain the stimulant caffeine and thus are excellent choices in the evening. One note of caution: tea itself contains no calories. However, lighteners or sweeteners added to it can add a significant amount of calories and fat. Minimize fat and calories by using skim milk. In addition, afternoon tea, an old but widespread tradition, often includes baked sweets such as scones or cookies. Keep your tea break healthful by limiting these sweets. Serve fruit or slices of whole-grain bread instead.

JUICE

Compared with the numerous beverages available today, juice remains a nutritious choice. It retains most of the nutrients (vitamins, minerals, and phytochemicals) in the original fruit or vegetable, although it may also be higher in calories and sugar than many suspect. There are 175 calories in 12 ounces of apple juice, for example, and 230 calories in grape juice. Apple juice has 45 grams of fructose, a naturally occurring form of sugar, and grape juice has 57 grams of this sugar. Tomato juice contains 62 calories in a 12-ounce serving; however, it also contains 1,314 milligrams of sodium. If you are monitoring your sodium intake, always check labels of vegetable-based juices.

In addition, be aware that not all juices sold today are all juice. Some may be 100 percent fruit juice, but other brands may have juice mixed with water or simply be water with added sugar and flavoring. The U.S. Food and Drug Administration regulates the terms used on juice labels and requires manufacturers to state the percentage of pure juice used in the product. Terms consumers may encounter include:

100 percent juice—These juices do not contain added water.

From concentrate—This juice has undergone processing to remove most of the water from the original fruit juice. The resulting concentrated liquid is frozen, and then rehydrated. It may be considered 100 percent juice if the amount of water added back to the concentrate does not exceed FDA guidelines for 100 percent juice.

Not from concentrate—The most expensive kind of juice in the supermarket, this is juice that does not have added water or sweeteners. Nor has it been reduced down to a concentrate and then rehydrated.

Other terms to look for on labels include cocktail, punch, drink, or beverage—These may signify that only a small percentage of actual juice was used. It also usually indicates that sweeteners were added.

Preparation Tips

Numerous juicing machines are available on the market, many of them relatively inexpensive. Try making your own juice at home with one of these machines. Not only do you guarantee that you are getting a pure product but also you can experiment with different combinations of fruits. Orange and pineapple, for example, make a tasty, tangy fruit juice. Carrot juice can be blended with juice from vegetables or fruits for an extra-nutritious drink. When choosing fruits at the supermarket, choose those that are sold during their peak season and are properly ripened.

Serving Suggestions

Juice makes an excellent addition to any meal. To reduce the amount of caffeine you consume, try substituting juice for coffee at coffee breaks. If you don't want to make your own juice at home, numerous juice combinations are available at the supermarket. Check the label to make sure you are getting fruit juice and not sweetened water, however.

SODA POP

Once soda pop was the exception rather than the rule when it came to choosing a beverage. Now, it is virtually the default beverage for adults and children alike. In fact, some studies indicate that soda pop may provide up to 20 percent of teens' calories each day. In addition, up to 20 percent of toddlers drink soda pop.

This increased consumption is unfortunate because soda pop is devoid of nutrition. Instead, it is high in calories and, typically, sugar. One can of regular, nondiet soda pop has between 150 and 200 calories. It also has between 9 and 11 teaspoons of sugar and 30 to 70 milligrams of caffeine. These values are only part of the issue, however.

By choosing soda pop, adults and children alike are missing out on the nutrients provided by beverages such as juice and milk. In fact, many physicians and nutritionists are concerned that today's teens may be placing themselves at high risk of osteoporosis—a bone-thinning disease—as they grow older because they are not getting enough calcium during their adolescent and early adult years. Phosphates in soda pop may cause the body to eliminate calcium during urination. There are also concerns about soda pop's effect on tooth decay because of its high sugar content and some of the acidic chemicals within it.

Preparation Tips and Serving Suggestions

Soda pop has little if any role in a healthful diet. If you do decide to drink it, limit it as much as possible. Selecting caffeine-free and diet soda pop can eliminate some of the nutritional hazards of these beverages, but not all of them. When eating out, encourage children to order milk. At home, make sure there's a selection of beverages available—including chilled water—that will help family members make more healthful choices.

SPORTS DRINKS

So-called sports drinks offer another beverage alternative, albeit one marketed mainly at athletes. The drinks contain mostly sugar, water, sodium, potassium, and flavorings. Whether they provide any additional benefit to athletes is debatable. For high-performance athletes, there may be some benefit from using these drinks because they replace sugar and sodium lost during extended periods of exercise. Water is still the preferred beverage for the average person and athlete. Plus, it has no calories or sugar, which most sports drinks do.

WINE

Wine is the naturally fermented juice of grapes or other fruit, vegetables, or grains. It has been a beverage used at meals and celebrations for thousands of years. Today, there are vineyards and wineries throughout the world.

General types of wine include the following:

Red wines—Typically dry and full-bodied, common red wines are made from a variety of grapes, including cabernet sauvignon, merlot, pinot noir, syrah, and sangiovese.

White wines—White wines range widely in flavor, from dry and tart to sweet and fruity. They also are made from various grapes, including chardonnay, sauvignon blanc, chenin blanc, and riesling.

WINE STORAGE

It is not necessary to have a specially built wine cellar to store wine. Anywhere with a temperature from 45° to 65° Fahrenheit is acceptable, as long as the temperature does not fluctuate. The warmer the temperature, the faster the wine will age. Cork-sealed wines should be stored on their sides to prevent the cork from drying.

Rosé or blush wines—These wines also vary in flavor from sweet to tart. These wines are made with red-skinned grapes but are processed in such a way that the juice's contact with grape skins is limited.

Sparkling wine—Sparkling is simply another way to say bubbly. Perhaps the best-known sparkling wine is champagne, although there are numerous other types.

The flavor of sparkling wines ranges from slightly sweet to dry. They usually have a lighter flavor than traditional wines.

Fortified wines—Brandy or other liquors may be added to sherry, port, or other dessert wines to increase alcohol in them.

Aromatic wines—So-named because they are flavored with herbs or spices. Vermouth is an example of a flavored wine.

Rice wine—Rice wine is a sweet, golden wine and usually has a low alcohol content. It is produced by fermenting steamed rice. Sake and mirin are two well-known rice wines. Rice wine is a frequent part of Asian cuisine.

Access to and varieties of wine have increased as vineyards and the art of winemaking have spread throughout the world. Whereas only a few regions of the world were thought to produce excellent wine decades ago, consumers now have an array of wines—not to mention prices—from which to choose. Wines from the traditional winemaking areas of France remain excellent, of course, but they now have competition from wines produced in Germany, Italy, Spain, the United States, and Australia. Less traditional winemaking areas such as South America and South Africa also offer enjoyable wines.

A wine myth that should be dispensed with is that wines automatically improve with age. Wines do need about a year to

Vineyard

FAST FACT

One of the sources for cream of tartar, an ingredient used in baking and candy making, comes from tartaric acid in wine that crystallizes on the inside of wine barrels.

age after their production. But after that, it is recommended that most white wines be drunk within 5 to 6 years. Many red wines improve with age. However, the extent of improvement depends largely on the quality of the wine. For less expensive varieties, age is not always a good thing.

Wine does complement the flavor of a variety of foods. However, it also can be high in calories. A 5-ounce glass of wine has 100 to 226 calories. Some studies show that phytochemicals in wine may have some health benefits (see sidebar: Alcohol and Health, page 387). However, moderation in wine consumption is still recommended.

Preparation Tips

Most red wines should be served at room temperature. White wine generally should

be served at a temperature between 50° and 55° Fahrenheit. Use caution when chilling white wine. Refrigerating it for more than 2 hours before serving can lower its temperature too much and blunt its flavor and aroma.

Serving Suggestions

Another wine myth is that only certain types of wine are served with particular foods. This is true in that certain types of wine complement the flavors of certain foods. Red wine, for example, is excellent with hearty or spicy meals or with a steak. White wines complement fish or poultry, and dry, tart wines are excellent with desserts. However, the vast majority of dinners are casual enough that it is not necessary to observe these guidelines.

BEER

Beer is another alcoholic beverage that has been enjoyed through the ages. In fact, historians believe that brewing began shortly after humans started to cultivate grains. Over the centuries, brewing has evolved into a highly scientific process.

Beer today generally contains about 5 percent alcohol and is brewed from malted barley and grains such as corn or rye and flavored with hops. Yeast also is used for fermentation. The quality of the water strongly influences beer's flavor and character because 90 percent of beer's volume is water.

The many varieties of beer include the following:

WHAT ARE HOPS?

Hops are plants that produce cone-like flowers. These flowers are dried and then used in brewing beer to give it a bitter but pleasant flavor. Hop shoots also can be cooked like asparagus and eaten.

Ale—Usually strongly flavored with a bitter taste, ale is popular in the pubs of England and Ireland. Its color ranges from light gold to amber.

Bock beer—A full-bodied beer with a dark color and somewhat sweet taste, bock beer is traditionally brewed in Germany in the fall and then drunk in spring celebrations.

Fruit beer—Fruit beers are mild beers with concentrated fruit juice added. A popular fruit beer in the United States is made with cranberries.

Lager—This clear, golden brew is an American favorite. It is stored in casks until sediment and residues left from the brewing process settle out. The residue is then removed.

Malt liquor—Despite the name, malt liquor is a beer, one that has a higher percentage of alcohol (up to 9 percent) than other types of beer. In comparison, most beers have an alcohol content, by weight, from 5 to 8 percent.

Pilsner, or pilsener—This is a pale, light lager beer that was originally brewed in Pilsen in the Czech Republic. The term is now used to describe most pale, mild-flavored lager beer.

Porter—The addition of roasted malt gives this beer its dark color and strong flavor. Porters may have a higher alcohol content than regular beer.

Stout—Another favorite of the pubs in England and Ireland, stout has a strong, bittersweet flavor and a dark-brown color. Roasted barley helps give this beer its character. Guinness is a well-known form of stout.

Wheat beer—Sometimes known as "weitzen," which is German for wheat beer, this type of beer has a pale-gold color and a mild flavor similar to that of a lager. It is made from malted wheat, which is why it is named wheat beer.

Light beer—Largely an American creation, light beer has fewer calories than regular beer and less alcohol. Nonalcoholic forms of beer also are available.

Unlike wine, beer is best soon after it is produced. Beer connoisseurs believe that beer older than 2 months should not be served. Beer manufacturers in the United

Bock beer

Fruit beer

Pilsner beer

ALCOHOL AND HEALTH

Heavy drinking always carries risks. But increasingly, studies are showing that light drinking (defined as 1 drink or less a day for women and 2 drinks or less a day for men) may have some health benefits.

Perhaps the most significant benefit is in cardiovascular health. Alcohol may help increase levels of high-density lipoprotein ("good") cholesterol and may help reduce the clotting that can lead to a heart attack or stroke. Red wine also is thought to contain phytochemicals (compounds occurring in plants) that also may help protect against cardiovascular disease.

Other studies suggest that light drinking may help protect against Alzheimer's disease, senility, and macular degeneration, an eye condition that is the leading cause of blindness in people age 65 years or older.

These benefits, however, are far from proved, and more study is needed to determine the role of light drinking in a healthful lifestyle. In addition, the potential benefits come with some substantial risks.

Any alcohol is hazardous for a pregnant woman and her developing fetus. In addition, it is risky for anyone with a family history of alcohol addiction to use alcohol.

Alcohol use has health risks for everyone else. Alcohol slows brain activity, which in turn affects alertness and coordination, increasing the risk of falls and accidents while driving. It also can affect sleep and sexual function, increase blood pressure, and play a role in heartburn. There is also the hazard of drug interaction, for both over-the-counter and prescription drugs. In addition, heavy, chronic drinking has been linked with an increased risk of obesity, high blood pressure, osteoporosis, and cancer of the throat, stomach, colon, and breast. Addiction is also a risk for anyone who uses alcohol.

For all of these reasons, moderation remains a key part of a healthful lifestyle. If you don't drink, there's no health reason to start doing so. If you already drink, there's no reason to stop. Just continue to enjoy wine, beer, or other spirits in moderation.

Type of Alcohol	One Drink Equals:
Wine	5 ounces
Beer	12 ounces
80-proof liquor	1.5 ounces

States recently have begun stamping cans or bottles with production dates to help consumers ensure they are getting fresh beer. Typically, beer is stored in a cool, dark place and then chilled before serving. Beer that is chilled, warmed, and then chilled again may lose its flavor.

Beer is relatively high in calories (between 120 and 150 calories for 12 ounces)—one reason it should be drunk in moderation. Other reasons for doing so include alcohol's other health hazards (see sidebar: Alcohol and Health, above).

Preparation Tips and Serving Suggestions
Beer is traditionally served in chilled glass steins or mugs. The temperature of the beer varies according to its type. Stouts and ales are often served at room temperature, and lagers typically are chilled. The most important consideration, however, is the personal preference of the person who is drinking it. Beer is often a beverage reserved for snacks and lighter meals, although in some European countries it is a staple at dinner.

Fats, oils, and sweeteners can enhance the flavor of the food you eat. However, virtually all health experts agree that intake of these foods should be limited, and certain types (saturated and partially hydrogenated) should be avoided. This chapter provides information on the various types of fats, oils, and sweeteners and the foods that contain them.

FATS, OILS, & SWEETENERS

I f there is any food group that we have a love-hate relationship with, it's fats, oils, and sweeteners.

On the one hand, these foods play a vital role in our enjoyment of what we eat. Fats and oils give a creaminess, richness, crispiness, or pleasing mouth-feel to foods. Sweeteners also satisfy a universal and natural craving. On the other hand, the pleasure that fats, oils, and sweeteners bring can come at a cost. These foods generally are high in calories, making it difficult for someone who eats a lot of them to maintain a healthful weight. They also have other health disadvantages. Too much of the wrong kinds of fats and oils can increase the blood cholesterol level, which in turn can increase risk for cardiovascular disease. Sugar and highly sweetened foods also are typically high in calories and provide few nutrients. For that reason, "empty calories" is a term often used to describe sweeteners or foods rich in them. Sugar and tooth decay are linked when sugar is eaten in excess and dental hygiene is poor.

The Food Guide Pyramid recognizes both the advantages and the disadvantages of these foods. It does not eliminate them. But it does place them at the very tip of the pyramid. It's okay to use these foods as long as they are eaten sparingly. Making sure fats, oils, and sweeteners play the proper role in your diet involves knowing more about them, their role in the diet, and what foods are rich in them. To help you do this, this chapter provides more detailed information on the types of fats, oils, and sweeteners you are likely to consume.

FATS & OILS

Basics

Fats, no matter what their source, play an important role in the food we eat. For centuries, every culture has taken advantage of the unique chemical properties of fats — for example, fats easily absorb other flavors. Fats also are used to cook foods, add a pleasing texture, impart tenderness to baked goods, and, in societies where food is scarce, increase the calorie content of a food. Fats, therefore, are a basic ingredient in cooking.

The term "fats," however, is broad, encompassing many different substances — from butter to lard to vegetable oil. All oils, for example, are fats. But not all fats

are oils. Although definitions vary, for cooking purposes fats are generally characterized as follows:

Fats — Generally defined as substances that are solid at room temperature, fats include butter, cocoa butter, lard, margarine, suet, and vegetable shortening.

Oils — Oils remain liquid at room temperature. Oils can be made from various plants and seeds: vegetables, olives, rapeseed (from which canola oil is made), sunflower seeds, corn, peanuts, soybeans, walnuts, almonds, hazelnuts, safflower seeds, grapeseed, sesame seed, mustard seed, and coconuts are among the most commonly used.

No matter what form they take, fats are made up of fatty acids, which are the

molecular building blocks of fats in the same way that amino acids combine to form protein. At their most basic, fatty acids are molecular chains of hydrogen, carbon, and oxygen atoms. The differences in the chemical structure of fats make some better for you than others (see Chapter 2, Sorting Out the Fats, page 26).

Sometimes a process called hydrogenation is used to make liquid oils solid at room temperature. This also converts unsaturated fat into trans fat. Trans fats are used in many processed and fast foods, such as doughnuts, crackers, chips, and french fries. Trans fats also give margarine its butter-like consistency. Trans fats tend to increase your cholesterol level and therefore should be eaten in only limited amounts (see Chapter 3, Limit Trans Fat, page 66).

Nutrition

Fat is an essential nutrient. Our bodies require small amounts of several fatty acids to build cell membranes and to support life-sustaining functions (see Chapter 2, Fat as a Nutrient, page 26). That said, the old adage about "too much of a good thing" is appropriate in discussing fat. Virtually all health experts agree that fat intake should be limited. The federal government, the American Heart Association, and other organizations recommend that fat intake for a healthy individual should be less than 30 percent of total daily calories. They also recommend that less than 8 to 10 percent of total calories come from saturated fat.

Although various kinds of fat have different effects on your blood cholesterol, all foods that are high in fat are high in calories. High-fat foods can easily increase your calories, making it difficult to maintain a healthful weight. Fat packs more calorie punch than any other type of nutrient group. Per gram, fat has 9 calories (about 100 calories per tablespoon, or 250 calories per ounce). In contrast, protein and carbohydrates have just 4 calories per gram. Because of this, you do not need to eat very much fat before reaching the 30 percent threshold.

If a low-fat diet is good, is an even lower-fat diet better? Not necessarily. Upper limits of fat intake have been established, but the same is not true for lower limits. Talk with your health care provider about the recommended fat intake that is best for you. Even a low-fat diet can lead to weight gain if you cut back on fat but take in excess calories by ignoring the rest of what you eat. Too many calories from any source result in added pounds. And if

they add up to obesity, you are at increased risk for health problems.

Selection

Be discriminating in the type of fat you consume. Limit animal fat (saturated) and trans fats (hydrogenated oils). Instead, use small, sensible amounts of plant-based (monounsaturated and polyunsaturated) fats.

When purchasing products, always check the expiration dates. Rancidity is a concern with any type of fat.

Purchasing oils, of which there are dozens of varieties, involves a close reading of the package label. Most cooks prefer to buy "cold pressed" oils, which means that minimal heat and pressure are used to extract the oil from the original plant or seed. This type of processing is considered important because it allows oil to maintain more of the plant's natural flavors and textures. With the exception of extra-virgin olive oil, however, it is difficult to find cold-pressed oils.

A good alternative is to use so-called unrefined oils. Unrefined oils are extracted with heat. Unlike other oils, unrefined oils undergo minimal processing after this point. The result is a more flavorful oil and, sometimes, a more darkly colored oil. Unrefined oils generally include virgin olive oil and corn, nut, soybean, canola, and sesame oils. Unrefined oils break down easily under heat and thus should not be used for deep-frying. Because deep-fried foods are not typically part of a healthful diet, you should not have to trade a flavorful oil for one with more cooking versatility.

Most oils in your supermarket, however, are not only extracted with heat but also undergo much more processing, including using chemicals to de-gum, refine, bleach, and deodorize the oil. The result? Less flavor. After experimenting with cold-pressed or

unrefined oils, you will likely appreciate the difference between these oils and their more highly processed counterparts.

Storage

The method of storage depends on whether a fat or oil is being stored. However, both become rancid given enough exposure to air, sunlight, and heat.

Fats such as butter, margarine, and lard should be tightly wrapped and refrigerated. They usually can be stored this way for up to 2 weeks. Extra butter or margarine can be stored in the freezer for up to 2 years. Hydrogenated vegetable shortening can be stored, tightly covered, at room temperature for as long as 3 months.

Oils require a slightly different strategy. They should be stored in airtight containers that are opaque to prevent light from penetrating. Refrigeration is also generally recommended for oils. Unopened oils can be kept this way for up to a year, although they should be used within a few months after they are opened.

Cooler temperatures may cause oil to look cloudy or congeal. Removing the oil from the refrigerator and allowing it to reach room temperature should resolve this problem. An important guideline for evaluating the freshness of oil is to trust your nose. If the oil smells fishy or musty, discard it.

The following sections provide more information on specific fats and oils.

FATS

Butter

Butter is made from the fat that comes from milk from cows, sheep, goats, horses, and other mammals. Most commercially produced butter in the United States is made from cow's milk.

Butter making occurs in several stages. Cream that separates from milk is pasteurized (heated at a high temperature) to kill any organisms that might be harmful to human health. Then the cream is placed in a ripening tank for 12 to 15 hours. There, it goes through another series of heat treatments that give butter a crystalline structure when it cools, helping it to solidify.

The next step is to churn the butter. This process breaks down the fat globules in the cream. The result is that the fat is coagulated into butter grains. The mixture is then separated, the remaining butter paste is worked until it is smooth, and, depending on the producer, it may be salted. Further variations in processing influence its characteristics, including aroma, taste, color, appearance, and quality.

There are numerous butter variations. Those you find in gourmet markets include a French butter known as beurre (butter) de Charentes. Beurre de Charentes has an ivory color and tastes very rich. Another European-style butter growing in popularity in the United States is ripened butter, traditionally made in Denmark and the Netherlands, which is softer than regular butter. It also has a slightly tangy taste because lactic acid is added to the cream from which it is made.

More common variations include the following:

Whipped butter — Whipped butter's name is self-explanatory. It has air beaten

WHAT IS AN OIL'S SMOKE POINT, AND WHY DOES IT MATTER?

An oil's smoke point is simply the point at which fat, when heated, starts to smoke, smell acrid, and, as a result, give an unpleasant flavor to food.

Each type of oil has a different smoke point. It is dependent on the free fatty acids that make up the oil. The higher the oil's smoke point, the higher the temperature it can withstand. Safflower and canola oils have the highest smoke point and are the most ideal for frying or sautéing (435° to 450° Fahrenheit). Olive oil has a lower smoke point and is best used in salad dressings (extra virgin, 250° Fahrenheit) or in baking (regular olive oil, 410° Fahrenheit).

Beyond the bad flavor imparted to foods, there are health reasons to avoid using an oil that has reached its smoke point. High temperatures can cause the oils to decompose, and this process, in turn, can irritate the lungs and cause gastrointestinal upset.

into it. The result is that it is slightly lower in fat and calories than regular butter. It is very soft and spreadable.

Light butter — Light butter usually has about half the calories of regular butter. It also generally has less fat and less salt because water is usually added to it.

Unsalted butter — This is butter to which no salt has been added.

Clarified butter — An ingredient in some recipes, clarified butter is butter that has the milk solids removed from it. The advantage is that it has a higher smoke point than regular butter, which increases its cooking versatility. It also keeps longer than butter and is thought to have a more pure flavor. Clarified butter is similar to a type of butter called ghee that is used in India.

Butter should be refrigerated and stored in opaque packaging that prevents light from entering. In addition, the packaging should seal in moisture to prevent the butter from becoming dehydrated, a process that intensifies its color and detracts from its flavor.

Preparation Tips

Should you use salted butter or unsalted? Although salted butter is the most common

type in supermarkets, many serious cooks prefer to use unsalted butter in cooking and baking. Unsalted butter is thought to have a sweeter flavor. In addition, many cooks prefer to control the salt they add to a dish or baked food.

Both light butter and whipped butter work well for toppings, but neither can be substituted for regular butter in recipes for baked goods because of the air or water they contain.

Serving Suggestions

Butter is one of the most versatile cooking ingredients and the foundation of numerous gourmet foods, sauces in particular. Its taste can be enhanced by mixing it with herbs and spices and then refrigerating it again. One popular flavored butter is garlic butter, which can be made by creaming the desired amount of butter and mashed garlic cloves to taste. Oregano, marjoram, basil, or parsley also can be added. Numerous recipes for flavored butter are available in cookbooks.

Butter's health drawbacks are well known, however, and thus it should be used selectively. If that special dish simply cannot be made without butter, don't try to sub-

How to "Clarify" Butter

To make clarified butter, start by cutting unsalted butter into small pieces. Melt the butter over low heat for 10 to 15 minutes. This allows the fat to separate from the milk solids. Then gently pour or spoon off the clarified butter fat and discard the solids. After the fat cools, it may appear to have a grainy texture. Clarified butter keeps longer than regular butter because the milk solids have been removed. It generally is used only for cooking because it has the advantage of having a higher smoke point than regular butter. Thus, it can be heated to higher temperatures, without burning, making it a good choice for frying or sautéing.

stitute something else. Instead, save the dish for special occasions.

Margarine

Hydrogenation, the process used to make liquid oils solid at room temperature, made possible a shift from animal fat to vegetable fat as a substitute for butter. The resulting product — which may be blended with other milk products or animal fats (such as lard or tallow) and salt for taste — is margarine. It has been used as a butter substitute since the late 19th century. Sometimes it is referred to as oleomargarine or oleo. Oleo means oil and refers to the vegetable oil base of margarine.

In addition to regular margarine, the dairy case may contain these variations:

Salted or unsalted margarine — As the name suggests, salt has been added for flavoring or left out.

Reduced-fat or nonfat margarine — These products contain 25 to 65 percent less fat than regular margarine. To reduce fat levels, modified margarines and spreads are created with varying amounts of water and thickening agents, such as gelatin, rice starch, and guar gum. Some margarines even make health claims (see Chapter 3, sidebar: Cholesterol-Lowering Margarine? page 66).

Butter-margarine blends — Designed to add butter flavor, these products are usually 40 percent butter and 60 percent margarine.

Soft margarine — These margarines are usually made from only vegetable oil and have been processed to stay soft and spreadable when cold.

Whipped margarine — This has had air beaten into it, making it fluffy and easy to spread.

Liquid margarine — Liquid margarine comes in squeezable bottles, making it convenient for picnics and other events away from home. It is specially blended so that it does not become too thick to squeeze out of the bottle. It is also handy for basting and for foods such as corn on the cob and waffles.

Like butter, regular margarine is about 80 percent fat (actually, law requires this percentage of fat for the product to be labeled margarine) and has the same number of calories. One tablespoon contains about 100 calories and 11 grams of fat.

For years, margarine was thought to be a much healthier choice than butter because it contains less saturated fat than butter and no cholesterol (because it is not made from animal fat). Although that's true, margarine is high in trans fats, which also are linked to cardiovascular disease. In addition, margarine is high in calories. Therefore, it should be used in moderation.

Softer margarines, such as those that are liquid or sold in tubs, are considered a healthier choice than harder "stick" margarines because the hydrogenation process used to make margarine hard adds trans fats to the product. When buying margarine, check the label and avoid products whose labels include these terms: "partially hydrogenated" or "hydrogenated." Instead, look for margarines whose main ingredient terms include "liquid" oils.

Preparation Tips

Generally, only regular margarine should be used for cooking and baking. Liquid, whipped, or reduced-fat versions of this product burn easily or contain too much water or air for these purposes. Margarine should be stored in the refrigerator and can be frozen for several months.

Serving Suggestions

Hard or stick margarine is a good choice for pastries, helping to make crusts light and flaky. It can be used to replace butter in most recipes, although its flavor may not be as rich as that of butter. Like butter, this type of margarine has health drawbacks — it is high in calories and trans fats that are linked with heart disease — and should be used in moderation.

Lard

Lard is simply a name for pure animal fat that has been processed, including filtering, bleaching, hydrogenation, and emulsification. Lard used in cooking is typically rendered from pork fat. Lard rendered from the fat around the pig's kidney is considered the best lard to use.

Lard is used worldwide in cooking, although it fell out of favor in the United States during the past several decades because of the amount of saturated fat in it.

Lard is usually sold at the supermarket in the dairy case or near the refrigerated meat section. When purchasing lard, make sure it is tightly wrapped to prevent the product from absorbing other flavors. Look at the label to determine whether it should be stored at room temperature or in the refrigerator. Lard also can be frozen if it is tightly wrapped.

Preparation Tips
Because lard is softer and oilier than butter and contains less water, some cooks believe it is the best choice for making pie crust. The reduced water content in lard helps make the crust especially light, flaky, and crumbly.

Serving Suggestions
Because lard is so high in saturated fat, it should be used sparingly at best.

Vegetable Shortening
Vegetable shortening, usually sold in coffee-can-sized canisters, is a solid fat made from vegetable oils. It gets its name from a property all fats have in common: they "shorten" gluten strands in flour-based products, which results in baked goods that have a tender texture.

Although the base of shortening is oil, it has undergone a process known as hydrogenation to make the shortening solid at room temperature. Trans fatty acids are created by hydrogenation. This type of fat is associated with increased risk for coronary artery disease.

Preparation Tips
Plain and butter-flavored vegetable shortenings are sold in supermarkets. Plain shortening has little taste. Both types can be used in place of other fats in baking and cooking.

Serving Suggestions
Shortening results in baked goods that are light and fluffy. Because of the health drawbacks of hydrogenation, however, shortening should be used sparingly.

OILS

Cooking Oils

Canola Oil
Canola oil is a bland-tasting oil made from rapeseed. Its health advantages over other oils has made it a popular choice in the United States, although it is also used around the world. Elsewhere, it may be referred to as lear oil or low erucic acid rapeseed oil.

Canola oil is the lowest in saturated fats of all oils. Only olive oil has more monoun-saturated fat. (See Chapter 2, A Comparison of Fats, page 27.) Another positive aspect is that canola oil contains omega-3 fatty acids, which are thought to play a role in reducing cardiovascular disease.

Preparation Tips
Because canola oil is relatively bland, some cooks combine it with olive oil to add additional flavor.

Serving Suggestions
Canola oil is suitable both for cooking and for salad dressings. It is also used as an ingredient in spreads that can be substituted for butter or margarine.

Coconut Oil
Pressed from the boiled nut meats of fresh or dried coconut, coconut oil is one of the few non-animal highly saturated fats.

Foods that contain trans fat.

Nearly 90 percent of coconut oil is saturated fat, topping even butter and lard in saturated fat content.

Although its high saturated fat content makes coconut oil an unhealthful fat choice, it helps coconut oil resist rancidity. Coconut oil is a common ingredient in commercial baked goods, ice cream, and salad dressings. It is particularly common in non-dairy coffee creamers and whipped toppings. The whiff of coconut that arises from an open bottle of suntan lotion should also tell you that coconut oil has nonfood uses.

Preparation Tips
Coconut oil is popular in cuisine from Southeast Asia, the Pacific, and the West Indies. For these cuisines, cooks often can substitute a more healthful oil or mix coconut oil with another oil to reduce saturated fat in a dish. Be aware that there is coconut oil in coconut cream and coconut milk.

Serving Suggestions
Because of its saturated fat content, coconut oil should be used sparingly.

Corn Oil

Corn oil is one of the most widely used cooking oils. It is pressed from the inside (endosperm) of corn kernels. It is more strongly flavored than other oils. Corn oil that has undergone less processing, known as unrefined corn oil, can be found in specialty stores. It is more dense than refined corn oil and has a darker gold color and a hint of popcorn flavor. Corn oil has a high level of the essential fatty acid linoleic acid and less saturated fat than many other oils. This is one reason it is used to make margarine.

Preparation Tips
Because of its high smoke point (410° Fahrenheit), corn oil is commonly used for sautéing and frying — two cooking methods that should be used in moderation. Many cooks believe corn oil helps make sautéed foods and fried foods crispier.

Serving Suggestions
Corn oil can have a strong flavor, which is why it is probably best used for cooking instead of serving "cold," such as in a salad dressing. Corn oil's stronger flavor works well in margarines, but many people prefer the taste of reduced-fat or light margarines made with corn oil.

Cottonseed Oil

Widely used at the turn of the century, cottonseed oil often is considered the original vegetable oil of the United States. Its fat is mostly polyunsaturated.

Preparation Tips
Cottonseed oil is used mainly as salad oil or in cooking. It is often used in processed foods because it is inexpensive.

Serving Suggestions
This oil has a neutral, clean taste that does not mask other flavors. It is used mainly by food manufacturers and is not common in home use.

Flaxseed (Linseed) Oil

The tall stems of the flax plant have many uses: they contain fibers that can be made into linen and high-grade paper. Oil also can be pressed from its shiny brown, oval-shaped seeds. Flaxseed is also sometimes known as linseed oil and is a common ingredient in paints, varnishes, and inks. This oil is low in saturated fat and high in omega-3 fatty acids, which may help protect against cardiovascular disease.

Preparation Tips
In the United States, flaxseed oil generally is not used for cooking purposes. Instead, the grassy-tasting oil is often sold at health stores as a supplement. The oil is more commonly used elsewhere in the world for cooking, particularly Eastern Europe. Flaxseed oil's low saturated fat content means that it is more prone to rancidity than oils with higher saturated fat content. Therefore, it should always be stored in the refrigerator.

Serving Suggestions
Baking with flaxseed meal is one way to incorporate this healthful oil into your diet. Flaxseed meal is available in the health foods section of most supermarkets.

Olive Oil

If there is one oil that gourmands and nutritionists agree on, it is olive oil. Low in saturated fat and rich in heart-healthy monounsaturated fats and flavor, olive oil is an excellent addition to any kitchen and meal. Sometimes it is even referred to as the "king of oils."

One of the first oils to be made by humans, olive oil is pressed from olives that are picked when their color turns to purplish black and their skin develops an oily sheen. The rich flavor of olive oil varies according to where the olives are grown and the type of tree that produces them. For example, Spanish olive oil has a strong, some would say overwhelming, flavor. Oils produced from California olive trees are said to have a mild, almost sweet flavor. Olive oil from Italy, however, is usually considered the best.

U.S. consumers can find a wide range of olive oils in supermarkets and specialty shops. Unrefined olive oil, which is less processed, is considered the most flavorful and has a greenish cast. Fresh, refined olive oil should have a sweet, somewhat nutty flavor and a golden color. A rule of thumb is that the more deeply colored the oil, the more flavorful it will be.

Olive oil is also graded by the International Olive Oil Council in these ways:

Extra virgin — Considered the finest olive oil, extra virgin is made without heat or solvents, from the first pressing of the olives. It is the most flavorful and the most expensive, and so it is most often used for seasoning.

Virgin — Also made without heat or solvents, virgin olive oil also comes from the first pressing of the olives. It is more acidic but has a less intense flavor than extra virgin. It is widely used in cooking.

Regular or pure — More subdued in flavor than virgin olive oil, regular (pure) olive oil is extracted with heat or solvent and may be made from pressed olives that have been washed and treated to extract more oil from them. This type of oil is often blended with virgin olive oil.

Light — Don't be misled by the title — light olive oil has the same amount of calories and fat as regular olive oil. "Light" refers simply to color, fragrance, and flavor. Light olive oil also may be referred to as "mild." Generally, this type of olive oil is dismissed by serious cooks. It does have the advantage of having the highest smoke point of all oils (468° Fahrenheit).

Preparation Tips

Generally, pressing and processing are done after the olives are collected in autumn. Shipments of freshly pressed olive oil begin arriving in specialty stores in the spring. Because most types of olive oil have a low smoke point, olive oil is not suited for deep-frying or sautéing. The best-tasting olive oil is the freshest.

Serving Suggestions

Olive oil is best used in sauces, salad dressings, and marinades, where its flavor can be put to full advantage.

Italian and Mediterranean cuisines take full advantage of olive oil's robust flavor and health benefits.

Palm Oil and Palm Kernel Oil

Often thought to be the same thing, palm oil and palm kernel oil are actually two separate oils. Palm oil, which is reddish brown, is extracted from the pulp of the fruit of the palm. Palm kernel oil, which is yellowish white and has a mild flavor, is extracted from the nut or kernel of the palm. Both, however, have this in common: they are among the highest of all oils in saturated fat.

Palm oil is often used in margarine and in commercially prepared gravies and soups.

Olives, avocados, peanuts, and the oils made from them are a source of monounsaturated fat.

It is also commonly used as a medium to fry potato chips. Palm kernel oil has various uses, including in nondairy creamers, dressings, whipped creams and toppings, baked goods, and candy.

Preparation Tips
Palm kernel oil and palm oil are generally used only for commercial food production purposes. They are not typically sold in supermarkets or specialty stores.

Serving Suggestions
Persons following a healthful diet would do well to avoid both palm oil and palm kernel oil. Check the list of ingredients on the products you buy. Avoid fatty products that contain these oils. Skim milk is a far healthier option for a coffee lightener.

Peanut Oil
One of the first native North American sources of vegetable oil, the peanut is comprised of about 50 percent oil. This oil is extracted by pressing steam-cooked peanuts.

Peanut oil is high in monounsaturated fat and has the added advantage of having a relatively high smoke point (410° Fahrenheit).

American peanut oils are often mild in flavor. Chinese peanut oils have a stronger peanut flavor because they are less processed.

Peanut oil keeps well when stored in a cool, dark place. However, its shelf life is longer when it is refrigerated.

Preparation Tips
Peanut oil's high smoke point makes it an excellent choice for deep-frying, a cooking technique always best used in moderation.

Serving Suggestions
A healthier use for peanut oil is to serve it as a salad dressing. It also adds a delicate flavor to mayonnaise recipes. The flavor of peanut oil also complements Indian and Asian cuisine.

Safflower Oil
Safflower oil is made from the seeds of the thistle-like safflower plant. Safflower oil has little flavor or color, but it is rich in polyunsaturated fat. It is also one of the oils with the least amount of saturated fat. Safflower is light in color and, like other oils that are rich in polyunsaturated fat, is a good all-around oil.

Preparation Tips
Safflower oil can be substituted for other oils in nearly any recipe. It has a high smoke point (450° Fahrenheit), making it a good choice for sautéing and deep-frying, two techniques best used in moderation.

Serving Suggestions
Safflower oil is often used in salad dressings because it does not solidify when chilled. It has an oily texture and a nutty flavor, making it a good substitute for peanut oil.

Soybean Oil
The soybean has the distinction of being the plant used most often to produce vegetable oil. The oil made from the yellow-

Olive oils vary in taste. Experiment to find the flavor you like.

Peanuts

ish brown legume is high in polyunsaturated fat and low in saturated fat. About 15 percent of the fat provided by soybean oil is saturated.

Soybean oil is one of the most commonly used oils for commercial food purposes. It has a smoke point of 410° Fahrenheit, making it useful for frying. However, sometimes manufacturers may hydrogenate the oil, decreasing its health attributes. It is commonly used in margarine and shortening.

Preparation Tips

Although soybean oil is versatile for cooking, some people find that it has a somewhat fishy taste and heavy texture. Use caution when using soybean oil by itself as a seasoning or dressing.

Serving Suggestions

When soybean oil is used as a cooking medium, its flavor complements Asian and Indian cuisines.

Sunflower Oil

Sunflower oil is pressed from the seeds of this well-known, towering yellow member of the daisy family. The oil itself is light and mild. It is low in saturated fat and high in polyunsaturated fat.

Preparation Tips

Sunflower oil is considered an excellent all-around oil. Its very mild taste makes it extremely versatile for both cooking and seasoning, particularly when cooks do not want to mask the taste of other foods.

Serving Suggestions

Because sunflower oil is relatively inexpensive and has little or no taste, it can be combined with more expensive specialty oils in foods such as salad dressings.

Specialty Oils

Almond Oil

A favorite for cakes, desserts, and candies, this clear, sweet, pale-yellow oil is pressed from either bitter or sweet almond kernels. Both flavor and consistency may depend on where the oil was made. In European countries, the oil comes from the "bitter almond," which is similar to a peach pit. When the bitter substance in the oil has been removed, the oil is then used to make candy. Almond oil from France, "huile

Sunflower seeds

d'amande," has a delicate flavor and smells like toasted almonds. It is also expensive. Oil produced in the United States is considered less pleasing by serious cooks, but it is more moderately priced.

Preparation Tips

Almond oil becomes rancid quickly and must be refrigerated. When using synthetic almond flavoring in recipes, be aware that it often is not made from almonds (instead peach pits may be used) and may have a bitter taste.

Serving Suggestions

Almond oil or flavoring is used in numerous desserts and sweets, including the famous German holiday specialty marzipan. Almond oil may be used as a flavoring in dishes in which you would have used the nut itself.

Grape-seed Oil

This pale-yellow oil has both a taste and an aroma that reflect where it came from — it is extracted from the tiny grape seed. Often, these seeds are left over after wine-making.

Grape-seed oil is produced mostly in France, Italy, and Switzerland, but it is easily found in the United States in both grocery and specialty stores. It is low in saturated fat and contains mostly polyunsaturated fat. Although light in consistency and color, grape-seed oil can have a relatively strong and distinctive taste.

Preparation Tips

Grape-seed oil's high smoke point (445° Fahrenheit) makes it a good all-purpose oil for deep-frying. It can be used to gently sauté foods. Grape-seed oil does not become rancid quickly and can be stored at room temperature.

Serving Suggestions

Grape-seed oil's distinctive flavor makes it a good candidate to serve cold. It is excellent as a salad dressing. A simple but elegant way to make use of it is as a dip for bread. Cut a baguette of French bread into cubes and serve them with a small bowl of grape-seed oil.

Hazelnut Oil

This delicious oil tastes like the nut it was pressed from and has a rich, strong flavor. Most hazelnut oil is imported from France, but it is easily found in most food stores. One caution: it can be fairly expensive. Hazelnut oil contains mostly monounsaturated fat.

Various oils are used to "dress" salads. Flavorful extra virgin olive oil, lighter-colored canola oil, and pale safflower oil are all low in saturated fat. Check labels for coconut oil, which is high in saturated fat and is not the healthiest choice.

Preparation Tips

Although it can be stored at room temperature, hazelnut oil lasts longer when kept refrigerated. Because it has a strong flavor, it is often combined with other, lighter oils. It also has a low smoke point, so it is not used for cooking.

Serving Suggestions

Hazelnut oil is excellent in salad dressing, drizzled over vegetables, and in cakes and pastries. It is also an unusual but pleasant addition to sauces.

Sesame Oil

Extracted from the oily seeds of the sesame plant, sesame oil is a flavorful, nutty-tasting, and aromatic oil. Although all sesame oils

GOURMET OILS

More expensive and hard-to-find oils that are also used in cooking or for flavoring include the following:

Pumpkin seed oil — As its name suggests, this oil is pressed from pumpkin seeds. Its distinctive flavor and expense mean that it is typically used sparingly as a condiment.

Mustard seed oil — This oil is popular in Indian cooking and often used as a substitute for ghee, an Indian form of clarified butter.

Poppy seed oil — In France, where it is a staple in cooking, poppy seed oil is referred to as "huile blanche," or "white oil."

Wheat germ oil — Made from the heart of the wheat seed, wheat germ oil is rich in vitamin E and low in saturated fat. It has a pleasing, nutty flavor, and is used by itself or mixed with other oils as a condiment.

are rich in flavor, taste and smell range in intensity. Lighter-colored oil is milder in flavor, and darker oil is made from toasted sesame seeds and has a more intense flavor. For serious cooks, darker oil is considered the most desirable.

Sesame oil is widely used as an accent in Middle Eastern, Asian, and Indian cuisine. It is high in monounsaturated and polyunsaturated fat, and it is low in saturated fat. It is used mostly to accent flavor and aroma and is used less often for frying because it burns easily.

Preparation Tips

Generally, sesame oil's strong flavor means that it is often added to other, less intense oils in cooking. Purchase sesame oil in glass or metal containers because it goes rancid more quickly in plastic. Sesame oil keeps for about a month in a cool place.

Serving Suggestions

Drizzle sesame oil lightly over dishes just after cooking. A small amount adds a distinct flavor and aroma.

Walnut Oil

Oil pressed from walnuts has a strong, nutty flavor. It is also high in polyunsaturated fat and low in saturated fat. Although widely available, walnut oil remains expensive. Blander, cheaper versions can be found.

Preparation Tips

Because of its intense flavor, walnut oil is often mixed with lighter oils. Walnut oil also has a low smoke point and will become rancid quickly if not refrigerated.

Serving Suggestions

Walnut oil is excellent as a salad dressing or drizzled over pasta or cooked vegetables. Like any nut oil, you can use walnut oil when the nut itself would be complementary.

WHAT ARE ESSENTIAL OILS?

Essential oils are extracted from various plants for use in perfumes, for aromatherapy purposes, or for use in medicines as flavorings. Common essential oils used for their scent include rose oil, geranium oil, and lavender oil. Essential oils used for flavorings are derived from lemons, cloves, peppermint leaves, and spearmint leaves.

SWEETENERS

Basics

Sweeteners come from various sources. They have been sought throughout history for their pleasing taste and many uses. Just stop and think of what you have eaten today. This morning you may have sweetened your tea with honey and put maple syrup on your pancakes. At lunch, you may have eaten a snack food made with corn syrup — a form of sugar — and, perhaps, you finished off dinner with a cake made with the most commonly used and best known sweetener of all, granulated sugar.

The science of sweetness, however, goes beyond the source of the foodstuff for the sweetener. At a molecular level, approximately 100 chemicals are sweet. They all are referred to as sugars. Common ones you may have heard of include the following:

Sucrose — Table sugar is the crystallized form of sucrose. Sucrose is referred to as a simple sugar. It is naturally occurring in all plants that depend on sunlight to produce energy. Sugar cane and sugar beets are among the most abundant producers of sucrose in the plant kingdom.

Glucose — A simple sugar that plays many key roles in the body, glucose is a simple sugar found in fruits, honey, cereal, flour, and nuts.

Fructose — The sweetest of all sugars is found in abundance in honey and fruit.

Lactose — Another simple sugar, lactose occurs only in milk. It is often added to other foods during processing to improve taste.

Maltose — The result of a chemical processing that uses starch and malt, maltose has numerous commercial food uses. It is often used in beer, bread, and baby food, among other things.

QUICK TIP

Although refrigeration helps to prevent oils from becoming rancid, the colder temperature may result in some oils becoming cloudy or thicker. This change does not affect the oil's nutritional value, quality, or taste. At room temperature, the oil will liquefy and, usually, clarify again.

Pectin — A complex sugar, pectin is found in apples, citrus fruits, and some vegetables. It is a form of fiber.

Nutrition

All sugars are carbohydrates and play a key role in providing the body with energy. The calorie content depends on the type of sweetener used. For example, table sugar has about 16 calories per teaspoon, and honey has about 21 calories per teaspoon.

Generally, sweeteners often are referred to as having "empty calories." They contain few or no vitamins, minerals, or other nutrients. And, because they are appealing, it can be easy to eat too much. This excess

Walnuts

could make it difficult to maintain a healthy weight. Heavy use of sweeteners also may increase the risk of tooth decay, which is why it is important to brush after eating a sweet food, particularly one that is sticky, such as caramel. For these reasons, it is best to minimize the amount of sweeteners in your diet.

Selection and Storage

These factors depend on the type of sweetener. Check the following sections, which provide more information on common types of sweeteners, for specifics.

Sugar

Sugar was once considered as valuable as gold because of its scarcity. Its use spread throughout the Western world after explorers, then armies, conquered parts of ancient Arabia. In early times, it was sold and traded in blocks, which were then ground into powder.

Although sugar is a carbohydrate that occurs naturally in every fruit and vegetable, it is found in the greatest quantities as sucrose in sugar cane, which is grown in the tropics, and sugar beets, which can be cultivated in colder climates. Juice extracted from the crushed cane or sliced beets is then processed to make sugar. Typically, the juice is boiled, and then chemicals are added to the solution to purify it. The resulting syrup is known as molasses. Continued processing separates crystals from the molasses and other by-products. The crystals are then dried and packaged as sugar.

The most common types of sugars found in supermarkets are as follows:

Granulated white sugar — Often referred to as table sugar, this is the most commonly used type of sugar. There are different grades of granulated white sugar, and the size of the sugar crystal determines how it is used. Regular, extra-fine, or fine sugar is the sugar found most commonly in the sugar bowl and called for in most cookbook recipes. Superfine sugar or ultra-fine sugar has the smallest crystal size and is often used in cakes and meringues and to sweeten fruits or iced drinks. Superfine sugar dissolves the most easily in water.

Brown sugar — Brown sugar is sold in dark and light varieties. It is simply white sugar crystals coated in a molasses syrup to add a natural mellow flavor and color. Dark brown sugar has more color and a stronger molasses flavor. Its fuller flavor is called for in recipes for gingerbread and baked beans. Lighter types are usually used in baking. Neither type of brown sugar is considered raw sugar, although they do look similar to it.

Confectioners' sugar — Also known as powdered sugar, this is granulated sugar that has been ground into a powder. A small amount of cornstarch can be added to prevent clumping. Confectioners' sugar typically is used to make icing, in whipping cream, and as a topping for desserts.

Decorating or coarse sugar — Also called sugar crystals, decorating sugar has granules about four times larger than those

There are many types of sweeteners.

of regular granulated sugar. It undergoes a special processing method to make it resistant to color change and breakdown at high temperatures. This makes it useful for making fondants or liqueurs.

Sanding sugar — Also called colored sugar, sanding sugar is used for decorating and is characterized by large crystals. This is desirable in decorating because it gives the food a sparkling appearance.

Flavored sugar — This is simply granulated sugar that has been combined or scented with various ingredients such as cinnamon or vanilla.

Fruit sugar — Slightly finer than "regular" sugar, fruit sugar is used in dry mixes such as gelatin desserts, pudding mixes, and drink mixes. The more uniform crystals prevent separation or settling of smaller crystals to the bottom of the box.

In addition to sweetening items, sugar plays an important role in making food. It is a critical ingredient in bread, in which it provides food for yeast and thus helps bread to rise. It also adds to the flavor and crust color of baked goods and helps extend shelf life.

In large amounts, sugar inhibits the growth of yeast and molds in jams and jellies. Sugar syrups protect frozen and canned foods from browning and withering. In ice cream, beverages, baked goods, and other products, sugar adds bulk, texture, and body. It is also used in many condiments, such as ketchup and salad dressing, where it blends flavors, reduces acidity, and helps create a smooth texture.

Sugar has a long shelf life. Kept tightly wrapped and in a cool, dark place, it will keep for months or even years.

Preparation Tips
Sugar adds flavor and calories but little else. Therefore, it is best for most of us to minimize its role in our diet. Make a little go a long way. To do so:
- Add spices, such as cinnamon or nutmeg, to foods to jazz up flavor while reducing sweeteners used in them.
- Add fruit or yogurt to foods such as cereal, instead of a sweetener.
- Avoid sweetened soft drinks, and minimize fruit juices with added sugar. Better yet, drink water.
- Check labels for sugar or any one of the chemical names for it: glucose, sucrose, lactose, or fructose, to name just a few. Also, watch for corn syrup or malt syrup, two more widely used sweeteners in food manufacturing.

Serving Suggestions
Manufacturers can reduce the fat in many foods, but it is hard to do without sugar. Sugar is a key component of baked goods and desserts and is used to enhance the flavor of everything from sweet-and-sour stir-fry to ham. The amount of sugar used in a recipe often can be reduced by up to half without compromising the flavor. However, this is not always the case, so a bit of trial and error is required.

Sugar cane

SUGAR IN THE RAW

Many so-called raw sugars are marketed in the United States. Two popular types of raw sugar are the coarse-textured dry Demerara sugar, which is produced in the South American country of Guyana, and the moist, fine-textured muscovado, or Barbados, sugar. Demerara sugar is light brown and is characterized by large golden crystals that are slightly sticky. It is excellent as a topping for hot cereals and is widely used in specialty coffee houses. Muscovado, or Barbados, sugar is a dark, rich brown sugar and retains a strong molasses flavor. Turbinado sugar is another variation of raw sugar. It has undergone steam processing to remove some of the lingering molasses in it. Its crystals are light-golden and generally are larger than those of regular sugar. Turbinado sugar tastes very similar to brown sugar.

Honey
In ancient times, this thick, sweet, golden liquid was thought of as a healing agent, a gift from the gods, and a symbol of wealth. Today, this sweetener is still revered, although not as highly, for being a natural source of sweet flavor.

Honey is made by bees. The basic ingredient is nectar gathered from flowers. Enzymes in the bee's saliva convert the nectar into honey. Essentially, this is a simple matter of chemistry, in which the sugar (sucrose) in nectar is converted into fructose and glucose.

As the phrase "busy as a bee" suggests, bees work hard to make honey. The bee must make up to 100,000 round trips from

hive to flower and back just to make a quart of honey.

Honey is divided into three basic categories:
- Liquid honey, which is extracted from the comb
- Chunk-style honey, a liquid honey with pieces of the honeycomb
- Comb honey, a square or round piece of the honeycomb, with the honey inside

Within these three categories are hundreds of different types of honey. Honey's color ranges from light to dark. The flavor ranges from mild to strong and depends on the type of flower from which the nectar was taken. In general, the darker the honey, the stronger the flavor.

One tablespoon of honey has about 64 calories. Although sugar has about 48 calories per tablespoon, honey does have some advantages over sugar. Its sweetening power is stronger. And honey, unlike other sweeteners, does contain trace amounts of vitamins and minerals.

If kept in a sealed container and a cool dark place, honey can be kept for a long time. Cooler temperatures, such as in a refrigerator, may cause honey to thicken. Warming it up, however, restores honey's appearance. Its taste is not altered. However, very warm temperatures can change honey's flavor.

Preparation Tips

Most honey sold in stores is pasteurized, filtered, and blended. Some cooks buy honey directly from an apiary because they believe that these processes alter or dull honey's delicate flavor. Honey also can be used as a substitute for sugar (about 1/2 cup honey for 1 cup of sugar) in many recipes. Keep in mind, however, that honey may cause food to brown more quickly. In addition, you will need to reduce the liquid in the recipe.

Serving Suggestions

Honey adds moisture to cakes, breads, and other confections. It is also an excellent topping for most baked goods. One other common use is as a glaze for meats such as ham.

Syrups

Sugar also comes in syrups — thick, viscous, sweet liquids that have various tastes and uses. The most common types include the following:

Cane syrup — Thick and extremely sweet, cane syrup is made from sugar cane. It is a common ingredient in Caribbean and Creole recipes.

Grain syrups — Sweet syrups can be made from several grains: barley, wheat, corn, or rice. They are not as sweet as sugar but are commonly used in food manufacturing because they do not readily form crystals. Corn syrup is perhaps the most widely used grain syrup. It is made by processing corn starch and is available in light and dark forms. Malt syrup, made from evaporated corn mash and sprouted barley, is another common grain syrup. It has a strong flavor and is used in bread making.

Golden syrup — Popular in England, golden syrup is similar in consistency to corn syrup and has a golden color. It is made from sugar cane juice and has a toasted flavor. It is also known as light treacle.

Maple syrup — The best known of all syrups, maple syrup is made by boiling the sap of certain species of maple trees (*Acer saccharum*) found mainly in Quebec, New York, and Vermont. This clear, subtly flavored syrup is sweeter than sugar and has a distinctive flavor.

Molasses — Molasses, a dark viscous syrup, is a by-product of the sugar-making process and is generally used for flavoring foods or as a glaze. It is poured over foods as a condiment in some regions of the United States. Light molasses is produced during the first stages of the sugar-extraction process. Dark molasses is made during the second stage and is referred to as unsulfured molasses. Blackstrap molasses, made during the final stage of sugar production, is darkly colored and has an intense flavor.

Black treacle — Black treacle is a thick, black, and sticky syrup. It is very similar to molasses and is a by-product of the sugar-production process. Black treacle is sweeter than molasses. A lighter-colored and lighter-flavored form is also available. Both are more common in England.

Palm syrup — This dark, thick, and intensely flavored syrup is made from palms. It is an ingredient in some Asian

HONEY AND INFANTS

Many parents are unaware that honey should not be served to children younger than 1 year because it may contain a small amount of botulism toxin. The amount of this toxin in honey is not enough to harm adults and children older than 1 year because their immune systems have matured. But in infants, this toxin can be life-threatening.

recipes and usually is sold only at specialty markets.

Preparation Tips

Most grain syrups are used commercially. In contrast, maple syrup is typically used at home. Pure maple syrup is found in supermarkets. However, pancake syrups commonly contain either a small portion of maple syrup or maple flavoring that is then mixed with a grain syrup. Many types of syrups are used to make candy. Keep in mind that syrups high in sugar have a higher boiling point than water.

Serving Suggestions

Maple syrup is typically used as a topping for waffles and pancakes. Numerous types of syrups can be used as glazes for meats (ham, in particular, and also poultry or fish) or on top of vegetables, such as carrots. Molasses also makes an excellent glaze.

Sugar syrups make an excellent glaze for pound cakes and bundt cakes.

Chocolate

If there is one flavoring that everyone seems to love, it's chocolate. Rich and sweet, with a distinctive taste that cannot be duplicated, chocolate is the universal favorite when it comes to flavor. In fact, ancient cultures even thought it was a gift from the gods, one with medicinal properties. It is perhaps the most popular sweet flavoring worldwide.

Chocolate is made from the beans of the cacao tree, which grows in the warm, humid weather of the equatorial regions. Like coffee beans, the beans of the cacao tree must be dried, chopped, and roasted before use. The processing of the beans results in a dark brown liquid called chocolate liquor. This fluid — which is 55 percent fat, 17 percent carbohydrate, and 11 percent protein — is used to make virtually all types of chocolate.

Different types of chocolate contain varying amounts of cocoa butter — a vegetable fat derived from the cacao bean — and solids from the cacao bean. Chocolate types include the following:

Unsweetened — Sometimes referred to as baker's chocolate, this dark, rich, and bitter chocolate does not have any sugar added to it. It is usually added to recipes in which sugar is an ingredient.

Bittersweet — This dark, rich chocolate is comprised mostly of chocolate liquor, meaning it is rich in cacao solids, but it may have some milk solids and other flavorings added to it.

Semisweet — A favorite of makers of chocolate chip cookies, semisweet chocolate contains more milk solids and other flavorings than bittersweet chocolate.

Milk — The sweetest of all chocolates, milk chocolate has a light-brown color and a mild chocolate flavor.

White — This ivory-colored chocolate contains no cacao bean solids, but it does contain cocoa butter, which gives it a rich, creamy mouth-feel. It differs from white almond bark or candy coating, which uses vegetable fat as a base instead of cocoa butter.

ARTIFICIAL SWEETENERS

Aspartame (better known as Nutra-Sweet), acesulfame-K, saccharin, and sucralose are synthetic substances that have been approved by the U.S. Food and Drug Administration. They are used commercially and in the home to sweeten products. They contain few or no calories but are several hundred times sweeter than sugar. For more information, see Chapter 2, page 21.

Imitation chocolate — Typically used in baking chips, imitation chocolate replaces some or all of the cocoa fat with other vegetable fats. It is high in fat (mostly saturated fat) and does contain caffeine.

Cocoa is another common type of chocolate. It is a powdered form of chocolate made from chocolate liquor. However, all cocoa butter has been removed from it. A tablespoon of unsweetened cocoa powder contains about 15 calories and just under 1 gram of fat.

Cocoa typically is not sweetened and is added to recipes in which sugar is used. There are two main types of cocoa: natural and "Dutch-process." Natural cocoa is light in color and has a strong chocolate flavor. In contrast, Dutch-process has a milder taste but is darker in color.

Chocolate's nutritional value varies. One ounce of unsweetened chocolate has 145 calories, 16 grams of fat, and 9 grams of saturated fat. In comparison, an ounce of semisweet chocolate has 135 calories, 9 grams of fat, and 5 grams of saturated fat. Chocolate also is a source of protein and contains trace amounts of vitamins and some minerals, such as potassium.

Chocolate should be stored in a cool, dark place and can be kept for several

Made from maple syrup, maple sugar is twice as sweet as regular table sugar.

WHAT IS LIQUID SUGAR?

Liquid sugars were first developed before processing made distribution of granulated sugars practical. It is generally used for commercial purposes in products in which dissolved granulated sugar is desired. A darker liquid sugar is also available. It is called amber liquid and is darker and has more of a cane sugar flavor.

Invert sugar is another type of liquid sugar. It helps prevent sugar crystallization and also helps the product it is in retain moisture. Invert sugar is sweeter than regular table sugar and is generally available only for commercial uses.

months. Dark chocolate can be stored for up to a year. Varying temperatures will cause lighter, whitish areas to appear on chocolate — something that does not affect taste and is resolved when the chocolate is melted. Chocolate also can be frozen, but it must be wrapped tightly to prevent moisture from damaging the chocolate when it is thawed.

Preparation Tips

Cocoa can be used as a substitute for chocolate in recipes. However, when this substitution is made, fat needs to be added to the cocoa to ensure that the final product will be moist. Each square of unsweetened chocolate can be replaced with 3 tablespoons of cocoa and 1 tablespoon of cooking oil.

Melting chocolate is difficult because it burns easily. For that reason, it is best to use a double boiler to melt chocolate. Avoid splashing any water into the melting chocolate because doing so can cause the chocolate to become hard and thus unusable. Chopping the chocolate into small bits before melting it helps achieve the smooth, even consistency that many recipes require.

Always look at the ingredient list of the chocolate you buy to ensure that you are not getting a substitute.

Serving Suggestions

Chocolate is the classic dessert ingredient, providing the flavoring power for cakes, tortes, frostings, mousses, creams, and other sweets too numerous to count.

Chocolate's high fat content and high calories mean it should be used in moderation. Use chocolate as an accent to a healthier food — such as a dip for strawberries — rather than as the main ingredient. Reduced-calorie hot cocoa mixes are an excellent way to feed a chocolate craving without the fat and calories. Or, simply save chocolate for special occasions.

GLOSSARY

Acute: Term used to describe disorders or symptoms that occur abruptly or that run a short course; opposite of *Chronic*

Aerobic: Requiring the presence of oxygen. Aerobic exercise, for example, requires increased oxygen consumption. Opposite of *Anaerobic*

Ambulatory: Able to walk

Amino acid: A component of protein, containing nitrogen. The body produces many amino acids; those it needs but cannot make are known as essential amino acids and must be obtained through the diet

Anaerobic: Able to live without oxygen (as certain bacteria), or a type of exercise in which short, vigorous bursts of activity requiring little additional oxygen are performed. Opposite of *Aerobic*

Anemia: Condition characterized by a reduced number of red blood cells, amount of hemoglobin, or amount of blood

Aneurysm: The localized bulging of a blood vessel, usually an artery, to form a bulge or sac

Anorexia: Loss of appetite, often due to depression, fever, illness, widespread cancer, or addiction to alcohol or drugs

Anorexia nervosa: An eating disorder characterized by aberrant eating patterns and disturbed ideas about body weight

Antibody: Protein of the immune system that counteracts or eliminates foreign substances known as antigens

Antigen: Substance foreign to the body that causes antibodies to form

Apnea: Temporary cessation of breathing

Arteriosclerosis: Condition in which the walls of arteries become hard and thick, sometimes interfering with blood circulation

Artery: Blood vessel that carries blood from the heart to other tissues of the body

Asymptomatic: Without symptoms

Atherosclerosis: Condition in which fatty deposits accumulate in the lining of the arteries, resulting in restricted, less flexible pathways for the blood

Atrophy: Wasting of tissue or an organ due to disease or lack of use

Autoimmune: Reaction of the body against one or some of its own tissues that are perceived as foreign substances, resulting in production of antibodies against that tissue

Bacteria: Single-celled microorganisms, some of which cause disease and some of which are beneficial to biological processes

Benign: Harmless; not progressive or recurrent

Blood pressure: Force placed on the walls of the arteries. See *Diastole* and *Systole*

Bowel: Small or large intestine. The small intestine is sometimes called the small bowel. The large intestine is also called the colon

Brand-name drug: A drug carrying a trademark name designated by its manufacturer

Cachexia: Malnutrition and wasting due to illness

Caffeine: A stimulant found naturally in coffee, tea, chocolate, and cocoa; may be added to, for example, soft drinks and over-the-counter drugs

Calorie: The amount of heat needed to raise the temperature of 1 gram of water by 1° Centigrade

Cancer: General term for various conditions characterized by abnormal growth of cells, forming malignant tumors that can develop in various parts of the body. See *Malignant* and *Benign*

Capillaries: Minute blood vessels connecting the smallest arteries to the smallest veins

Carbohydrate: A group of compounds composed of starches or sugars, found primarily in breads and cereals and in fruits and vegetables

Carcinogen: A potential cancer-causing agent

Cardiac: Pertaining to the heart

Cardiopulmonary: Pertaining to both heart and lungs

Cardiovascular: Pertaining to the heart and blood vessels

Carotid artery: Main (right and left) artery of the neck which carries blood to the head and brain

Cerebrovascular: Pertaining to the blood vessels of the brain

Chemotherapy: Treatment of disease by chemicals that have a direct effect on the disease-causing organism or disease cells; widely used in the treatment of cancer

Cholesterol: A fat-like substance made in the liver and found in the blood, brain, liver, and bile and as deposits in the walls of blood vessels. Essential to the production of sex hormones. Found in foods of animal sources

Chromosome: One of 46 rod-shaped structures in the nucleus that carry genetic information to each cell

Chronic: Term used to describe long-lasting disease or conditions. Opposite of *Acute*

Clinical: Pertaining to information gathered from direct observation of patients, as distinct from laboratory findings

Coagulate: To solidify or change from a liquid to a semisolid, as when blood clots

Colon: The large intestine extending from the small intestine and ending in the anus. It is responsible for extracting water from undigested food and storing the waste, which is eliminated in bowel movements

Colorectal: Pertaining to the colon and rectum

Complex carbohydrate: A substance that contains several sugar units linked together, such as starch

Constipation: The difficult or infrequent passage of stool

Coronary: Pertaining to the arteries that supply blood to the heart

Coronary artery disease: Narrowing or blockage of one or more of the coronary arteries, resulting in decreased blood supply to the heart (ischemia). Also called "ischemic heart disease"

Corticosteroids: Hormones produced by the cortex of the adrenal glands; also, synthetic hormones used as medications

Debility: A state of physical weakness

Dehydration: A lack of an adequate amount of fluid in the body. Dehydration may be accompanied by dry mouth, thirst, constipation, dizziness, concentrated urine, or fever

Deoxyribonucleic acid (DNA): A substance found in the nucleus of cells that carries genetic information

Dextrose: A simple sugar that is found in the blood

Diabetes mellitus: Disorder characterized by high levels of glucose in the blood. Diabetes mellitus may be caused by a failure of the pancreas to produce sufficient insulin or by resistance of the body to the action of insulin

Diabetic ketoacidosis: A serious condition that develops in persons with diabetes when there is not enough insulin and the body begins breaking down fat, producing ketones (acids)

Diagnosis: Identification of a disease or disorder

Diarrhea: An increase in the number or liquidity of bowel movements

Diastole: Period during the heart cycle in which the muscle relaxes, followed by contraction (*Systole*). In a blood pressure reading, the lower number is the diastolic measurement

Diastolic pressure: The lowest blood pressure reached during the relaxation of your heart. Recorded as the second number in a blood pressure measurement

Digestion: Breakdown of food so it can be absorbed

Duodenum: The part of the small intestine next to the stomach

Edema: Swelling of body tissues due to excessive fluid

Endocardium: The thin, inner membrane that lines the heart. See *Epicardium* and *Myocardium*

Enzyme: A complex protein that stimulates a chemical reaction

Epicardium: The thin membrane on the surface of the heart. See *Endocardium* and *Myocardium*

Epinephrine: Adrenal hormone that increases heart rate and blood pressure and affects other body functions

Esophagus: The muscular tube that connects the throat to the stomach

Estrogen: Hormone produced primarily in women that contributes to the development of female secondary sex characteristics and cyclic changes such as menstruation and pregnancy. An oral replacement dose of estrogen is often used to lessen the effects of menopause, among other effects. The hormone is also produced in small quantities in men

Ethanol: Grain or ethyl alcohol

Fats: A group of organic compounds that are composed of fatty acids. Fats are either saturated or unsaturated. Unsaturated fats are classified further as either monounsaturated or polyunsaturated

Fatty acids: Substances that occur in foods; different fatty acids have different effects on cholesterol and triglyceride levels

Fiber: As applies to food, a substance that resists digestion and passes through the system essentially unchanged. Fiber adds bulk to the diet and aids in the passage of bowel movements

Flatulence: Excessive gas in the stomach or intestine

Fracture: To break or crack a bone; or, a break or a crack in a bone

Fructose: A sugar found in fruit, corn syrup, and honey

Gallbladder: Structure located under the liver that stores bile and then releases it into the small intestine

Gastric: Pertaining to the stomach

Gastroenteritis: An inflammatory condition of the stomach and intestines leading to nausea, vomiting, abdominal pain, and diarrhea. Usually of bacterial or viral origin

Gastrointestinal tract: The stomach and intestines

Gene: Structure within a chromosome that is responsible for inheritance of a particular characteristic

Generalized: Overall, not limited to one area of the body

Genetic engineering: Manufacture, alteration, or repair of genetic material by synthetic means

Geriatrics: The branch of medicine that specializes in the care of problems related to aging

Germ: A microorganism that causes disease

Gestational diabetes: Diabetes that develops during pregnancy, resulting in improper regulation of glucose levels in the blood

Gland: Any organ or tissue that releases a substance to be used elsewhere in the body; endocrine glands release hormones directly into the bloodstream

Glucose: A form of sugar. All of carbohydrate and part of fat can be changed by the body into glucose; used by the body for energy

Gluten: Protein found in grains such as wheat, rye, oats, and barley. Gluten helps hold in the gas bubbles when flour dough rises

Glycogen: Stored form of carbohydrate in the liver and muscles

Goiter: Enlargement of the thyroid gland

Gout: A condition in which excess uric acid may lead to arthritis and kidney stones

HDL cholesterol: High-density lipoprotein cholesterol; a type of cholesterol thought to help protect against atherosclerosis; known as "good" cholesterol

Heart attack: Descriptive term for a myocardial infarction: an incident caused by the blockage of one or more of the coronary arteries, resulting in interruption of blood flow to a part of the heart

Heartburn: Pain due to regurgitation (reflux) of juices from the stomach into the esophagus; pyrosis

Hemoglobin: A iron-containing protein found in the red blood cells. Hemoglobin transports oxygen to body tissues

Hemorrhage: Loss of blood from a blood vessel

Hemorrhoid: Swollen vein in and around the anus that may bleed

Hepatic: Pertaining to the liver

Heredity: Genetic transmission of traits from parent to offspring

Hernia: Protrusion of an organ or part of an organ into surrounding tissues

High blood pressure: See *Hypertension*

Hormone: A substance secreted in the body and carried through the bloodstream to various tissues of the body, where it serves a regulatory function

Hydrogenation: A process that changes an unsaturated fat to a more saturated one

Hyper-: Prefix meaning "excessive" or "increased"

Hyperactivity: Condition of disturbed behavior characterized by constant overactivity, distractibility, impulsiveness, inability to concentrate, and aggressiveness

Hypercholesterolemia: Increased level of cholesterol in the bloodstream

Hyperglycemia: Increased level of sugar (glucose) in the bloodstream

Hyperlipidemia: Excess of fats (lipids) in the bloodstream

Hyperplasia: Excessive growth of tissues

Hypertension: Condition in which the blood is pumped through the body under abnormally high pressure; also known as high blood pressure

Hypo-: Prefix meaning "inadequate" or "insufficient"

Hypoglycemia: Condition in which the sugar (glucose) in the bloodstream decreases below normal levels

Hypotension: Low blood pressure

Iatrogenic disease: Disorder or disease resulting as a side effect of a prescribed treatment

Idiopathic: Pertaining to a condition or disease of unknown cause

Ileum: Lower portion of the small intestine

Immobilize: To make a limb or part immovable in order to promote healing

Immunity: State of being resistant to a disease, particularly an infectious one

Indigestion: Impaired digestion, commonly refers to abdominal pain after meals

Infarct: An area of tissue that dies because of lack of blood supply

Infection: Disease caused by invasion of body tissue by bacteria, viruses, or fungi

Infectious: Ability to transmit a disease caused by microorganisms

Inferior vena cava: Large vein returning blood from your legs and abdomen to your heart

Inflammation: Body tissue's reaction to injury that leads to swelling, pain, heat, and redness

Insulin: A hormone made by the pancreas or taken by injection that regulates the amount of sugar (glucose) in the bloodstream

Insulin pump: A device that delivers a predetermined amount of insulin into the body

Insulin reaction: A condition in insulin-taking diabetics resulting in low blood sugar (hypoglycemia) due to excess insulin or inadequate carbohydrate intake

Intestines: Portion of the digestive tract extending from stomach to anus and responsible for much of the absorption of nutrients. See *Duodenum*, *Ileum*, *Jejunum*, and *Colon*

Intolerance: Inability to endure, as with pain or a drug therapy

Involuntary: Not controlled through will

Irradiation of food: A process of exposing food to low-dose radiation in order to extend shelf life by killing microorganisms and insects

Ischemia: Deficiency of blood flow within an organ or part of an organ. Often refers to the situation in which an artery is narrowed or blocked by spasm or atherosclerosis and cannot deliver sufficient blood to the organ it supplies

Jejunum: The portion of the small intestine located between the duodenum and ileum

Joint: The point of juncture between two or more bones where movement occurs

Ketoacidosis: A disturbance of body chemistry that occurs in starvation or as a complication of type 1 (insulin-dependent) diabetes

Ketone: An acidic substance produced when the body must use fat for energy

Kidneys: The two bean-shaped organs located in the back portion of the upper abdomen that are responsible for excreting urine and regulating the water and chemical contents of the blood

Kilogram: A metric unit of weight; 1 kilogram equals 2.2 pounds or 1,000 grams

Kyphosis: Excessive curvature of the upper spine, resulting in humpback, hunchback, or rounding of the shoulders. May result from diseases such as osteoarthritis or rheumatoid arthritis, osteoporosis, or rickets, from conditions such as compression fracture, or from a congenital abnormality

Lactation: The production of breast milk

Lactose: The sugar found in milk

Laparoscopy: Examination of the inside of the abdominal cavity by means of a laparoscope (a viewing instrument) inserted through a small incision

LDL cholesterol: Low-density lipoprotein cholesterol; provides cholesterol for necessary body functions, but in excessive amounts it tends to accumulate in artery walls; known as "bad" cholesterol

Lesion: Area of tissue that is injured or diseased such as a wound, abscess, sore, tumor, mole, or cyst

Lipid: Description term for a fat or fat-like substance found in the blood, such as cholesterol

Lipoproteins: Proteins that combine with lipids to make them dissolve in blood

Liver: A large organ in the upper abdomen that is the site of many metabolic functions, including the secretion of bile, the manufacture of proteins, and the storage of glycogen and certain vitamins

Lungs: The two organs of respiration that bring air and blood into close contact so that oxygen can be added to and carbon dioxide removed from the blood

Malabsorption: Inadequate absorption of nutrients from the small intestine. Symptoms and signs of malabsorption syndrome include loose, fatty stools, diarrhea, and weight-loss, but anemia is not a symptom

Malignant: Harmful, as in cancerous tissue that can grow uncontrollably and spread (metastasize)

Malnutrition: Deficiency of nourishment in the body due to lack of healthful food or improper digestion and distribution of nutrients

Masticate: Chew

Melanoma: A pigmented tumor of the skin and, in rare instances, of the mucous membranes. A malignant melanoma can be invasive and spread to lymph nodes and other sites more frequently than other skin cancers

Membrane: A thin layer of tissue that lines, separates, or covers organs or structures

Menopause: The age-related, permanent cessation of menstruation

Menstruation: Monthly shedding of blood and tissue from the lining of the uterus

Metabolism: Physical and chemical processes by which food is transformed into energy and tissues are broken down into waste products

Metastasis: Spreading of a disease from one part of the body to another, usually refers to movement of malignant cells (as in cancer) or bacteria through the lymph or blood

Microbes: Microscopic one-celled organisms such as bacteria, many of which cause disease

Mineral: A class of nutrients made from inorganic compounds

Mitosis: Type of cell division in which the new cells have the same number of chromosomes as the parent cell

mm Hg (millimeters of mercury): Unit used for measuring blood pressure

Mono-: Prefix meaning "one"

Muscle: Tissue that produces movement by its ability to contract

Musculoskeletal: Pertaining to the muscles and the skeleton

Myalgia: Muscle tenderness or pain

Myocardial infarction: Heart attack; death of an area of heart muscle due to lack of blood supply

Myocardium: The heart muscle. See *Endocardium* and *Epicardium*

Nausea: An unpleasant sensation in the stomach, often followed by vomiting

Necrosis: Changes due to death of cells or organs

Nerve: A bundle of nerve fibers through which nerve impulses pass

Neuropathy: A functional or structural change in nerves

Nucleus: Center portion of cells essential for cell growth, nourishment, and reproduction

Nutrients: Substances supplied by food that provide nourishment for the body

Nutrition: A combination of processes by which the body receives and uses the substances necessary for its function, for energy, and for growth and repair of the body

Obesity: Abnormal body weight, usually defined as more than 30 percent above average for age, height, and bone structure

Occlusion: Closure of a passage such as ducts or blood vessels. In dentistry, the alignment of upper and lower teeth when the jaws are closed

Olfactory: Pertaining to the sense of smell

-oma: Suffix meaning "tumor"; generally not a cancer

Oncology: The study of cancer

Organic food: Food that is grown and processed without the use of chemicals, including fertilizers, insecticides, artificial coloring, and additives

Orthostatic hypotension: Decrease in blood pressure upon standing; may lead to light-headedness or fainting
-osis: suffix meaning "diseased state"

Osteoporosis: Reduction in bone that can result in weak bones and fractures

Over-the-counter (OTC): Sold without a prescription

Pancreas: Gland that produces enzymes essential to the digestion of food. The islets of Langerhans within the pancreas secrete insulin into the blood

Parasite: An organism that lives on or within another organism at the expense of the host

Parathyroid gland: Endocrine glands located behind the thyroid gland that maintain the level of calcium in the blood

Parenteral: Method of administering medication or nutrition other than via the digestive tract, such as intravenous, subcutaneous, or intramuscular

Pareve: A term describing food made without animal or dairy ingredients, according to kosher dietary regulations

Pasteurization: A method of killing bacteria in milk and other liquids by heating to moderately high temperatures for a short time

Pathogen: Disease-producing microorganism

Pathology: Study of the cause and nature of a disease

Pernicious: Destructive, sometimes fatal. Pernicious anemia is caused by the inability to absorb vitamin B_{12} from the intestinal tract

Pharmacology: Study of drugs and their effects on living beings

Phytochemicals: Plant chemicals that when eaten may have an effect on health

Pica: An uncommon urge to eat nonfood items such as laundry starch, dirt, baking powder, or frost from the freezer

Pinch: A measure of dry ingredients equivalent to approximately 1/16 of a teaspoon

Placebo: Substance given for psychological benefit or as part of a clinical research study; it has no specific pharmacologic activity against illness

Plaque: A film or deposit of bacteria and other material on the surface of a tooth that may lead to tooth decay or periodontal disease

Plasma: Fluid part of the blood and lymph

Poly-: Prefix meaning "multiple"

Polyp: A protruding growth, often on a stalk

Primary care physician: Physician responsible for a person's general health care

Progesterone: Female sex hormone responsible for, among other things, preparation of the uterine lining for implantation of the fertilized egg

Prognosis: Prediction of the course or outcome of a disease

Prostate gland: Gland located at the base of the bladder in men that contributes to production of seminal fluid

Protein: One of many complex nitrogen-containing compounds, composed of amino acids; essential for the growth and repair of tissue

Puberty: The time when body changes particular to the sex occur and when reproduction becomes possible

Pulmonary: Pertaining to the lungs

Pulse: Expansion of an artery after each contraction of the heart

Radiation therapy: The use of high-energy penetrating waves to treat disease. Sources of radiation used in radiation therapy include x-ray, cobalt, and radium

Rectum: The lowest portion of the large intestine. Stores stool until it is emptied

Renal: Pertaining to the kidneys

Renal failure: The inability of the kidneys to excrete wastes, concentrate urine, and maintain electrolyte balance

Retinopathy: Abnormality of the retina that may cause deterioration of eyesight

Risk factors: A factor that increases the chance of developing or aggravating a condition

Roughage: Indigestible fiber of fruits, vegetables, and cereals

Saline: Salt (sodium chloride) solution

Saliva: Fluid secreted by the salivary and mucous glands of the mouth that moistens food and begins the process of digestion

Sauté: To cook food quickly in a small amount of oil or seasoned liquid over a high heat

Sclerosis: Hardening or thickening of an organ or tissue, usually due to abnormal growth of fibrous tissue

Screening: Tests or observations applied to a large cohort of individuals to identify disease or risk of disease

Secretion: The process of producing a substance by a gland; also, the substance produced

Sedentary: Lacking exercise; inactive

Sepsis: Infection with disease-causing microorganisms or other toxins in the bloodstream

Side effects: Undesirable effect of a medication or other treatment

Spleen: The largest organ in the lymphatic system. Located near the stomach, it has a role in the production, storage, and breakdown of blood cells

Squamous cell carcinoma: A malignant tumor arising from cells known as squamous epithelium; a common form of skin cancer

Stenosis: The narrowing or closure of an opening or passageway in the body

Sterilization: The process by which all microorganisms are killed, as in sterilization of surgical instruments

Steroids: See *Corticosteroids*

Stomach: A sac-like organ to which food is delivered by the esophagus. After the food is processed mechanically by a churning action and chemically with gastric acids, it passes from the stomach to the small intestine

Stool: Body waste excreted from the bowel; feces

Stroke: An injury of the brain due to bleeding or to an interruption of the blood supply

Sucrose: The simple sugar processed from sugarcane and sugar beets

Syndrome: A constellation of symptoms that characterize an ailment

Systemic: Affecting or pertaining to the entire body rather than one of its parts

Systole: The portion of the heart cycle during which the heart muscle is contracting

Systolic pressure: The highest blood pressure produced by the contraction of the heart. Recorded as the first number in a blood pressure measurement

Thoracic: Having to do with the chest (thorax)

Thyroid gland: The endocrine gland that produces thyroid hormone

Thyroxine: One of the forms of thyroid hormone that is involved in the control of the pace of chemical activity (metabolism) in the body

Tissue: A collection of similar cells that form a body structure

Toxin: A poison

Transient ischemic attack: Symptoms caused by temporary lack of circulation to part of the brain

Transplantation: The surgical transfer of an organ or tissue from one position (or person) to another

Trauma: The process or event leading to an injury or wound

Triglyceride: A form of fat that the body can make from sugar, alcohol, or excess calories

Truncal obesity: Fat deposited in the thorax and abdomen, instead of the hips and thighs

Tumor: A new growth of tissue; a neoplasm

Ulcer: An open sore on the skin or a mucous membrane

Ulcerative colitis: A disease characterized by inflammation of the lining of the colon and rectum

Urine: Fluid waste produced in the kidneys, stored in the bladder, and released through the urethra

Vascular: Pertaining to blood vessels; includes veins and arteries

Vegans: People who do not eat any food of animal origin

Vein: A blood vessel that returns blood to the heart

Venous: Pertaining to veins

Viral: Pertaining to or caused by a virus

Virus: Tiny organism that causes disease; viruses range from minor (common cold) to potentially deadly (AIDS)

Vital signs: Respiration, heart rate, and body temperature

Vitamins: Organic substances that are essential for most metabolic functions of the body; they are fat-soluble (A, D, E, K) and water-soluble (B vitamins and C).

Vomit: The ejection of contents of the stomach through the mouth; also, the material itself

X-ray: Electromagnetic vibrations of short wavelength that penetrate most matter and produce an image on film; also called roentgen ray

READING LIST

American Diabetes Association Complete Guide to Diabetes: the Ultimate Home Diabetes Reference. Second edition. Alexandria, VA, American Diabetes Association, 1999

Calhoun S, Bradley J: Nutrition, Cancer, and You: What You Need to Know, and Where to Start. Lenexa, KS, Addax Publishing Group, 1997

Collazo-Clavell M (editor): Mayo Clinic on Managing Diabetes. Rochester, MN, Mayo Clinic, 2001

Corriher SO: Cookwise: The Hows and Whys of Successful Cooking. New York, William Morrow, 1997

Donovan MD (editor): The Professional Chef's Techniques of Healthy Cooking. New York, Van Nostrand Reinhold, 1997

Duyff RL: The American Dietetic Association's Complete Food & Nutrition Guide. Minneapolis, Chronimed Publishing, 1998

Ensminger AH, Ensminger ME, Konlande JE, Robson JRK: Foods & Nutrition Encyclopedia. Second edition. Boca Raton, CRC Press, 1994

Ensminger AH, Ensminger ME, Konlande JE, Robson JRK: The Concise Encyclopedia of Foods and Nutrition. Second edition. Boca Raton, CRC Press, 1995

Fortin F, D'Amico S: The Visual Food Encyclopedia: the Definitive Practical Guide to Food and Cooking. New York, Macmillan Publishing, 1996

Gersh BJ (editor): Mayo Clinic Heart Book: the Ultimate Guide to Heart Health. Second edition. New York, William Morrow & Company, 2000

Hagen PT (editor): Mayo Clinic Guide to Self-Care: Answers for Everyday Health Problems. Second edition. Rochester, MN, Mayo Clinic, 1999

Heinerman J: Heinerman's New Encyclopedia of Fruits & Vegetables. West Nyack, NY, Parker, 1995

Hensrud DD (editor): Mayo Clinic on Healthy Weight. Rochester, MN, Mayo Clinic, 2000

Herbst ST: The Food Lover's Tiptionary: an A to Z Culinary Guide With More Than 4,500 Food and Drink Tips, Secrets, Shortcuts, and Other Things Cookbooks Never Tell You. New York, Hearst Books, 1994

Herbst ST: The New Food Lover's Companion: Comprehensive Definitions of Nearly 6,000 Food, Drink, and Culinary Terms. Third edition. Hauppauge, NY, Barron's Educational Series, 2001

Hoffman M, Joachim D (editors): Prevention's the Healthy Cook: the Ultimate Illustrated Kitchen Guide to Great Low-Fat Food: Featuring 450 Homestyle Recipes and Hundreds of Time-Saving Tips. Emmaus, PA, Rodale Press, 1997

Johnson RV (editor): Mayo Clinic Complete Book of Pregnancy & Baby's First Year. New York, William Morrow & Company, 1994

Larson DE (editor): Mayo Clinic Family Health Book. Second edition. New York, William Morrow & Company, 1996

Margen S: The Wellness Encyclopedia of Food and Nutrition: How to Buy, Store, and Prepare Every Fresh Food. New York, Rebus, 1992

Rinzler CA: The New Complete Book of Food: A Nutritional, Medical and Culinary Guide. New York, Checkmark Books, 1999

Robbers JE, Tyler VE: Tyler's Herbs of Choice: The Therapeutic Use of Phytomedicinals. New York, Haworth Herbal Press, 1999

Sarubin A: The Health Professional's Guide to Popular Dietary Supplements. Chicago, American Dietetic Association, 2000

Sheps SG (editor): Mayo Clinic on High Blood Pressure. Rochester, MN, Mayo Clinic, 1999

Shils ME, Olson JA, Shike M, Ross AC (editors): Modern Nutrition in Health and Disease. Ninth edition. Baltimore, Williams & Wilkins, 1999

Stare FJ, Whelan EM: Fad-Free Nutrition. Alameda, CA, Hunter House Publishers, 1998

WEB SITES

Cancer
American Cancer Society: http://www.cancer.org

American Institute for Cancer Research: http://www.aicr.org

CancerNet: http://cancernet.nci.nih.gov/index.html

National Cancer Institute: http://rex.nci.nih.gov

Diabetes
American Diabetes Association: http://www.diabetes.org/

National Institute of Diabetes & Digestive & Kidney Diseases:
http://www.niddk.nih.gov/index.htm

Heart Disease
American Heart Association: http://www.americanheart.org/

National Heart, Lung, and Blood Institute: http://www.nhlbi.nih.gov/index.htm

Hypertension
Dietary Approaches to Stop Hypertension (DASH): http://dash.bwh.harvard.edu

Nutrition
American Dietetic Association: http://www.eatright.org

Dietary Guidelines for Americans: http://warp.nal.usda.gov:80/fnic/dga/index.html

Food Guide Pyramid: http://www.nal.usda.gov:8001/py/pmap.htm

Healthy People 2010: http://web.health.gov/healthypeople/

Mayo Clinic (health Web site): http://www.MayoClinic.com

Tuft's Nutrition Navigator: http://navigator.tufts.edu/

USDA Food and Nutrition Information Center: http://warp.nal.usda.gov:80/fnic/

Obesity
National Institutes of Health: Aim for a Healthy Weight:
http://www.nhlbi.nih.gov/health/public/heart/obesity/lose_wt/index.htm

Shape Up America!: http://www.shapeup.org

Osteoporosis
National Osteoporosis Foundation: http://www.nof.org

Osteoporosis and Related Bone Diseases National Resource Center: http://www.osteo.org

Supplements
National Institutes of Health Office of Dietary Supplements:
http://ods.od.nih.gov/databases/ibids.html

(All Web sites were retrieved on June 29, 2001.)

APPENDIX

Dietary Reference Intakes
and
Nutrients in Foods

The tables on pages 422 to 429 are from Dietary Reference Intakes, National Academy of Sciences, National Academy Press, Washington, DC, 2001.

The data on pages 434 to 483 are from U.S. Department of Agriculture, Agricultural Research Service. 1999. USDA Nutrient Database for Standard Reference, Release 13. Nutrient Data Laboratory Home Page, http://www.nal.usda.gov/fnic/foodcomp. In some instances, manufacturers' data also were used. This does not indicate an endorsement of the product. "0" value = negligible level. (-) = value not available. Values were rounded to nearest decimal point.

TABLE 1: DIETARY REFERENCE INTAKES (DRIs): RECOMMENDED INTAKES FOR INDIVIDUALS. VITAMINS: FOOD AND NUTRITION BOARD, INSTITUTE OF MEDICINE, NATIONAL ACADEMIES

Life Stage Group	Vitamin A (µg/d)[a]	Vitamin C (mg/d)	Vitamin D (µg/d)[b,c]	Vitamin E (mg/d)[d]	Vitamin K (µg/d)	Thiamin (mg/d)	Riboflavin (mg/d)
Infants							
0–6 mo	400*	40*	5*	4*	2.0*	0.2*	0.3*
7–12 mo	500*	50*	5*	5*	2.5*	0.3*	0.4*
Children							
1–3 y	300	15	5*	6	30*	0.5	0.5
4–8 y	400	25	5*	7	55*	0.6	0.6
Males							
9–13 y	600	45	5*	11	60*	0.9	0.9
14–18 y	900	75	5*	15	75*	1.2	1.3
19–30 y	900	90	5*	15	120*	1.2	1.3
31–50 y	900	90	5*	15	120*	1.2	1.3
51–70 y	900	90	10*	15	120*	1.2	1.3
> 70 y	900	90	15*	15	120*	1.2	1.3
Females							
9–13 y	600	45	5*	11	60*	0.9	0.9
14–18 y	700	65	5*	15	75*	1.0	1.0
19–30 y	700	75	5*	15	90*	1.1	1.1
31–50 y	700	75	5*	15	90*	1.1	1.1
51–70 y	700	75	10*	15	90*	1.1	1.1
> 70 y	700	75	15*	15	90*	1.1	1.1
Pregnancy							
≤ 18 y	750	80	5*	15	75*	1.4	1.4
19–30 y	770	85	5*	15	90*	1.4	1.4
31–50 y	770	85	5*	15	90*	1.4	1.4
Lactation							
≤ 18 y	1,200	115	5*	19	75*	1.4	1.6
19–30 y	1,300	120	5*	19	90*	1.4	1.6
31–50 y	1,300	120	5*	19	90*	1.4	1.6

NOTE: This table (taken from the Dietary Reference Intake reports, see www.nap.edu) presents Recommended Dietary Allowances (RDA) in **bold type** and Adequate Intakes (AIs) in ordinary type followed by an asterisk (*). RDAs and AIs may both be used as goals for individual intake. RDAs are set to meet the needs of almost all (97 to 98 percent) individuals in a group. For healthy breastfed infants, the AI is the mean intake. The AI for other life stage and gender groups is believed to cover needs of all individuals in the group, but lack of data or uncertainty in the data prevent being able to specify with confidence the percentage of individuals covered by this intake.

[a]As retinol activity equivalents (RAEs). 1 RAE = 1 µg retinol, 12 µg β-carotene, 24 µg α-carotene, or 24 µg β-cryptoxanthin. To calculate RAEs from REs of provitamin A carotenoids in food, divide the REs by 2. For preformed vitamin A in foods or supplements and for provitamin A carotenoids in supplements, 1 RE = 1 RAE.

[b]Calciferol. 1 µg Calciferol = 40 IU vitamin D.

[c]In the absence of adequate exposure to sunlight.

[d]As α-tocopherol. α-tocopherol includes RRR-α-tocopherol, the only form of α-tocopherol that occurs naturally in foods, and the 2R-stereoisomeric forms of α-tocopherol (RRR-, RSR-, RRS-, and RSS-α-tocopherol) that occur in fortified foods and supplements. It does not include the 2S-stereoisomeric forms of α-tocopherol (SRR-, SSR-, SRS-, and SSS-α-tocopherol), also found in fortified foods and supplements.

Niacin (mg/d)[e]	Vitamin B_6 (mg/d)	Folate (µg/d)[f]	Vitamin B_{12} (µg/d)	Pantothenic Acid (mg/d)	Biotin (µg/d)	Choline (mg/d)[g]
2*	0.1*	65*	0.4*	1.7*	5*	125*
4*	0.3*	80*	0.5*	1.8*	6*	150*
6	0.5	150	0.9	2*	8*	200*
8	0.6	200	1.2	3*	12*	250*
12	1.0	300	1.8	4*	20*	375*
16	1.3	400	2.4	5*	25*	550*
16	1.3	400	2.4	5*	30*	550*
16	1.3	400	2.4	5*	30*	550*
16	1.7	400	2.4[h]	5*	30*	550*
16	1.7	400	2.4[h]	5*	30*	550*
12	1.0	300	1.8	4*	20*	375*
14	1.2	400[i]	2.4	5*	25*	400*
14	1.3	400[i]	2.4	5*	30*	425*
14	1.3	400[i]	2.4	5*	30*	425*
14	1.5	400	2.4[h]	5*	30*	425*
14	1.5	400	2.4[h]	5*	30*	425*
18	1.9	600[j]	2.6	6*	30*	450*
18	1.9	600[j]	2.6	6*	30*	450*
18	1.9	600[j]	2.6	6*	30*	450*
17	2.0	500	2.8	7*	35*	550*
17	2.0	500	2.8	7*	35*	550*
17	2.0	500	2.8	7*	35*	550*

[e]As niacin equivalents (NE). 1 mg of niacin = 60 mg of tryptophan; 0-6 months = preformed niacin (not NE).

[f]As dietary folate equivalents (DFE). 1 DFE = 1 µg food folate = 0.6 µg of folic acid from fortified food or as a supplement consumed with food = 0.5 µg of a supplement taken on an empty stomach.

[g]Although AIs have been set for choline, there are few data to assess whether a supply of choline is needed at all stages of the life cycle, and it may be that the choline requirement can be met by endogenous synthesis at some of these stages.

[h]Because 10-30 percent of older people may malabsorb food-bound B_{12}, it is advisable for those older than 50 years to meet their RDA mainly by consuming foods fortified with B_{12} or a supplement containing B_{12}.

[i]In view of evidence linking folate intake with neural tube defects in the fetus, it is recommended that all women capable of becoming pregnant consume 400 µg from supplements or fortified foods in addition to intake of food folate from a varied diet.

[j]It is assumed that women will continue consuming 400 µg of folic acid from supplements or fortified food until their pregnancy is confirmed and they enter prenatal care, which ordinarily occurs after the end of the periconception period—the critical time for formation of the neural tube.

TABLE 2: DIETARY REFERENCE INTAKES (DRIs): RECOMMENDED INTAKES FOR INDIVIDUALS. MINERALS: FOOD AND NUTRITION BOARD, INSTITUTE OF MEDICINE, NATIONAL ACADEMIES

Life Stage Group	Calcium (mg/d)	Chromium (μg/d)	Copper (μg/d)	Fluoride (mg/d)	Iodine (μg/d)	Iron (mg/d)
Infants						
0–6 mo	210*	0.2*	200*	0.01*	110*	0.27*
7–12 mo	270*	5.5*	220*	0.5*	130*	11*
Children						
1–3 y	500*	11*	340	0.7*	90	7
4–8 y	800*	15*	440	1*	90	10
Males						
9–13 y	1,300*	25*	700	2*	120	8
14–18 y	1,300*	35*	890	3*	150	11
19–30 y	1,000*	35*	900	4*	150	8
31–50 y	1,000*	35*	900	4*	150	8
51–70 y	1,200*	30*	900	4*	150	8
> 70 y	1,200*	30*	900	4*	150	8
Females						
9–13 y	1,300*	21*	700	2*	120	8
14–18 y	1,300*	24*	890	3*	150	15
19–30 y	1,000*	25*	900	3*	150	18
31–50 y	1,000*	25*	900	3*	150	18
51–70 y	1,200*	20*	900	3*	150	8
> 70 y	1,200*	20*	900	3*	150	8
Pregnancy						
≤ 18 y	1,300*	29*	1,000	3*	220	27
19–30 y	1,000*	30*	1,000	3*	220	27
31–50 y	1,000*	30*	1,000	3*	220	27
Lactation						
≤ 18 y	1,300*	44*	1,300	3*	290	10
19–30 y	1,000*	45*	1,300	3*	290	9
31–50 y	1,000*	45*	1,300	3*	290	9

NOTE: This table presents Recommended Dietary Allowances (RDAs) in **bold type** and Adequate Intakes (AIs) in ordinary type followed by an asterisk (*). RDAs and AIs may both be used as goals for individual intake. RDAs are set to meet the needs of almost all (97 to 98 percent) individuals in a group. For healthy breastfed infants, the AI is the mean intake. The AI for other life stage and gender groups is believed to cover needs of all individuals in the group, but lack of data or uncertainty in the data prevent being able to specify with confidence the percentage of individuals covered by this intake.

Sources: Dietary Reference Intakes for Calcium, Phosphorus, Magnesium, Vitamin D, and Fluoride (1997); Dietary Reference Intakes for Thiamin, Riboflavin, Niacin, Vitamin B_6, Folate, Vitamin B_{12}, Pantothenic Acid, Biotin, and Choline (1998); Dietary Reference Intakes for Vitamin C, Vitamin E, Selenium, and Carotenoids (2000); and Dietary Reference Intakes for Vitamin A, Vitamin K, Arsenic, Boron, Chromium, Copper, Iodine, Iron, Manganese, Molybdenum, Nickel, Silicon, Vanadium, and Zinc (2001). These reports may be accessed via www.nap.edu.

Magnesium (mg/d)	Manganese (mg/d)	Molybdenum (µg/d)	Phosphorus (mg/d)	Selenium (µg/d)	Zinc (mg/d)
30*	0.003*	2*	100*	15*	2*
75*	0.6*	3*	275*	20*	3
80	1.2*	17	460	20	3
130	1.5*	22	500	30	5
240	1.9*	34	1,250	40	8
410	2.2*	43	1,250	55	11
400	2.3*	45	700	55	11
420	2.3*	45	700	55	11
420	2.3*	45	700	55	11
420	2.3*	45	700	55	11
240	1.6*	34	1,250	40	8
360	1.6*	43	1,250	55	9
310	1.8*	45	700	55	8
320	1.8*	45	700	55	8
320	1.8*	45	700	55	8
320	1.8*	45	700	55	8
400	2.0*	50	1,250	60	13
350	2.0*	50	700	60	11
360	2.0*	50	700	60	11
360	2.6*	50	1,250	70	14
310	2.6*	50	700	70	12
320	2.6*	50	700	70	12

Table 3: Dietary Reference Intakes (DRIs): Tolerable Upper Intake Levels (UL[a]). Vitamins: Food and Nutrition Board, Institute of Medicine, National Academies

Life Stage Group	Vitamin A (µg/d)[b]	Vitamin C (mg/d)	Vitamin D (µg/d)	Vitamin E (mg/d)[c,d]	Vitamin K	Thiamin	Riboflavin
Infants							
0–6 mo	600	ND[f]	25	ND	ND	ND	ND
7–12 mo	600	ND	25	ND	ND	ND	ND
Children							
1–3 y	600	400	50	200	ND	ND	ND
4–8 y	900	650	50	300	ND	ND	ND
Males, females							
9–13 y	1,700	1,200	50	600	ND	ND	ND
14–18 y	2,800	1,800	50	800	ND	ND	ND
19–70 y	3,000	2,000	50	1,000	ND	ND	ND
> 70 y	3,000	2,000	50	1,000	ND	ND	ND
Pregnancy							
≤ 18 y	2,800	1,800	50	800	ND	ND	ND
19–50 y	3,000	2,000	50	1,000	ND	ND	ND
Lactation							
≤ 18 y	2,800	1,800	50	800	ND	ND	ND
19–50 y	3,000	2,000	50	1,000	ND	ND	ND

[a]UL = The maximum level of daily nutrient intake that is likely to pose no risk of adverse effects. Unless otherwise specified, the UL represents total intake from food, water, and supplements. Due to lack of suitable data, ULs could not be established for vitamin K, thamin, riboflavin, vitamin B_{12}, pantothenic acid, biotin, or carotenoids. In the absence of ULs, extra caution may be warranted in consuming levels above recommended intakes.

[b]As preformed vitamin A only.

[c]As α-tocopherol; applies to any form of supplemental α-tocopherol.

[d]The ULs for vitamin E, niacin, and folate apply to synthetic forms obtained from supplements, fortified foods, or a combination of the two.

[e]β-Carotene supplements are advised only to serve as a provitamin A source for individuals at risk of vitamin A deficiency.

[f]ND = Not determinable due to lack of data on adverse effects in this age group and concern with regard to lack of ability to handle excess amounts. Source of intake should be from food only to prevent high levels of intake.

Sources: Dietary Reference Intakes for Calcium, Phosphorus, Magnesium, Vitamin D, and Fluoride (1997); Dietary Reference Intakes for Thiamin, Riboflavin, Niacin, Vitamin B_6, Folate, Vitamin B_{12}, Pantothenic Acid, Biotin, and Choline (1998); Dietary Reference Intakes for Vitamin C, Vitamin E, Selenium, and Carotenoids (2000); and Dietary Reference Intakes for Vitamin A, Vitamin K, Arsenic, Boron, Chromium, Copper, Iodine, Iron, Manganese, Molybdenum, Nickel, Silicon, Vanadium, and Zinc (2001). These reports may be accessed via www.nap.edu.

Niacin (mg/d)[d]	Vitamin B$_6$ (mg/d)	Folate (µg/d)[d]	Vitamin B$_{12}$	Pantothenic Acid	Biotin	Choline (g/d)	Carotenoids[e]
ND	ND	ND	ND	ND	ND	ND	ND
ND	ND	ND	ND	ND	ND	ND	ND
10	30	300	ND	ND	ND	1.0	ND
15	40	400	ND	ND	ND	1.0	ND
20	60	600	ND	ND	ND	2.0	ND
30	80	800	ND	ND	ND	3.0	ND
35	100	1,000	ND	ND	ND	3.5	ND
35	100	1,000	ND	ND	ND	3.5	ND
30	80	800	ND	ND	ND	3.0	ND
35	100	1,000	ND	ND	ND	3.5	ND
30	80	800	ND	ND	ND	3.0	ND
35	100	1,000	ND	ND	ND	3.5	ND

TABLE 4: DIETARY REFERENCE INTAKES (DRIs): TOLERABLE UPPER INTAKE LEVELS (UL[a]). MINERALS: FOOD AND NUTRITION BOARD, INSTITUTE OF MEDICINE, NATIONAL ACADEMIES

Life Stage Group	Arsenic[b]	Boron (mg/d)	Calcium (g/d)	Chromium	Copper (µg/d)	Fluoride (mg/d)	Iodine (µg/d)	Iron (mg/d)
Infants								
0–6 mo	ND[f]	ND	ND	ND	ND	0.7	ND	40
7–12 mo	ND	ND	ND	ND	ND	0.9	ND	40
Children								
1–3 y	ND	3	2.5	ND	1,000	1.3	200	40
4–8 y	ND	6	2.5	ND	3,000	2.2	300	40
Males, females								
9–13 y	ND	11	2.5	ND	5,000	10	600	40
14–18 y	ND	17	2.5	ND	8,000	10	900	45
19–70 y	ND	20	2.5	ND	10,000	10	1,100	45
> 70 y	ND	20	2.5	ND	10,000	10	1,100	45
Pregnancy								
≤ 18 y	ND	17	2.5	ND	8,000	10	900	45
19–50 y	ND	20	2.5	ND	10,000	10	1,100	45
Lactation								
≤ 18 y	ND	17	2.5	ND	8,000	10	900	45
19–50 y	ND	20	2.5	ND	10,000	10	1,100	45

[a]UL = The maximum level of daily nutrient intake that is likely to pose no risk of adverse effects. Unless otherwise specified, the UL represents total intake from food, water, and supplements. Due to lack of suitable data, ULs could not be established for arsenic, chromium, and silicon. In the absence of ULs, extra caution may be warranted in consuming levels above recommended intakes.

[b]Although the UL was not determined for arsenic, there is no justification for adding arsenic to food or supplements.

[c]The ULs for magnesium represent intake from a pharmacologic agent only and do not include intake from food and water.

[d]Although silicon has not been shown to cause adverse effects in humans, there is no justification for adding silicon to supplements.

[e]Although vanadium in food has not been shown to cause adverse effects in humans, there is no justification for adding vanadium to food, and vanadium supplements should be used with caution. The UL is based on adverse effects in laboratory animals and this data could be used to set a UL for adults but not children and adolescents.

[f]ND = Not determinable due to lack of data on adverse effects in this age group and concern with regard to lack of ability to handle excess amounts. Source of intake should be from food only to prevent high levels of intake.

Sources: Dietary Reference Intakes for Calcium, Phosphorus, Magnesium, Vitamin D, and Fluoride (1997); Dietary Reference Intakes for Thiamin, Riboflavin, Niacin, Vitamin B_6, Folate, Vitamin B_{12}, Pantothenic Acid, Biotin, and Choline (1998); Dietary Reference Intakes for Vitamin C, Vitamin E, Selenium, and Carotenoids (2000); and Dietary Reference Intakes for Vitamin A, Vitamin K, Arsenic, Boron, Chromium, Copper, Iodine, Iron, Manganese, Molybdenum, Nickel, Silicon, Vanadium, and Zinc (2001). These reports may be accessed via www.nap.edu.

Magnesium (mg/d)[c]	Manganese (mg/d)	Molybdenum (µg/d)	Nickel (mg/d)	Phosphorus (g/d)	Selenium (µg/d)	Silicon[d]	Vanadium (mg/d)[e]	Zinc (mg/d)
ND	ND	ND	ND	ND	45	ND	ND	4
ND	ND	ND	ND	ND	60	ND	ND	5
65	2	300	0.2	3	90	ND	ND	7
110	3	600	0.3	3	150	ND	ND	12
350	6	1,100	0.6	4	280	ND	ND	23
350	9	1,700	1.0	4	400	ND	ND	34
350	11	2,000	1.0	4	400	ND	1.8	40
350	11	2,000	1.0	3	400	ND	1.8	40
350	9	1,700	1.0	3.5	400	ND	ND	34
350	11	2,000	1.0	3.5	400	ND	ND	40
350	9	1,700	1.0	4	400	ND	ND	34
350	11	2,000	1.0	4	400	ND	ND	40

TABLE 5: A QUICK LOOK – VITAMINS, THEIR FUNCTIONS AND FOOD SOURCES

Below are recommended vitamin intakes for adults aged 19 or older.

(For infants, children, or women who are pregnant or lactating, see pages 422-423 and 426-427.)

Vitamin	Food Sources	Functions
Fat-Soluble Vitamins		
Vitamin A *Men:* 19 years or older – 900 micrograms *Women:* 19 years or older – 700 micrograms *Upper limit:* 3,000 micrograms/day	Vitamin A (retinol): eggs, liver, fortified dairy products, vitamin A-fortified foods. Beta-carotenes (converted by the body into vitamin A): dark green, yellow, and red vegetables and fruit	Growth, reproduction, maintenance of body tissues, immune function, vision
Vitamin D *Men and women:* 19–50 years – 5 micrograms 51–70 years – 10 micrograms 71 years or older – 15 micrograms *Upper limit:* 50 micrograms/day	Fortified dairy products, egg yolk, fatty fish; also made by skin exposed to sunlight	Building and maintenance of bones and teeth, calcium and phosphorus metabolism
Vitamin E *Men and women:* 19 years or older – 15 milligrams *Upper limit:* 1,000 milligrams/day from supplement/fortified foods	Vegetable oil, wheat germ, margarine, nuts, green leafy vegetables, beans	Antioxidant: protects cell membranes and red blood cells from oxidation damage; immune function
Vitamin K *Men:* 19 years or older – 120 micrograms *Women:* 19 years or older – 90 micrograms *Upper limit:* Not established*	Green leafy vegetables, milk, dairy products, meats, eggs, cereals, fruits; also made by bacteria in gut	Formation of blood clotting substances and building of bones
Water-Soluble Vitamins		
Vitamin C *Men:* 19 years or older – 90 milligrams *Women:* 19 years or older – 75 milligrams *Upper limit:* 2,000 milligrams/day	Citrus fruits, strawberries, melon, tomatoes, green and red peppers, collard greens, broccoli, spinach, potatoes	Maintains collagen (intracellular cement): blood vessel integrity; enhances immunity, wound healing; antioxidant; increases absorption of iron from plant foods
Thiamin (B_1) *Men:* 19 years or older – 1.2 milligrams *Women:* 19 years or older – 1.1 milligrams *Upper limit:* Not established*	Wheat germ, whole and enriched grains, brewer's yeast, organ meats, pork, legumes, seeds, nuts	Carbohydrate metabolism, nerve function, growth, and muscle tone

Riboflavin (B$_2$) *Men:* 19 years or older – 1.3 milligrams *Women:* 19 years or older – 1.1 milligrams *Upper limit:* Not established*	Dairy products, whole and enriched grain products, animal proteins	Energy release in cells, maintenance of tissues
Niacin (B$_3$) *Men:* 19 years or older – 16 milligrams *Women:* 19 years or older – 14 milligrams *Upper limit:* 35 milligrams/day from supplement/ fortified foods	Animal protein, enriched grains, dried beans and peas	Energy release in cells, growth hormone production, skin and gut maintenance, nerve function

Pyridoxine (B$_6$)
Men and women:
19–50 years – 1.3 milligrams
51 years or older –
   Men: 1.7 milligrams
   Women: 1.5 milligrams
 	
Upper limit:
100 milligrams/day
	Fish, poultry, meat, liver, whole grains, potato	Energy release in cells, red blood cell formation, nerve function

Folate *Men and women:* 19 years or older – 400 micrograms *Upper limit:* 1,000 micrograms/day from supplement/fortified foods	Legumes, green leafy vegetables, fortified grain products, yeast, oranges, nuts	Prevention of birth defects, red blood cell formation, growth and cell division
Cobalamin (B$_{12}$) *Men and women:* 19 years or older – 2.4 micrograms *Upper limit:* Not established*	Animal and dairy products	Red blood cell formation, nerve function, energy release in cells
Biotin *Men and women:* 19 years or older – 30 micrograms *Upper limit:* Not established*	Egg yolk, organ meats (kidney, liver), milk, dark green vegetables	Formation of fatty acids, utilization of B vitamins, nerve maintenance
Choline *Men:* 19 years or older – 550 milligrams *Women:* 19 years or older – 425 milligrams *Upper limit:* 3,500 milligrams	Eggs, liver, soybeans, cauliflower, lettuce, fats that are emulsified such as margarine and salad dressings	Growth and development, nerve transmission, component of lipoproteins and cell membranes

(continues)

A Quick Look – Vitamins, Their Functions and Food Sources *(continued)*

Pantothenic Acid *Men and women:* 19 years or older – 5 milligrams *Upper limit:* Not established*	Animal products, whole-grain cereals, legumes	Conversion of energy into blood glucose, hormone synthesis, vitamin utilization, nerve function

These values are from the tables on pages 422-423 and 426-427.
*Upper limits not determinable. Caution is advised in consuming amounts above recommended intakes.

Table 6: A Quick Look – Minerals, Their Functions and Food Sources

Below are recommended mineral intakes for adults aged 19 or older.

(For infants, children, or women who are pregnant or lactating, see pages 424-425 and 428-429.)

Mineral	Food Sources	Functions
Calcium *Men and women:* 19–50 years – 1,000 milligrams 51 years or older – 1,200 milligrams *Upper limit:* 2,500 milligrams/day	Milk and milk products, fish with edible bones, dark green vegetables, fortified foods	Bone development and maintenance, nerve function, blood clotting, muscle contraction
Chromium *Men:* 19–50 years – 35 micrograms 51 years or older – 30 micrograms *Women:* 19–50 years – 25 micrograms 51 years or older – 20 micrograms *Upper limit:* Not established*	Brewer's yeast, wheat germ, cheese, whole grains	Glucose regulation, muscle function
Copper *Men and women:* 19 years or older – 900 micrograms *Upper limit:* 10,000 micrograms	Liver, seafoods, nuts and seeds, cocoa powder	Formation of red blood cells, pigmentation, bone maintenance
Fluoride *Men:* 19 years and older – 4 milligrams *Women:* 19 years or older – 3 milligrams *Upper limit:* 10 milligrams/day	Fluoridated water, tea, ocean fish with edible bones	Reduction of dental caries, bone maintenance
Iodine *Men and women:* 19 years or older – 150 micrograms *Upper limit:* 1,100 micrograms	Iodized salt, seafood	Thyroid function, growth, mental development, energy metabolism

Iron *Men:* 19 years or older – 8 milligrams *Women:* 19–50 years – 18 milligrams 51 years or older – 8 milligrams *Upper limit:* 45 milligrams	Meat, liver, egg yolk, dark green vegetables, whole and enriched grain products	Formation of hemoglobin in blood and myoglobin in muscle, which helps with utilization of oxygen
Magnesium *Men:* 19–30 years – 400 milligrams 31 years or older – 420 milligrams *Women:* 19–30 years – 310 milligrams 31 years or older – 320 milligrams *Upper limit:* 350 milligrams from supplement only	Nuts, seeds, whole grains, wheat germ, bran, green vegetables, bananas	Enzyme, nerve and muscle function, bone growth
Manganese *Men:* 19 years or older – 2.3 milligrams *Women:* 19 years or older – 1.8 milligrams *Upper limit:* 11 milligrams	Whole grains, fruits, vegetables, tea	Reproduction, growth, bone formation, glucose regulation
Molybdenum *Men and women:* 19 years or older – 45 micrograms *Upper limit:* 2,000 micrograms	Milk, beans, grain products	Enzyme systems, nerve function, mental development
Phosphorus *Men and women:* 19 years or older – 700 milligrams *Upper limit:* 19–70 years – 4,000 milligrams/day 71 years or older – 3,000 milligrams/day	Animal and high-protein vegetable products, whole grains	Bone development and maintenance, energy release
Selenium *Men and women:* 19 years or older – 55 micrograms *Upper limit:* 400 milligrams/day	Seafood, meats, liver and kidney, onions, grains	Antioxidant, fat utilization, heart muscle maintenance
Zinc *Men:* 19 years or older – 11 milligrams *Women:* 19 years or older – 8 milligrams *Upper limit:* 40 milligrams/day	Meat, liver, eggs, seafoods, whole grains	Growth, wound healing, taste and smell sensitivity

These values are from the tables on pages 424-425 and 428-429.

*Upper limits not determinable. Caution is advised in consuming amounts above recommended intakes.

TABLE 7: NUTRIENTS IN FOODS

FRUITS

	Serving size	Weight (g)	Calories	Protein (g)	Fat (g)	Carbohydrates (g)	Fiber (g)	Calcium (mg)	Iron (mg)	Magnesium (mg)	Phosphorus (mg)
Acerola (West Indian cherry), raw	1 cup	98	31	0	0	8	1	12	0	18	11
Apple, raw, with skin	1 medium (2 3/4″ diameter)	138	81	0	0	21	4	10	0	7	10
Apricot, raw	2 apricots	70	34	1	0	8	2	10	0	6	13
Apricot, dried, uncooked	9 halves	32	75	1	0	19	3	14	1	15	37
Avocado, raw	1	200	324	4	31	15	10	22	2	78	82
Banana, raw	1 medium (7″ to 7 7/8″ long)	118	109	1	1	28	3	7	0	34	24
Blackberry, raw	1/2 cup	72	37	1	0	9	4	23	0	14	15
Blueberry, raw	1/2 cup (about 53 berries)	73	41	0	0	10	2	4	0	4	7
Breadfruit, raw	1/4 small fruit	96	99	1	0	26	5	16	1	24	29
Cantaloupe, raw	1 medium wedge (1/4 of medium melon)	138	48	1	0	12	1	15	0	15	23
Carambola (starfruit), raw	1 large (4 1/2″ long)	127	42	1	0	10	3	5	0	11	20
Casaba melon, raw	1 cup, cubes	170	44	2	0	11	1	9	1	14	12
Cherimoya, raw	1/8 fruit, without skin and seeds	68	64	1	0	16	2	16	0	(-)	27
Cherry, sweet, raw	1/2 cup, without pits (11 fruits)	73	52	1	1	12	2	11	0	8	14
Cherry, sour, red, raw	1/2 cup, without pits (11 fruits)	78	39	1	0	9	1	12	0	7	12
Coconut meat (nuts), raw	1 piece (1 1/4″ x 1″ diameter)	23	80	1	8	3	2	3	1	7	25
Cranberry, raw	1/2 cup, whole	48	23	0	0	6	2	3	0	2	4
Currant, red and white, raw	1/2 cup	56	31	1	0	8	2	18	1	7	25
Date, domestic, natural, and dry	5 dates, dried	42	114	1	0	31	3	13	0	15	17
Durian, raw or frozen	1/4 cup, chopped or diced	61	89	0	3	16	2	4	0	18	23
Elderberry, raw	1/2 cup	73	53	0	0	13	5	28	1	4	28
Feijoa, raw	3 fruits, without peel	150	74	2	1	16	(-)	26	0	14	30
Fig, raw	3 medium (2 1/4″ diameter)	150	111	1	0	29	5	53	1	26	21
Fig, dried, uncooked	2 figs	38	97	1	0	25	5	55	1	22	26
Gooseberry, raw	1/2 cup	75	33	1	0	8	3	19	0	8	20
Grapefruit, raw, pink, red, and white	1/2 medium (approx 4″ diameter), without peel	128	41	1	0	10	1	15	0	10	10
Grapes, American type (slip skin), raw	1/2 cup (18 fruits)	46	31	0	0	8	0	6	0	2	5
Grapes, red or green, seedless, raw	1/2 cup, seedless (18 fruits)	80	57	1	0	14	1	9	0	5	10
Guava, common, raw	1 fruit, without peel	90	46	1	1	11	5	18	0	9	23
Honeydew melon, raw	1 wedge (1/8 of 5 1/4″ diameter melon)	125	44	1	0	11	1	8	0	9	13

Potassium (mg)	Sodium (mg)	Zinc (mg)	Copper (mg)	Manganese (mg)	Selenium (µg)	Vitamin C (mg)	Thiamin (mg)	Riboflavin (mg)	Niacin (mg)	Pantothenic acid (mg)	Vitamin B$_6$ (mg)	Folate (µg)	Vitamin B$_{12}$ (µg)	Vitamin A (IU)	Vitamin A (RE)	Vitamin E (mg)	Saturated fat (g)	Monounsaturated fat (g)	Polyunsaturated fat (g)	Cholesterol (mg)
143	7	0	0	(-)	1	1,644	0.0	0.1	0	0	0.0	14	0.0	752	75	0	0	0	0	0
159	0	0	0	0	0	8	0.0	0.0	0	0	0.1	4	0.0	73	7	0	0	0	0	0
207	1	0	0	0	0	7	0.0	0.0	0	0	0.0	6	0.0	1,828	183	1	0	0	0	0
434	3	0	0	0	1	1	0.0	0.0	1	0	0.0	3	0.0	2,281	228	0	0	0	0	0
1,204	20	1	0	0	1	16	0.2	0.2	3.8	2	0.5	124	0.0	1,230	123	3	5	19	4	0
467	1	0	0	0	1	11	0.1	0.1	1	0	0.7	23	0.0	96	9	0	0	0	0	0
141	0	0	0	1	0	15	0.0	0.0	0	0	0.0	24	0.0	119	12	1	0	0	0	0
65	4	0	0	0	0	9	0.0	0.0	0	0	0.0	5	0.0	73	7	1	0	0	0	0
470	2	0	0	0	1	28	0.1	0.0	1	0	0.1	13	0.0	38	4	1	0	0	0	0
426	12	0	0	0	0	58	0.1	0.1	1	0	0.2	23	0.0	4,449	444	0	0	0	0	0
207	3	0	0	0	1	27	0.0	0.0	1	(-)	0.1	18	0.0	626	62	0	0	0	0	0
357	20	0	0	(-)	1	27	0.1	0.0	1	(-)	0.2	29	0.0	51	5	0	0	0	0	0
(-)	(-)	(-)	(-)	(-)	(-)	6	0.1	0.1	1	(-)	(-)	(-)	0.0	7	1	(-)	(-)	(-)	(-)	0
162	0	0	0	0	0	5	0.0	0.0	0	0	0.0	3	0.0	155	15	0	0	0	0	0
134	2	0	0	0	0	8	0.0	0.0	0	0	0.0	6	0.0	994	99	0	0	0	0	0
80	5	0	0	0	2	1	0.0	0.0	0	0	0.0	6	0.0	0	0	0	7	0	0	0
34	0	0	0	0	0	6	0.0	0.0	0	0	0.0	1	0.0	22	2	0	0	0	0	0
154	1	0	0	0	0	23	0.0	0.0	0	0	0.0	4	0.0	67	7	0	0	0	0	0
271	1	0	0	0	1	0	0.0	0.0	1	0	0.1	5	0.0	21	2	0	0	0	0	0
265	0	0	0	0	(-)	12	0.0	0.0	0	0	0.0	(-)	0.0	27	3	(-)	(-)	(-)	(-)	0
203	4	0	0	(-)	0	26	0.1	0.0	0	0	0.2	4	0.0	435	44	1	0	0	0	0
233	5	0	0	0	(-)	30	0.0	0.0	0	0	0.1	57	0.0	0	0	(-)	(-)	(-)	(-)	0
348	2	0	0	0	1	3	0.1	0.1	1	0	0.2	9	0.0	213	21	1	0	0	0	0
271	4	0	0	0	0	0	0.0	0.0	0	0	0.1	3	0.0	51	5	0	0	0	0	0
149	1	0	0	0	0	21	0.0	0.0	0	0	0.1	5	0.0	218	22	0	0	0	0	0
178	0	0	0	0	2	44	0.0	0.0	0	0	0.1	13	0.0	159	15	0	0	0	0	0
88	1	0	0	0	0	2	0.0	0.0	0	0	0.1	2	0.0	46	5	0	0	0	0	0
148	2	0	0	0	0	9	0.1	0.0	0	0	0.1	3	0.0	58	6	1	0	0	0	0
256	3	0	0	0	1	165	0.0	0.0	1	0	0.1	13	0.0	713	71	1	0	0	0	0
339	13	0	0	0	1	31	0.1	0.0	1	0	0.1	8	0.0	50	5	0	0	0	0	0

(continues)

FRUITS (*continued*)

	Serving size	Weight (g)	Calories	Protein (g)	Fat (g)	Carbohydrates (g)	Fiber (g)	Calcium (mg)	Iron (mg)	Magnesium (mg)	Phosphorus (mg)
Jackfruit, raw	1/2 cup, sliced	83	78	1	0	20	1	28	0	31	30
Jujube, raw	3.5 oz	100	79	1	0	20	(-)	21	0	10	23
Jujube, dried	1 oz	28	82	1	0	21	(-)	22	0	10	28
Kiwi fruit, fresh, raw	1 large, without skin	91	56	1	0	14	3	24	0	27	36
Kumquat, raw	4 fruits, without peel	76	48	1	0	12	5	33	0	10	14
Lemon, raw, with peel	1 fruit, without seeds	108	22	1	0	12	5	66	1	13	16
Lime, raw	1 fruit (2″ diameter)	67	20	0	0	7	2	22	0	4	12
Longan, raw	10 fruits, without peel	32	19	0	0	5	0	0	0	3	7
Longan, dried	10 g	10	29	0	0	7	(-)	5	1	5	20
Loquat, raw	1/2 cup, cubed	75	35	0	0	9	1	12	0	10	20
Lychee, raw	10 fruits, without peel	96	63	1	0	16	1	5	0	10	30
Lychee, dried	10 fruits	25	69	1	0	18	1	8	0	11	45
Mango, raw	1/2 fruit, without peel	104	67	1	0	18	2	10	0	9	11
Mulberry, raw	1/2 cup	70	30	1	0	7	1	27	1	13	27
Nectarine, raw	1 fruit (2 1/2″ diameter)	136	67	1	1	16	2	7	0	11	22
Olives, ripe, canned	10 large (1/3 cup)	44	51	0	5	3	1	39	1	2	1
Orange, raw, all commercial varieties	1 fruit (2 5/8″ diameter)	131	62	1	0	15	3	52	0	13	18
Papaya, raw	1/4 medium (about 1/2 cup cubes)	76	30	0.5	0	8	2	18	0	8	4
Passion fruit, purple, raw	4 fruits, without peel	72	68	2	0	16	8	8	0	20	48
Peach, raw	1 medium (2 1/2″ diameter)	98	42	1	0	11	2	5	0	7	12
Pear, raw	1 medium (2 1/2 per pound)	166	98	1	1	25	4	18	0	10	18
Pear, Asian, raw	1 fruit (2 1/4" high × 2-1/2" diameter)	122	51	1	0	13	4	5	0	10	13
Persimmon, Japanese, raw	1/2 fruit (2 1/2″ diameter)	84	59	0	0	16	3	7	0	8	14
Pineapple, raw	1/2 cup, diced	78	38	0	0	10	1	5	0	11	5
Plantain, cooked	1/2 cup, mashed	100	116	1	0	31	2	2	1	32	28
Plum, raw	1 fruit (2 1/8″ diameter)	66	36	1	0	9	1	3	0	5	7
Pomegranate, raw	1 fruit (3 3/8″ diameter)	154	105	1	0	26	1	5	0	5	12
Prickly pear, raw	1 fruit, without peel	103	42	1	1	10	4	58	0	88	25
Prune (dried plum), uncooked	1/4 cup, pitted (about 5 fruits)	43	102	1	0	27	3	22	1	19	34
Pummelo, raw	1/2 cup sections	95	36	1	0	9	1	4	0	6	16
Quince, raw	1 fruit, without peel	92	52	0	0	14	2	10	1	7	16
Raisins, seedless	1/4 cup, packed	41	124	1	0	33	2	20	1	14	40

Potassium (mg)	Sodium (mg)	Zinc (mg)	Copper (mg)	Manganese (mg)	Selenium (µg)	Vitamin C (mg)	Thiamin (mg)	Riboflavin (mg)	Niacin (mg)	Pantothenic acid (mg)	Vitamin B₆ (mg)	Folate (µg)	Vitamin B₁₂ (µg)	Vitamin A (IU)	Vitamin A (RE)	Vitamin E (mg)	Saturated fat (g)	Monounsaturated fat (g)	Polyunsaturated fat (g)	Cholesterol (mg)
250	2	0	0	0	0	6	0.0	0.1	0	(-)	0.1	12	0.0	245	25	0	0	0	0	0
250	3	0	0	0	(-)	69	0.0	0.0	1	(-)	0.1	(-)	0.0	40	4	(-)	(-)	(-)	(-)	0
149	2	0	0	0	(-)	4	0.0	0.1	0	(-)	(-)	(-)	0.0	(-)	(-)	(-)	(-)	(-)	(-)	0
302	5	(-)	(-)	(-)	(-)	89	0.0	0.0	0	(-)	0.1	35	0.0	159	16	1	0	0	0	0
148	5	0	0	0	0	28	0.1	0.1	0	(-)	0.0	12	0.0	230	23	0	0	0	0	0
157	3	0	0	(-)	(-)	83	0.1	0.0	0	0	0.1	(-)	0.0	32	3	(-)	0	0	0	0
68	1	0	0	0	0	19	0.0	0.0	0	0	0.0	5	0.0	7	1	0	0	0	0	0
85	0	0	0	0	(-)	27	0.0	0.0	0	(-)	(-)	(-)	0.0	(-)	(-)	(-)	(-)	(-)	(-)	0
66	5	0	0	0	(-)	3	0.0	0.0	0	(-)	(-)	(-)	0.0	0	0	(-)	(-)	(-)	(-)	0
198	1	0	0	0	0	1	0.0	0.0	0	(-)	0.1	10	0.0	1,138	114	1	0	0	0	0
164	1	0	0	0	1	69	0.0	0.1	1	(-)	0.1	13	0.0	0	0	1	0	0	0	0
278	1	0	0	0	0	46	0.0	0.1	1	(-)	0.0	3	0.0	0	0	0	0	0	0	0
161	2	0	0	0	1	29	0.1	0.1	1	0	0.1	14	0.0	4,030	403	1	0	0	0	0
136	7	0	0	(-)	0	25	0.0	0.1	0	(-)	0.0	4	0.0	18	2	0	0	0	0	0
288	0	0	0	0	1	7	0.0	0.1	1	0	0.0	5	0.0	1,001	101	1	0	0	0	0
4	384	0	0	0	0	0	0.0	0.0	0	0	0.0	0	0.0	177	18	1	1	3	0	0
237	0	0	0	0	1	70	0.1	0.1	0	0	0.1	40	0.0	269	28	0	0	0	0	0
196	3	0	0	0	1	47	0.0	0.0	1	0	0.0	29	0.0	216	22	1	0	0	0	0
252	20	0	0	(-)	0	20	0.0	0.0	1	(-)	0.0	12	0.0	504	52	1	0	0	0	0
193	0	0	0	0	0	6	0.0	0.0	1	0	0.0	3	0.0	524	53	1	0	0	0	0
208	0	0	0	0	2	7	0.0	0.1	0	0	0.0	12	0.0	33	3	1	0	0	0	0
148	0	0	0	0	1	5	0.0	0.0	0	0	0.0	10	0.0	0	0	1	0	0	0	0
135	1	0	0	0	1	6	0.0	0.0	0	(-)	0.1	6	0.0	1,820	182	0	0	0	0	0
89	1	0	0	1	0	12	0.1	0.0	0	0	0.1	8	0.0	18	2	0	0	0	0	0
465	5	0	0	(-)	1	11	0.0	0.1	1	0	0.2	26	0.0	909	91	0	0	0	0	(-)
114	0	0	0	0	0	6	0.0	0.1	0	0	0.1	1	0.0	213	21	0	0	0	0	0
399	5	0	0	(-)	1	9	0.0	0.0	0	1	0.2	9	0.0	0	0	1	0	0	0	0
227	5	0	0	(-)	1	14	0.0	0.1	0	(-)	0.1	6	0.0	53	5	0	0	0	0	0
317	2	0	0	0	1	1	0.0	0.1	1	0	0.1	2	0.0	844	85	1	0	0	0	0
205	1	0	0	0	(-)	58	0.0	0.0	0	(-)	0.0	(-)	0.0	0	0	(-)	(-)	(-)	(-)	0
181	4	0	0	(-)	1	14	0.0	0.0	0	0	0.0	3	0.0	37	4	1	0	0	0	0
310	5	0	0	0	0	1	0.1	0.0	0	0	0.1	1	0.0	3	0	0	0	0	0	0

(continues)

FRUITS (*continued*)

	Serving size	Weight (g)	Calories	Protein (g)	Fat (g)	Carbohydrates (g)	Fiber (g)	Calcium (mg)	Iron (mg)	Magnesium (mg)	Phosphorus (mg)
Rambutan, canned, syrup pack	1/2 cup, drained	75	62	0	0	16	1	17	0	5	7
Raspberry, raw	1/2 cup (30 fruits)	62	30	1	0	7	4	14	0	11	7
Rhubarb, raw	1/2 cup, diced	61	13	1	0	3	1	52	0	7	9
Sapodilla, raw	1 fruit	170	141	1	2	34	9	36	1	20	20
Strawberry, raw	1/2 cup, whole (5 fruits)	72	22	0	0	5	2	10	0	7	14
Tamarind, raw	10 fruits	20	48	1	0	13	1	15	1	18	23
Tangerine, raw	1 medium (2 3/8″ diameter)	84	37	1	0	9	2	12	0	10	8
Watermelon, raw	1/2 cup, diced	76	24	0	0	5	0	6	0	8	7

VEGETABLES

	Serving size	Weight (g)	Calories	Protein (g)	Fat (g)	Carbohydrates (g)	Fiber (g)	Calcium (mg)	Iron (mg)	Magnesium (mg)	Phosphorus (mg)	Potassium (mg)
Amaranth leaves, cooked	1/2 cup	66	14	1	0	3	(-)	138	2	36	48	423
Artichoke, cooked	1 medium globe	120	60	4	0	13	6	54	2	72	103	425
Arugula, raw	1 cup	20	5	1	0	1	0	32	0	9	10	74
Asparagus, raw	4 medium spears (5 1/4″ to 7″ long)	64	14	1	0	3	1	13	1	11	36	175
Asparagus, cooked, boiled, drained, without salt	1/2 cup (6 spears, 1/2″ base)	90	22	2	0	4	1	18	1	9	49	144
Bamboo (shoots), cooked	1/2 cup (1/2″ slices)	60	7	1	0	1	1	7	0	2	12	320
Beans, snap, green, cooked, boiled, drained, without salt	1/2 cup	67	22	1	0	5	2	29	1	16	24	187
Beans, snap, yellow, cooked, boiled, drained, without salt	1/2 cup	67	22	1	0	5	2	29	1	16	24	187
Beet, cooked, boiled, drained	1/2 cup slices	85	37	1	0	8	2	14	1	20	32	259
Beet greens, cooked	1/2 cup	72	19	2	0	4	2	82	1	49	30	655

Potassium (mg)	Sodium (mg)	Zinc (mg)	Copper (mg)	Manganese (mg)	Selenium (µg)	Vitamin C (mg)	Thiamin (mg)	Riboflavin (mg)	Niacin (mg)	Pantothenic acid (mg)	Vitamin B$_6$ (mg)	Folate (µg)	Vitamin B$_{12}$ (µg)	Vitamin A (IU)	Vitamin A (RE)	Vitamin E (mg)	Saturated fat (g)	Monounsaturated fat (g)	Polyunsaturated fat (g)	Cholesterol (mg)
32	8	0	0	0	(-)	4	0.0	0.0	1	0	0.0	6	0.0	2	0	(-)	(-)	(-)	(-)	0
93	0	0	0	1	0	15	0.0	0.1	1	0	0.0	16	0.0	80	8	0	0	0	0	0
176	2	0	0	0	1	5	0.0	0.0	0	0	0.0	4	0.0	61	6	0	0	0	0	0
328	20	0	0	(-)	1	25	0.0	0.0	0	0	0.1	24	0.0	102	10	0	0	1	0	0
120	1	0	0	0	1	41	0.0	0.0	0	0	0.0	13	0.0	19	2	0	0	0	0	0
126	6	0	0	(-)	0	1	0.1	0.0	0	0	0.0	3	0.0	6	1	0	0	0	0	0
132	1	0	0	0	0	26	0.1	0.0	0	0	0.1	17	0.0	773	77	0	0	0	0	0
88	2	0	0	0	0	7	0	0.0	0	0	0.1	2	0.0	278	28	0	0	0	0	0

Sodium (mg)	Zinc (mg)	Copper (mg)	Manganese (mg)	Selenium (µg)	Vitamin C (mg)	Thiamin (mg)	Riboflavin (mg)	Niacin (mg)	Pantothenic acid (mg)	Vitamin B$_6$ (mg)	Folate (µg)	Vitamin B$_{12}$ (µg)	Vitamin A (IU)	Vitamin A (RE)	Vitamin E (mg)	Saturated fat (g)	Monounsaturated fat (g)	Polyunsaturated fat (g)	Cholesterol (mg)
14	1	0	1	1	27	0.0	0.1	0	0	0.1	37	0.0	1,830	183	(-)	0	0	0	0
114	1	0	0	0	12	0.1	0.1	1	0	0.1	61	0.0	212	22	0	0	0	0	0
5	0	0	0	0	3	0.0	0.0	0	0	0.0	19	0.0	475	47	0	0	0	0	0
1	0	0	0	1	8	0.1	0.1	1	0	0.1	82	0.0	373	37	1	0	0	0	0
10	0	0	0	2	10	0.1	0.1	1	0	0.1	131	0.0	485	49	0	0	0	0	0
2	0	0	0	0	0	0.0	0.0	0	0	0.1	1	0.0	0	0	(-)	0	0	0	0
2	0	0	0	0	6	0.0	0.1	0	0	0.0	21	0.0	416	42	0	0	0	0	0
2	0	0	0	0	6	0.0	0.1	0	0	0.0	21	0.0	51	5	0	0	0	0	0
65	0	0	0	1	3	0.0	0.0	0	0	0.1	68	0.0	30	3	0	0	0	0	0
174	0	0	0	1	18	0.1	0.2	0	0	0.1	10	0.0	3,672	367	0	0	0	0	0

(continues)

VEGETABLES *(continued)*

	Serving size	Weight (g)	Calories	Protein (g)	Fat (g)	Carbohydrates (g)	Fiber (g)	Calcium (mg)	Iron (mg)	Magnesium (mg)	Phosphorus (mg)	Potassium (mg)
Bitter melon (balsam-pear), leafy tips, cooked, boiled, drained, without salt	1/2 cup	29	10	1	0	2	0	12	0	27	22	175
Bitter melon (balsam-pear), pods, cooked, boiled, drained, without salt	1/2 cup (1/2″ pieces)	62	12	1	0	3	1	6	0	10	22	198
Broccoli, cooked, boiled, drained, without salt	1/2 cup (about 2 spears)	78	22	2	0	4	2	36	1	19	46	228
Broccoli, raw	1/2 cup (about 3 florets)	33	9	1	0	2	(-)	15	0	8	22	107
Brussels sprouts, cooked, boiled, drained, without salt	1/2 cup (about 4 medium)	78	30	2	0	7	2	28	1	16	44	247
Cabbage, cooked, boiled, drained, without salt	1/2 cup, shredded	75	17	1	0	3	2	23	0	6	11	73
Cabbage, raw	1 cup, shredded	70	18	1	0	4	2	33	0	11	16	172
Carrot, cooked	1/2 cup	78	35	1	0	8	3	24	0	10	23	177
Carrot, raw	1 medium	61	26	1	0	6	2	16	0	9	27	140
Cassava, raw	1/4 cup	51	82	1	0	20	1	8	0	11	14	140
Cauliflower, cooked, boiled, drained, without salt	1/2 cup (1″ pieces)	62	14	1	0	3	2	10	0	6	20	88
Cauliflower, raw	1/2 cup	50	13	1	0	3	1	11	0	8	22	152
Celeriac, cooked, boiled, drained, without salt	1/2 cup, pieces	77	21	1	0	5	1	20	0	9	51	134
Celeriac, raw	1/2 cup	78	33	1	0	7	1	34	0	16	90	234
Celery, cooked, boiled, drained, without salt	2 stalks	75	14	1	0	3	1	32	0	9	19	213
Celery, raw	2 medium stalks (7 1/2″–8″ long)	80	13	1	0	3	1	32	0	9	20	230
Chayote, fruit, cooked, boiled, drained, without salt	1/2 cup (1″ pieces)	80	17	1	0	4	2	10	0	10	23	138
Chicory greens, raw	1 cup, chopped	180	41	3	1	8	7	180	2	54	85	756
Chicory roots, raw	1/2 cup (1″ pieces)	45	33	1	0	8	(-)	18	0	10	27	131
Chinese cabbage (bok choy), cooked, boiled, drained, without salt	1/2 cup, shredded	85	10	1	0	2	1	79	1	9	25	316
Chinese cabbage (bok choy), raw	1 cup, shredded	70	9	1	0	2	1	74	1	13	26	176
Chinese cabbage (napa), cooked	1/2 cup	55	7	1	0	1	(-)	16	0	4	10	47
Collards, cooked, boiled, drained, without salt	1/2 cup, chopped	95	25	2	0	5	3	113	0	16	25	247
Corn, sweet, yellow, cooked, boiled, drained, without salt	1/2 cup, cut	82	89	3	1	21	2	2	1	26	84	204

Sodium (mg)	Zinc (mg)	Copper (mg)	Manganese (mg)	Selenium (µg)	Vitamin C (mg)	Thiamin (mg)	Riboflavin (mg)	Niacin (mg)	Pantothenic acid (mg)	Vitamin B$_6$ (mg)	Folate (µg)	Vitamin B$_{12}$ (µg)	Vitamin A (IU)	Vitamin A (RE)	Vitamin E (mg)	Saturated fat (g)	Monounsaturated fat (g)	Polyunsaturated fat (g)	Cholesterol (mg)
4	0	0	0	0	16	0.0	0.1	0	0	0.2	25	0.0	503	50	0	0	0	0	0
4	0	0	0	0	20	0.0	0.0	0	0	0.0	32	0.0	70	7	0	0	0	0	0
20	0	0	0	1	58	0.0	0.1	0	0	0.1	39	0.0	1,082	108	1	0	0	0	0
9	0	0	0	1	30	0.0	0.0	0	0	0.0	23	0.0	990	99	0	0	0	0	0
16	0	0	0	1	48	0.1	0.1	0	0	0.1	47	0.0	561	56	1	0	0	0	0
6	0	0	0	0	15	0.0	0.0	0	0	0.1	15	0.0	99	10	0	0	0	0	0
13	0	0	0	1	23	0.0	0.0	0	0	0.1	30	0.0	93	9	0	0	0	0	0
51	0	0	1	1	2	0.0	0.0	0	0	0.2	11	0.0	19,152	1,915	0	0	0	0	0
21	0	0	0	1	6	0.1	0.0	1	0	0.1	9	0.0	17,159	1,716	0	0	0	0	0
7	0	0	0	0	11	0.0	0.0	0	0	0.0	14	0.0	13	1	0	0	0	0	0
9	0	0	0	0	27	0.0	0.0	0	0	0.1	27	0.0	11	1	0	0	0	0	0
15	0	0	0	0	23	0.0	0.0	0	0	0.1	29	0.0	10	1	0	0	0	0	0
47	0	0	0	0	3	0.0	0.0	0	0	0.1	3	0.0	0	0	(-)	(-)	(-)	(-)	0
78	0	0	0	1	6	0.0	0.0	1	0	0.1	6	0.0	0	0	0	0	0	0	0
68	0	0	0	1	5	0.0	0.0	0	0	0.1	17	0.0	99	10	0	0	0	0	0
70	0	0	0	0	6	0.0	0.0	0	0	0.1	22	0.0	107	10	0	0	0	0	0
1	0	0	0	0	6	0.0	0.0	0	0	0.1	14	0.0	37	4	(-)	0	0	0	0
81	1	1	1	1	43	0.1	0.2	1	2	0.2	197	0.0	7,200	720	4	0	0	0	0
23	0	0	0	0	2	0.0	0.0	0	0	0.1	10	0.0	3	0	(-)	0	0	0	0
29	0	0	0	0	22	0.0	0.0	0	0	0.1	35	0.0	2,183	218	0	0	0	0	0
46	0	0	0	(-)	32	0.0	0.0	0	0	0.1	46	0.0	2,100	210	0	0	0	0	0
6	2	0	0	0	2	0.0	0.0	0	0	0.0	23	0.0	48	5	(-)	(-)	(-)	(-)	0
9	0	0	1	1	17	0.0	0.1	1	0	0.1	88	0.0	2,973	297	1	0	0	0	0
14	0	0	0	1	5	0.2	0.1	1	1	0.0	38	0.0	178	18	0	0	0	0	0

(continues)

VEGETABLES (*continued*)

	Serving size	Weight (g)	Calories	Protein (g)	Fat (g)	Carbohydrates (g)	Fiber (g)	Calcium (mg)	Iron (mg)	Magnesium (mg)	Phosphorus (mg)	Potassium (mg)
Corn, sweet, yellow, cooked, boiled, drained, without salt	1 ear kernels	77	83	3	1	19	2	2	0	25	79	192
Cress (watercress), raw	1 cup	34	4	1	0	0	1	41	0	7	20	112
Cucumber, with peel, raw	1/2 cup, slices	52	7	0	0	1	0	7	0	6	10	75
Eggplant, cooked, boiled, drained, without salt	1/2 cup (1″ cubes)	50	14	0	0	3	1	3	0	6	11	123
Fennel, bulb, raw	1/2 cup, sliced	87	13	0	0	3	1	21	0	7	22	180
Fennel seed (spice)	1 teaspoon	2	7	0	0	1	1	24	0	8	10	34
Fenugreek seed (spice)	1 teaspoon	4	12	1	0	2	1	6	1	7	11	28
Garlic, raw	1 teaspoon (about 1 clove)	3	4	0	0	1	0	5	0	1	4	11
Gingerroot, raw	5 slices (1″ diameter)	11	8	0	0	2	0	2	0	5	3	46
Gingerroot, raw	1 teaspoon	2	1	0	0	0	0	0	0	0	0	8
Horseradish, prepared	1 teaspoon	5	2	0	0	1	0	3	0	1	2	12
Jerusalem artichoke, raw	1/2 cup, slices	75	57	2	0	13	1	11	3	13	59	322
Jicama (yam bean), raw	1/2 cup, slices	60	23	0	0	5	3	7	0	7	11	90
Kelp (seaweed), raw	1/8 cup	10	4	0	0	1	0	17	0	12	4	9
Kohlrabi, cooked, boiled, drained, without salt	1/2 cup, slices	83	24	1	0	6	1	21	0	16	37	281
Kohlrabi, raw	1/2 cup	70	18	1	0	4	2	16	0	13	31	236
Leek (bulb and lower leaf portion), cooked, boiled, drained, without salt	1/2 cup	52	16	0	0	4	1	16	1	7	9	45
Leek (bulb and lower leaf portion), raw, chopped	1/2 cup	45	27	1	0	6	1	26	1	12	16	80
Lettuce, butterhead (includes Boston and Bibb types), raw	1 cup, shredded	55	7	1	0	1	1	18	0	7	13	141
Lettuce, iceberg (includes crisphead types), raw	1 cup, shredded	55	7	1	0	1	1	10	0	5	11	87
Lettuce, looseleaf, raw	1 cup, shredded	56	10	1	0	2	1	38	1	6	14	148
Lettuce, romaine, raw	1 cup, shredded	56	8	1	0	1	1	20	1	3	25	162
Mushroom, cooked, boiled, drained, without salt	7 medium	84	22	2	0	4	2	5	1	10	73	299
Mushroom, raw	1/2 cup, pieces	35	9	1	0	1	0	2	0	4	36	130
Okra, cooked, boiled, drained, without salt	1/2 cup slices	80	26	2	0	6	2	50	0	46	45	258

Sodium (mg)	Zinc (mg)	Copper (mg)	Manganese (mg)	Selenium (µg)	Vitamin C (mg)	Thiamin (mg)	Riboflavin (mg)	Niacin (mg)	Pantothenic acid (mg)	Vitamin B$_6$ (mg)	Folate (µg)	Vitamin B$_{12}$ (µg)	Vitamin A (IU)	Vitamin A (RE)	Vitamin E (mg)	Saturated fat (g)	Monounsaturated fat (g)	Polyunsaturated fat (g)	Cholesterol (mg)
13	0	0	0	1	5	0.2	0.1	1	1	0.0	36	0	167	17	0	0	0	0	0
14	0	0	0	0	15	0.0	0.0	0	0	0.0	3	0.0	1,598	159	0	0	0	0	0
1	0	0	0	0	3	0.0	0.0	0	0	0.0	7	0.0	112	11	0	0	0	0	0
1	0	0	0	0	1	0.0	0.0	0	0	0.0	7	0.0	32	3	0	0	0	0	0
23	0	0	0	0	5	0.0	0.0	0	0	0.0	12	0.0	58	6	(-)	(-)	(-)	(-)	0
2	0	0	0	(-)	0	0.0	0.0	0	(-)	(-)	(-)	0.0	3	0	(-)	0	0	0	0
2	0	0	0	0	0	0.0	0.0	0	(-)	(-)	2	0.0	2	0	(-)	0	(-)	(-)	0
0	0	0	0	0	1	0.0	0.0	0	0	0.0	0	0.0	0	0	0	0	0	0	0
1	0	0	0	0	1	0.0	0.0	0	0	0.0	1	0.0	0	0	0	0	0	0	0
0	0	0	0	0	0	0.0	0	0	0	0.0	0	0.0	0	0	0	0	0	0	0
16	0	0	0	0	1	0.0	0.0	0	0	0.0	3	0.0	0	0	0	0	0	0	0
3	0	0	0	1	3	0.2	0.0	1	1	0.1	10	0.0	15	2	0	0	0	0	0
2	0	0	0	0	12	0.0	0.0	0	0	0.1	7	0.0	13	1	0	0	0	0	0
23	0	0	0	0	0	0.0	0.0	0	0	0.0	18	0.0	12	1	0	0	0	0	0
17	0	0	0	1	45	0.0	0.0	0	0	0.1	10	0.0	29	3	1	0	0	0	0
14	0	0	0	0	42	0.0	0.0	0	0	0.1	11	0.0	24	3	0	0	0	0	0
5	0	0	0	0	2	0.0	0.0	0	0	0.1	13	0.0	24	3	(-)	0	0	0	0
9	0	0	0	0	5	0.0	0.0	0	0	0.1	29	0.0	42	4	0	0	0	0	0
3	0	0	0	0	4	0.0	0.0	0	0	0.0	40	0.0	534	53	0	0	0	0	0
5	0	0	0	0	2	0.0	0.0	0	0	0.0	31	0.0	182	18	0	0	0	0	0
5	0	0	0	0	10	0.0	0.0	0	0	0.0	28	0.0	1,064	106	0	0	0	0	0
4	0	0	0	0	13	0.1	0.1	0	0	0.0	76	0.0	1,456	146	0	0	0	0	0
2	1	0	0	10	3	0.1	0.3	4	2	0.1	15	0.0	0	0	0	0	0	0	0
1	0	0	0	3	1	0.0	0.1	1	1	0.0	4	0.0	0	0	0	0	0	0	0
4	0	0	1	1	13	0.1	0.0	1	0	0.2	37	0.0	460	46	1	0	0	0	0

(continues)

VEGETABLES *(continued)*

	Serving size	Weight (g)	Calories	Protein (g)	Fat (g)	Carbohydrates (g)	Fiber (g)	Calcium (mg)	Iron (mg)	Magnesium (mg)	Phosphorus (mg)	Potassium (mg)
Okra, cooked, boiled, drained, without salt	8 pods (3″ long)	85	27	2	0	6	2	54	0	48	48	273
Onion, cooked	1 medium	94	41	1	0	10	1	21	0	10	33	156
Onion, raw	1/2 cup, chopped	80	30	1	0	7	1	16	0	8	26	126
Parsley, raw	1 tablespoon	4	1	0	0	0	0	5	0	2	2	21
Parsnip, cooked, boiled, drained, without salt	1/2 cup, slices	78	63	1	0	15	3	29	0	23	54	286
Parsnip, raw	1/2 cup	67	50	1	0	12	3	24	0	19	47	249
Pepper, hot chili, green, raw	1 pepper	45	18	1	0	4	1	8	1	11	21	153
Pepper, hot chili, red, raw	1 pepper	45	18	1	0	4	1	8	1	11	21	153
Pepper, sweet, green, raw	1/2 cup, chopped	75	20	1	0	5	1	7	0	7	14	131
Pepper, sweet, red, raw	1/2 cup, chopped	75	20	1	0	5	1	7	0	7	14	132
Pepper, sweet, yellow, raw	1/2 cup, chopped	75	20	1	0	5	1	8	0	9	18	159
Potato, baked, flesh, without salt	1/2 cup	61	57	1	0	13	1	3	0	15	31	239
Potato, baked, flesh and skin, without salt	1 potato (2 1/3″ x 4 3/4″)	202	220	5	0	51	5	20	3	54	115	844
Radish, raw	1/2 cup, slices (13 medium)	58	12	0	0	2	1	12	0	5	10	135
Rutabaga, cooked	1/2 cup, cubes	85	33	1	0	7	2	41	0	20	48	277
Rutabaga, raw	1/2 cup, cubes	70	25	1	0	6	2	33	0	16	41	236
Salsify (vegetable oyster), cooked, boiled, drained, without salt	1/2 cup, slices	68	46	2	0	10	2	32	0	12	38	191
Scallion (including top and bulb)	1/2 cup	50	16	1	0	4	1	36	1	10	19	138
Spinach, cooked, boiled, drained, without salt	1/2 cup	90	21	3	0	3	2	122	3	78	50	419
Spinach, raw	1 cup	30	7	1	0	1	1	30	1	24	15	167
Squash, summer, all varieties, cooked, boiled, drained, without salt	1/2 cup, slices	90	18	1	0	4	1	24	0	22	35	173
Squash, winter, all varieties, cooked, baked, without salt	1/2 cup, cubes	103	40	1	1	9	3	14	0	8	21	448
Sweet potato, cooked, baked in skin, without salt	3/4 cup	150	155	3	0	36	5	42	1	30	83	522
Taro, cooked, without salt	1/2 cup	66	94	0	0	23	3	12	1	20	50	319
Taro leaves, cooked, steamed, without salt	1/2 cup	72	17	2	0	3	1	62	1	15	20	333
Tomatillo, raw	1/2 cup, chopped	66	21	1	1	4	1	5	0	13	26	177

Sodium (mg)	Zinc (mg)	Copper (mg)	Manganese (mg)	Selenium (µg)	Vitamin C (mg)	Thiamin (mg)	Riboflavin (mg)	Niacin (mg)	Pantothenic acid (mg)	Vitamin B_6 (mg)	Folate (µg)	Vitamin B_{12} (µg)	Vitamin A (IU)	Vitamin A (RE)	Vitamin E (mg)	Saturated fat (g)	Monounsaturated fat (g)	Polyunsaturated fat (g)	Cholesterol (mg)
4	0	0	1	1	14	0.1	0.0	1	0	0.2	39	0	489	49	1	0	0	0	0
3	0	0	0	1	5	0.0	0.0	0	0	0.1	14	0.0	0	0	0	0	0	0	0
2	0	0	0	0	5	0.0	0.0	0	0	0.1	15	0.0	0	0	0	0	0	0	0
2	0	0	0	0	5	0.0	0.0	0	0	0.0	6	0.0	198	20	0	0	0	0	0
8	0	0	0	1	10	0.1	0.0	1	0	0.1	45	0.0	0	0	1	0	0	0	0
7	0	0	0	1	11	0.1	0.0	0	0	0.0	44	0.0	0	0	(-)	0	0	0	0
3	0	0	0	0	109	0.0	0.0	0	0	0.1	11	0.0	347	35	0	0	0	0	0
3	0	0	0	0	109	0.0	0.0	0	0	0.1	11	0.0	4,838	484	0	0	0	0	0
1	0	0	0	0	66	0.0	0.0	0	0	0.2	16	0.0	471	47	1	0	0	0	0
1	0	0	0	0	142	0.0	0.0	0	0	0.2	16	0.0	4,247	425	1	0	0	0	0
2	0	0	0	0	138	0.0	0.0	1	0	0.1	20	0.0	179	18	(-)	0	(-)	(-)	0
3	0	0	0	0	8	0.1	0.0	1	0	0.2	6	0.0	0	0	0	0	0	0	0
16	1	1	0	2	26	0.2	0.1	3	1	0.7	22	0.0	0	0	0.1	0	0	0	0
14	0	0	0	0	13	0.0	0.0	0	0	0.0	16	0.0	5	1	0	0	0	0	0
17	0	0	0	1	16	0.1	0.0	1	0	0.1	13	0.0	477	48	0	0	0	0	0
14	0	0	0	0	18	0.1	0.0	0	0	0.1	15	0.0	406	41	0	0	0	0	0
11	0	0	0	0	3	0.0	0.1	0	0	0.1	10	0.0	0	0	0	(-)	(-)	(-)	0
8	0	0	0	0	9	0.0	0.0	0	0	0.0	32	0.0	193	20	0	0	0	0	0
63	1	0	1	1	9	0.1	0.2	0	0	0.2	131	0.0	7,371	737	1	0	0	0	0
24	0	0	0	0	8	0.0	0.1	0	0	0.1	58	0.0	2,015	202	1	0	0	0	0
1	0	0	0	0	5	0.0	0.0	0	0	0.1	18	0.0	258	26	0	0	0	0	0
1	0	0	0	0	10	0.1	0.0	1	0	0.1	29	0.0	3,646	365	0	0	0	1	0
15	0	0	1	1	37	0.1	0.2	1	1	0.0	34	0.0	32,733	3,273	0	0	0	0	0
10	0	0	0	0	3	0.1	0.0	0	0	0.2	13	0.0	0	0	0	0	0	0	0
1	0	0	0	1	26	0.1	0.3	1	0	0.1	35	0.0	3,073	307	(-)	0	0	0	0
1	0	0	0	0	8	0.0	0.0	0	0	0.0	5	0.0	75	7	0	0	0	0	0

(continues)

VEGETABLES (*continued*)

	Serving size	Weight (g)	Calories	Protein (g)	Fat (g)	Carbohydrates (g)	Fiber (g)	Calcium (mg)	Iron (mg)	Magnesium (mg)	Phosphorus (mg)	Potassium (mg)
Tomato, red, ripe, raw	1 medium, whole (2 3/5″ diameter)	123	26	1	0	6	1	6	1	14	30	273
Turnip, cooked, boiled, drained, without salt	1/2 cup, cubes	78	16	1	0	4	2	17	0	6	15	105
Turnip, raw	1/2 cup, cubes	65	18	1	0	4	1	20	0	7	18	124
Water chestnut, cooked	1/2 cup, slices	70	35	1	0	9	2	3	1	4	13	83
Water chestnut, raw	1/2 cup, slices	62	60	1	0	15	2	7	0	14	39	362
Yam, cooked, boiled, drained, or baked, without salt	1/2 cup, cubes	68	79	1	0	19	3	10	0	12	33	456

GRAINS (INCLUDING BREADS, CEREALS, FLOUR, AND PASTA)

BREADS	Serving size	Weight (g)	Calories	Protein (g)	Fat (g)	Carbohydrates (g)	Fiber (g)	Calcium (mg)	Iron (mg)	Magnesium (mg)	Phosphorus (mg)	Potassium (mg)
Bagels, plain, enriched, without calcium propionate (includes onion, poppy, sesame)	1/2 bagel (3 1/2″ diameter)	35	96	4	1	19	1	6	1	10	34	36
Biscuits, plain or buttermilk, commercially baked	1 biscuit	35	127	2	6	17	0	17	1	6	151	78
Bread, cornbread, dry mix, enriched (includes corn muffin mix)	1 oz (about 3 tablespoons)	28	119	2	3	20	2	16	1	7	139	32
Bread, cornbread, prepared (made with 2% milk)	1 piece	65	173	4	5	28	(-)	162	2	16	110	95
Bread, cracked-wheat	1 slice	25	65	2	1	12	1	11	1	13	38	44
Bread, French or Vienna (includes sourdough)	1 slice (4 3/4″ × 4″ × 1/2″)	25	69	2	1	13	1	19	1	7	26	28
Bread, mixed-grain (includes whole-grain, 7-grain)	1 slice	26	65	3	1	12	2	24	1	14	46	53
Bread, oat bran	1 slice	30	71	3	1	12	1	20	1	11	42	44

Sodium (mg)	Zinc (mg)	Copper (mg)	Manganese (mg)	Selenium (µg)	Vitamin C (mg)	Thiamin (mg)	Riboflavin (mg)	Niacin (mg)	Pantothenic acid (mg)	Vitamin B$_6$ (mg)	Folate (µg)	Vitamin B$_{12}$ (µg)	Vitamin A (IU)	Vitamin A (RE)	Vitamin E (mg)	Saturated fat (g)	Monounsaturated fat (g)	Polyunsaturated fat (g)	Cholesterol (mg)
11	0	0	0	0	23	0.1	0.1	1	0	0.1	18	0.0	766	76	0	0	0	0	0
39	0	0	0	0	9	0.0	0.0	0	0	0.1	7	0.0	0	0	0	0	0	0	0
44	0	0	0	0	14	0.0	0.0	0	0	0.1	9	0.0	0	0	0	0	0	0	0
6	0	0	0	0	1	0.0	0.0	0	0	0.1	4	0.0	3	0	0	0	0	0	0
9	0	0	0	0	2	0.1	0.1	1	0	0.2	10	0.0	0	0	1	0	0	0	0
5	0	0	0	0	8	0.1	0.0	0	0	0.2	11	0.0	0	0	0	0	0	0	0

Sodium (mg)	Zinc (mg)	Copper (mg)	Manganese (mg)	Selenium (µg)	Vitamin C (mg)	Thiamin (mg)	Riboflavin (mg)	Niacin (mg)	Pantothenic acid (mg)	Vitamin B$_6$ (mg)	Folate (µg)	Vitamin B$_{12}$ (µg)	Vitamin A (IU)	Vitamin A (RE)	Vitamin E (mg)	Saturated fat (g)	Monounsaturated fat (g)	Polyunsaturated fat (g)	Cholesterol (mg)
190	0	0	0	0	0	0.2	0.1	2	0	0.0	31	0.0	0	0	0	0	0	0	0
368	0	0	0	7	0	0.1	0.1	1	0	0.0	21	0.0	1	0	1	1	2	2	0
315	0	0	0	2	0	0.1	0.1	1	0	0.0	30	0.0	33	3	0	1	2	0	1
428	0	0	0	0	0	0.2	0.2	1	0	0.0	42	0.0	180	35	(-)	1	1	2	26
135	0	0	0	6	0	0.1	0.1	1	0	0.1	15	0.0	0	0	0	0	0	0	0
152	0	0	0	8	0	0.1	0.1	1	0	0.0	24	0.0	0	0	0	0	0	0	0
127	0	0	0	8	0	0.1	0.1	1	0	0.1	21	0.0	0	0	0	0	0	0	0
122	0	0	0	9	0	0.2	0.1	1	0	0.0	24	0.0	2	0	0	0	0	1	0

(*continues*)

GRAINS (INCLUDING BREADS, CEREALS, FLOUR, AND PASTA) *(continued)*

BREADS *(continued)*	Serving size	Weight (g)	Calories	Protein (g)	Fat (g)	Carbohydrates (g)	Fiber (g)	Calcium (mg)	Iron (mg)	Magnesium (mg)	Phosphorus (mg)	Potassium (mg)
Bread, pita, white, enriched	1 small pita (4″ diameter)	28	77	3	0	16	1	24	1	7	27	34
Bread, pumpernickel	1 slice	26	65	2	1	12	2	18	1	14	46	54
Bread, raisin, enriched	1 slice	26	71	2	1	14	1	17	1	7	28	59
Bread, reduced-calorie, white	1 slice	23	48	2	1	10	2	22	1	5	28	17
Bread, rye	1 slice	32	83	3	1	15	2	23	1	13	40	53
Bread, wheat (includes wheat berry)	1 slice	25	65	2	1	12	1	26	1	12	38	50
Bread, wheat bran	1 slice	36	89	3	1	17	1	27	1	29	67	82
Bread, white, commercially prepared (includes soft bread crumbs)	1 slice	25	67	2	1	12	1	27	1	6	24	30
Bread, whole-wheat, commercially prepared	1 slice	28	69	3	1	13	2	20	1	24	64	71
Croissants, butter	1 medium croissant	57	231	5	12	26	1	21	1	9	60	67
English muffins, raisin-cinnamon (includes apple-cinnamon)	1 muffin	57	139	4	2	28	2	84	1	9	39	119
Rolls, dinner, plain, commercially prepared (includes brown-and-serve)	1 roll (1 oz)	28	85	2	2	14	1	34	1	7	33	38
Rolls, dinner, whole-wheat	1 roll (1 oz)	28	75	2	1	14	2	30	1	24	64	77
Rolls, hamburger or hotdog, plain	1 roll	43	123	4	2	22	1	60	1	9	38	61
Taco shells, baked	2 medium (approx 5″ diameter)	26	122	2	6	16	2	42	1	27	64	46
Tortillas, ready-to-bake or -fry, corn	1 medium tortilla (approx 6″ diameter)	26	58	1	1	12	1	46	0	17	82	40
Tortillas, ready-to-bake or -fry, flour	1 medium tortilla (approx 6″ diameter)	32	104	3	2	18	1	40	1	8	40	42
CEREALS												
Cereals, corn grits, white, regular, quick, enriched, cooked with water, with salt (corn)	1/2 cup	121	73	2	0	16	0	0	1	5	15	27
Cereals, cream of rice, cooked with water, without salt	1/2 cup	122	63	1	0	14	0	4	0	4	21	24
Cereals, cream of wheat, regular, cooked with water, without salt	1/2 cup	126	67	2	0	14	1	25	5	5	21	21
Cereals, farina, enriched, cooked with water, without salt (wheat)	1/2 cup	117	58	2	0	12	2	2	1	2	14	15

Sodium (mg)	Zinc (mg)	Copper (mg)	Manganese (mg)	Selenium (µg)	Vitamin C (mg)	Thiamin (mg)	Riboflavin (mg)	Niacin (mg)	Pantothenic acid (mg)	Vitamin B$_6$ (mg)	Folate (µg)	Vitamin B$_{12}$ (µg)	Vitamin A (IU)	Vitamin A (RE)	Vitamin E (mg)	Saturated fat (g)	Monounsaturated fat (g)	Polyunsaturated fat (g)	Cholesterol (mg)
150	0	0	0	8	0	0.2	0.1	1	0	0.0	27	0.0	0	0	0	0	0	0	0
174	0	0	0	6	0	0.1	0.1	1	0	0.0	21	0.0	0	0	0	0	0	0	0
101	0	0	0	5	0	0.1	0.1	1	0	0.0	23	0.0	0	0	0	0	1	0	0
104	0	0	0	5	0	0.1	0.1	1	0	0.0	22	0.1	1	0	0	0	0	0	0
211	0	0	0	10	0	0.1	0.1	1	0	0.0	28	0.0	2	0	0	0	0	0	0
133	0	0	0	8	0	0.1	0.1	1	0	0.0	19	0.0	0	0	0	0	0	0	0
175	0	0	1	11	0	0.1	0.1	2	0	0.1	25	0.0	0	0	0	0	1	0	0
135	0	0	0	7	0	0.1	0.1	1	0	0.0	24	0.0	0	0	0	0	0	0	0
148	1	0	1	10	0	0.1	0.1	1	0	0.1	14	0.0	0	0	0	0	0	0	0
424	0	0	0	13	0	0.2	0.1	1	0	0.0	35	0.1	424	106	0	7	3	1	38
255	1	0	0	9	0	0.2	0.2	2	0	0.0	46	0.0	1	0	0	0	0	1	0
148	0	0	0	8	0	0.1	0.1	1	0	0.0	27	0.0	0	0	0	0	1	0	0
136	1	0	1	14	0	0.1	0.0	1	0	0.1	9	0.0	0	0	0	0	0	1	0
241	0	0	0	11	0	0.2	0.1	2	0	0.0	41	0.0	0	0	1	1	0	1	0
4	0	0	0	(-)	0	0.1	0.0	0	0	0.0	27	0.0	0	0	1	1	2	2	0
42	0	0	0	1	0	0.0	0.0	0	0	0.1	30	0.0	0	0	0	0	0	0	0
153	0	0	0	7	0	0.2	0.1	1	0	0.0	39	0.0	0	0	0	1	1	0	0
270	0	0	0	0	0	0.1	0.1	1	0	0.0	38	0.0	0	0	0	0	0	0	0
1	0	0	0	4	0	0.0	0.0	0	0	0.0	4	0.0	0	0	0	0	0	0	0
1	0	0	0	0	0	0.1	0.0	1	0	0.0	23	0.0	0	0	0	0	0	0	0
0	0	0	0	11	0	0.1	0.1	1	0	0.0	27	0.0	0	0	0	0	0	0	0

(*continues*)

GRAINS (INCLUDING BREADS, CEREALS, FLOUR, AND PASTA) (*continued*)

CEREALS (*continued*)	Serving size	Weight (g)	Calories	Protein (g)	Fat (g)	Carbohydrates (g)	Fiber (g)	Calcium (mg)	Iron (mg)	Magnesium (mg)	Phosphorus (mg)	Potassium (mg)
Cereals, oatmeal, regular or quick or instant, fortified, plain, cooked with water, without salt	1/2 cup	117	73	3	1	13	2	9	1	28	89	66
Cereals, ready-to-eat, 40% Bran Flakes, Ralston Purina	2/3 cup (about 1 ounce)	33	106	4	0	26	5	15	5	78	182	191
Cereals, ready-to-eat, corn flakes (Kellogg's)	1 cup (1 ounce)	28	102	2	0	24	1	1	9	3	11	25
Cereals, ready-to-eat, rice, puffed, fortified	2 cups (1 ounce)	28	113	2	0	25	0	2	9	7	27	32
Cereals, ready-to-eat, wheat, puffed, fortified	2 cups (about 1 ounce)	24	87	3	0	19	1	7	8	35	85	83
CRACKERS												
Crackers, crispbread, rye	3 crispbreads or crackers	30	110	2	0	25	5	9	1	23	81	96
Crackers, matzo, plain	1 matzo (1 ounce)	28	112	3	0	24	1	4	1	7	25	32
Crackers, melba toast, plain	3 pieces	15	58	2	0	11	1	14	0	9	29	30
Crackers, saltines	6 small crackers	18	78	2	4	13	0	21	1	5	19	23
Crackers, saltines, fat-free, low-sodium	6 small saltines	30	118	3	0	25	1	7	2	8	34	34
Crackers, wheat, regular	6 small thin square crackers	12	57	1	2	8	0	6	0	7	26	22
Crackers, whole-wheat	6 small crackers	24	106	2	4	16	2	12	1	24	71	71
FLOUR												
Buckwheat flour, whole-groat	1 cup	120	402	15	4	85	12	49	5	301	404	692
Corn flour, masa, enriched, yellow	1 cup	114	416	11	4	87	0	161	8	125	254	340
Cornmeal, whole-grain, yellow	1 cup	122	442	10	4	94	9	7	4	155	294	350
Peanut flour, defatted	1 cup	60	196	31	0	21	9	84	1	222	456	774
Potato flour	1 cup	160	571	11	1	133	9	104	2	104	269	1,602
Rice flour, white	1 cup	158	578	9	2	127	4	16	1	55	155	120
Rye flour, medium	1 cup	102	361	10	2	79	15	24	2	77	211	347
Semolina (wheat), enriched	1 cup	167	601	21	2	122	7	28	7	78	227	311
Soy flour, defatted	1 cup	100	329	47	1	38	18	241	9	290	674	2,384
Wheat flour, durum	1 cup	192	651	26	5	137	(-)	65	7	276	975	827
Wheat flour, white, all-purpose, enriched, bleached	1 cup	125	455	13	1	95	3	19	6	28	135	134
Wheat flour, white, all-purpose, self-rising, enriched	1 cup	125	442	12	1	93	3	422	6	24	745	155

Sodium (mg)	Zinc (mg)	Copper (mg)	Manganese (mg)	Selenium (µg)	Vitamin C (mg)	Thiamin (mg)	Riboflavin (mg)	Niacin (mg)	Pantothenic acid (mg)	Vitamin B$_6$ (mg)	Folate (µg)	Vitamin B$_{12}$ (µg)	Vitamin A (IU)	Vitamin A (RE)	Vitamin E (mg)	Saturated fat (g)	Monounsaturated fat (g)	Polyunsaturated fat (g)	Cholesterol (mg)
1	1	0	1	9	0	0.1	0.0	0	0	0.0	5	0.0	19	2	0	0	0	0	0
304	1	0	1	(-)	17	0.4	0.5	6	0	0.1	115	2.0	1,440	432	0	0	0	0	0
298	0	0	0	1	14	0.4	0.4	5	0	0.5	99	0.0	700	210	0	0	0	0	0
1	0	0	0	3	0	0.7	0.5	10	0	0.0	5	0.0	0	0	0	0	0	0	0
1	1	0	0	29	0	0.6	0.4	8	0	0.0	8	0.0	0	0	0	0	0	0	0
79	1	0	1	11	0	0.0	0.0	0	0	0.0	9	0.0	0	0	0	0	0	0	0
1	0	0	0	10	0	0.1	0.1	1	0	0.0	33	0.0	0	0	0	0	0	0	0
124	0	0	0	5	0	0.1	0.0	1	0	0.0	19	0.0	0	0	0	0	0	0	0
234	0	0	0	2	0	0.1	0.1	1	0	0.0	22	0.0	0	0	0	0	0	0	0
191	0	0	0	6	0	0.1	0.2	2	0	0.0	37	0.0	0	0	0	0	0	0	0
95	0	0	0	1	0	0.1	0.0	1	0	0.0	5	0.0	0	0	0	0	0	0	0
158	0	0	0	3	0	0.0	0.0	1	0	0.0	7	0.0	0	0	0	0	0	0	0
13	4	1	2	7	0	0.5	0.2	7	1	0.7	65	0.0	0	0	1	1	1	1	0
6	2	0	1	0	0	1.6	0.9	11	1	0.4	213	0.0	535	54	0	1	1	2	0
43	2	0	1	19	0	0.5	0.2	4	1	0.4	31	0.0	572	57	1	1	1	2	0
108	3	1	3	4	0	0.4	0.3	16	2	0.3	149	0.0	0	0	0	0	0	0	0
88	1	0	1	2	6	0.4	0.1	6	1	1.2	40	0.0	0	0	0	0	0	0	0
0	1	0	2	24	0	0.2	0.0	4	1	0.7	6	0.0	0	0	0	1	1	1	0
3	2	0	6	36	0	0.3	0.1	2	1	0.3	19	0.0	0	0	1	0	0	1	0
2	2	0	1	0	0	1.4	1.0	10	1	0.2	257	0.0	0	0	0	0	0	1	0
20	2	4	3	2	0	0.7	0.3	3	2	0.6	305	0.0	40	4	0	0	0	1	0
4	8	1	6	172	0	0.8	0.2	13	2	1.0	83	0.0	0	0	0	1	1	2	0
3	1	0	1	42	0	1.0	0.6	7	1	0.1	193	0.0	0	0	0	0	0	1	0
1,587	1	0	1	43	0	1.0	0.5	7	0	0.0	192	0.0	0	0	0	0	0	1	0

(continues)

GRAINS (INCLUDING BREADS, CEREALS, FLOUR, AND PASTA) *(continued)*

FLOUR *(continued)*	Serving size	Weight (g)	Calories	Protein (g)	Fat (g)	Carbohydrates (g)	Fiber (g)	Calcium (mg)	Iron (mg)	Magnesium (mg)	Phosphorus (mg)	Potassium (mg)
Wheat flour, white, cake, enriched	1 cup	137	496	11	1	107	2	19	10	22	116	144
Wheat flour, whole-grain	1 cup	120	407	16	2	87	15	41	5	166	415	486
GRAINS												
Amaranth	1/2 cup	98	365	14	6	65	15	149	7	259	444	357
Barley, pearled, cooked	1/2 cup	79	97	2	0	22	3	9	1	17	42	73
Buckwheat groats, roasted, cooked	1/2 cup	84	77	3	1	17	2	6	1	43	59	74
Bulgur (wheat)	1/2 cup	91	76	3	0	17	4	9	1	29	36	62
Hominy, canned, white (corn)	1/2 cup	83	59	1	1	12	2	8	1	13	29	7
Millet, cooked	1/2 cup	87	104	3	1	21	1	3	1	38	87	54
Flax seed	1/2 cup	77	381	15	26	26	22	154	5	280	386	528
Popcorn, air-popped	2 cups	16	61	2	1	12	2	2	0	21	48	48
Quinoa	1/2 cup	85	318	11	5	59	5	51	8	179	349	629
Rice bran, crude	1/2 cup	59	186	8	12	29	12	34	11	461	989	876
Rice, brown, long-grain, cooked	1/2 cup	98	108	3	1	22	2	10	0	42	81	42
Rice, white, long-grain, regular, cooked	1/2 cup	79	103	2	0	22	0	8	1	9	34	28
Rye	1/2 cup	85	283	12	2	59	12	28	2	102	316	223
Sorghum	1/2 cup	96	325	11	3	72	0	27	4	0	276	336
Triticale	1/2 cup	96	323	13	2	69	0	36	2	125	344	319
Wheat bran, crude	1/2 cup	29	63	5	1	19	12	21	3	177	294	343
Wild rice, cooked	1/2 cup	82	83	3	0	17	1	2	0	26	67	83
PASTA												
Couscous, cooked	1/2 cup	79	88	3	0	18	1	6	0	6	17	46
Macaroni, cooked, enriched	1/2 cup	70	99	3	0	20	1	5	1	13	38	22
Noodles, egg, cooked, enriched	1/2 cup	80	106	4	1	20	1	10	1	15	55	22
Noodles, Japanese, soba, cooked	1/2 cup	57	56	3	0	12	0	2	0	5	14	20
Noodles, Japanese, somen, cooked	1/2 cup	88	115	4	0	24	0	7	0	2	24	26
Pasta, fresh-refrigerated, plain, cooked	1/2 cup	70	92	4	1	17	0	4	1	13	44	17
Spaghetti, cooked, enriched, without added salt	1/2 cup	70	99	3	0	20	1	5	1	13	38	22
Spaghetti, whole-wheat, cooked	1/2 cup	70	87	4	0	19	3	11	1	21	62	31

Sodium (mg)	Zinc (mg)	Copper (mg)	Manganese (mg)	Selenium (µg)	Vitamin C (mg)	Thiamin (mg)	Riboflavin (mg)	Niacin (mg)	Pantothenic acid (mg)	Vitamin B$_6$ (mg)	Folate (µg)	Vitamin B$_{12}$ (µg)	Vitamin A (IU)	Vitamin A (RE)	Vitamin E (mg)	Saturated fat (g)	Monounsaturated fat (g)	Polyunsaturated fat (g)	Cholesterol (mg)
3	1	0	1	7	0	1.0	0.6	9	1	0.0	211	0.0	0	0	0	0	0	1	0
6	4	0	5	85	0	0.5	0.3	8	1	0.4	53	0.0	0	0	1	0	0	1	0
20	3	1	2	0	4	0.1	0.2	1	1	0.2	48	0.0	0	0	1	2	1	3	0
2	1	0	0	7	0	0.1	0.0	2	0	0.1	13	0.0	5	1	0	0	0	0	0
3	1	0	0	2	0	0.0	0.0	1	0	0.1	12	0.0	0	0	0	0	0	0	0
5	1	0	1	1	0	0.1	0.0	1	0	0.1	16	0.0	0	0	0	0	0	0	0
173	1	0	0	2	0	0.0	0.0	0	0	0.0	1	0.0	0	0	0	0	0	0	0
2	1	0	0	1	0	0.1	0.1	1	0	0.1	17	0.0	0	0	0	0	0	0	0
26	3	1	2	4	1	0.1	0.1	1	1	0.7	215	0.0	0	0	4	2	5	19	0
1	0	0	0	2	0	0.0	0.0	0	0	0.0	4	0.0	31	3	0	0	0	1	0
18	3	1	2	0	0	0.2	0.3	2	1	0.2	42	0.0	0	0	0	0	1	2	0
3	4	0	8	9	0	1.6	0.2	20	4	2.4	37	0.0	0	0	4	2	4	4	0
5	1	0	1	10	0	0.1	0.0	1	0	0.1	4	0.0	0	0	1	0	0	0	0
1	0	0	0	6	0	0.1	0.0	1	0	0.1	46	0.0	0	0	0	0	0	0	0
5	3	0	2	30	0	0.3	0.2	4	1	0.2	51	0.0	0	0	2	0	0	1	0
6	0	0	0	0	0	0.2	0.1	3	0	0.0	0	0.0	0	0	0	0	1	1	0
5	3	0	3	0	0	0.4	0.1	1	1	0.1	70	0.0	0	0	1	0	0	1	0
1	2	0	3	23	0	0.2	0.2	4	1	0.4	23	0.0	0	0	1	0	0	1	0
2	1	0	0	1	0	0.0	0.1	1	0	0.1	21	0.0	0	0	0	0	0	0	0
4	0	0	0	22	0	0.0	0.0	1	0	0.0	12	0.0	0	0	0	0	0	0	0
1	0	0	0	15	0	0.1	0.1	1	0	0.0	49	0.0	0	0	0	0	0	0	0
6	0	0	0	17	0	0.1	0.1	1	0	0.0	51	0.1	16	5	0	0	0	0	26
34	0	0	0	0	0	0.1	0.0	0	0	0.0	4	0.0	0	0	0	0	0	0	0
142	0	0	0	0	0	0.0	0.0	0	0	0.0	2	0.0	0	0	0	0	0	0	0
4	0	0	0	0	0	0.1	0.1	1	0	0.0	45	0.1	14	4	0	0	0	0	23
1	0	0	0	15	0	0.1	0.1	1	0	0.0	49	0.0	0	0	0	0	0	0	0
2	1	0	1	18	0	0.1	0.0	0	0	0.1	4	0.0	0	0	0	0	0	0	0

HIGH-PROTEIN FOODS

MEAT	Serving Size	Weight (g)	Calories	Protein (g)	Fat (g)	Carbohydrates (g)	Calcium (mg)	Iron (mg)	Magnesium (mg)	Phosphorus (mg)	Potassium (mg)
Beef											
Chuck, arm pot roast, trimmed to 1/4" fat, choice, braised	3 oz	85	296	23	22	0	9	3	16	184	207
Cured, corned brisket, cooked	3 oz	85	213	15	16	0	7	2	10	106	123
Flank, trimmed to 0" fat, choice, braised	3 oz	85	224	23	14	0	5	3	20	218	286
Ground, extra lean, baked, medium	3 oz	85	213	21	14	0	6	2	14	105	190
Ground, regular, baked, medium	3 oz	85	244	20	18	0	9	2	13	116	188
Rib, eye, small end (ribs 10-12), trimmed to 1/4" fat, choice, broiled	3 oz	85	261	21	19	0	11	2	20	156	292
Rib, shortribs, choice, braised	3 oz	85	400	18	36	0	10	2	13	138	190
Rib, whole (ribs 6-12), trimmed to 1/4" fat, choice, broiled	3 oz	85	306	19	25	0	10	2	16	149	262
Round, bottom round, trimmed to 1/4" fat, choice, braised	3 oz	85	241	24	15	0	5	3	19	208	240
Round, eye of round, trimmed to 1/4" fat, choice, roasted	3 oz	85	205	23	12	0	5	2	20	175	305
Round, full cut, trimmed to 1/4" fat, choice, broiled	3 oz	85	204	23	12	0	5	2	21	202	333
Round, top round, trimmed to 1/4" fat, choice, broiled	3 oz	85	190	26	9	0	6	2	25	199	356
Short loin, porterhouse steak, separable lean only, trimmed to 1/4" fat, choice, broiled	3 oz	85	183	22	10	0	6	3	23	179	312
Short loin, T-bone steak, trimmed to 1/4" fat, choice, broiled	3 oz	85	263	20	20	0	7	2	20	156	273
Tenderloin, trimmed to 1/4" fat, choice, broiled	3 oz	85	258	21	19	0	7	3	22	178	310
Top sirloin, trimmed to 1/4" fat, choice, broiled	3 oz	85	229	23	14	0	9	3	24	187	309
Lamb											
Ground, broiled	3 oz	85	241	21	17	0	19	2	20	171	288
Leg, shank half, trimmed to 1/4" fat, choice, roasted	3 oz	85	191	22	11	0	9	2	21	168	277
Loin, trimmed to 1/4" fat, choice, broiled	3 oz	85	269	21	20	0	17	2	20	167	278
Shoulder, whole (arm and blade), trimmed to 1/4" fat, choice, braised	3 oz	85	292	24	21	0	21	2	20	158	211
Ostrich											
Top loin, cooked	3 oz	85	130	24	3	0	5	3	(-)	(-)	(-)

Nutrients in Foods 455

Sodium (mg)	Zinc (mg)	Copper (mg)	Manganese (mg)	Selenium (µg)	Vitamin C (mg)	Thiamin (mg)	Riboflavin (mg)	Niacin (mg)	Pantothenic acid (mg)	Vitamin B$_6$ (mg)	Folate (µg)	Vitamin B$_{12}$ (µg)	Vitamin A (IU)	Vitamin A (RE)	Vitamin E (mg)	Saturated fat (g)	Monounsaturated fat (g)	Polyunsaturated fat (g)	Cholesterol (mg)
50	6	0	0	(-)	0	0.1	0.2	3	0	0.2	8	2.5	0	0	0	9	9	1	84
964	4	0	0	(-)	0	0.0	0.1	3	0	0.2	5	1.4	0	0	0	5	8	1	83
60	5	0	0	(-)	0	0.1	0.2	4	0	0.3	8	2.8	0	0	(-)	6	6	0	61
42	5	0	0	(-)	0	0.0	0.2	4	0	0.2	8	1.5	0	0	(-)	5	6	1	70
51	4	0	0	16	0	0.0	0.1	4	0	0.2	8	2.0	0	0	(-)	7	8	1	74
54	5	0	0	(-)	0	0.1	0.2	4	0	0.3	6	2.6	0	0	(-)	8	8	1	71
43	4	0	0	(-)	0	0.0	0.1	2	0	0.2	4	2.2	0	0	0	15	16	1	80
53	4	0	0	(-)	0	0.1	0.1	3	0	0.2	5	2.4	0	0	(-)	10	11	1	70
43	4	0	0	(-)	0	0.1	0.2	3	0	0.3	9	2.0	0	0	0	6	7	1	82
50	4	0	0	(-)	0	0.1	0.1	3	0	0.3	5	1.8	0	0	(-)	5	5	0	61
52	4	0	0	(-)	0	0.1	0.2	3	0	0.3	8	2.6	0	0	0	4	5	0	68
51	4	0	0	(-)	0	0.1	0.2	5	0	0.5	9	2.1	0	0	(-)	3	4	0	72
59	4	0	0	(-)	0	0.1	0.2	4	0	0.3	7	1.9	0	0	0	3	4	0	59
54	4	0	0	(-)	0	0.1	0.2	3	0	0.3	6	1.8	0	0	0	8	9	1	57
50	4	0	0	(-)	0	0.1	0.2	3	0	0.3	5	2.0	0	0	(-)	7	8	1	73
53	5	0	0	(-)	0	0.1	0.2	3	0	0.3	8	2.3	0	0	0	6	6	1	77
69	4	0	0	(-)	0	0.1	0.2	6	1	0.1	16	2.2	0	0	0	7	7	1	82
55	4	0	0	(-)	0	0.1	0.2	6	1	0.1	19	2.3	0	0	0	4	4	1	77
65	3	0	0	(-)	0	0.1	0.2	6	1	0.1	15	2.1	0	0	0	8	8	1	85
64	5	0	0	(-)	0	0.1	0.2	5	1	0.1	14	2.4	0	0	0	9	9	2	99
65	(-)	(-)	(-)	(-)	(-)	(-)	(-)	(-)	(-)	(-)	(-)	(-)	(-)	(-)	(-)	1	1	1	79

(*continues*)

HIGH-PROTEIN FOODS *(continued)*

	Serving Size	Weight (g)	Calories	Protein (g)	Fat (g)	Carbohydrates (g)	Calcium (mg)	Iron (mg)	Magnesium (mg)	Phosphorus (mg)	Potassium (mg)
Pork											
Backribs, roasted	3 oz	85	315	21	25	0	38	1	18	166	268
Bacon, cooked: broiled, pan-fried, or roasted	3 strips	19	109	6	9	0	2	0	5	64	93
Canadian-style bacon, grilled	3 oz	70	129	17	6	1	7	1	15	206	271
Feet, cured, pickled	3 oz	85	173	11	14	0	27	0	3	29	260
Ground, cooked	3 oz	85	252	22	18	0	19	1	20	192	308
Ham, boneless, regular (approximately 11% fat), roasted	3 oz	85	151	19	8	0	7	1	19	239	348
Leg (ham), whole, roasted	3 oz	85	232	23	15	0	12	1	19	224	299
Loin, center rib (chops), bone-in, braised	3 oz	85	213	23	13	0	21	1	15	150	329
Loin, sirloin (chops), bone-in, braised	3 oz	85	208	22	13	0	15	1	16	148	276
Loin, tenderloin, broiled	3 oz	85	171	25	7	0	4	1	30	247	377
Loin, whole, braised	3 oz	85	203	23	12	0	18	1	16	154	318
Veal											
Ground, broiled	3 oz	85	146	21	6	0	14	1	20	184	286
Leg (top round), roasted	3 oz	85	136	24	4	0	5	1	24	199	331
Loin, roasted	3 oz	85	184	21	10	0	16	1	21	180	276
Rib, roasted	3 oz	85	194	20	12	0	9	1	19	167	251
Sirloin, roasted	3 oz	85	172	21	9	0	11	1	22	190	298
Variety meats											
Brain, simmered	3 oz	85	136	9	11	0	8	2	12	299	204
Heart (beef), simmered	3 oz	85	149	24	5	0	5	6	21	212	198
Kidney (beef), cooked	3 oz	85	122	22	3	1	14	6	15	260	152
Liver (beef), braised	3 oz	85	137	21	4	3	6	6	17	343	200
Pancreas, braised	3 oz	85	230	23	15	0	14	2	18	385	209
Thymus, braised	3 oz	85	271	19	21	0	9	1	9	309	368
Sausages/luncheon/deli meats											
Beef, lunch meat, thin sliced	3 oz	85	150	24	3	5	9	2	16	143	365
Bologna, beef and pork	3 oz	85	269	10	24	2	10	1	9	77	153
Bratwurst, pork	3 oz	85	256	12	22	2	37	1	13	127	180
Frankfurter, beef	1 frankfurter (5")	45	142	5	13	1	9	1	1	39	75
Frankfurter, turkey	1 frankfurter	45	102	6	8	1	48	1	6	60	81
Italian sausage, pork	1 link, 4/lb	83	268	17	21	1	20	1	15	141	252

Sodium (mg)	Zinc (mg)	Copper (mg)	Manganese (mg)	Selenium (µg)	Vitamin C (mg)	Thiamin (mg)	Riboflavin (mg)	Niacin (mg)	Pantothenic acid (mg)	Vitamin B₆ (mg)	Folate (µg)	Vitamin B₁₂ (µg)	Vitamin A (IU)	Vitamin A (RE)	Vitamin E (mg)	Saturated fat (g)	Monounsaturated fat (g)	Polyunsaturated fat (g)	Cholesterol (mg)
86	3	0	0	33	0	0.4	0.2	3	0	0.3	3	0.5	8	3	(-)	9	11	2	100
303	1	0	0	5	0	0.1	0.1	1	0	0.1	1	0.3	0	0	0	3	5	1	16
1,074	1	0	0	17	0	0.6	0.1	5	0	0.3	3	0.5	0	0	0	2	3	1	40
785	1	0	0	6	0	0	0	0	0	0.3	3	0	0	0	0	5	6	1	78
62	3	0	0	30	1	0.6	0.2	4	0	0.3	5	0.5	7	2	0	7	8	2	80
1,275	2	0	0	17	0	0.6	0.3	5	1	0.3	3	0.6	0	0	0	3	4	1	50
51	3	0	0	39	0	0.5	0.3	4	1	0.3	9	0.6	9	3	0	5	7	1	80
34	2	0	0	35	0	0.5	0.2	4	0	0.3	2	0.5	6	2	(-)	5	6	1	62
43	2	0	0	34	1	0.6	0.2	3	1	0.3	3	0.5	6	2	(-)	5	6	1	70
54	2	0	0	41	1	0.8	0.3	4	1	0.4	5	0.8	6	2	(-)	2	3	1	80
41	2	0	0	39	1	0.5	0.2	4	1	0.3	3	0.5	6	2	0	4	5	1	68
71	3	0	0	(-)	0	0.1	0.2	7	1	0.3	9	1.1	0	0	0	3	2	0	88
58	3	0	0	(-)	0	0.1	0.3	8	1	0.3	14	1.0	0	0	0	2	1	0	88
79	3	0	0	(-)	0	0.0	0.2	8	1	0.3	13	1.1	0	0	0	4	4	1	88
78	3	0	0	(-)	0	0.0	0.2	6	1	0.2	11	1.2	0	0	0	5	5	1	94
71	3	0	0	(-)	0	0.1	0.3	8	1	0.3	13	1.2	0	0	0	4	3	1	87
102	1	0	0	(-)	1	0.1	0.1	2	0	0.2	6	7.3	0	0	2	2	2	1	1,746
53	3	1	0	33	1	0.1	0.3	3	1	0.2	2	12	0	0	1	1	1	1	164
114	4	1	0	239	1	0.2	3.4	5	1	0.4	83	44	1,055	317	0	1	1	1	329
60	5	4	0	(-)	20	0.2	3.5	9	4	0.8	184	60.4	30,327	9,012	(-)	2	1	1	331
51	4	0	0	(-)	17	0.2	0.4	3	4	0.2	3	14.1	0	0	(-)	5	5	3	223
99	2	0	0	(-)	26	0.1	0.2	2	2	0.1	1	1.3	0	0	(-)	7	7	4	250
1,224	3	0	0	24	0	0.1	0.2	4	0	0.3	9	2	0	0	0	1	1	0	35
867	2	0	0	10	0	0.1	0.1	2	0	0.2	4	1.1	0	0	0	9	11	2	47
474	2	0	0	18	1	0.4	0.2	3	0	0.2	2	0.8	0	0	0	8	10	2	51
462	1	0	0	6	0	0.0	0.0	1	0	0.1	2	0.7	0	0	0	5	6	1	27
642	1	0	0	7	0	0.0	0.1	2	0	0.1	4	0.1	0	0	0	3	3	2	48
765	2	0	0	18	2	0.5	0.2	3	0	0.3	4	1.1	0	0	0	8	10	3	65

(continues)

HIGH-PROTEIN FOODS (*continued*)

	Serving Size	Weight (g)	Calories	Protein (g)	Fat (g)	Carbohydrates (g)	Calcium (mg)	Iron (mg)	Magnesium (mg)	Phosphorus (mg)	Potassium (mg)
Sausages/luncheon/deli meats (*continued*)											
Luncheon meat, jellied, beef	3 oz	85	94	16	3	0	9	3	15	118	342
Mortadella, pork	3 oz	85	265	14	22	3	15	1	9	83	139
Pastrami, beef	3 oz	85	297	15	25	3	8	2	15	128	194
Pastrami, turkey	3 oz	85	120	16	5	1	8	1	12	170	221
Pepperoni, pork, beef	3 oz	85	422	18	37	2	8	1	14	101	295
Pork, lunch meat, thin sliced	3 oz	85	300	11	27	2	8	1	12	73	172
Pork sausage, fresh, cooked	3 oz	85	314	17	27	1	27	1	14	156	307
Salami, beef	3 oz	85	223	13	18	2	8	2	12	96	191
Turkey roll, light meat	3 oz	85	125	16	6	0	34	1	14	156	213
Game meats											
Antelope, roasted	3 oz	85	128	25	2	0	3	4	24	179	316
Bear, simmered	3 oz	85	220	28	11	0	4	9	20	145	224
Beaver, roasted	3 oz	85	180	30	6	0	19	9	25	248	343
Beefalo, roasted	3 oz	85	160	26	5	0	20	3	0	213	390
Bison, roasted	3 oz	85	122	24	2	0	7	3	22	178	307
Boar, wild, roasted	3 oz	85	136	24	4	0	14	1	23	114	337
Buffalo, water, roasted	3 oz	85	111	23	2	0	13	2	28	187	266
Caribou, roasted	3 oz	85	142	25	4	0	19	5	23	198	264
Deer, roasted	3 oz	85	134	26	3	0	6	4	20	192	285
Elk, roasted	3 oz	85	124	26	2	0	4	3	20	153	279
Goat, roasted	3 oz	85	122	23	3	0	14	3	0	171	344
Hare, stewed	3 oz	85	147	28	3	0	15	4	26	204	291
Moose, roasted	3 oz	85	114	25	1	0	5	4	20	150	284
Muskrat, roasted	3 oz	85	199	26	10	0	31	6	22	230	272
Opossum, roasted	3 oz	85	188	26	9	0	14	4	29	236	372
Rabbit, domesticated, roasted	3 oz	85	167	25	7	0	16	2	18	224	326
Raccoon, roasted	3 oz	85	217	25	12	0	12	6	26	222	338
Squirrel, roasted	3 oz	85	147	26	4	0	3	6	24	179	299
POULTRY											
Chicken											
Broilers or fryers, breast, meat and skin, roasted	3 oz	85	168	25	7	0	12	1	23	182	209
Broilers or fryers, breast, meat only, roasted	3 oz	85	141	27	3	0	13	1	25	196	220

Sodium (mg)	Zinc (mg)	Copper (mg)	Manganese (mg)	Selenium (µg)	Vitamin C (mg)	Thiamin (mg)	Riboflavin (mg)	Niacin (mg)	Pantothenic acid (mg)	Vitamin B_6 (mg)	Folate (µg)	Vitamin B_{12} (µg)	Vitamin A (IU)	Vitamin A (RE)	Vitamin E (mg)	Saturated fat (g)	Monounsaturated fat (g)	Polyunsaturated fat (g)	Cholesterol (mg)
1,124	3	0	0	14	0	0.1	0.2	4	1	0.2	6	4.4	0	0	(-)	1	1	0	29
1,060	2	0	0	19	0	0.1	0.1	2	0	0.1	3	1.3	0	0	0	8	10	3	48
1,044	4	0	0	9	0	0.1	0.1	4	0	0.2	6	1.5	0	0	0	9	12	1	79
890	2	0	0	14	0	0.0	0.2	3	0	0.2	4	0.2	0	0	0	2	2	1	46
1,733	2	0	0	20	0	0.3	0.2	4	2	0.2	3	2.1	0	0	0	14	18	4	67
1,100	1	0	0	24	0	0.3	0.1	2	0	0.2	5	1	0	0	0	10	13	3	47
1,101	2	0	0	15	2	0.6	0.2	4	1	0.3	2	1.5	0	0	0	9	12	3	71
1,000	2	0	0	12	0	0.1	0.2	3	1	0.2	2	2.6	0	0	0	8	8	1	55
416	1	0	0	19	0	0.1	0.2	6	0	0.3	3	0.2	0	0	0	2	2	1	37
46	1	0	0	(-)	0	0.2	0.6	(-)	(-)	(-)	(-)	(-)	0	0	(-)	1	1	0	107
60	9	0	(-)	(-)	0	0.1	0.7	3	(-)	0.2	5	2.1	0	0	0	3	5	2	83
50	2	0	(-)	(-)	3	0.0	0.3	2	1	0.4	9	7.1	0	0	1	2	2	1	99
70	5	(-)	(-)	(-)	8	0.0	0.1	4	0	(-)	15	2.2	0	0	(-)	2	2	0	49
48	3	0	0	(-)	0	0.1	0.2	3	(-)	0.3	7	2.4	0	0	0	1	1	0	70
51	3	0	(-)	(-)	0	0.3	0.1	4	(-)	0.4	5	0.6	0	0	0	1	1	1	65
48	2	0	(-)	(-)	0	0.0	0.2	5	0	0.4	8	1.5	0	0	(-)	1	0	0	52
51	4	0	0	(-)	3	0.2	0.8	5	2	0.3	4	5.6	0	0	0	1	1	1	93
46	2	0	0	(-)	0	0.2	0.5	6	(-)	(-)	(-)	(-)	0	0	(-)	1	1	1	95
52	3	0	0	(-)	0	(-)	(-)	(-)	(-)	(-)	(-)	(-)	0	0	(-)	1	1	0	62
73	4	0	0	(-)	0	0.1	0.5	3	(-)	0.0	4	1.0	0	0	0	1	1	0	64
38	2	0	13	0	0	0.0	0.1	5	(-)	0.3	7	5	0	0	1	1	1	1	104
59	3	0	0	(-)	4	0.0	0.3	4	(-)	0.3	3	5.4	0	0	0	0	0	0	66
81	2	0	0	(-)	6	0.1	0.6	6	1	0.4	9	7.1	0	0	(-)	0	0	0	103
49	2	0	(-)	(-)	0	0.1	0.3	7	(-)	0.4	9	7.1	0	0	1	1	3	3	110
40	2	0	0	(-)	0	0.1	0.2	7	1	0.4	9	7.1	0	0	(-)	2	2	1	70
67	2	0	(-)	(-)	0	0.5	0.4	4	(-)	0.4	9	7.1	0	0	1	3	4	2	82
101	2	0	0	(-)	0	0.1	0.2	4	1	0.3	8	5.5	0	0	1	1	1	1	103
61	1	0	0	(-)	0	0.1	0.1	11	1	0.5	3	0.3	79	23	0	2	3	1	72
64	1	0	0	24	0	0.0	0.1	12	1	0.5	3	0.3	18	5	0	1	1	1	73

(*continues*)

High-Protein Foods (*continued*)

	Serving Size	Weight (g)	Calories	Protein (g)	Fat (g)	Carbohydrates (g)	Calcium (mg)	Iron (mg)	Magnesium (mg)	Phosphorus (mg)	Potassium (mg)
Chicken (*continued*)											
Broilers or fryers, dark meat, meat and skin, roasted	3 oz	85	215	22	13	0	13	1	19	143	187
Broilers or fryers, dark meat, meat only, roasted	3 oz	85	174	23	8	0	13	1	21	166	206
Broilers or fryers, drumstick, meat and skin, roasted	3 oz	85	183	23	9	0	10	1	19	148	194
Broilers or fryers, meat only, roasted	3 oz	85	161	25	6	0	13	1	21	166	206
Broilers or fryers, thigh, meat and skin, roasted	3 oz	85	210	21	13	0	10	1	19	148	189
Broilers or fryers, wing, meat and skin, roasted	3 oz	85	246	23	17	0	13	1	16	128	156
Cornish game hens, meat and skin, roasted	3 oz	85	221	19	16	0	11	1	15	124	209
Liver, simmered	3 oz	85	133	21	5	1	12	7	18	265	119
Pâté, chicken liver, canned	3 oz	85	171	11	11	6	9	8	11	149	81
Turkey											
Breast, meat and skin, roasted	3 oz	85	161	24	6	0	18	1	23	179	245
Breast, meat only, roasted	3 oz	85	93	19	1	0	6	0	17	195	263
Dark meat, roasted	3 oz	85	159	24	6	0	27	2	20	173	246
Ground, cooked	3 oz	85	200	23	11	0	21	2	20	167	229
Leg, meat and skin, roasted	3 oz	85	177	24	8	0	27	2	20	169	238
Thigh, pre-basted, meat and skin, roasted	3 oz	85	133	16	7	0	7	1	14	145	205
Wing, meat and skin, roasted	3 oz	85	195	23	11	0	20	1	21	167	226
Duck											
Meat and skin, roasted	3 oz	85	286	16	24	0	9	2	14	133	173
Meat only, roasted	3 oz	85	171	20	10	0	10	2	17	173	214
Goose											
Meat and skin, roasted	3 oz	85	259	21	19	0	11	2	19	229	280
Meat only, roasted	3 oz	85	202	25	11	0	12	2	21	263	330
Pâté de foie gras, canned (goose liver pâté), smoked	3 oz	85	393	10	37	4	60	5	11	170	117
EGGS											
Egg substitute, liquid	1/2 cup	126	105	15	4	1	67	3	11	152	414
Egg, white, raw	2 large whites	67	33	7	0	1	4	0	7	9	95
Egg, whole, raw, fresh	1 large egg	50	75	6	5	1	25	1	5	89	61
Egg. yolk, raw	1 large yolk	17	59	3	6	0	23	1	1	81	16

Sodium (mg)	Zinc (mg)	Copper (mg)	Manganese (mg)	Selenium (µg)	Vitamin C (mg)	Thiamin (mg)	Riboflavin (mg)	Niacin (mg)	Pantothenic acid (mg)	Vitamin B$_6$ (mg)	Folate (µg)	Vitamin B$_{12}$ (µg)	Vitamin A (IU)	Vitamin A (RE)	Vitamin E (mg)	Saturated fat (g)	Monounsaturated fat (g)	Polyunsaturated fat (g)	Cholesterol (mg)
74	2	0	0	(-)	0	0.1	0.2	5	1	0.3	6	0.2	171	49	(-)	4	5	3	78
79	2	0	0	(-)	0	0.1	0.2	6	1	0.3	7	0.3	61	19	0	2	3	2	79
76	2	0	0	(-)	0	0.1	0.2	5	1	0.3	7	0.3	85	25	0	3	4	2	77
73	2	0	(-)	0	0	0.1	0.2	8	1	0.4	5	0.3	45	14	0	2	2	1	76
71	2	0	0	(-)	0	0.1	0.2	5	1	0.3	6	0.2	140	41	0	4	5	3	79
70	2	0	0	(-)	0	0.0	0.1	6	1	0.4	3	0.2	134	40	0	5	6	4	71
54	1	0	0	(-)	0	0.1	0.2	5	1	0.3	2	0.2	90	27	0	4	7	3	112
43	4	0	0	(-)	13	0.1	1.5	4	5	0.5	655	16.5	13,919	4,176	1	2	1	1	536
328	2	0	0	39	9	0.0	1.2	6	2	0.2	273	6.9	616	185	1	3	4	2	333
54	2	0	0	(-)	0	0.0	0.1	5	1	0.4	5	0.3	0	0	(-)	2	2	2	63
1,216	1	0	0	26	0	0.0	0.1	7	0	0.3	3	2	0	0	0	0	0	0	35
67	4	0	0	35	0	0.1	0.2	3	1	0.3	8	0.3	0	0	1	2	1	2	72
91	2	0	0	(-)	0	0.0	0.1	4	1	0.3	6	0.3	0	0	0	3	4	3	87
65	4	0	0	(-)	0	0.1	0.2	3	1	0.3	8	0.3	0	0	0	3	2	2	72
371	4	0	0	(-)	0	0.1	0.2	2	1	0.2	5	0.2	0	0	(-)	2	2	2	53
52	2	0	0	(-)	0	0.0	0.1	5	0	0.4	5	0.3	0	0	0	3	4	3	69
50	2	0	0	(-)	0	0.1	0.2	4	1	0.2	5	0.3	178	54	1	8	11	3	71
55	2	0	0	(-)	0	0.2	0.4	4	1	0.2	9	0.3	65	20	1	4	3	1	76
59	2	0	0	(-)	0	0.1	0.3	4	1	0.3	2	0.3	59	18	1	6	9	2	77
65	3	0	0	(-)	0	0.1	0.3	3	2	0.4	10	0.4	34	10	(-)	4	4	1	82
593	1	0	0	(-)	2	0.1	0.3	2	1	0.1	51	8.0	2,835	851	(-)	12	22	1	128
222	2	0	0	31	0	0.1	0.4	0	3	0.0	19	0.4	2,711	271	1	1	1	2	1
109	0	0	0	12	0	0.0	0.3	0	0	0.0	2	0.1	0	0	0	0	0	0	0
63	1	0	0	15	0	0.0	0.3	0	1	0.1	24	0.5	318	96	1	2	2	1	213
7	1	0	0	7	0	0.0	0.1	0	1	0.1	24	0.5	323	97	1	2	0	1	213

(*continues*)

High-Protein Foods (*continued*)

FINFISH AND SHELLFISH	Serving Size	Weight (g)	Calories	Protein (g)	Fat (g)	Carbohydrates (g)	Calcium (mg)	Iron (mg)	Magnesium (mg)	Phosphorus (mg)	Potassium (mg)
Finfish (lean)											
Bass, sea, mixed species, cooked, dry heat	3 oz	85	105	20	2	0	11	0	45	211	279
Cod, Atlantic, cooked, dry heat	3 oz	85	89	19	1	0	12	0	36	117	207
Flounder (and sole species), cooked, dry heat	3 oz	85	99	21	1	0	15	0	49	246	292
Halibut, Atlantic and Pacific, cooked, dry heat	3 oz	85	119	23	2	0	51	1	91	242	490
Ocean perch, Atlantic, cooked, dry heat	3 oz	85	103	20	2	0	116	1	33	235	298
Perch, mixed species, cooked, dry heat	3 oz	85	99	21	1	0	87	1	32	218	292
Pike, northern, cooked, dry heat	3 oz	85	96	21	1	0	62	1	34	240	281
Roughy, orange, cooked, dry heat	3 oz	85	76	16	1	0	32	0	32	218	327
Snapper, mixed species, cooked, dry heat	3 oz	85	109	22	1	0	34	0	31	171	444
Tuna, light, canned in water, without salt, drained solids	3 oz	85	99	22	1	0	9	1	23	139	201
Finfish (moderate fat)											
Bass, freshwater, mixed species, cooked, dry heat	3 oz	85	124	21	4	0	88	2	32	218	388
Catfish, channel, wild, cooked, dry heat	3 oz	85	93	15	3	0	8	0	24	260	356
Salmon, chinook, smoked (lox), regular	3 oz	85	99	16	4	0	9	1	15	139	149
Salmon, coho, wild, cooked, moist heat	3 oz	85	156	23	6	0	39	1	30	253	387
Salmon, pink, cooked, dry heat	3 oz	85	127	22	4	0	14	1	28	251	352
Salmon, sockeye, canned, drained solids with bone	3 oz	85	130	17	6	0	203	1	25	277	320
Swordfish, cooked, dry heat	3 oz	85	132	22	4	0	5	1	29	286	314
Trout, rainbow, wild, cooked, dry heat	3 oz	85	128	19	5	0	73	0	26	229	381
Tuna, fresh, bluefin, cooked, dry heat	3 oz	85	156	25	5	0	9	1	54	277	275
Tuna, light, canned in oil, without salt, drained solids	3 oz	85	168	25	7	0	11	1	26	264	176
Finfish (high fat)											
Mackerel, Atlantic, cooked, dry heat	3 oz	85	223	20	15	0	13	1	82	236	341
Salmon, Atlantic, wild, cooked, dry heat	3 oz	85	155	22	7	0	13	1	31	218	534
Salmon, chinook, cooked, dry heat	3 oz	85	196	22	11	0	24	1	104	315	429
Salmon, sockeye, cooked, dry heat	3 oz	85	184	23	9	0	6	0	26	235	319
Sardine, Atlantic, canned in oil, drained solids with bone	3 oz	85	177	21	10	0	325	2	33	417	338
Smelt, rainbow, cooked, dry heat	3 oz	85	105	19	3	0	65	1	32	251	316
Trout, mixed species, cooked, dry heat	3 oz	85	162	23	7	0	47	2	24	267	394

Sodium (mg)	Zinc (mg)	Copper (mg)	Manganese (mg)	Selenium (µg)	Vitamin C (mg)	Thiamin (mg)	Riboflavin (mg)	Niacin (mg)	Pantothenic acid (mg)	Vitamin B$_6$ (mg)	Folate (µg)	Vitamin B$_{12}$ (µg)	Vitamin A (IU)	Vitamin A (RE)	Vitamin E (mg)	Saturated fat (g)	Monounsaturated fat (g)	Polyunsaturated fat (g)	Cholesterol (mg)
74	0	0	0	40	0	0.1	0.1	2	1	0.4	5	0.3	181	54	(-)	1	0	1	45
66	0	0	0	32	1	0.1	0.1	2	0	0.2	7	0.9	39	12	0	0	0	0	47
89	1	0	0	49	0	0.1	0.1	2	0	0.2	8	2.1	32	9	2	0	0	1	58
59	0	0	0	40	0	0.1	0.1	6	0	0.3	12	1.2	152	46	1	0	1	1	35
82	1	0	0	47	1	0.1	0.1	2	0	0.2	9	1.0	39	12	(-)	0	1	0	46
67	1	0	1	14	1	0.1	0.1	2	1	0.1	5	1.9	27	9	(-)	0	0	0	98
42	1	0	0	14	3	0.1	0.1	2	1	0.1	15	2.0	69	20	(-)	0	0	0	43
69	1	0	0	40	0	0.1	0.2	3	1	0.3	7	2.0	69	20	(-)	0	1	0	22
48	0	0	0	42	1	0.0	0.0	0	1	0.4	5	3.0	98	30	(-)	0	0	1	40
43	1	0	0	68	0	0.0	0.1	11	0	0.3	3	2.5	48	14	0	0	0	0	26
77	1	0	1	14	2	0.1	0.1	1	1	0.1	14	1.9	98	30	(-)	1	2	1	74
42	0	0	0	12	1	0.1	0.0	2	1	0.0	9	2.4	42	13	(-)	1	1	1	62
1,700	0	0	0	32	0	0.0	0.1	4	1	0.2	2	2.8	75	22	(-)	1	2	1	20
45	0	0	0	39	1	0.1	0.1	7	1	0.5	8	3.8	92	27	(-)	1	2	2	48
73	1	0	0	49	0	0.2	0.1	7	1	0.2	4	2.9	116	35	(-)	1	1	1	57
457	1	0	0	30	0	0.0	0.2	5	0	0.3	8	0.3	150	45	1	1	3	2	37
98	1	0	0	52	1	0.0	0.1	10	0	0.3	2	1.7	116	35	(-)	1	2	1	43
48	0	0	0	11	2	0.1	0.1	5	1	0.3	16	5.4	43	13	(-)	1	1	2	59
43	1	0	0	40	0	0.2	0.3	9	1	0.4	2	9.2	2,142	643	(-)	1	2	2	42
43	1	0	0	65	0	0.0	0.1	11	0	0.1	5	1.9	66	20	(-)	1	3	2	15
71	1	0	0	44	0	0.1	0.4	6	1	0.4	1	16.2	153	46	(-)	4	6	4	64
48	1	0	0	40	0	0.2	0.4	9	2	0.8	25	2.6	37	11	(-)	1	2	3	60
51	0	0	0	40	3	0.0	0.1	9	1	0.4	30	2.4	422	127	(-)	3	5	2	72
56	0	0	0	32	0	0.2	0.1	6	1	0.2	4	4.9	178	54	(-)	2	4	2	74
430	1	0	0	45	0	0.1	0.2	4	1	0.1	10	7.6	191	57	0	1	3	4	121
65	2	0	1	40	0	0.0	0.1	2	1	0.1	4	3.4	49	14	(-)	0	1	1	77
57	1	0	1	14	0	0.4	0.4	5	2	0.2	13	6.3	54	16	(-)	1	4	2	63

(continues)

HIGH-PROTEIN FOODS (*continued*)

	Serving Size	Weight (g)	Calories	Protein (g)	Fat (g)	Carbohydrates (g)	Calcium (mg)	Iron (mg)	Magnesium (mg)	Phosphorus (mg)	Potassium (mg)
Shellfish											
Abalone, mixed species, cooked, fried	3 oz	85	161	17	6	9	31	3	48	184	241
Clams, breaded and fried	3 oz	85	333	9	20	29	15	2	23	176	196
Crab, Alaska king, cooked, moist heat	3 oz	85	82	16	1	0	50	1	54	238	223
Crab, blue, canned	1/3 cup	45	44	9	0	0	45	0	17	117	168
Crayfish, mixed species, wild, cooked, moist heat	3 oz	85	70	14	1	0	51	1	28	230	252
Lobster, northern, cooked, moist heat	3 oz	85	83	17	1	1	52	0	30	157	299
Mussel, blue, cooked, moist heat	3 oz	85	146	20	4	6	28	6	31	242	228
Octopus, common, cooked, moist heat	3 oz	85	139	25	2	4	90	8	51	237	536
Oysters, battered or breaded, and fried	3 oz	85	225	8	11	24	17	3	14	120	111
Scallops, breaded and fried	3 oz	85	228	9	11	23	11	1	19	173	173
Shrimp, mixed species, breaded and fried	3 oz (about 11 shrimp)	85	206	18	10	10	57	1	34	185	191
Shrimp, mixed species, canned	1/4 cup	32	38	7	1	0	19	1	13	75	67
Shrimp, mixed species, cooked, moist heat	3 oz	85	84	18	1	0	33	3	29	116	155
Spiny lobster, mixed species, cooked, moist heat	3 oz	85	122	22	2	3	54	1	43	195	177

Sodium (mg)	Zinc (mg)	Copper (mg)	Manganese (mg)	Selenium (µg)	Vitamin C (mg)	Thiamin (mg)	Riboflavin (mg)	Niacin (mg)	Pantothenic acid (mg)	Vitamin B$_6$ (mg)	Folate (µg)	Vitamin B$_{12}$ (µg)	Vitamin A (IU)	Vitamin A (RE)	Vitamin E (mg)	Saturated fat (g)	Monounsaturated fat (g)	Polyunsaturated fat (g)	Cholesterol (mg)
502	1	0	0	44	2	0.2	0.1	2	2	0.1	12	0.6	4	2	(-)	1	2	1	80
616	1	0	0	7	0	0.2	0.2	2	0	0.0	31	0.8	90	27	(-)	5	8	5	65
911	6	1	0	34	6	0.0	0.0	1	0	0.2	43	9.8	25	8	(-)	0	0	0	45
150	2	0	0	14	1	0.0	0.0	1	0	0.1	19	0.2	2	1	0	0	0	0	40
80	1	1	0	31	1	0.0	0.1	2	0	0.1	37	1.8	43	13	1	0	0	0	113
323	2	2	0	36	0	0.0	0.1	1	0	0.1	9	2.6	74	22	1	0	0	0	61
314	2	0	6	76	12	0.3	0.4	3	1	0.1	64	20.4	258	77	(-)	1	1	1	48
391	3	1	0	76	7	0.0	0.1	3	1	0.6	20	30.6	230	69	1	0	0	0	82
414	10	0	0	56	3	0.2	0.2	3	1	0.0	19	0.6	222	66	(-)	3	4	3	66
542	1	0	0	23	0	0.1	0.5	0	0	0.0	31	0.3	82	25	(-)	3	7	0	64
292	1	0	0	35	1	0.1	0.1	3	0	0.1	7	1.6	161	48	(-)	2	3	4	150
54	0	0	0	13	1	0.0	0.0	1	0	0.1	1	0.4	19	6	0	0	0	0	55
190	1	0	0	34	2	0.0	0.0	2	0	0.1	3	1.3	186	56	0	0	0	0	166
193	6	0	0	50	2	0.0	0.0	4	0	0.1	1	3.4	17	5	(-)	0	0	1	77

(*continues*)

HIGH-PROTEIN FOODS (*continued*)

LEGUMES	Serving Size	Weight (g)	Calories	Protein (g)	Fat (g)	Carbohydrates (g)	Fiber (g)	Calcium (mg)	Iron (mg)	Magnesium (mg)	Phosphorus (mg)	Potassium (mg)
Alfalfa seeds, sprouted, raw	1 cup	33	10	1	0	1	1	11	0	9	23	26
Broad beans (fava beans), mature seeds, cooked, boiled, without salt	1/2 cup	85	94	6	0	17	5	31	1	37	106	228
Chickpeas (garbanzo beans, Bengal gram), mature seeds, cooked, boiled, without salt	1/2 cup	82	134	7	2	22	6	40	2	39	138	239
Common beans												
Beans, black, mature seeds, cooked, boiled, without salt	1/2 cup	86	114	8	0	20	7	23	2	60	120	305
Beans, cranberry (Roman), mature seeds, cooked, boiled, without salt	1/2 cup	89	120	8	0	22	9	44	2	44	119	342
Beans, Great Northern, mature, boiled	1/2 cup	88	104	7	0	19	6	60	2	44	146	346
Beans, kidney, all types, mature seeds, cooked, boiled, without salt	1/2 cup	89	112	8	0	20	6	25	3	40	126	357
Beans, navy, mature, boiled, without salt	1/2 cup	91	129	8	0	24	6	64	2	54	143	335
Beans, pinto, mature seeds, cooked, boiled, without salt	1/2 cup	86	117	7	0	22	7	41	2	47	137	400
Beans, small white, mature seeds, cooked, boiled, without salt	1/2 cup	90	127	8	1	23	9	65	3	61	151	414
Dolichos beans												
Beans, adzuki, mature seeds, cooked, boiled, without salt	1/2 cup	115	147	9	0	28	0	32	2	60	193	612
Black-eyed peas, mature seeds, cooked, boiled, without salt	1/2 cup	86	100	7	1	17	3	22	3	82	121	321
Lablab bean (hyacinth bean), mature, boiled	1/2 cup	97	113	8	1	20	(-)	39	4	79	116	327
Mung beans, mature seeds, cooked, boiled, without salt	1/2 cup	101	106	7	0	19	8	27	1	48	100	269
Mung beans, mature seeds, sprouted, cooked, boiled, drained, without salt	1/2 cup	62	13	1	0	3	0	7	0	9	17	63
Mung beans, mature seeds, sprouted, raw	1 cup	104	31	3	0	6	2	14	1	22	56	155
Other lentils & legumes												
Lentils, mature seeds, cooked, boiled, without salt	1/2 cup	99	115	9	0	20	8	19	3	36	178	365
Lima beans, large, mature seeds, cooked, boiled, without salt	1/2 cup	94	108	7	0	20	7	16	2	40	104	478
Peanuts, all types, raw	1/3 cup	48	273	12	24	8	4	44	2	81	181	340
Peanuts, all types, oil-roasted, with salt (halves and whole)	1/3 cup	48	276	13	23	9	4	42	1	88	246	324
Peanuts, all types, cooked, boiled, with salt (shelled)	1/3 cup	59	189	8	13	13	5	33	1	61	118	107

Sodium (mg)	Zinc (mg)	Copper (mg)	Manganese (mg)	Selenium (µg)	Vitamin C (mg)	Thiamin (mg)	Riboflavin (mg)	Niacin (mg)	Pantothenic acid (mg)	Vitamin B₆ (mg)	Folate (µg)	Vitamin B₁₂ (µg)	Vitamin A (IU)	Vitamin A (RE)	Vitamin E (mg)	Saturated fat (g)	Monounsaturated fat (g)	Polyunsaturated fat (g)	Cholesterol (mg)
2	0	0	0	0	3	0.0	0.0	0	0	0.0	12	0.0	51	5	0	0	0	0	0
4	1	0	0	2	0	0.1	0.1	1	0	0.1	88	0.0	13	2	0	0	0	0	0
6	1	0	1	3	1	0.1	0.1	0	0	0.1	141	0.0	22	2	0	0	0	1	0
1	1	0	0	1	0	0.2	0.1	0	0	0.1	128	0.0	5	1	0	0	0	0	0
1	1	0	0	1	0	0.2	0.1	0	0	0.1	183	0.0	0	0	0	0	0	0	0
2	1	0	0	4	1	0.1	0.0	1	0	0.1	90	0.0	2	0	(-)	0	0	0	0
2	1	0	0	1	1	0.1	0.1	1	0	0.1	115	0.0	0	0	0	0	0	0	0
1	1	0	0	5	1	0.2	0.0	0	0	0.1	127	0.0	2	0	(-)	0	0	0	0
2	1	0	0	6	2	0.2	0.1	0	0	0.1	147	0.0	2	0	1	0	0	0	0
2	1	0	0	1	0	0.2	0.1	0	0	0.1	123	0.0	0	0	0	0	0	0	0
9	2	0	1	1	0	0.1	0.1	1	0	0.1	139	0.0	7	1	0	0	0	0	0
16	2	0	0	2	0	0.1	0.0	1	0	0.1	121	0.0	9	1	0	0	0	0	0
7	3	0	0	3	0	0.3	0.0	0	0	0.0	4	0.0	0	0	(-)	0	0	0	0
2	1	0	0	3	1	0.2	0.1	1	0	0.1	160	0.0	24	2	1	0	0	0	0
6	0	0	0	0	7	0.0	0.1	1	0	0.0	18	0.0	9	1	0	0	0	0	0
6	0	0	0	1	14	0.1	0.1	1	0	0.1	63	0.0	22	2	0	0	0	0	0
2	1	0	0	3	1	0.2	0.1	1	1	0.2	179	0.0	8	1	0	0	0	0	0
2	1	0	0	4	0	0.2	0.1	0	0	0.2	78	0.0	0	0	0	0	0	0	0
9	2	1	1	3	0	0.3	0.1	6	1	0.2	116	0.0	0	0	4	3	12	7	0
206	3	1	1	4	0	0.1	0.1	7	1	0.1	60	0.0	0	0	4	3	12	7	0
446	1	0	1	3	0	0.2	0.0	3	0	0.1	44	0.0	0	0	2	2	6	4	0

(*continues*)

HIGH-PROTEIN FOODS (*continued*)

	Serving Size	Weight (g)	Calories	Protein (g)	Fat (g)	Carbohydrates (g)	Fiber (g)	Calcium (mg)	Iron (mg)	Magnesium (mg)	Phosphorus (mg)	Potassium (mg)
Other lentils & legumes (*continued*)												
Peanut butter, chunk style, with salt	2 Tblsp	32	188	8	16	7	2	13	1	51	101	239
Peanut butter, smooth style, with salt	2 Tblsp	32	190	8	16	6	2	12	1	51	118	214
Peas, edible-podded, boiled without salt	1/2 cup	80	34	3	0	6	2	34	2	21	44	192
Peas, green, cooked without salt	1/2 cup	80	67	4	0	12	4	22	1	31	94	217
Peas, split, mature, boiled without salt	1/2 cup	98	116	8	0	21	8	14	1	35	97	355
Soybeans, mature, cooked, boiled, without salt	1/2 cup	86	149	14	8	9	5	88	4	74	211	443
Soybeans, mature seeds, sprouted, cooked, steamed	1/2 cup	47	38	4	2	3	0	28	1	28	63	167
Soybeans, mature seeds, sprouted, raw	1 cup	70	84	9	5	7	1	47	1	50	115	339
Soybeans, miso	1/2 cup	138	283	16	8	38	7	91	4	58	210	226
Soybeans, tempeh	1/2 cup	83	165	16	6	14	0	77	2	58	171	305
Soybeans, tofu, firm	1/2 cup	126	97	10	6	4	1	204	2	58	185	222
Soybeans, tofu, lite, firm	1/2 cup	125	46	8	1	1	0	45	1	12	101	79
Soybeans, tofu, soft	1/2 cup	124	76	8	5	2	0	138	1	33	114	149
NUTS & SEEDS												
Almonds, dry roasted, without salt added	1/3 cup whole kernels	46	275	10	24	9	5	122	2	132	225	343
Beechnuts, dried	1 oz	28	163	2	14	9	(-)	0	1	0	0	288
Brazil nuts, dried, blanched	1/3 cup shelled (11 kernels)	47	306	7	31	6	3	82	2	105	280	280
Cashew nuts, dry roasted, without salt added	1/3 cup halves and whole	46	262	7	21	15	1	21	3	119	224	258
Chestnuts, European, roasted	1/3 cup	48	117	2	1	25	2	14	0	16	51	282
Ginkgo nuts, dried	1 oz	28	99	3	1	21	(-)	6	0	15	76	283
Hazelnuts (filberts)	1/3 cup whole	45	283	7	27	8	4	51	2	73	131	306
Lotus seeds, dried	1/4 cup	8	27	1	0	4	(-)	13	0	17	50	109
Macadamia nuts, dry roasted, without salt added	1/3 cup whole or halves	45	321	3	34	6	4	31	1	53	88	162
Pecans	1/3 cup halves	36	249	3	26	5	3	25	1	44	100	148
Pistachio nuts, dry roasted, without salt added	1/3 cup	43	243	9	20	12	4	46	2	51	207	441
Safflower seed kernels, dried	1 oz	28	147	5	11	10	(-)	22	1	100	183	195
Sesame seeds, whole, dried	1/4 cup	36	206	6	18	8	4	351	5	126	226	168
Squash and pumpkin seed kernels, dried	1/4 cup	34	187	8	16	6	1	15	5	185	405	278
Squash and pumpkin seeds, whole, roasted, without salt added	1/4 cup	16	71	3	3	9	(-)	9	0	42	15	147
Sunflower seed kernels, dry roasted, without salt added	1/4 cup	32	186	6	16	8	3	22	1	41	370	272
Walnuts, English	1/3 cup halves	33	218	5	22	5	2	35	1	53	115	147

Sodium (mg)	Zinc (mg)	Copper (mg)	Manganese (mg)	Selenium (µg)	Vitamin C (mg)	Thiamin (mg)	Riboflavin (mg)	Niacin (mg)	Pantothenic acid (mg)	Vitamin B_6 (mg)	Folate (µg)	Vitamin B_{12} (µg)	Vitamin A (IU)	Vitamin A (RE)	Vitamin E (mg)	Saturated fat (g)	Monounsaturated fat (g)	Polyunsaturated fat (g)	Cholesterol (mg)
156	1	0	1	2	0	0.0	0.0	4	0	0.1	29	0.0	0	0	3	3	8	5	0
149	1	0	0	2	0	0.0	0.0	4	0	0.1	24	0.0	0	0	3	3	8	4	0
3	0	0	0	0	38	0.1	0.0	0	0	0.1	23	0.0	104	11	0	0	0	0	0
2	1	0	0	1	11	0.2	0.0	2	0	0.2	51	0.0	478	48	0	0	0	0	0
2	1	0	0	1	0	0.2	0.0	1	1	0.0	64	0.0	7	1	0	0	0	0	0
1	1	0	1	6	1	0.1	0.2	0	0	0.2	46	0.0	8	1	2	1	2	4	0
5	0	0	0	0	4	0.1	0.0	1	0	0.0	38	0.0	5	0	0	0	0	1	0
10	1	0	0	0	11	0.2	0.1	1	1	0.1	120	0.0	8	1	0	1	1	3	0
5,015	5	1	1	2	0	0.1	0.3	1	0	0.3	45	0.0	120	12	0	1	2	5	0
5	2	1	1	7	0	0.1	0.1	4	0	0.2	43	0.8	569	57	0	1	1	4	0
10	1	0	1	12	0	0.1	0.1	0	0	0.1	42	0.0	10	1	0	1	1	3	0
106	0	0	(-)	(-)	0	0.0	0.0	0	(-)	0.0	(-)	0.0	0	0	0	0	0	1	0
10	1	0	0	11	0	0.1	0.0	1	0	0.1	55	0.0	9	1	0	1	1	3	0
0	2	1	1	4	0	0.0	0.4	2	0	0.1	15	0.0	0	0	12	2	15	6	0
11	0	0	0	(-)	4	0.1	0.1	0	0	0.2	32	0.0	0	0	(-)	2	6	6	0
1	2	1	0	1,381	0	0.5	0.1	1	0	0.1	2	0.0	0	0	1	8	11	11	0
7	3	1	0	5	0	0.1	0.1	1	1	0.1	32	0.0	0	0	0	4	12	6	0
1	0	0	1	1	12	0.1	0.1	1	0	0.2	33	0.0	11	0	1	0	0	0	0
4	0	0	0	(-)	8	0.1	0.1	3	0	0.2	30	0.0	309	31	(-)	0	0	0	0
0	1	1	3	2	3	0.3	0.1	1	0	0.3	51	0.0	18	2	7	3	21	4	0
0	0	0	0	(-)	0	0.0	0.0	0	0	0.0	8	0.0	4	0	(-)	0	0	0	0
2	1	0	1	2	0	0.3	0.0	1	0	0.2	4	0.0	0	0	0	5	26	1	0
0	2	0	2	2	0	0.2	0.0	0	0	0.1	8	0.0	28	3	1	2	15	8	0
4	1	1	1	3	1	0.4	0.1	1	0	0.7	21	0.0	227	23	2	2	10	6	0
1	1	0	1	(-)	0	0.3	0.1	1	1	0.3	45	0.0	14	1	(-)	1	1	8	0
4	3	1	2	2	0	0.3	0.1	2	0	0.3	35	0.0	3	0	(-)	2	7	8	0
6	3	0	1	2	1	0.1	0.1	1	0	0.1	20	0.0	131	13	0	3	5	7	0
3	2	0	0	(-)	0	0.0	0.0	0	0	0.0	1	0.0	10	1	(-)	1	1	1	0
1	2	1	1	25	0	0.0	0.1	2	2	0.2	76	0.0	0	0	16	2	3	10	0
1	1	1	1	2	0	0.1	0.1	1	0	0.2	33	0.0	14	1	1	2	3	16	0

Dairy Foods

CHEESE	Serving Size	Weight (g)	Calories	Protein (g)	Fat (g)	Carbohydrates (g)	Fiber (g)	Calcium (mg)	Iron (mg)	Magnesium (mg)	Phosphorus (mg)	Potassium (mg)
American, pasteurized processed	2 oz	57	213	13	18	1	0	349	0	13	422	92
Blue	1 1/2 oz	43	150	9	12	1	0	224	0	10	165	109
Brick	1 1/2 oz	43	158	10	13	1	0	286	0	10	192	58
Brie	1 1/2 oz	43	142	9	12	0	0	78	0	9	80	65
Camembert	1 1/2 oz	43	170	8	10	0	0	165	0	8	147	79
Caraway	1 1/2 oz	43	160	11	12	1	0	286	0	9	208	40
Cheddar, low fat	1 1/2 oz	43	74	10	3	1	0	176	0	7	206	28
Cheddar, whole	1 1/2 oz	57	171	11	14	1	0	307	0	12	218	42
Cheese food	2 oz	57	186	11	14	4	0	326	0	17	428	158
Cheese spread	2 oz	57	165	9	12	5	0	319	0	16	496	137
Cheshire	1 1/2 oz	43	165	10	13	2	0	273	0	9	197	40
Colby	1 1/2 oz	43	167	10	14	1	0	291	0	11	194	54
Cottage, creamed, large curd	2 cups	420	434	52	19	11	0	252	1	22	554	354
Cottage, creamed, small curd	2 cups	450	465	56	20	12	0	270	1	24	593	379
Cottage, nonfat, uncreamed, large, small, or dry curd	2 cups	290	245	50	1	5	0	92	1	11	302	94
Cottage, 1%	2 cups	452	327	56	5	12	0	275	1	24	605	386
Cottage, 2%	2 cups	452	405	62	9	16	0	310	1	27	680	435
Cream cheese	2 Tblsp (1 oz)	28	99	2	10	1	0	23	0	2	34	12
Cream cheese, fat free	2 Tblsp (1 oz)	28	28	4	0	2	0	54	0	4	126	47
Edam	1 1/2 oz	43	152	11	12	1	0	311	0	13	228	80
Farmer's	1 1/2 oz	43	150	9	12	1	0	300	0	12	225	38
Feta	1 1/2 oz	43	112	6	9	2	0	209	0	8	143	26
Fontina	1 1/2 oz	43	165	11	13	1	0	234	0	6	147	27
Goat	1 1/2 oz	43	155	9	13	1	0	127	1	12	159	67
Gorgonzola	1 1/2 oz	43	150	9	14	0	0	225	0	(-)	(-)	(-)
Gouda	1 1/2 oz	43	152	11	12	1	0	298	0	12	232	51
Gruyère	1 1/2 oz	43	176	13	14	0	0	430	0	15	257	34
Limburger	1 1/2 oz	43	139	9	12	0	0	211	0	9	167	54
Monterey Jack	1 1/2 oz	43	159	10	13	0	0	317	0	11	189	34
Mozzarella, part skim	1 1/2 oz	43	108	10	7	1	0	275	0	10	197	36
Mozzarella, whole	1 1/2 oz	43	120	8	9	1	0	220	0	8	158	29
Muenster	1 1/2 oz	43	157	10	13	0	0	305	0	12	199	57
Neufchâtel	1 1/2 oz	43	111	4	10	1	0	32	0	3	58	49
Parmesan	1 Tblsp	5	23	2	2	0	0	69	0	3	40	5
Provolone	1 1/2 oz	43	149	11	11	1	0	321	0	12	211	59
Ricotta, part skim	1/2 cup	124	171	14	10	6	0	337	1	18	226	155
Ricotta, whole	1/2 cup	124	216	14	16	4	0	257	0	14	196	130
Romano	1 1/2 oz	43	164	14	11	2	0	452	0	17	323	37

Sodium (mg)	Zinc (mg)	Copper (mg)	Manganese (mg)	Selenium (µg)	Vitamin C (mg)	Thiamin (mg)	Riboflavin (mg)	Niacin (mg)	Pantothenic acid (mg)	Vitamin B$_6$ (mg)	Folate (µg)	Vitamin B$_{12}$ (µg)	Vitamin A (IU)	Vitamin A (RE)	Vitamin D (IU)	Vitamin E (mg)	Saturated fat (g)	Monounsaturated fat (g)	Polyunsaturated fat (g)	Cholesterol (mg)
811	2	0	0	8	0	0.0	0.2	0	0	0.0	4	0.4	686	164	(-)	0	11	5	1	54
593	1	0	0	6	0	0.0	0.2	0	1	0.1	15	0.5	307	97	(-)	0	8	3	0	32
238	1	0	0	6	0	0.0	0.2	0	0	0.0	9	0.5	460	128	(-)	0	8	4	0	40
268	1	0	0	6	0	0.0	0.2	0	0	0.1	28	0.7	284	77	(-)	0	7	3	0	43
358	1	0	0	6	0	0.0	0.2	0	1	0.1	26	0.6	393	107	(-)	0	6	3	0	31
293	1	0	0	6	0	0.0	0.2	0	0	0.0	8	0.1	448	123	(-)	(-)	8	4	0	40
260	1	0	0	6	0	0.0	0.1	0	0	0.0	5	0.2	99	27	(-)	0	2	1	0	9
264	1	0	0	6	0	0.0	0.2	0	0	0.0	8	0.4	450	118	(-)	0	9	4	0	45
905	2	0	0	9	0	0.0	0.3	0	0	0.1	4	0.6	518	124	(-)	(-)	9	4	0	36
921	1	0	0	6	0	0.0	0.2	0	0	0.1	4	0.2	447	107	(-)	(-)	8	4	0	31
298	1	0	0	6	0	0.0	0.1	0	0	0.0	8	0.4	419	104	(-)	(-)	8	4	0	44
257	1	0	0	6	0	0.0	0.2	0	0	0.0	8	0.4	440	117	(-)	0	9	4	0	40
1,700	2	0	0	38	0	0.1	0.7	1	1	0.3	51	2.6	685	202	(-)	1	12	5	1	63
1,822	2	0	0	41	0	0.1	0.7	1	1	0.3	55	2.8	734	216	(-)	1	13	6	1	67
37	1	0	0	31	0	0.1	0.4	0	0	0.2	43	2.4	87	22	(-)	0	1	0	0	19
1,835	2	0	0	41	0	0.1	0.7	1	1	0.3	56	2.9	167	50	(-)	0	3	1	0	20
1,835	2	0	0	46	0	0.1	0.8	1	1	0.3	59	3.2	316	90	(-)	0	6	2	0	38
84	0	0	0	1	0	0.0	0.1	0	0	0.0	7	0.1	405	108	(-)	0	6	3	0	31
158	0	0	0	1	0	0.0	0.0	0	0	0.0	11	0.1	270	81	(-)	0	0	0	0	0
410	2	0	0	6	0	0.0	0.2	0	0	0.0	7	0.7	390	108	(-)	0	7	3	0	38
285	1	(-)	(-)	(-)	0	(-)	0.1	(-)	(-)	(-)	(-)	0.7	450	129	(-)	(-)	(-)	(-)	(-)	(-)
475	1	0	0	6	0	0.1	0.4	0	0	0.2	14	0.7	190	54	(-)	0	6	2	0	38
340	1	0	0	6	0	0.0	0.1	0	0	0.0	3	0.7	499	123	(-)	0	8	4	1	49
219	0	0	0	2	0	0.0	0.3	0	0	0.0	1	0.1	567	170	(-)	0	9	3	0	34
585	(-)	(-)	(-)	(-)	0	(-)	(-)	(-)	(-)	(-)	(-)	(-)	450	90	(-)	(-)	9	(-)	(-)	38
348	2	0	0	6	0	0.0	0.1	0	0	0.0	9	0.7	274	74	(-)	0	7	3	0	48
143	2	0	0	6	0	0.0	0.1	0	0	0.0	4	0.7	518	128	(-)	0	8	4	1	47
340	1	0	0	4	0	0.0	0.2	0	1	0.0	24	0.4	545	134	(-)	0	7	4	0	38
228	1	0	0	6	0	0.0	0.2	0	0	0.0	8	0.4	404	108	(-)	0	8	4	0	38
198	1	0	0	6	0	0.0	0.1	0	0	0.0	4	0.4	248	75	(-)	0	4	2	0	25
159	1	0	0	6	0	0.0	0.1	0	0	0.0	3	0.3	337	102	(-)	0	6	3	0	33
267	1	0	0	6	0	0.0	0.1	0	0	0.0	5	0.6	476	134	(-)	0	8	4	0	41
170	0	0	0	1	0	0.0	0.1	0	0	0.0	5	0.1	482	128	(-)	(-)	6	3	0	32
93	0	0	0	1	0	0.0	0.0	0	0	0.0	0	0.1	35	9	(-)	0	1	0	0	4
372	1	0	0	6	0	0.0	0.1	0	0	0.0	4	0.6	347	112	(-)	0	7	3	0	29
155	2	0	0	21	0	0.0	0.2	0	0	0.0	16	0.4	536	140	(-)	0	6	3	0	38
104	1	0	0	18	0	0.0	0.2	0	0	0.1	15	0.4	608	166	(-)	0	10	4	0	63
510	1	0	0	6	0	0.0	0.2	0	0	0.0	3	0.5	162	60	(-)	0	7	3	0	44

(continues)

DAIRY FOODS (continued)

	Serving Size	Weight (g)	Calories	Protein (g)	Fat (g)	Carbohydrates (g)	Fiber (g)	Calcium (mg)	Iron (mg)	Magnesium (mg)	Phosphorus (mg)	Potassium (mg)
CHEESE (continued)												
Roquefort	1 1/2 oz	43	157	9	13	1	0	281	0	13	167	39
Swiss	1 1/2 oz	43	160	12	12	1	0	409	0	15	257	47
Tilsit	1 1/2 oz	43	145	10	11	1	0	298	0	6	213	27
CREAM												
Half & half	1 fl oz	30	39	1	3	1	0	32	0	3	29	39
Light	1 fl oz	30	59	1	6	1	0	29	0	3	24	37
Sour	1 Tblsp	12	26	0	3	1	0	14	0	1	10	17
Substitute, dry (nondairy creamer)	2 Tblsp[a]	12	66	1	4	7	0	3	0	0	51	97
Substitute, liquid (nondairy creamer)	1 fl oz[a]	30	41	0	3	3	0	3	0	0	19	57
Whipped, pressurized	1 fl oz	6	15	0	1	1	0	6	0	1	5	9
Whipping, heavy	1 fl oz	30	103	1	11	1	0	19	0	2	19	22
Whipping, light	1 fl oz	30	88	1	9	1	0	21	0	2	18	29
FROZEN DESSERT												
Ice cream, chocolate	1 1/2 cup	198	428	8	22	56	2	216	2	57	212	493
Ice cream, vanilla	1 1/2 cup	198	398	7	22	47	0	253	0	28	208	394
Ice milk, vanilla	1 1/2 cup	198	276	8	8	45	0	276	0	30	216	417
Tofu, frozen dessert, vanilla	1 1/2 cup	225	600	6	33	63	0	(-)	(-)	(-)	(-)	90
Yogurt, softserve (chocolate)	1 cup	144	230	6	9	39	3	212	2	39	200	376
MILK												
Acidophilus milk, low fat (1%)	1 cup	244	90	8	3	12	0	300	0	34	234	381
Buttermilk	1 cup	245	99	8	2	12	0	285	0	27	219	371
Chocolate, low fat	1 cup	250	158	8	3	26	1	287	1	33	257	426
Condensed, sweetened	1 fl oz[a]	38	123	3	3	21	0	108	0	10	97	142
Dry, instant	1/3 cup[b]	23	82	8	0	12	0	283	0	27	226	392
Dry, nonfat	1/3 cup[a]	30	109	11	0	16	0	377	0	33	290	538
Evaporated, nonfat	1 fl oz[a]	32	25	2	0	4	0	92	0	9	62	106
Evaporated, whole	1 fl oz[a]	32	42	2	2	3	0	82	0	8	64	95
Goat	1 cup	244	168	9	10	11	0	326	0	34	270	499
Low fat (1%)	1 cup	244	102	8	3	12	0	300	0	34	235	381
Low lactose milk, low fat (1%)	1 cup	246	103	9	3	12	0	303	0	34	236	384
Nonfat (skim)	1 cup	245	86	8	0	12	0	302	0	28	247	406
Reduced fat (2%)	1 cup	244	121	8	5	12	0	297	0	33	232	377
Rice milk	1 cup	245	120	0.4	2	25	0	20[c]	0	10	34	69

Sodium (mg)	Zinc (mg)	Copper (mg)	Manganese (mg)	Selenium (µg)	Vitamin C (mg)	Thiamin (mg)	Riboflavin (mg)	Niacin (mg)	Pantothenic acid (mg)	Vitamin B$_6$ (mg)	Folate (µg)	Vitamin B$_{12}$ (µg)	Vitamin A (IU)	Vitamin A (RE)	Vitamin D (IU)	Vitamin E (mg)	Saturated fat (g)	Monounsaturated fat (g)	Polyunsaturated fat (g)	Cholesterol (mg)
769	1	0	0	6	0	0.0	0.2	0	1	0.1	21	0.3	445	127	(-)	(-)	8	4	1	38
111	2	0	0	5	0	0.0	0.2	0	0	0.0	3	0.7	359	108	(-)	0	8	3	0	39
320	1	0	0	6	0	0.0	0.2	0	0	0.0	9	0.9	444	124	(-)	0	7	3	0	43
12	0	0	0	1	0	0.0	0.1	0	0	0.0	1	0.1	131	32	(-)	0	2	1	0	11
12	0	0	0	0	0	0.0	0.0	0	0	0.0	1	0.1	190	55	(-)	0	4	2	0	20
6	0	0	0	0	0	0.0	0.0	0	0	0.0	1	0.0	95	23	(-)	0	2	1	0	5
22	0	0	0	0	0	0.0	0.0	0	0	0.0	0	0.0	24	2	(-)	0	4	0	0	0
24	0	0	0	0	0	0.0	0.0	0	0	0.0	0	0.0	27	3	(-)	(-)	3	0	0	0
8	0	0	0	0	0	0.0	0.0	0	0	0.0	0	0.0	51	12	(-)	0	1	0	0	5
11	0	0	0	0	0	0.0	0.0	0	0	0.0	1	0.1	438	125	(-)	0	7	3	0	41
10	0	0	0	0	0	0.0	0.0	0	0	0.0	1	0.1	169	89	(-)	0	6	3	0	33
150	1	0	0	5	1	0.1	0.4	0	1	0.1	32	0.6	824	236	(-)	1	13	6	1	67
158	1	0	0	5	1	0.1	0.5	0	1	0.1	10	0.8	810	232	(-)	0	13	6	1	7
168	1	0	0	5	1	0.1	0.4	0	1	0.1	12	1.3	327	93	0	0	5	2	0	27
270	(-)	(-)	(-)	(-)	(-)	(-)	(-)	(-)	(-)	(-)	(-)	(-)	(-)	(-)	(-)	(-)	4	9	19	0
141	1	0	0	3	0	0.1	0.3	0	1	0.1	16	0.4	230	62	(-)	0	5	3	0	7
122	1	0	0	5	2	0.1	0.4	0	1	0.1	12	0.9	500	144	96	0	2	1	0	10
257	1	0	0	5	2	0.1	0.4	0	1	0.1	12	0.5	81	20	(-)	0	1	1	0	9
152	1	0	0	5	2	0.1	0.4	0	1	0.1	12	0.9	500	148	100	0	2	1	0	7
49	0	0	0	6	1	0.0	0.2	0	0	0.0	4	0.2	125	31	(-)	0	2	1	0	13
126	1	0	0	6	1	0.1	0.4	0	1	0.1	11	0.9	545	163	133	0	0	0	0	4
161	1	0	0	8	2	0.1	0.5	0	1	0.1	15	1.2	659	198	101	0	0	0	0	6
37	0	0	0	1	0	0.0	0.1	0	0	0.0	3	0.1	125	37	25	0	0	0	0	1
33	0	0	0	1	1	0.0	0.1	0	0	0.0	2	0.1	77	17	(-)	0	1	1	0	9
122	1	0	0	3	3	0.1	0.3	1	1	0.1	1	0.2	451	137	29	0	7	3	0	28
123	1	0	0	5	2	0.1	0.4	0	1	0.1	12	0.9	500	144	98	0	2	1	0	10
123	1	0	0	5	2	0.1	0.4	0	1	0.1	12	0.9	504	145	100	(-)	2	1	0	10
126	1	0	0	5	2	0.1	0.3	0	1	0.1	13	0.9	500	149	98	0	0	0	0	4
122	1	0	0	5	2	0.1	0.4	0	1	0.1	12	0.9	500	139	98	0	3	1	0	18
86	0	0	0	(-)	1	0.1	0.0	1	0	0.0	91	0.0	5	0	0	(-)	0	1	0	0

(continues)

DAIRY FOODS (*continued*)

	Serving Size	Weight (g)	Calories	Protein (g)	Fat (g)	Carbohydrates (g)	Fiber (g)	Calcium (mg)	Iron (mg)	Magnesium (mg)	Phosphorus (mg)	Potassium (mg)
MILK (*continued*)												
Soy	1 cup	245	81	7	5	4	3	10ᶜ	1	47	120	345
Whole	1 cup	244	150	8	8	11	0	291	0	33	228	370
YOGURT												
Low fat, plain	1 cup	245	155	13	4	17	0	447	0	43	352	573
Low fat, fruit	8-oz container	227	225	9	3	42	0	314	0	30	247	402
Skim, plain	1 cup	245	137	14	0	19	0	488	0	47	383	625
Whole milk, plain	1 cup	245	150	9	8	11	0	296	0	28	233	379

ᵃNo serving size specified for this item by the Food Guide Pyramid.
ᵇMakes 1 cup reconstituted milk.
ᶜNot calcium-fortified. Calcium-fortified soy and rice milks, which provide the same amount of calcium as cow's milk, are a
 better choice.

FATS AND OILS

FATS (PLANT & ANIMAL)	Serving Size	Weight (g)	Calories	Protein (g)	Fat (g)	Carbohydrates (g)	Fiber (g)	Calcium (mg)	Iron (mg)	Magnesium (mg)	Phosphorus (mg)	Potassium (mg)	Sodium (mg)
Butter, whipped, with salt	1 Tblsp	9	67	0	8	0	0	2	0	0	2	2	78
Butter, with salt	1 Tblsp	14	102	0	12	0	0	3	0	0	3	4	117
Butter, without salt	1 Tblsp	14	102	0	12	0	0	3	0	0	3	4	1
Lard	1 Tblsp	13	115	0	13	0	0	0	0	0	0	0	0
Margarine-butter blend, 60% corn oil margarine and 40% butter	1 Tblsp	14	102	0	11	0	0	4	0	0	3	5	127
Margarine, hard, corn oil (hydrogenated and regular)	1 Tblsp	14	101	0	11	0	0	4	0	0	3	6	133
Margarine, hard, unspecified oil, without salt	1 Tblsp	14	99	0	11	0	0	2	0	0	2	9	1
Margarine, liquid, soybean and cottonseed oils (hydrogenated)	1 Tblsp	14	102	0	11	0	0	9	0	1	7	13	111

Sodium (mg)	Zinc (mg)	Copper (mg)	Manganese (mg)	Selenium (µg)	Vitamin C (mg)	Thiamin (mg)	Riboflavin (mg)	Niacin (mg)	Pantothenic acid (mg)	Vitamin B$_6$ (mg)	Folate (µg)	Vitamin B$_{12}$ (µg)	Vitamin A (IU)	Vitamin A (RE)	Vitamin D (IU)	Vitamin E (mg)	Saturated fat (g)	Monounsaturated fat (g)	Polyunsaturated fat (g)	Cholesterol (mg)
29	1	0	0	3	0	0.4	0.2	0	0	0.1	4	0.0	78	7	(-)	0	1	1	2	0
120	1	0	0	5	2	0.1	0.4	0	1	0.1	12	0.9	307	76	98	0	5	2	0	33
172	2	0	0	8	2	0.1	0.5	0	1	0.1	27	1.4	162	39	(-)	0	2	1	0	15
121	2	0	0	6	1	0.1	0.4	0	1	0.1	19	1.0	111	27	(-)	0	2	1	0	10
187	2	0	0	9	2	0.1	0.6	0.3	2	0.1	30	1.5	17	5	(-)	0	0	0	0	4
114	1	0	0	5	1	0.1	0.4	0	1	0.1	18	0.9	301	74	(-)	0	5	2	0	31

Zinc (mg)	Copper (mg)	Manganese (mg)	Selenium (µg)	Vitamin C (mg)	Thiamin (mg)	Riboflavin (mg)	Niacin (mg)	Pantothenic acid (mg)	Vitamin B$_6$ (mg)	Folate (µg)	Vitamin B$_{12}$ (µg)	Vitamin A (IU)	Vitamin A (RE)	Vitamin E (mg)	Saturated fat (g)	Monounsaturated fat (g)	Polyunsaturated fat (g)	Cholesterol (mg)	Phytosterol (mg)
0	0	0	0	0	0.0	0.0	0	0	0.0	0	0.0	287	71	0	5	2	0	20	(-)
0	0	0	0	0	0.0	0.0	0	0	0.0	0	0.0	434	107	0	7	3	0	31	0
0	0	0	0	0	0.0	0.0	0	0	0.0	0	0.0	434	107	0	7	3	0	31	(-)
0	0	0	0	0	0.0	0.0	0	0	0.0	0	0.0	0	0	0	5	6	1	12	0
0	0	0	0	0	0.0	0.0	0	0	0.0	0	0.0	507	113	1	4	5	2	12	49
0	0	0	0	0	0.0	0.0	0	0	0.0	0	0.0	504	113	2	2	5	3	0	80
0	0	0	0	0	0.0	0.0	0	0	0.0	0	0.0	501	113	2	2	5	3	0	37
0	0	0	0	0	0.0	0.0	0	0	0.0	0	0.0	507	113	1	2	4	5	0	25

(*continues*)

FATS AND OILS (*continued*)

	Serving Size	Weight (g)	Calories	Protein (g)	Fat (g)	Carbohydrates (g)	Fiber (g)	Calcium (mg)	Iron (mg)	Magnesium (mg)	Phosphorus (mg)	Potassium (mg)	Sodium (mg)
Fats (Plant & Animal) (*continued*)													
Margarine, reduced fat (40% fat) corn	1 Tblsp	14	50	0	6	0	0	3	0	0	2	4	139
Margarine, soft, corn (hydrogenated)	1 Tblsp	14	101	0	11	0	0	4	0	0	3	6	153
Shortening, household, soybean and cottonseed oils (hydrogenated)	1 Tblsp	13	113	0	13	0	0	0	0	0	0	0	0
PLANT OILS													
Canola	1 Tblsp	14	124	0	14	0	0	0	0	0	0	0	0
Coconut	1 Tblsp	14	117	0	14	0	0	0	0	0	0	0	0
Corn	1 Tblsp	14	120	0	14	0	0	0	0	0	0	0	0
Cottonseed	1 Tblsp	14	120	0	14	0	0	0	0	0	0	0	0
Olive	1 Tblsp	14	119	0	14	0	0	0	0	0	0	0	0
Palm	1 Tblsp	14	120	0	14	0	0	0	0	0	0	0	0
Palm kernel	1 Tblsp	14	117	0	14	0	0	0	0	0	0	0	0
Peanut	1 Tblsp	14	119	0	14	0	0	0	0	0	0	0	0
Safflower, linoleic (over 70%)	1 Tblsp	14	120	0	14	0	0	0	0	0	0	0	0
Soybean	1 Tblsp	14	120	0	14	0	0	0	0	0	0	0	0
Sunflower, linoleic (60% and over)	1 Tblsp	14	120	0	14	0	0	0	0	0	0	0	0
SPECIALTY OILS													
Almond	1 Tblsp	14	120	0	14	0	0	0	0	0	0	0	0
Grape-seed	1 Tblsp	14	120	0	14	0	0	0	0	0	0	0	0
Hazelnut	1 Tblsp	14	120	0	14	0	0	0	0	0	0	0	0
Poppyseed	1 Tblsp	14	120	0	14	0	0	0	0	0	0	0	0
Sesame	1 Tblsp	14	120	0	14	0	0	0	0	0	0	0	0
Walnut	1 Tblsp	14	120	0	14	0	0	0	0	0	0	0	0
Wheat germ	1 Tblsp	14	120	0	14	0	0	0	0	0	0	0	0

Zinc (mg)	Copper (mg)	Manganese (mg)	Selenium (µg)	Vitamin C (mg)	Thiamin (mg)	Riboflavin (mg)	Niacin (mg)	Pantothenic acid (mg)	Vitamin B$_6$ (mg)	Folate (µg)	Vitamin B$_{12}$ (µg)	Vitamin A (IU)	Vitamin A (RE)	Vitamin E (mg)	Saturated fat (g)	Monounsaturated fat (g)	Polyunsaturated fat (g)	Cholesterol (mg)	Phytosterol (mg)
0	(-)	(-)	0	0	0.0	0.0	0	0	0.0	0	0.0	518	116	(-)	1	2	2	0	39
0	(-)	(-)	0	0	0.0	0.0	0	0	0.0	0	0.0	506	113	(-)	2	4	4	0	68
0	0	0	0	0	0.0	0.0	0	0	0.0	0	0.0	0	0	1	3	6	3	0	26
0	0	0	0	0	0.0	0.0	0	0	0.0	0	0.0	0	0	3	1	8	4	0	0
0	0	0	0	0	0.0	0.0	0	0	0.0	0	0.0	0	0	0	12	1	0	0	12
0	0	0	0	0	0.0	0.0	0	0	0.0	0	0.0	0	0	3	2	3	8	0	132
0	0	0	0	0	0.0	0.0	0	0	0.0	0	0.0	0	0	5	4	2	7	0	44
0	0	0	0	0	0.0	0.0	0	0	0.0	0	0.0	0	0	2	2	10	1	0	30
0	0	0	0	0	0.0	0.0	0	0	0.0	0	0.0	0	0	3	7	5	1	0	0
0	0	0	0	0	0.0	0.0	0	0	0.0	0	0.0	0	0	1	11	2	0	0	13
0	0	0	0	0	0.0	0.0	0	0	0.0	0	0.0	0	0	2	2	6	4	0	28
0	0	0	0	0	0.0	0.0	0	0	0.0	0	0.0	0	0	6	1	2	10	0	60
0	0	0	0	0	0.0	0.0	0	0	0.0	0	0.0	0	0	2	2	3	8	0	34
0	0	0	0	0	0.0	0.0	0	0	0.0	0	0.0	0	0	7	1	3	9	0	14
0	0	0	0	0	0.0	0.0	0	0	0.0	0	0.0	0	0	5	1	10	2	0	36
0	0	0	0	0	0.0	0.0	0	0	0.0	0	0.0	0	0	0	1	2	10	0	24
0	0	0	0	0	0.0	0.0	0	0	0.0	0	0.0	0	0	0	1	11	1	0	16
0	(-)	(-)	0	0	0.0	0.0	0	0	0.0	0	0.0	0	0	(-)	2	3	8	0	36
0	0	0	0	0	0.0	0.0	0	0	0.0	0	0.0	0	0	1	2	5	6	0	118
0	0	0	0	0	0.0	0.0	0	0	0.0	0	0.0	0	0	0	1	3	9	0	24
0	0	0	0	0	0.0	0.0	0	0	0.0	0	0.0	0	0	26	3	2	8	0	75

SWEETENERS

	Serving Size	Weight (g)	Calories	Protein (g)	Fat (g)	Carbohydrates (g)	Calcium (mg)	Iron (mg)	Magnesium (mg)	Phosphorus (mg)	Potassium (mg)	Sodium (mg)
Chocolate, baking, unsweetened squares	1 square (1 oz)	28	148	3	16	8	21	2	88	118	236	4
Chocolate, milk chocolate	1 bar (1.5 oz)	44	226	3	14	26	84	1	26	95	169	36
Chocolate, semisweet, chips	60 pieces (1 oz)	28	136	1	9	18	9	1	33	37	103	3
Cocoa powder, unsweetened	1 Tblsp	5	11	1	1	3	6	1	25	37	76	1
Honey, strained or extracted	1 Tblsp	21	64	0	0	17	1	0	0	1	11	1
Sugar, brown	1 tsp, packed	5	17	0	0	4	4	0	1	1	16	2
Sugar, maple	1 tsp	3	11	0	0	3	3	0	1	0	8	0
Sugar, powdered (confectioners')	1 tsp	3	10	0	0	2	0	0	0	0	0	0
Sugar, white, granulated	1 tsp	4	16	0	0	4	0	0	0	0	0	0
Sugar, white, granulated	1 individual packet	6	23	0	0	6	0	0	0	0	0	0
Syrup, corn, dark	1 Tblsp	20	56	0	0	15	4	0	2	2	9	31
Syrup, corn, high-fructose	1 Tblsp	19	53	0	0	14	0	0	0	0	0	0
Syrup, corn, light	1 Tblsp	20	56	0	0	15	1	0	0	0	1	24
Syrup, malt	1 Tblsp	24	76	1	0	17	15	0	17	56	77	8
Syrup, maple	1 Tblsp	20	52	0	0	13	13	0	3	0	41	2
Syrup, molasses	1 Tblsp	20	53	0	0	14	41	1	48	6	293	7
Syrup, sorghum	1 Tblsp	21	61	0	0	16	32	1	21	12	210	2
Syrup, table blends, corn, refiner, and sugar	1 Tblsp	20	64	0	0	17	5	0	2	2	13	14
Syrup, table blends, pancake, with 2% maple	1 Tblsp	20	53	0	0	14	1	0	0	2	1	12

Zinc (mg)	Copper (mg)	Manganese (mg)	Selenium (µg)	Vitamin C (mg)	Thiamin (mg)	Riboflavin (mg)	Niacin (mg)	Pantothenic acid (mg)	Vitamin B$_6$ (mg)	Folate (µg)	Vitamin B$_{12}$ (µg)	Vitamin A (IU)	Vitamin A (RE)	Vitamin E (mg)	Saturated fat (g)	Monounsaturated fat (g)	Polyunsaturated fat (g)	Cholesterol (mg)
1	1	0	2	0	0.0	0.0	0	0	0.0	2	0.0	28	2.8	0	9	5	0	0
1	0	0	2	0	0.0	0.1	0	0	0.0	4	0.2	81	24	0.5	8	4	0	10
0	0	0	1	0	0.0	0.0	0	0	0.0	1	0.0	6	0.5	0	5	3	0	0
0	0	0	(-)	0	0.0	0.0	0.1	0	0.0	2	0.0	1	0	(-)	0	0	0	0
0	0	0	0	0	0.0	0.0	0	0	0.0	0	0.0	0	0	0	0	0	0	0
0	0	0	0	0	0.0	0.0	0	0	0.0	0	0.0	0	0	0	0	0	0	0
0	0	0	0	0	0.0	0.0	0	0	0.0	0	0.0	1	0	0	0	0	0	0
0	0	0	0	0	0.0	0.0	0	0	0.0	0	0.0	0	0	0	0	0	0	0
0	0	0	0	0	0.0	0.0	0	0	0.0	0	0.0	0	0	0	0	0	0	0
0	0	0	0	0	0.0	0.0	0	0	0.0	0	0.0	0	0	0	0	0	0	0
0	0	0	0	0	0.0	0.0	0	0	0.0	0	0.0	0	0	0	0	0	0	0
0	0	0	0	0	0.0	0.0	0	0	0.0	0	0.0	0	0	0	0	0	0	0
0	0	0	0	0	0.0	0.0	2	0	0.1	3	0.0	0	0	0	0	0	0	0
1	0	1	0	0	0.0	0.0	0	0	0.0	0	0.0	0	0	0	0	0	0	0
0	0	0	4	0	0.0	0.0	0	0	0.1	0	0.0	0	0	0	0	0	0	0
0	0	0	0	0	0.0	0.0	0	0	0.1	0	0.0	0	0	0	0	0	0	0
0	0	0	0	0	0.0	0.0	0	0	0.0	1	0.0	0	0	(-)	0	0	0	0
0	0	0	0	0	0.0	0.0	0	0	0.0	0	0.0	0	0	0	0	0	0	0

BEVERAGES

ALCOHOLIC BEVERAGES	Serving Size	Weight (g)	Calories	Fat (g)	Carbohydrates (g)	Fiber (g)	Alcohol (g)	Calcium (mg)	Iron (mg)	Magnesium (mg)	Phosphorus (mg)	Potassium (mg)
Beer												
Light	1 can or bottle (12 fl oz)	354	99	0	5	0	11	18	0	18	42	64
Regular	1 can or bottle (12 fl oz)	356	146	0	13	1	13	18	0	21	43	89
Distilled spirits												
80 proof (gin, rum, vodka, whiskey)	1.5 fl oz	42	97	0	0	0	14	0	0	0	2	1
86 proof (gin, rum, vodka, whiskey)	1.5 fl oz	42	105	0	0	0	15	0	0	0	2	1
94 proof (gin, rum, vodka, whiskey)	1.5 fl oz	42	116	0	0	0	17	0	0	0	2	1
100 proof (gin, rum, vodka, whiskey)	1.5 fl oz	42	124	0	0	0	18	0	0	0	2	1
Wine												
Dessert, dry	5 fl oz	147	186	0	6	0	23	12	0	13	13	136
Dessert, sweet	5 fl oz	147	226	0	17	0	23	12	0	13	13	136
Table, red	5 fl oz	147	106	0	2	0	14	12	1	19	21	165
Table, rosé	5 fl oz	147	105	0	2	0	14	12	1	15	22	146
Table, white	5 fl oz	147	100	0	1	0	14	13	0	15	21	118

Sodium (mg)	Zinc (mg)	Copper (mg)	Manganese (mg)	Selenium (µg)	Vitamin C (mg)	Thiamin (mg)	Riboflavin (mg)	Niacin (mg)	Pantothenic acid (mg)	Vitamin B$_6$ (mg)	Folate (µg)	Vitamin B$_{12}$ (µg)	Vitamin A (IU)	Vitamin A (RE)	Vitamin E (mg)	Saturated fat (g)	Monounsaturated fat (g)	Polyunsaturated fat (g)	Cholesterol (mg)
11	0	0	0	4	0	0.0	0.1	1	0	0.1	15	0.0	0	0	0	0	0	0	0
18	0	0	0	4	0	0.0	0.1	2	0	0.2	21	0.1	0	0	0	0	0	0	0
0	0	0	0	0	0	0.0	0.0	0	0	0.0	0	0.0	0	0	0	0	0	0	0
0	0	0	0	0	0	0.0	0.0	0	0	0.0	0	0.0	0	0	0	0	0	0	0
0	0	0	0	0	0	0.0	0.0	0	0	0.0	0	0.0	0	0	0	0	0	0	0
0	0	0	0	0	0	0.0	0.0	0	0	0.0	0	0.0	0	0	0	0	0	0	0
13	0	0	0	0	0	0.0	0.0	0	0	0.0	0	0.0	0	0	0	0	0	0	0
13	0	0	0	1	0	0.0	0.0	0	0	0.0	0	0.0	0	0	0	0	0	0	0
7	0	0	1	0	0	0.0	0.0	0	0	0.0	3	0.0	0	0	0	0	0	0	0
7	0	0	0	0	0	0.0	0.0	0	0	0.0	2	0.0	0	0	0	0	0	0	0
7	0	0	1	0	0	0.0	0.0	0	0	0.0	0	0.0	0	0	0	0	0	0	0

(*continues*)

BEVERAGES (*continued*)

COFFEE	Serving Size	Weight (g)	Calories	Protein (g)	Fat (g)	Carbohydrates (g)	Fiber (g)	Calcium (mg)	Iron (mg)	Magnesium (mg)	Phosphorus (mg)	Potassium (mg)	Sodium (mg)
Brewed, prepared with tap water	6 fl oz	178	4	0	0	1	0	4	0	9	2	96	4
Instant, decaffeinated, powder, prepared with water	6 fl oz	179	4	0	0	1	0	5	0	7	5	63	5
Instant, regular, prepared with water	6 fl oz	179	4	0	0	1	0	5	0	7	5	64	5
Substitute, cereal grain beverage, prepared with water	6 fl oz	180	9	0	0	2	0	5	0	7	13	43	7
SODA													
Club soda	12 fl oz	355	0	0	0	0	0	18	0	3.55	0	7	75
Cola, contains caffeine	12 fl oz	370	152	0	0	38	0	11	0	4	44	4	15
Cola, low calorie, with aspartame, contains caffeine	12 fl oz	355	4	0	0	0	0	14	0	4	32	0	21
Cream soda	12 fl oz	371	189	0	0	49	0	19	0	4	0	4	45
Ginger ale	12 fl oz	366	124	0	0	32	0	11	1	4	0	4	26
Grape soda	12 fl oz	372	160	0	0	42	0	11	0	4	0	4	56
Lemon-lime soda	12 fl oz	368	147	0	0	38	0	7	0	4	0	4	40
Orange, contains caffeine	12 fl oz	372	179	0	0	46	0	19	0	4	4	7	45
Pepper-type	12 fl oz	368	151	0	0	38	0	11	0	0	40	4	37
Root beer	12 fl oz	370	152	0	0	39	0	19	0	4	0	4	48
TEA													
Black (unspecified), brewed, prepared with tap water	6 fl oz	178	2	0	0	1	0	0	0	5	2	66	5
Herb, chamomile, brewed	6 fl oz	178	2	0	0	1	0	4	0	2	0	16	2
Herb, not chamomile, brewed	6 fl oz	178	2	0	0	0	0	4	0	2	0	16	2

Zinc (mg)	Copper (mg)	Manganese (mg)	Selenium (µg)	Vitamin C (mg)	Thiamin (mg)	Riboflavin (mg)	Niacin (mg)	Pantothenic acid (mg)	Vitamin B_6 (mg)	Folate (µg)	Vitamin B_{12} (µg)	Vitamin A (IU)	Vitamin A (RE)	Vitamin E (mg)	Saturated fat (g)	Monounsaturated fat (g)	Polyunsaturated fat (g)	Cholesterol (mg)	Caffeine (mg)
0	0	0	0	0	0.0	0.0	0	0	0.0	0	0.0	0	0	0	0	0	0	0	103
0	0	0	0	0	0.0	0.0	1	0	0.0	0	0.0	0	0	0	0	0	0	0	2
0	0	0	0	0	0.0	0.0	1	0	0.0	0	0.0	0	0	0	0	0	0	0	57
0	0	0	0	0	0.0	0.0	0	0	0.0	1	0.0	0	0	0	0	0	0	0	0
0	0	0	0	0	0.0	0.0	0	0	0.0	0	0.0	0	0	0	0	0	0	0	0
0	0	0	0	0	0.0	0.0	0	0	0.0	0	0.0	0	0	0	0	0	0	0	37
0	0	0	0	0	0.0	0.1	0	0	0.0	0	0.0	0	0	0	0	0	0	0	50
0	0	0	0	0	0.0	0.0	0	0	0.0	0	0.0	0	0	0	0	0	0	0	0
0	0	0	0	0	0.0	0.0	0	0	0.0	0	0.0	0	0	0	0	0	0	0	0
0	0	0	0	0	0.0	0.0	0	0	0.0	0	0.0	0	0	0	0	0	0	0	0
0	0	0	0	0	0.0	0.0	0	0	0.0	0	0.0	0	0	0	0	0	0	0	0
0	0	0	0	0	0.0	0.0	0	0	0.0	0	0.0	0	0	0	0	0	0	0	0
0	0	0	0	0	0.0	0.0	0	0	0.0	0	0.0	0	0	0	0	0	0	0	37
0	0	0	0	0	0.0	0.0	0	0	0.0	0	0.0	0	0	0	0	0	0	0	0
0	0	0	0	0	0.0	0.0	0	0	0.0	9	0.0	0	0	0	0	0	0	0	36
0	0	0	0	0	0.0	0.0	0	0	0.0	1	0.0	36	4	0	0	0	0	0	0
0	0	0	0	0	0.0	0.0	0	0	0.0	1	0.0	0	0	0	0	0	0	0	0

TABLE 8: PHYTOCHEMICAL CONTENTS OF SELECTED FOODS

| | Carotenoids (µg) | | | | | Isoflavones (mg) | |
	Alpha-carotene	Beta-carotene	Beta-cryptoxanthin	Lutein & zeaxanthin	Lycopene	Daidzein	Genistein
Fruits							
Apples, raw, with skin	30						
Apricots, raw	0	2,554	0	0	5		
Apricots, dried, uncooked							
Avocados, raw	28	53	36				
Bananas, raw	5	21	0	0	0		
Blackberries, raw							
Blueberries, raw	0	35					
Breadfruit, raw							
Cantaloupe, raw	27	1,595	0	40	0		
Carambola (starfruit), raw (2)		42	36		42		
Casaba melon, raw							
Cherimoya, raw							
Cherries, sweet, raw	28						
Cherries, sour, red, raw							
Coconut meat (nuts), raw							
Cranberries, raw							
Currants, red and white, raw							
Dates, domestic, natural and dry							
Durian, raw or frozen	6	23	0				
Elderberries, raw							
Feijoa, raw							
Figs, raw							
Figs, dried, uncooked							
Gooseberries, raw							
Grapes, American type (slip skin), raw							
Grapes, red or green, seedless, raw							
Grapefruit, raw, pink and red, all areas (not white)	5	603	12	13	1,462		
Guavas, common, raw		984	66		1,150		
Honeydew melon, raw							
Jackfruit, raw (2)		360	36		17		
Jujube, raw							
Kiwi fruit, fresh, raw							
Kumquats, raw		0					
Lemons, raw, with peel							
Limes, raw							
Longans, raw							
Longans, dried							

Glycitein	Total isoflavones	Other phytochemicals										
		Terpenes	Ellagic acid	Inositol phosphates (phytates)	Lignans	Coumestrol	Formononetin	Biochanin A	Indoles	Isothiocyanates	Phenols & cyclic compounds	Sulfides & thiols
			X									
					X							
					X							
		X										
		X										
		X										

PHYTOCHEMICAL CONTENTS OF SELECTED FOODS (*continued*)

	Carotenoids (µg)					Isoflavones (mg)	
	Alpha-carotene	Beta-carotene	Beta-cryptoxanthin	Lutein & zeaxanthin	Lycopene	Daidzein	Genistein
Loquats, raw							
Lychees, raw							
Lychees, dried							
Mangos, raw	17	445	11				
Mangosteen, canned, syrup pack	1	16	9				
Mulberries, raw							
Nectarines, raw	0	101	59				
Olives, ripe, canned							
Oranges, raw, all commercial varieties	16	51	122	187	0		
Papayas, raw	0	276	761	75	0		
Passion-fruit, yellow, raw	35	525	46				
Peaches, raw	1	97	24	57	0		
Pears, raw	6	27					
Pears, Asian, raw							
Persimmons, Japanese, raw	253	1,447	834	158			
Pineapple, raw (2)		230	89		399		
Plantains, cooked							
Plums, raw	98	16					
Pomegranates, raw							
Prickly pears, raw	0	24	3				
Prunes (dried plums), uncooked							
Pummelo, raw	0	0					
Quinces, raw							
Raisins, seedless							
Rambutan, canned, syrup pack	0	2	0				
Raspberries, raw	12	8	0				
Rhubarb, raw							
Sapodilla, raw							
Strawberries, raw	5						
Tamarinds, raw	0	8					
Tangerines, raw	14	71	485	243	0		
Watermelon, raw	0	295	103	17	4,868		
Vegetables							
Amaranth leaves, cooked							
Artichokes, cooked							
Arugula, raw							
Asparagus, raw	12	493					
Asparagus, cooked, boiled, drained							

Glycitein	Total isoflavones	Other phytochemicals										
		Terpenes	Ellagic acid	Inositol phosphates (phytates)	Lignans	Coumestrol	Formononetin	Biochanin A	Indoles	Isothiocyanates	Phenols & cyclic compounds	Sulfides & thiols
		X			X							
					X							
					X							
			X									
			X		X							

PHYTOCHEMICAL CONTENTS OF SELECTED FOODS *(continued)*

	Carotenoids (μg)					Isoflavones (mg)	
	Alpha-carotene	Beta-carotene	Beta-cryptoxanthin	Lutein & zeaxanthin	Lycopene	Daidzein	Genistein
Bamboo (shoots), cooked							
Beans, snap, green, cooked, boiled, drained, without salt	92	552	0	700	0		
Beans, snap, yellow, cooked, boiled, drained, without salt							
Beet greens, cooked, boiled, drained without salt		2,560					
Beets, cooked, boiled, drained							
Bitter melon (balsam pear), leafy tips, cooked, boiled, drained, without salt							
Bitter melon (balsam pear), pods, cooked, boiled, drained, without salt							
Broccoli, cooked, boiled, drained, without salt	0	1,042	0	2,226	0		
Broccoli, raw	1	779	0	2,445	0		
Brussels sprouts, cooked, boiled, drained, without salt	0	465	0	1,290	0		
Cabbage, cooked, boiled, drained, without salt		90					
Cabbage, raw	0	65	0	310	0		
Carrots, cooked	4,109	8,015					
Carrots, baby, raw	4,425	7,275	0	358	0		
Cassava, raw	0	8	0				
Cauliflower, cooked, boiled, drained, without salt							
Cauliflower, raw							
Celeriac, cooked, boiled, drained, without salt							
Celeriac, raw							
Celery, cooked, boiled, drained, without salt	0	210	0	250	0		
Celery, raw	0	150	0	232	0		
Chayote, fruit, cooked, boiled, drained, without salt	0	0	0				
Chicory greens, raw							
Chicory roots, raw							
Cabbage, Chinese (bok choy), boiled, drained, without salt							
Cabbage, Chinese (bok choy), raw							
Cabbage, napa, cooked	49	133	0				
Collards, cooked, boiled, drained, without salt	90	4,418	20	8,091	0		
Corn, sweet, yellow, canned, whole kernel, drained solids	33	30	0	884	0		
Cress, water, raw							
Cucumber, with peel, raw		138					
Eggplant, cooked, boiled, drained, without salt							
Fennel, bulb, raw							
Fennel seed (spice)							
Fenugreek seed (spice)							
Garlic, raw							

Glycitein	Total isoflavones	Terpenes	Ellagic acid	Inositol phosphates (phytates)	Lignans	Coumestrol	Formononetin	Biochanin A	Indoles	Isothiocyanates	Phenols & cyclic compounds	Sulfides & thiols
					X				X	X		
					X				X	X		
									X	X		
					X				X	X		
					X				X	X		
					X							
					X							
										X		
										X		
					X				X	X		
					X				X	X		
					X				X	X		
												X

Phytochemical Contents of Selected Foods *(continued)*

	Carotenoids (µg)					Isoflavones (mg)	
	Alpha-carotene	Beta-carotene	Beta-cryptoxanthin	Lutein & zeaxanthin	Lycopene	Daidzein	Genistein
Gingerroot, raw							
Horseradish, prepared	0	0	0				
Jerusalem artichokes, raw							
Jicama (yam bean), raw							
Kelp (seaweed), raw							
Kohlrabi, cooked, boiled, drained, without salt							
Kohlrabi, raw							
Leeks (bulb and lower leaf-portion), cooked, boiled, drained, without salt							
Leeks (bulb and lower leaf-portion), raw, chopped							
Lettuce, iceberg (includes crisphead types), raw	2	192	0	352	0		
Mushrooms, black, dried	0	0	0				
Mushrooms, raw							
Okra, cooked, boiled, drained, without salt	0	170	0	390	0		
Onion, cooked							
Onions, raw							
Parsley, raw							
Parsnips, cooked, boiled, drained, without salt							
Parsnips, raw							
Peppers, hot chili, green, raw							
Peppers, hot chili, red, raw							
Peppers, sweet, green, raw	22	198					
Peppers, sweet, red, raw	59	2,379	2,205				
Peppers, sweet, yellow, raw		120					
Potatoes, baked, flesh, without salt							
Radishes, raw							
Rutabagas, cooked							
Rutabagas, raw							
Salsify (vegetable oyster), cooked, boiled, drained, without salt							
Scallions (including tops and bulbs)							
Spinach, cooked, boiled, drained, without salt	0	5,242	0	7,043	0		
Spinach, raw	0	5,597	0	11,938	0		
Squash, summer, crookneck and straightneck, raw	0	90	0	290	0		
Squash, winter, acorn, cooked, boiled, mashed, without salt	0	490	0	66	0		
Sweet potato, cooked, baked in skin, without salt	0	9,488	0	0	0		
Taro, cooked, without salt							
Taro, raw							
Taro leaves, cooked, steamed, without salt							

Glycitein	Total isoflavones	Terpenes	Ellagic acid	Inositol phosphates (phytates)	Lignans	Coumestrol	Formononetin	Biochanin A	Indoles	Isothiocyanates	Phenols & cyclic compounds	Sulfides & thiols
												X
												X
					X							
					X							X
					X							X
					X							
					X							
					X							
					X							

PHYTOCHEMICAL CONTENTS OF SELECTED FOODS (*continued*)

	Carotenoids (µg)					Isoflavones (mg)	
	Alpha-carotene	Beta-carotene	Beta-cryptoxanthin	Lutein & zeaxanthin	Lycopene	Daidzein	Genistein
Taro leaves, raw							
Tomatillo							
Tomatoes, red, ripe, raw	112	393	0	130	3,025		
Turnip greens, cooked, boiled, drained, without salt	0	4,575	0	8,440	0		
Turnips, raw							
Water chestnuts, cooked							
Water chestnuts, raw							
Yam, cooked, boiled, drained, or baked, without salt		0					
Grains							
Amaranth							
Barley, bran							
Buckwheat groats, roasted, cooked							
Bulgur, cooked							
Hominy, canned, white							
Millet, cooked							
Oat bran							
Quinoa							
Rice bran, crude							
Rice, brown, long-grain, cooked							
Rice, white, long-grain, regular, cooked							
Rye							
Sorghum							
Triticale							
Wheat bran							
Wild rice, cooked							
Lentils and Legumes							
Alfalfa seeds, sprouted, raw						0	0
Beans, adzuki, mature seeds, cooked, boiled, without salt							
Beans, black, mature seeds, raw						0	0
Beans, cranberry, mature seeds, cooked, boiled, without salt							
Beans, kidney, all types, mature seeds, cooked, boiled, without salt						0	0
Beans, pinto, mature seeds, raw						0.01	0.26
Beans, small white, mature seeds, raw						0	0.74
Broadbeans (fava beans), mature seeds, raw						0.02	0

Glycitein	Total isoflavones	Terpenes	Ellagic acid	Inositol phosphates (phytates)	Lignans	Coumestrol	Formononetin	Biochanin A	Indoles	Isothiocyanates	Phenols & cyclic compounds	Sulfides & thiols
					X							
									X			
					X							
					X							
											X	
					X							
	0											
	0											
0				X	0	0	0.41					
	0.27					3.61	trace	0.56				
	0.74					0	0.82	0				
	0.03											

PHYTOCHEMICAL CONTENTS OF SELECTED FOODS (*continued*)

	Carotenoids (μg)					Isoflavones (mg)	
	Alpha-carotene	Beta-carotene	Beta-cryptoxanthin	Lutein & zeaxanthin	Lycopene	Daidzein	Genistein
Chickpeas (garbanzo beans, bengal gram), mature seeds, raw						0.04	0.06
Cowpeas, catjang, mature seeds, cooked, boiled, without salt						0.01	0.02
Lentils, mature seeds, raw						0	0
Lima beans, large, mature seeds, cooked, boiled, without salt						0	0
Miso						16.13	24.56
Mung beans, mature seeds, raw						0.01	0.18
Natto						21.85	29.04
Peanuts, all types, raw						0.03	0.24
Peas, split, mature seeds, raw						2.42	0
Soy sauce made from soy (tamari)							
Soybeans, mature cooked, boiled, without salt						26.95	27.71
Soybeans, mature seeds, sprouted, raw						19.12	21.6
Tempeh						17.59	24.85
Tofu, firm, prepared with calcium sulfate & nigari						9.44	13.35
Tofu, soft, prepared with calcium sulfate & nigari						11.99	18.23
Other							
Tea, green, Japanese						0.01	0.04
Tea, jasmine, Twinings						0.01	0.03
Wine, red							

Nutrient values are in 100 grams of edible food unless otherwise noted. Data on the phytochemical contents of foods are limited. "0" value = phytochemicals are not present. "X" = the amount of phytochemicals is unspecified. Blanks indicate that no data are available.

This table was compiled from the following sources: Meagher LP, Beecher GR: Assessment of Data on the Lignan Content of Foods. J Food Composition Analysis 13:935-947, 2000; Milner JA: Nonnutritive components in foods as modifiers of the cancer process. In Preventive Nutrition: the Comprehensive Guide for Health Professionals. 2nd ed. Edited by A Bendich, RJ Deckelbaum. Totowa, New Jersey, Humana Press, 2001, pp 131-154; Setiawan B, Sulaeman A, Giraud DW, Driskell JA: Carotenoid content of selected Indonesian fruits. J Food Composition Analysis 14:169-176, 2001; USDA–Iowa State University Database on the Isoflavone Content of Foods—1999: Table of analytical isoflavone values and table of analytical Coumestrol, Biochanin A, and Formononetin values (http://www.nal.usda.gov/fnic/foodcomp/Data/isoflav/isoflav.html); USDA–NCC Carotenoid Database for US Foods—1998 (http://www.nal.usda.gov/fnic/foodcomp/Data/car98/car98.html).

Glycitein	Total isoflavones	Other phytochemicals										
		Terpenes	Ellagic acid	Inositol phosphates (phytates)	Lignans	Coumestrol	Formononetin	Biochanin A	Indoles	Isothiocyanates	Phenols & cyclic compounds	Sulfides & thiols
	1					0	0	1.52				
	0.03											
	0.01				X							
	0											
2.87	42.55											
	0.19											
8.17	58.93											
	0.26				X							
	2.42											
	54.66				X							
	40.71				X							
2.10	43.52											
2.08	24.74											
2.03	31.10											
	0.05				X	0.03					X	
	0.04				X	0.03					X	
											X	

CREDITS

Top 10 Causes of Death (U.S. Population) (page 6)

Modified from Hoyert DL, Kochanek KD, Murphy SL: Deaths: final data for 1997. National Vital Statistics Reports 47:1-104, 1999.

Healthy People 2010 (page 7)

From the U.S. Department of Health and Human Services. Healthy People 2010 (Conference Edition, in two volumes). Washington, DC, January 2000.

Eat 5 A Day for Better Health (page 8)

By permission of the Produce for Better Health Foundation.

Dietary Guidelines for Americans (page 9)

From the U.S. Department of Agriculture, Agricultural Research Service, Dietary Guidelines Advisory Committee, 2000. Nutrition and Your Health: Dietary Guidelines for Americans, 2000.

The Food Guide Pyramid (pages 11 and 79)

Modified from the U.S. Department of Agriculture and the U.S. Department of Health and Human Services, The Food Guide Pyramid.

How Many Servings Do You Need Each Day? (page 13)

Modified from the International Food Information Council Foundation, U.S. Department of Agriculture Center for Nutrition Policy and Promotion, and the Food Marketing Institute: The Food Guide Pyramid: Beyond the Basic 4. Revised 1996.

Pyramids (page 15)

Mayo Clinic Healthy Weight Pyramid: From Energy density: how to eat more and achieve a healthy weight. Mayo Clinic Women's HealthSource, January 2001, p 2. By permission of Mayo Foundation for Medical Education and Research.

California Pyramid: From Heber D: The Resolution Diet. Garden City Park, New York, Avery Publishing Group, 1999, p 80. By permission of the author.

Asian Pyramid and Mediterranean Pyramid: From Putting the whole back in wholesome. Supplement to Mayo Clinic Women's HealthSource, November 1998. By permission of Mayo Foundation for Medical Education and Research.

Body Mass Index Table (page 49)

Modified from National Institutes of Health Clinical Guidelines on the Identification, Evaluation, and Treatment of Overweight and Obesity in Adults, 1998.

Illustration (top) on page 50
From Weight control: what works and why. Medical Essay (Supplement to Mayo Clinic Health Letter), June 1994, p 1. By permission of Mayo Foundation for Medical Education and Research.

Classification of High Blood Pressure (page 54)
From National Institutes of Health. The Sixth Report of the Joint National Committee on Prevention, Detection, Evaluation, and Treatment of High Blood Pressure, 1997.

Illustration on page 55
From Sheps SG: Mayo Clinic on High Blood Pressure. Rochester, Minnesota, Mayo Clinic, 1999, p 11. By permission of Mayo Foundation for Medical Education and Research.

The Combination Diet From the DASH Study (page 56)
Modified from National Institutes of Health. The DASH Diet (Publication No. 98-4082), 1998.

Illustration on page 62
From Larson DE: Mayo Clinic Family Health Book. New York, William Morrow and Company, 1996, p B-3. By permission of Mayo Foundation for Medical Education and Research.

Mini-Glossary of Lipid-Related Terms (page 63)
Modified from McGoon MD: Mayo Clinic Heart Book. New York, William Morrow and Company, 1993, pp 347-355. By permission of Mayo Foundation for Medical Education and Research.

Your Blood Lipid Test Results—What Do Those Numbers Mean? (page 65)
From Cholesterol: put knowledge behind your numbers to lower your confusion level. Medical Essay (Supplement to Mayo Clinic Health Letter), June 1993, p 4. By permission of Mayo Foundation for Medical Education and Research.

How You Grow Shorter (page 68)
From Osteoporosis: it's never too late to protect your bones. Medical Essay (Supplement to Mayo Clinic Health Letter), October 1997, p 2. By permission of Mayo Foundation for Medical Education and Research.

Osteoporosis in Men (page 70)
From Osteoporosis: it's never too late to protect your bones. Medical Essay (Supplement to Mayo Clinic Health Letter), October 1997, p 3. By permission of Mayo Foundation for Medical Education and Research.

How Much Vitamin D Is Enough? (page 71)

Data from Standing Committee on the Scientific Evaluation of Dietary Reference Intakes: Dietary Reference Intakes for Calcium, Phosphorus, Magnesium, Vitamin D, and Fluoride. National Academy Press, Washington, D.C., 1997, pp 250-287.

Tips for Selecting and Taking a Calcium Supplement (page 72)

From Osteoporosis: it's never too late to protect your bones. Medical Essay (Supplement to Mayo Clinic Health Letter), October 1997, p 7. By permission of Mayo Foundation for Medical Education and Research.

Logo on page 91

From US Food and Drug Administration: Irradiation: A Safe Measure for Safer Food. May-June 1998 FDA Consumer (revised June 1998).

Health Claims (page 93)

Modified from US Food and Drug Administration Center for Food Safety and Applied Nutrition: A food labeling guide, September 1994 (editorial revisions June 1999).

Sources of Bacteria (page 148)

Data from USDA Food Safety and Inspection Service: Four simple steps to food safety, September 1999.

Cooking It Safe (page 149)

Modified from USDA Food Safety and Inspection Service: Cook it safely! It's a matter of degrees, September 1999.

*Portions of the text of **Grains** (pages 269-289) and **Beverages** (pages 377-387) are from the following articles on http://www.MayoClinic.com (retrieved May 1, 2001):*

Celiac disease: when food becomes the enemy

Palm oil in breakfast cereals

What is hominy?

Facts on flaxseed

The wonder of water: a drink to your health

Water, water everywhere

INDEX